CLASSICAL MECHANICS
Volume 1

CLASSICAL MECHANICS

Volume 1

EDWARD A. DESLOGE
Department of Physics
Florida State University

A WILEY-INTERSCIENCE PUBLICATION

JOHN WILEY & SONS
New York · Chichester · Brisbane · Toronto · Singapore

Copyright © 1982 by John Wiley & Sons, Inc.

All rights reserved. Published simultaneously in Canada.

Reproduction or translation of any part of this work beyond that permitted by Section 107 or 108 of the 1976 United States Copyright Act without the permission of the copyright owner is unlawful. Requests for permission or further information should be addressed to the Permissions Department, John Wiley & Sons, Inc.

Library of Congress Cataloging in Publication Data:

Desloge, Edward A., 1926–
 Classical mechanics.

  "A Wiley-Interscience publication."
  Includes index.
  1. Mechanics. I. Title.
QC122.D47 531 81-11402

ISBN 0-471-09144-8 (v. 1)AACR2
ISBN 0-471-09145-6 (v. 2)

Printed in the United States of America

10 9 8 7 6 5 4 3 2 1

Preface

This book is the product of teaching classical mechanics on both the undergraduate and the graduate levels intermittently over the past 20 years. It covers mechanics in a unified fashion from the foundations, through elementary and intermediate mechanics, to a moderately advanced graduate level. A knowledge of calculus is presumed. Any other mathematics needed is provided in the appendices.

Volume 1 can be used as a text for an undergraduate course. Volume 2 can be used as a text for a graduate course. By judicious deletion of material, courses of almost any length can be accommodated. The breadth and detail of the coverage is such that the book can also be used by students wanting to learn mechanics on their own, or by instructors wanting to direct students through self-paced programs. All the topics customarily found in an undergraduate or graduate text are covered, though frequently, in greater depth or detail, or from a slightly different point of view. In addition there are many interesting and useful topics that are seldom found in the standard texts. Hence this book can also serve as a source book for an instructor in mechanics, or as an aid to a researcher whose need for mechanics exceeds what is provided by the usual text.

Much material is covered, but no attempt has been made to write an encyclopedic text on classical mechanics; rather, the subject is arranged in such a way that additional material can be inserted easily and naturally.

A knowledge of mechanics that will continue to mature beyond the termination of a formal course requires a clear and accurate grasp of the organization of the subject. Throughout this book, I have tried to stress organization, clarity, and accuracy. This emphasis may sometimes appear to result in an approach that is overly stiff, detailed, and formal. The niceties of charm, elegance, and warmth, where lacking, can be supplied by a good instructor. A weakness in organization is almost impossible to repair.

One of the most crucial stages in the exposition of any subject in physics is the choice of notation. A good notation is comprehensive; that is, it carries all the information that is required to avoid misinterpretation, and yet it is simple.

Slight differences are often quite significant. For example it is both more meaningful and more useful to express the components of a vector **A** with respect to a frame S' as $A_{i'}$ rather than A_i'. I have thought a great deal about the notation. If I have erred, it is more often than not in using notation that is somewhat overloaded. The appearance of many of the discussions and derivations could be improved by stripping the notation of some of its appendages. The loss however would generally outweigh the gain. For example, the appearance of arguments involving partial derivatives, such as $(\partial f/\partial x)_y$ or equivalently $\partial f(x, y)/\partial y$ could be improved by writing such terms simply $\partial f/\partial x$. I have at times done this myself. However, anyone who has taught a course in thermodynamics will verify that the consequences can be disastrous, if one is not careful.

In writing this book, I have used a number of organizational and pedagogical devices that I have found over the years to be particularly helpful. (1) *The material is highly subdivided*, to give emphasis to the organization and to facilitate reorganization. (2) *The chapters are short*, to make assimilation easier and to aid in the addition, deletion, or rearrangement of material. (3) *Formal definitions, postulates, and theorems are frequently used*, to emphasize and clarify important concepts and to foster exactness. (4) *Most chapters contain one or more examples* designed to illustrate an idea, or to provide general techniques for the solution of problems, rather than to serve as models to be slavishly imitated. (5) *There are a large number of problems*, since the ability to systematically solve a wide variety of problems is one of the goals of a course in mechanics. (6) *All mathematical developments needed are provided in appendices*. In this way the continuity of a physical argument is not broken, and yet the mathematical tools are readily available.

A number of benefits, other than problem solving, can be derived from a course in classical mechanics. Three in particular have influenced me strongly in writing this book.

1. Since physics as we know it today had its origins in classical mechanics, the very structure and language of physics is permeated by ideas that come from classical mechanics; hence a *knowledge of the foundations of mechanics can provide a student with a deeper grasp of all of physics*. I have therefore devoted considerable effort to developing the foundations of classical mechanics.
2. *A course in classical mechanics is not only a course in physics but a course in applied mathematics*. The inclusion of certain topics, and the extensive mathematical appendices in this book, reflect this aspect of mechanics. As much consideration has been given to the writing of the appendices as to the text proper. Though relegated to appendices, the mathematical topics are an integral part of the text.
3. The human brain contains two hemispheres whose characters have been shown to be different but complementary. In most individuals, the right

hemisphere, which is associated directly with the left hand, the left field of vision, and so forth, is superior in handling geometrical concepts, and the left hemisphere, which is associated directly with the right hand, the right field of vision, and so forth, is superior in handling formal analytical concepts. Learning physics involves an interesting interplay between these two abilities. The left hemisphere provides the analytical map that takes us from one point to another, while the right hemisphere provides the geometrical vision necessary to see the goal and landmarks along the way. *A course in classical mechanics offers a marvelous vehicle for developing and integrating both the geometrical and the analytical powers of the brain.* The Newtonian approach to mechanics is strongly geometrical, whereas the Lagrangian and Hamiltonian approaches are strongly analytical. It follows that a course designed to exercise both sides of the brain should, as this text does, include a thorough foundation in Newtonian mechanics before proceeding to Lagrangian and Hamiltonian mechanics. Too many modern courses in classical mechanics are built on a weak foundation of Newtonian mechanics, with the resultant complaint by instructors that a particular student is good at mathematics but does not seem to have any physical intuition.

To make Volume 1 adequate for an undergraduate course and Volume 2 adequate for a graduate course, certain topics are covered in both volumes. Central force motion, the differential scattering cross section, and small oscillations are introduced in Volume 1, and reviewed and extended in Volume 2. Rigid body motion is covered in great detail from the Newtonian point of view in Volume 1, and briefly reviewed and treated from the Lagrangian point of view in Volume 2. An introduction to Lagrangian and Hamiltonian mechanics is given in Volume 1, to prepare the student for the full treatment in Volume 2.

To conclude this preface, I point out some aspects of the book that are unique or are, in my opinion, treated better or more thoroughly here than elsewhere.

In Chapters 1–4, 85, and 86 the basic principles of Newtonian and relativistic kinematics are derived starting from the assumption of the existence and equivalence of inertial frames of reference. With this assumption, the law of transformation between inertial frames arises naturally and contains only one undetermined parameter, which is identified as the upper speed with which a particle can move with respect to an inertial frame. By choosing this parameter to be infinite we obtain the Galilean transformation, and by choosing it to be finite we obtain the Lorentz transformation.

In Chapters 8, 9, 88, and 89 an investigation of quantities that might be conserved in a collision leads naturally to the momentum conservation laws of Newtonian and relativistic mechanics, and from these laws to the definitions of force in both Newtonian and relativistic mechanics.

The treatment of the foundations of Newtonian and relativistic mechanics contained in the two above-mentioned sets of chapters clearly brings out the close relation between Newtonian and relativistic mechanics, and the primacy of the law of conservation of momentum over Newton's equation of motion and its counterpart in relativistic mechanics. It is the most thorough treatment that can be found anywhere. Many of the details are new and unique.

Chapter 6 presents very carefully and thoroughly the definition and properties of the angular velocity of one coordinate system with respect to another. By starting with the analytical definition of angular velocity rather than the geometrical definition, as is customarily done, many of the difficulties associated with this concept are avoided. Even though a good grasp of this concept is a prerequisite to an ability to express the laws of motion in rotating frames, and to an understanding of the kinematics and dynamics of rigid bodies, it is amazing to me how many of the standard texts are weak on this subject, and frequently contain spurious definitions of angular velocity.

Chapter 23 contains one of the simplest and most organized treatments of central force motion of which I am aware.

Chapters 24, 26, 60, and 61 contain a very complete treatment of the differential scattering cross section. Chapter 61 derives the relationship between the center of mass and laboratory cross sections for a collision in which both particles are moving, the collision is inelastic, and the product particles differ from the incident particles. No other text that I know of contains this complete result.

The treatment of rigid body motion in Chapters 34–40, with the possible addition of Chapters 41, 63, and 64, is sufficiently complete and detailed to form a course by itself. Chapter 38 is a thorough treatment of the inertia tensor from both the analytical and the geometric points of view. Chapter 39 contains an extremely detailed treatment of Euler angles, and the kinematics of rigid body motion. The expression in Chapter 39 of the equations of transformation between different reference frames and the relative angular velocities between these frames in terms of Euler angles is a very useful aid to anyone interested in solving rigid body motion problems.

Chapters 47–54 give a thorough treatment of Lagrange's equations of motion and include a detailed treatment of constraints both holonomic and anholonomic. Since most undergraduate courses do not have the time for such a complete treatment, an introductory abbreviated treatment appears in Chapters 42–44.

The treatment of small vibrations in Chapters 45 and 65 contains a number of results that cannot be found elsewhere.

The material in Chapter 66 together with the material in Appendices 28–30 represents a complete introduction to group theory and its application to symmetrical vibrating systems. Although the complexity of the notation makes the going a little tedious at times, the text contains none of the gaps and guesswork that seem to mar the treatments I have found elsewhere.

Quasi-coordinates and the Gibbs-Appell equations of motion covered in Chapters 67–70 are probably unknown to most physicists. There are only a few mechanics books in which they are even mentioned. Because I think they deserve more attention and are quite useful in solving certain problems, I have included them.

The treatment of Hamilton's equations of motion, canonical transformations, and Hamilton-Jacobi theory in Chapters 71–78 is quite unusual in that no use is made of the calculus of variations. Although this results in many of the proofs being a little longer than usual, I feel that it provides a more secure foundation in the subject, since most students who are encountering this subject for the first time are not completely sure of themselves in the use of the calculus of variations. This unsureness is usually compounded by the casual and sometimes erroneous use of the calculus of variations by some authors.

Although I prefer not to use the calculus of variations in a student's first encounter with Hamiltonian mechanics, I certainly believe it to be an extremely useful tool, hence have devoted Part 8 and Appendix 32 to this subject. The initial separation of the calculus of variations from Hamiltonian mechanics has the added organizational advantage of allowing the instructor who so desires to pursue the subject beyond and apart from what is required in Hamiltonian mechanics.

The presentation of the foundation and basic principles of relativistic mechanics in Part 9 is I believe one of the most logical and straightforward to be found anywhere. Though some of the proofs are a little ponderous, the general flow of concepts does not require an extraordinary distortion of a student's imagination. As a consequence relativistic mechanics seems almost inevitable.

The nature and importance of constants of the motion is stressed time and again throughout the text, starting with Chapter 17 and proceeding through Chapters 33, 46, 56, 71, and 81. Chapter 56 considers in detail the relationship between constants of the motion and the invariance of the Lagrangian under certain transformations, and Noether's theorem is presented without making use of variational techniques as is usually done in those few sources where this theorem can be found.

The subject of impulse, which usually causes students a great deal of unnecessary grief, is developed in detail in Chapters 18, 41, and 57.

Many of the appendices are quite useful in themselves, apart from their value in the body of the text. However since this book is intended primarily as a mechanics text, and only secondarily as a course in applied mathematics, many theorems in the appendices are stated without proof. Proofs that are particularly pertinent or cannot easily be found elsewhere are given.

In Appendices 4–6, and 23, the reader is led systematically from the geometric concept of a vector as a directed line segment to the highly analytical concept of general tensors. With a little amplification and completion of proofs, the material would make a good course in vectors and tensors.

Similarly the material in Appendices 12 and 24–26, with the possible addition of the material on quadratic forms in Appendix 27, forms a good outline for a course in matrices.

Appendix 32 provides an excellent introduction to the calculus of variations.

Appendices 11, 15, 19, and 22 cover a number of topics very important to classical mechanics in a manner that is both simpler and clearer than can be found elsewhere.

While writing this book I have not had in mind a hypothetical audience, but rather have written as if I were to be the reader. The book is in a sense a reflection of myself. Its exposure to numerous students over the years has sharpened rather than altered this reflection. I suspect that this is how most texts are written. Interestingly, I am dominantly a right hemisphere thinker—that is, I think in terms of pictures—but the first impression one gains of the text is that it is dominantly analytical. The probable explanation of this apparent anomaly is that one tends to emphasize the things that are personally difficult while at the same time ignoring what comes easily, with the result that the material is more analogous to a photographic negative than to a positive print. In any case the success of this book will depend on how many others share the difficulties, problems, loves, and hates that I experienced in learning classical mechanics. I hope that there are many, and that through this book they will derive some of the pleasures I have found in my encounter with classical mechanics.

EDWARD A. DESLOGE

Tallahassee, Florida
December 1981

Contents

VOLUME 1

Introduction

PART 1. THE NEWTONIAN MECHANICS OF PARTICLES

Section 1. The Basic Principles of Newtonian Kinematics	5

    1. Space and Time, 7
    2. Inertial Frames, 11
    3. Transformation Between Inertial Frames, 13
    4. Absolute Space and Time, 4

Section 2. Auxiliary Principles of Newtonian Kinematics	25

    5. Relative Motion of Particles, 27
    6. Relative Motion of Frames of Reference, 32
    7. The Description of Motion Using Orthogonal Curvilinear Coordinates, 43

Section 3. The Basic Principles of Newtonian Dynamics	47

    8. Mass and Momentum, 49
    9. Force, 59
  10. Center of Mass, 65

Section 4. Elementary Applications of Newton's Law	71

  11. Some Basic Forces, 73
  12. Statics of a Particle, 83
  13. Dynamics Problems, 88

Section 5. Auxiliary Principles of Newtonian Dynamics   99

14. Torque and Angular Momentum, 101
15. Work and Kinetic Energy, 107
16. Potential Energy, 113
17. Constants of the Motion, 120
18. Impulse, 126
19. The Equations of Motion in Noninertial Reference Frames, 136
20. The Equations of Motion in Orthogonal Curvilinear Coordinate Systems, 144

Section 6. Applications   149

21. One Dimensional Motion in an Arbitrary Potential, 151
22. The Harmonic Oscillator, 158
23. Central Force Motion, 172
24. The Differential Scattering Cross Section, 187
25. Two Particle Systems, 196
26. Two Particle Collisions, 203

PART 2. THE NEWTONIAN MECHANICS OF SYSTEMS OF PARTICLES

Section 1. Basic Principles   213

27. Dynamical Systems, 215
28. Force and Linear Momentum, 218
29. Torque and Angular Momentum, 226
30. Work and Kinetic Energy, 234
31. Cartesian Configuration Space, 240
32. Potential Energy, 243
33. Constants of the Motion, 247

Section 2. Rigid Body Motion   251

34. Rigid Bodies, 253
35. Equivalent Systems of Forces, 255
36. Statics of a Rigid Body, 264
37. Uniplanar Motion of a Rigid Body, 269
38. The Inertia Tensor, 286

39. Rigid Body Kinematics, 303
40. Rigid Body Dynamics, 320
41. Impulsive Motion of a Rigid Body, 338

PART 3.   AN INTRODUCTION TO LAGRANGIAN AND HAMILTONIAN MECHANICS

Section 1.   Lagrangian Mechanics   347

42. Holonomic Constraint Forces, 349
43. Generalized Coordinates for Holonomic Systems, 352
44. Lagrange's Equations of Motion for a Holonomic System, 361

Section 2.   Applications of Lagrangian Mechanics   371

45. Vibrating Systems, 373

Section 3.   Hamiltonian Mechanics   385

46. Hamilton's Equations of Motion, 387

SUPPLEMENTARY MATERIAL

Appendices   395

1. Analytical Representation of a Sine Function, 397
2. Partial Differentiation, 398
3. Jacobians, 401
4. Vector Algebra, 402
5. Vector Calculus, 408
6. Cartesian Tensors, 417
7. Orthogonal Curvilinear Coordinates, 427
8. Ordinary Differential Equations, 432
9. Linear Differential Equations, 436
10. Differentiation of an Integral, 441
11. Exact Differentials, 442
12. Matrices, 447
13. Systems of Linear Equations, 458
14. Functional Dependence, 460
15. The Method of Lagrange Multipliers, 464
16. Elliptic Functions, 467
17. Coordinate Transformations, 470

Tables 475

1. Abbreviations of Units, 477
2. Constants, 477
3. Conversion Factors, 478
4. Centers of Mass, 479
5. Moments of Inertia, 482
6. Vector Identities, 486

Answers 489

Combined Index I-1

VOLUME 2

PART 4. LAGRANGIAN MECHANICS

Section 1. Lagrange's Equations of Motion   511

47. Generalized Coordinates, 513
48. Lagrange's Equations of Motion for Elementary Systems, 521
49. Constraint Forces, 528
50. Lagrange's Equations of Motion for Holonomic Systems, 538
51. The Determination of Holonomic Constraint Forces, 549
52. Lagrange's Equations of Motion for Anholonomic Systems, 554
53. Generalized Force Functions, 558
54. Lagrange's Equations of Motion for Lagrangian Systems, 564
55. Lagrange's Equations of Motion and Tensor Analysis, 570

Section 2. Auxiliary Principles of Lagrangian Mechanics   573

56. Constants of the Motion in the Lagrangian Formulation, 575
57. Lagrange's Equations of Motion for Impulsive Forces, 588

PART 5. APPLICATIONS OF LAGRANGIAN MECHANICS

Section 1. Central Force Motion   595

58. Central Force Motion, 597
59. Bertrand's Theorem, 606
60. The Differential Scattering Cross Section, 614
61. Two Particle Collisions, 622
62. The Restricted Three Body Problem, 634

Section 2. Rigid Body Motion — 645

63. Rigid Body Kinematics, 647
64. Rigid Body Dynamics, 653

Section 3. Small Oscillations — 663

65. Vibrating Systems, 665
66. Symmetrical Vibrating Systems, 682

PART 6. QUASI-COORDINATES

67. Quasi-Coordinates, 711
68. Lagrange's Equations for Quasi-Coordinates, 715
69. The Gibbs-Appell Equations of Motion, 720
70. The Gibbs-Appell Equations and Rigid Body Motion, 727

PART 7. HAMILTONIAN MECHANICS

Section 1. Hamilton's Equations of Motion — 735

71. Hamilton's Equations of Motion, 737
72. Equations of Motion of the Hamiltonian Type, 744
73. Point Transformations, 750

Section 2. Canonical Transformations — 753

74. Canonical Transformations, 755
75. A Condensed Notation, 768
76. Hamilton's Canonical Equations of Motion, 773

Section 3. Hamilton-Jacobi Theory — 787

77. Generating Functions, 789
78. The Hamilton-Jacobi Equations, 797
79. Action and Angle Variables, 807

Section 4. Poisson Formulation of the Equations of Motion — 821

80. Poisson Brackets, 823
81. The Poisson Formulation of the Equations of Motion, 828

PART 8. VARIATIONAL PRINCIPLES IN CLASSICAL MECHANICS

82. D'Alembert's Principle, 835
83. Hamilton's Principle, 838
84. The Modified Hamilton's Principle, 843

PART 9. RELATIVISTIC MECHANICS

Section 1. Relativistic Kinematics   855

85. The Basic Postulates of Relativistic Kinematics, 857
86. The Lorentz Transformation, 859
87. Some Consequences of the Lorentz Transformation, 867

Section 2. Relativistic Dynamics of a Particle   873

88. Mass and Momentum, 875
89. Force, 885
90. Work and Energy, 888

Section 3. Four Dimensional Formulation of Relativistic Mechanics   891

91. Four Dimensional Formulation of Relativistic Mechanics, 893

Section 4. Relativistic Lagrangian and Hamiltonian Mechanics   899

92. Relativistic Lagrangian Mechanics, 901
93. Relativistic Hamiltonian Mechanics, 903

SUPPLEMENTARY MATERIAL

Appendices   907

18. Inverse Transformations, 909
19. Change of Variables in an Integral, 912
20. Euler's Theorem, 914
21. The Gram-Schmidt Orthogonalization Process, 915
22. Legendre Transformations, 918
23. General Tensors, 921

24. Matrix Transformations, 928
25. Eigenvalues of Matrices, 931
26. Diagonalization of Matrices, 935
27. Quadratic Forms, 937
28. Group Theory, 945
29. Vector Spaces, 951
30. Representations of a Group, 955
31. Multiply Periodic Functions, 967
32. Calculus of Variations, 968

Answers   **981**

Combined Index   **I-1**

CLASSICAL MECHANICS

Volume 1

Introduction

Mechanics may be defined as the study of the motion of material objects and the circumstances that influence this motion.

The objects in the universe can be roughly divided into three size categories: microscopic, macroscopic, and astronomic. Microscopic objects are objects of atomic or subatomic size; astronomic objects are objects of stellar or galactic size, and macroscopic objects are objects whose sizes fall in between the microscopic and the astronomic.

We can also distinguish between high speed motion and low speed motion. By high speed motion we mean motion in which the speeds one encounters are comparable to the speed of light. By low speed motion we mean motion in which the speeds encountered are small compared to the speed of light.

In this text we are primarily, though not exclusively, interested in the low speed motion of macroscopic objects. This area of mechanics is referred to as classical mechanics. The results we obtain are *not* generally valid for objects of microscopic or astronomic size or objects that are moving with speeds comparable to the speed of light. The mechanics of objects of microscopic size is the realm of quantum mechanics; the mechanics of high speed motion is the realm of special relativity; and the mechanics of objects of astronomic size is the realm of general relativity.

The foregoing categorization of classical mechanics is not meant to rule out the possibility of applying the laws of classical mechanics to objects of microscopic or astronomic size or to objects moving with high speeds; but if we do make such an application we should not be surprised to find some discrepancy between theory and experiment.

PARTICLES

If we are considering the low speed motion of an object of macroscopic size we can begin our study either by standing so far away from the object that it appears to be localized at a point, or by directing our attention to some small portion of the object, the dimensions of which are so small that it can be considered to be concentrated at a point. An object or a portion of an object that can for all practical purposes be localized at a point is called a particle.

Once we have determined the laws that govern the motion of a particle, we can determine the motion of larger objects by considering them to be made up of a finite number of particles.

It should be noted that the particles of classical mechanics are *not* molecules or atoms or electrons or nucleons. The motions of such objects are beyond the scope of classical mechanics. The particles of classical mechanics are generally large compared to such objects.

THE BASIC POSTULATES OF CLASSICAL MECHANICS

Experiment has shown that all the basic phenomena of classical mechanics can be derived from a few simple postulates. There are many different possible postulational formulations of classical mechanics. We have chosen a set of postulates that are close to the original historical postulates of classical mechanics, plausible in terms of our experience, easy to understand, easy to apply analytically, and easy to generalize into the relativistic domain. In the chapters that follow we consider these postulates and some of their consequences.

PART 1

THE NEWTONIAN MECHANICS OF PARTICLES

SECTION 1

The Basic Principles of Newtonian Kinematics

1

Space and Time

INTRODUCTION

Mechanics was defined in the Introduction as the study of the motion of material objects and the circumstances that influence this motion. To describe the motion of objects, it is necessary to have a clear notion of the meaning of space and time.

SPACE

The set of all particles and possible particles that comprise the universe can be broken down into subsets of particles whose members are at rest with respect to one another. Each of these subsets constitutes a three dimensional continuum of points, which we refer to as a space or a reference frame.

MEASUREMENT OF LENGTH

If a particle moves from a point P_1 to a point P_2 in a given space, it traces or defines a curve C in the space. A very important property of the curve is its length, or equivalently the distance traveled by the particle in going from P_1 to P_2.

To define length we first choose as a reference a short rigid rod AB whose ends can be treated approximately as points. We then put the end A of the rod on point P_1 and the end B on the curve C, move the rod along the curve in a succession of steps in which the end A is always moved to the prior position of B; when we cannot make a step without the end B going beyond P_2 and off the curve, we stop. We then define the length of the curve as the number of times the rod is contained in the curve, or equivalently the number of steps taken plus one. The same rod is used to measure lengths in different frames of reference.

The foregoing definition of length does not depend on the shape of the rod. The rod serves only to define two points whose relative positions are assumed to be fixed. A pair of dividers fixed in one configuration would serve the same purpose.

The accuracy of this method of defining length depends on the closeness to each other of the ends of the rod. In studying the macroscopic motion of a particular object it is always possible to obtain reasonable accuracy by simply choosing a rod that is suitably small.

NOTE: There is a certain amount of circularity in our definition of length. We have assumed that the shape of the reference rod remains fixed, or that the rod remains rigid, and yet we cannot strictly define what we mean by this until we have defined "length." One can partially escape this dilemma by picking an arbitrary rod as reference, using it to measure length, and then determining whether the results obtained can be interpreted in a simple fashion.

GEOMETRY

Starting with the fundamental concepts of a point and distance, mathematicians are able to define or construct different geometries. Which of these geometries, if any, is applicable to a particular physical space must be determined experimentally, by using the reference measuring rod to measure the distances between points, and then examining the relationships between these measurements.

NOTE: The determination of the geometry of space depends on our physical definition of length. If we choose some method of defining length other than the one introduced in the preceding section, we may obtain different results.

TIME

If the internal state of a particle or a localized object is changing, its successive states define a one dimensional continuum, which we call the time of that particle.

MEASUREMENT OF TIME INTERVALS

As the internal state of a particle changes, it passes from one point t_1 in its time to another point t_2 in its time. An important property of this change is the length or interval of time that elapses between times t_1 and t_2.

To define this time interval, we choose as reference a particle whose internal state is changing in a cyclic fashion. We then juxtapose the reference particle and the particle we are studying and define the time interval between t_1 and t_2 as the number of cycles the reference particle goes through between times t_1 and t_2.

NOTE: Just as there was a certain circularity in our definition of length, there is also a certain circularity in our definition of a time interval. To measure a time interval, we need to find a particle whose internal state is periodic in time. But we cannot determine whether a state is periodic in time until we have determined how to measure time intervals. The only way to remove ourselves from this dilemma is to pick a reference particle that we judge to be suitable, use it, and then see if the results we obtain are able to be interpreted in a simple fashion.

CLOCKS

Using the reference particle, we can divide the time line of a given particle into equal intervals. If we then number the successive points between adjacent intervals consecutively, the particle constitutes a clock.

Each particle in the universe can be thought of as a clock. We have no a priori knowledge of the relationship between the times of different particles, or equivalently the reading on different clocks.

SYNCHRONIZATION OF CLOCKS

To determine the relationship, if any, between clocks at different points in a particular space S, we start with our reference clock at a particular point P in the space S and synchronize the readings on the reference clock with the readings on the clock at P. We then proceed infinitesimally slowly from point to point in the space S and at each point synchronize the reading on the clock at the given point with the readings on the reference clock. If we then find at any future time that when we move slowly from point to point along any path in S, the clocks remain synchronized with the reference clock without the necessity of any adjustment, we say that the clocks in S are synchronized with one another.

It may not be possible to synchronize the clocks in a given frame. However if this can be done, we can speak of a single time associated with that frame.

REFERENCE FRAMES AND COORDINATE SYSTEMS

In the analysis of the motion of a particle, it is necessary to carefully distinguish between what is meant by a reference frame and what is meant by a coordinate system.

A reference frame is a real physical entity. A coordinate system is a convenient mathematical entity that can be used to describe the position of a point with respect to a reference frame. There are an infinite number of coordinate systems that can be used to describe the position of a point in a reference frame. We can even have coordinate systems that describe the position of a particle in a given reference frame that are in the frame but are moving with respect to the frame. For example, we can mathematically consider a coordinate system that at each instant of time is congruent to a given fixed coordinate system, but whose position is changing with time. One must be careful to distinguish between a reference frame that is moving with respect to another frame, and a coordinate system that is in a frame but moving with respect to that frame.

2

Inertial Frames

INTRODUCTION

It is possible to describe the motion of a particle with respect to any frame of reference. Experiment reveals, however, that out of all possible frames of reference, there are those with respect to which the formulation of the laws of motion is particularly simple. In this chapter we consider these frames.

INERTIAL FRAMES

If we consider all possible frames of reference, we find the following postulate to be true, if not universally, at least locally.

Postulate I. There exists at least one frame of reference, called an inertial frame, having the following properties:
  a. The geometry is Euclidean.
  b. The clocks can be synchronized.
  c. All positions, directions, and instants of time are equivalent with respect to the formulation of the laws of motion.

THE PRINCIPLE OF RELATIVITY

Most of us know that the relative motions of bodies inside planes, ships, and trains that are moving with a constant velocity seem to obey the same laws irrespective of the value of the constant velocity. This behavior is a reflection of an extremely important law of nature that we formalize in the following postulate.

Postulate II. Frames have the following properties:
  a. Any frame all of whose points are moving with the same constant velocity with respect to an inertial frame is also an inertial frame.
  b. All inertial frames are equivalent with respect to the formulation of the laws of motion.

The preceding postulate is sometimes called the principle of relativity. It was first practically used by Christian Huygens (1629–1695), who adopted it as a basic law of motion, and it is also one of the fundamental principles of the special theory of relativity as developed by Albert Einstein (1879–1955).

THE LAW OF INERTIA

If a particle that is far removed from the influence of all other particles in the universe is at rest with respect to an inertial frame, we expect it to remain at rest. If it moved spontaneously, it would have to move in some direction, and this would violate the condition that all directions are equivalent in an inertial frame. A particle that is at rest with respect to one inertial frame will be moving with a constant velocity with respect to a second inertial frame; hence if a particle that is far removed from the influence of all other particles in the universe is moving with a constant velocity with respect to an inertial frame, we expect it to continue moving with the same constant velocity. These observations are summarized in the following theorem.

Theorem 1. If a particle is far removed from the influence of all other particles in the universe, the particle will move with a constant velocity with respect to an inertial frame.

The theorem above is sometimes called the law of inertia. It is frequently attributed to Isaac Newton (1642–1727), who stated it as the first of his three laws of motion, or to Galileo Galilei (1564–1642), to whom Newton gave credit. However modern historians have shown that Galileo did not have a complete understanding of the law of inertia and that the source of Newton's first law was probably René Descartes (1596–1650) rather than Galileo. In any case the law was commonly accepted prior to Newton's use of it, primarily because Descartes had publicized it.

3

Transformation Between Inertial Frames

INTRODUCTION

Frequently, it is desirable to solve for the motion of a particle with respect to one inertial frame, but to express the results with respect to a second inertial frame. In this chapter we consider how this can be done.

The Galilean transformation we will ultimately obtain seems so obvious that it is customary in texts on Newtonian mechanics to write it down as immediately evident. It is instructive, however, to derive this transformation on the basis of a careful analysis, to prepare ourselves for the modifications of Newtonian theory that are required in relativistic mechanics. Readers who are solely interested in Newtonian mechanics may prefer to skip the proof of the following theorem and to move to the section entitled "The Galilean Transformation."

LAW OF TRANSFORMATION

Theorem 1. Let R be an inertial reference frame and R' a second inertial frame that is moving with respect to R. Let O be the origin and OX, OY, and OZ the axes of a Cartesian coordinate system S fixed in frame R. Let O' be the origin and $O'X'$, $O'Y'$, and $O'Z'$ the axes of a similar Cartesian coordinate system S' fixed in frame R'. The coordinates of a point with respect to S will be designated x, y and z, respectively; and the coordinates of a point with respect to S' will be designated x', y', and z', respectively. The time reading on a clock in the R frame will be designated by t, and the time reading on a clock in the R' frame will be designated t'. If we choose the coordinate systems and the zero setting on the clocks in such a way that at zero time in both frames the origins coincide, the respective axes are collinear, and the relative motion of S' with respect to S is in the positive x direction, the coordinates x, y, z, t of a particle and the coordinates x', y', z', t' of the same particle are related as follows:

$$x' = \Gamma(x - Vt) \qquad (1)$$

$$y' = y \qquad (2)$$

$$z' = z \qquad (3)$$

$$t' = \Gamma\left(t - \frac{Vx}{c^2}\right) \qquad (4)$$

where V is the speed with which S' is moving with respect to S, c is the upper limit on the speed with which a particle can move with respect to an inertial frame, and

$$\Gamma \equiv \left(1 - \frac{V^2}{c^2}\right)^{-1/2} \qquad (5)$$

If there is no upper limit on the speed of a particle, then $c = \infty$ and $\Gamma = 1$.

NOTE: The choice of coordinate axes and the setting on the clocks puts no physical restriction on the results.

PROOF: If we let $x_1 \equiv x$, $x_2 \equiv y$, $x_3 \equiv z$, $x_4 \equiv t$, and $\mathbf{x} \equiv x_1, x_2, x_3, x_4$, then the equations for transforming from R to R' can be written in the form

$$x_i' = f_i(\mathbf{x}) \qquad (6)$$

where the $f_i(\mathbf{x})$ are unknown functions of $\mathbf{x}$. If $\mathbf{a}$ and $\mathbf{b} \equiv \mathbf{a} + \mathbf{c}$ are two arbitrary points in R, then

$$b_i' - a_i' = f_i(\mathbf{b}) - f_i(\mathbf{a}) = f_i(\mathbf{a} + \mathbf{c}) - f_i(\mathbf{a}) \qquad (7)$$

Since there are no preferred spatial positions or instants of time, we expect $b_i' - a_i'$ to depend only on $\mathbf{c} \equiv \mathbf{b} - \mathbf{a}$ and not on $\mathbf{a}$. Hence we expect

$$f_i(\mathbf{a} + \mathbf{c}) - f_i(\mathbf{a}) = \text{fcn}(\mathbf{c}) \qquad (8)$$

If we take the derivative of Eq. (8) with respect to a_j, we obtain

$$\frac{\partial f_i(\mathbf{a} + \mathbf{c})}{\partial a_j} - \frac{\partial f_i(\mathbf{a})}{\partial a_j} = 0 \qquad (9)$$

which can be written equivalently

$$\left[\frac{\partial f_i(\mathbf{x})}{\partial x_j}\right]_{\mathbf{x} = \mathbf{a} + \mathbf{c}} = \left[\frac{\partial f_i(\mathbf{x})}{\partial x_j}\right]_{\mathbf{x} = \mathbf{a}} \qquad (10)$$

The equation above must be valid for arbitrary $\mathbf{a}$ and $\mathbf{c}$. Letting $\mathbf{a} = 0$ we obtain

$$\frac{\partial f_i(\mathbf{c})}{\partial c_j} = \text{constant} \qquad (11)$$

It follows that all the partial derivatives of f_i must be constants, hence $f_i(\mathbf{x})$ must be a linear function of x_1, x_2, x_3, and x_4. We conclude that the transformation between R and R' must be of the form

$$x' = A_1 x + A_2 y + A_3 z + A_4 t + A_5 \tag{12}$$

$$y' = B_1 x + B_2 y + B_3 z + B_4 t + B_5 \tag{13}$$

$$z' = C_1 x + C_2 y + C_3 z + C_4 t + C_5 \tag{14}$$

$$t' = D_1 x + D_2 y + D_3 z + D_4 t + D_5 \tag{15}$$

where the A_i, B_i, C_i, and D_i are constants.

At $t = 0$ in R and $t' = 0$ in R' the origins of S and S' coincide, hence the point $x = y = z = t = 0$ in R coincides with the point $x' = y' = z' = t' = 0$ in R'. Using these values in Eqs. (12)–(15) we obtain

$$A_5 = B_5 = C_5 = D_5 = 0 \tag{16}$$

Substituting Eq. (16) in Eqs. (12)–(15) we obtain

$$x' = A_1 x + A_2 y + A_3 z + A_4 t \tag{17}$$

$$y' = B_1 x + B_2 y + B_3 z + B_4 t \tag{18}$$

$$z' = C_1 x + C_2 y + C_3 z + C_4 t \tag{19}$$

$$t' = D_1 x + D_2 y + D_3 z + D_4 t \tag{20}$$

Since there are no preferred positions, or directions, we expect the transformations Eqs. (17)–(20) to be valid if we rotate about the x axis, or reflect in a plane contining the x axis. In particular we expect Eqs. (17)–(20) to be valid if: (a) we replace y by z, z by $-y$, y' by z', and z' by $-y'$, which corresponds to a rotation through $\pi/2$ about the x axis; or (b) we replace y by $-y$, and y' by $-y'$, which corresponds to a reflection in the x-z plane; or (c) we replace z by $-z$ and z' by $-z'$, which corresponds to a reflection in the x-y plane. Replacing y by z, z by $-y$, y' by z', and z' by $-y'$ in Eqs. (17), (19), and (20) we obtain

$$x' = A_1 x + A_2 z - A_3 y + A_4 t \tag{21}$$

$$z' = B_1 x + B_2 z - B_3 y + B_4 t \tag{22}$$

$$t' = D_1 x + D_2 z - D_3 y + D_4 t \tag{23}$$

Replacing y by $-y$ and y' by $-y'$ in Eq. (18) we obtain

$$-y' = B_1 x - B_2 y + B_3 z + B_4 t \tag{24}$$

Replacing z by $-z$ and z' by $-z'$ in Eq. (19) we obtain

$$-z' = C_1 x + C_2 y - C_3 z + C_4 t \tag{25}$$

Equations (17) and (21) can be simultaneously true for arbitrary x, y, z, t only if

$$A_2 = A_3 = 0 \tag{26}$$

16 Transformation Between Inertial Frames

Equations (20) and (23) can be simultaneously true for arbitrary x, y, z, t only if
$$D_2 = D_3 = 0 \tag{27}$$
Equations (18) and (24) can be simultaneously true for arbitrary x, y, z, t only if
$$B_1 = B_3 = B_4 = 0 \tag{28}$$
Equations (19) and (25) can be simultaneously true for arbitrary x, y, z, t only if
$$C_1 = C_2 = C_4 = 0 \tag{29}$$
Equations (19) and (22) can be simultaneously true for arbitrary x, y, z, t only if
$$B_2 = C_3 \equiv \beta \tag{30}$$
Substituting Eqs. (26)–(30) into Eqs. (17)–(20) we obtain
$$x' = A_1 x + A_4 t \tag{31}$$
$$y' = \beta y \tag{32}$$
$$z' = \beta z \tag{33}$$
$$t' = D_1 x + D_4 t \tag{34}$$
From Eqs. (31) and (34) it follows that
$$\frac{dx'}{dt'} = \frac{A_1 dx + A_4 dt}{D_1 dx + D_4 dt} = \frac{A_1 (dx/dt) + A_4}{D_1 (dx/dt) + D_4} \tag{35}$$
If we consider the motion of a point that is fixed in frame R', then $dx'/dt' = 0$ and $dx/dt = +V$. Substituting these values in Eq. (35) we obtain
$$0 = \frac{A_1 V + A_4}{D_1 V + D_4} \tag{36}$$
If we consider the motion of a point that is fixed in frame R, then $dx/dt = 0$ and $dx'/dt' = -V$. Substituting these values in Eq. (35) we obtain
$$-V = \frac{A_4}{D_4} \tag{37}$$
If a particle is moving in the positive x' direction, its speed with respect to R must be greater than or equal to its speed with respect to R'. Hence if a particle is moving in the positive x' direction with speed c with respect to R', where c is the upper limit, finite or infinite, on the speed of the particle with respect to R', then the speed of the particle with respect to R must be greater than or equal to c. But because of the equivalence of inertial frames, the upper limit c of the speed of a particle must have the same value in all inertial frames. Therefore the speed of the particle above with respect to R cannot exceed c, hence must be equal to c. Thus a particle that is moving in the

positive x' direction with a speed c with respect to R' must also be moving in the positive x direction with a speed c with respect to R. Setting $dx/dt = c$ and $dx'/dt' = c$ in Eq. (35) we obtain

$$c = \frac{A_1 c + A_4}{D_1 c + D_4} \tag{38}$$

From Eqs. (36)–(38) it follows that

$$A_4 = -A_1 V \tag{39}$$

$$D_4 = A_1 \tag{40}$$

$$D_1 = \frac{-A_1 V}{c^2} \tag{41}$$

Substituting Eqs. (39)–(41) in Eqs. (31)–(34) and redefining $A_1 \equiv \Gamma$ we obtain

$$x' = \Gamma(x - Vt) \tag{42}$$

$$y' = \beta y \tag{43}$$

$$z' = \beta z \tag{44}$$

$$t' = \Gamma\left(t - \frac{Vx}{c^2}\right) \tag{45}$$

We now let R'' represent an inertial frame fixed with respect to R', hence basically the same inertial frame as R', but in which the position of a particle is given with respect to the coordinate system S'' obtained by inverting the x' and y' axes of S'. We assume further that the time t'' in R'' is measured by the same set of clocks used in R'. The transformation between R'' and R' is given by

$$x'' = -x' \tag{46}$$

$$y'' = -y' \tag{47}$$

$$z'' = z' \tag{48}$$

$$t'' = t' \tag{49}$$

Combining Eqs. (42)–(45) and Eqs. (46)–(49) we obtain for the transformation between R'' and R

$$x'' = -\Gamma(x - Vt) \tag{50}$$

$$y'' = -\beta y \tag{51}$$

$$z'' = \beta z \tag{52}$$

$$t'' = \Gamma\left(t - \frac{Vx}{c^2}\right) \tag{53}$$

Because of the symmetrical relationship between R and R'', that is, R is moving with speed V in the positive x'' direction with respect to R'', and similarly R'' is moving with speed V in the positive x direction with respect to

R, the inverse of the transformation Eqs. (50)–(53) is given by

$$x = -\Gamma(x'' - Vt'') \tag{54}$$

$$y = -\beta y'' \tag{55}$$

$$z = \beta z'' \tag{56}$$

$$t = \Gamma\left(t'' - \frac{Vx''}{c^2}\right) \tag{57}$$

Combining Eqs. (51) and (55) we obtain

$$\beta = \pm 1 \tag{58}$$

When $V = 0$, $y'' = -y$, hence the positive sign must be chosen in Eq. (58). Thus

$$\beta = +1 \tag{59}$$

Substituting Eqs. (54) and (57) in Eq. (50) we obtain

$$x'' = -\Gamma\left\{[-\Gamma(x'' - Vt'')] - V\Gamma\left(t'' - \frac{Vx''}{c^2}\right)\right\}$$

$$= \Gamma^2\left(1 - \frac{V^2}{c^2}\right)x'' \tag{60}$$

From Eq. (60) it follows that

$$\Gamma = \pm\left(1 - \frac{V^2}{c^2}\right)^{-1/2} \tag{61}$$

When $V = 0$, $x'' = -x$, hence the positive sign must be chosen in Eq. (61). Thus

$$\Gamma = \left(1 - \frac{V^2}{c^2}\right)^{-1/2} \tag{62}$$

This completes the proof of the theorem.

NOTE: In the proof above it would appear that we could obtain the same result without introducing the frame R'', by first obtaining the inverse of Eqs. (42)–(45) by interchanging the primed and unprimed coordinates, and replacing V by $-V$, and then combining the transformation Eqs. (42)–(45) with their inverse to obtain β and Γ. However before one could do this it would be necessary to show somehow that the dependence of β and Γ on V was such that $\beta(-V) = \beta(V)$ and $\Gamma(-V) = \Gamma(V)$. The introduction of the frame R'' avoids this difficulty.

VECTOR FORMULATION OF THE LAW OF TRANSFORMATION

In the preceding section we chose coordinate systems S and S' in such a way as to make the results come out as simply as possible. This puts no physical restriction on our results. Nevertheless, it is sometimes useful to write the transformation in a form that allows a wider choice of coordinate systems S and S'.

Corollary. Let R and R' be two inertial frames. Let $\mathbf{V}$ be the velocity in R of R' with respect to R and $\mathbf{V}'$ the velocity in R' of R with respect to R'. Let $\mathbf{r}$ be the position in R of a particle with respect to a point O fixed in R and $\mathbf{r}'$ the position in R' of the same particle with respect to a point O' fixed in R'. Let S and S' be Cartesian coordinate systems fixed in R and R' respectively with origins at O and O' respectively and such that the orientation of S' with respect to $-\mathbf{V}'$ is the same as the orientation of S with respect to $\mathbf{V}$, and the points O and O' coincide at times $t = t' = 0$. Then the relationship between the coordinates x, y, z, t and x', y', z', t' are given by the vector equation

$$\mathbf{r}' = \mathbf{r} + \frac{(\Gamma - 1)(\mathbf{r} \cdot \mathbf{V})\mathbf{V}}{V^2} - \Gamma \mathbf{V} t \tag{63}$$

$$t' = \Gamma t - \frac{\Gamma(\mathbf{r} \cdot \mathbf{V})}{c^2} \tag{64}$$

where c is the upper limit on the speed with which a particle may move with respect to an inertial frame, and

$$\Gamma \equiv \left(1 - \frac{V^2}{c^2}\right)^{-1/2} \tag{65}$$

If there is no upper limit on the speed of a particle, then $c = \infty$ and $\Gamma = 1$.

NOTE: Ordinarily, one thinks of a vector equation as a relation that is independent of the choice of coordinate system. This is true in the ordinary vector equation where the vectors are vectors in the same space, and when the vectors are always resolved with respect to the same coordinate system. In Eqs. (63) and (64), however, the vectors $\mathbf{r}$ and $\mathbf{r}'$ are vectors in different spaces, and when resolved the components of $\mathbf{r}$ are expressed with respect to a coordinate system fixed in R and the components of $\mathbf{r}'$ are expressed with respect to a coordinate system fixed in R'.

PROOF: If we let $\mathbf{r}_\parallel$ and $\mathbf{r}'_\parallel$ be the components of $\mathbf{r}$ and $\mathbf{r}'$ parallel to $\mathbf{V}$, and $\mathbf{r}_\perp$ and $\mathbf{r}'_\perp$ be the components of $\mathbf{r}$ and $\mathbf{r}'$ perpendicular to $\mathbf{V}$, then from Theorem 1 we can write

$$\mathbf{r}'_\parallel = \Gamma\left[\mathbf{r}_\parallel - \mathbf{V}t\right] \tag{66}$$

$$\mathbf{r}'_\perp = \mathbf{r}_\perp \tag{67}$$

$$t' = \Gamma\left(t - \frac{\mathbf{V} \cdot \mathbf{r}}{c^2}\right) \tag{68}$$

Summing Eqs. (66) and (67) we obtain

$$\mathbf{r}'_\parallel + \mathbf{r}'_\perp = \Gamma\left[\mathbf{r}_\parallel - \mathbf{V}t\right] + \mathbf{r}_\perp \tag{69}$$

But
$$\mathbf{r}'_\parallel + \mathbf{r}'_\perp = \mathbf{r}' \tag{70}$$
and
$$\mathbf{r}_\perp = \mathbf{r} - \mathbf{r}_\parallel \tag{71}$$
Substituting Eqs. (70) and (71) in Eq. (69) we obtain
$$\mathbf{r}' = \mathbf{r} + (\Gamma - 1)\mathbf{r}_\parallel - \Gamma \mathbf{V} t \tag{72}$$
But
$$\mathbf{r}_\parallel = \left(\mathbf{r} \cdot \frac{\mathbf{V}}{V}\right) \frac{\mathbf{V}}{V} \tag{73}$$

Substituting Eq. (73) in Eq. (72) we obtain Eq. (63). Thus Eqs. (66) and (67) give us Eq. (63), and Eq. (68) gives us Eq. (64). All we have actually proved so far is that Eqs. (63) and (64) are valid for the coordinate systems S and S' described in Theorem 1. However, the coordinate systems described in the present theorem can be obtained by rotating each of the coordinate systems in Theorem 1 by the same amount, hence for such pairs of coordinate systems the relations expressed in Eqs. (63) and (64) remain valid.

THE UPPER LIMIT ON THE SPEED OF A PARTICLE

Experiment reveals that there is a finite upper limit c to the speed with which a particle can move, and this speed is equal to the speed with which light moves with respect to an inertial frame. In classical mechanics, one deals with speeds that are small compared to c; hence in the realm where classical mechanics is valid, one may assume that there is effectively no upper limit. We formalize this fact in the following postulate.

Postulate III. There is no limit to the speed with which a particle can move with respect to an inertial frame.

THE GALILEAN TRANSFORMATION

If Postulate III is true, $c = \infty$ and $\Gamma = 1$ in Theorem 1. Hence in classical mechanics the law of transformation from the frame S to the frame S' in Theorem 1 is given by
$$x' = x - Vt \tag{74}$$
$$y' = y \tag{75}$$
$$z' = z \tag{76}$$
$$t' = t \tag{77}$$

or in vector notation

$$\mathbf{r}' = \mathbf{r} - \mathbf{V}t \tag{78}$$
$$t' = t \tag{79}$$

This transformation is called the Galilean transformation.

RELATIVISTIC MECHANICS

In relativistic mechanics, one deals with speeds approaching the upper limit c; hence one cannot accept Postulate III. In relativistic mechanics, the upper limit c is recognized to be the speed of light. The transformation one obtains by setting c in Theorem 1 equal to the speed of light is called the Lorentz transformation.

PROBLEM

1. A frame of reference S' moves with velocity v in the x direction of frame S. The x and x' axes are parallel. If a particle moves at a speed u and at an angle θ with the x axis in frame S, show that its speed and direction in frame S' are given by $\tan\theta' = \sin\theta/(\cos\theta - v/u)$, $u'^2 = u^2 + v^2 - 2uv\cos\theta$.

4

Absolute Space and Time

INTRODUCTION

In the preceding chapter we introduced the idea of inertial reference frames and showed how results can be transformed from one inertial frame to another. In the present chapter we lay the groundwork for generalization of this result to noninertial frames.

ABSOLUTE TIME

If the Galilean transformation is valid, it is possible to synchronize the clocks in different inertial frames. This is equivalent to saying that there is effectively a single time for all inertial frames of reference. To be able to handle noninertial as well as inertial frames, we shall make a stronger assumption, namely, that there is a single time for all frames of reference, whether inertial or noninertial. This is equivalent to assuming that there is an absolute time. From now on the word "time" has reference to this absolute time.

ABSOLUTE SPACE

Let R be an inertial frame and R' a second inertial frame moving with speed V with respect to R. If we choose a coordinate system S in frame R, and a similar coordinate system S' in frame R' in such a way that at time zero the origins coincide and the respective coordinate axes are collinear, and R' is moving in the positive x direction with respect to R, then the coordinates of a particle with respect to S' are related to its coordinates with respect to S by the equations

$$x' = x - Vt \tag{1}$$
$$y' = y \tag{2}$$
$$z' = z \tag{3}$$

This result was proved in the preceding chapter. Now let us let $\bar{S}$ be a coordinate system that is in R, coincides point for point with S at time $t = 0$, and is moving in the positive x direction with a speed V. The statement that the coordinate system is in R means that at every instant it coincides point for point with a coordinate system that is fixed in R and differs from S only in the location of its origin. From simple geometrical considerations we can show that the coordinates of the particle with respect to $\bar{S}$ are related to the coordinates with respect to S by the equations

$$\bar{x} = x - Vt \tag{4}$$
$$\bar{y} = y \tag{5}$$
$$\bar{z} = z \tag{6}$$

The two transformations above are formally identical. It follows that we can treat the coordinate system S' as if it were simply a coordinate system in R moving with a speed V in the positive x direction. Hence coordinate systems in different inertial frames can be treated as if they are coordinate systems in the same space, moving with respect to one another. To be able to handle noninertial as well as inertial frames, we shall assume that it is possible to do this with noninertial frames as well as inertial frames. This assumption is equivalent to assuming that as far as an observer in one frame is concerned, space is absolute, since the measurements of any other observer in any other frame can be interpreted as if they were measurements in the first observer's frame but simply referred to a moving coordinate system. From now on, unless stated otherwise, the word "space" refers to the space of an observer in an inertial frame, treated as if it were absolute.

REFERENCE FRAMES AND COORDINATE SYSTEMS

If we assume that space and time are absolute in the sense described in the preceding sections, the distinctions we have made between reference frames and coordinate systems are not terribly critical. Therefore in future chapters we are rather loose in our use of these terms and in fact use them interchangeably.

SECTION 2

Auxiliary Principles
of Newtonian Kinematics

5

The Relative Motion of Particles

INTRODUCTION

In this chapter we introduce a few simple but useful techniques for analyzing the relative motion of particles and demonstrate their use with a number of examples.

POSITION, VELOCITY, AND ACCELERATION

The position of a point or a particle with respect to another point or particle can be represented by a vector joining the two particles. We designate the vector joining a particular particle i to a particular particle j as $\mathbf{r}(ij)$, and we speak of $\mathbf{r}(ij)$ as the vector position of particle j with respect to particle i. Thus

$$\mathbf{r}(ij) \equiv \text{vector position of particle } j \text{ with respect to particle } i \qquad (1)$$

Note that the first term in the argument (ij) designates the position of the tail of the vector and the second term the position of the head of the vector. If we know $\mathbf{r}(ij)$, then the velocity of particle j with respect to particle i is given by

$$\mathbf{v}(ij) \equiv \dot{\mathbf{r}}(ij) \qquad (2)$$

and the acceleration of particle j with respect to particle i is given by

$$\mathbf{a}(ij) \equiv \dot{\mathbf{v}}(ij) \equiv \ddot{\mathbf{r}}(ij) \qquad (3)$$

If we have a system of particles $i, j, k, \ldots$, and we are given sufficient information in the form of the vectors $\mathbf{r}(ij), \mathbf{r}(jk), \ldots$, to uniquely specify the relative configuration of the system, the position of any one of the particles with respect to any other particle can be obtained by exploiting the following properties of the position vectors.

Property 1

$$\mathbf{r}(ij) \equiv -\mathbf{r}(ji) \qquad (4)$$

Property 2

$$\mathbf{r}(ij) + \mathbf{r}(jk) = \mathbf{r}(ik) \tag{5}$$

Similar relations hold for the velocity and acceleration. Thus

Property 1a

$$\mathbf{v}(ij) \equiv -\mathbf{v}(ji) \tag{6}$$

Property 2a

$$\mathbf{v}(ij) + \mathbf{v}(jk) \equiv \mathbf{v}(ik) \tag{7}$$

Property 1b

$$\mathbf{a}(ij) \equiv -\mathbf{a}(ji) \tag{8}$$

Property 2b

$$\mathbf{a}(ij) + \mathbf{a}(jk) \equiv \mathbf{a}(ik) \tag{9}$$

EXAMPLES

Example 1. A ship that is heading due south steams across a current running northwest. At the end of 2 hours it is found that the ship has gone 20 km in the direction 30° west of south. Find the speed of the ship relative to the water, and the speed of the current.

Solution. Let $\mathbf{i}$ be a unit vector pointing east and $\mathbf{j}$ be a unit vector pointing north. Let o be a point fixed in the earth, w a point fixed in the water, and s a point fixed on the ship. We then have from the information given

$$\mathbf{v}(ws) = -A\mathbf{j} \tag{10}$$
$$\mathbf{v}(ow) = -B\mathbf{i} + B\mathbf{j} \tag{11}$$
$$\mathbf{v}(os) = 10(-\sin 30°\,\mathbf{i} - \cos 30°\,\mathbf{j}) \tag{12}$$

where A and B are constants and the units are kilometers per hour. The velocities above are not independent but must satisfy the condition

$$\mathbf{v}(os) = \mathbf{v}(ow) + \mathbf{v}(ws) \tag{13}$$

Substituting Eqs. (10)–(12) in Eq. (13) we obtain

$$-B\mathbf{i} + (B - A)\mathbf{j} = -5\mathbf{i} - 5\sqrt{3}\,\mathbf{j} \tag{14}$$

Equating the components and solving for A and B we obtain

$$A = 5 + 5\sqrt{3} \tag{15}$$
$$B = 5 \tag{16}$$

Substituting Eqs. (15) and (16) in Eqs. (10) and (11) we obtain

$$\mathbf{v}(ws) = -(5 + 5\sqrt{3})\mathbf{j} \tag{17}$$

$$\mathbf{v}(ow) = -5\mathbf{i} + 5\mathbf{j} \tag{18}$$

Hence the speed of the ship relative to the water is 13.66 km/hr and the speed of the current is 7.07 km/hr.

Example 2. A ship a, steaming due east at 15 km/hr, sights a second ship b, steaming with constant velocity, at a distance of 10 km and in a direction 30° east of north. Ten minutes later b lies in a direction 45° east of north, and after a further 5 minutes in a direction 60° east of north. Find the distance ab at the last observation, and determine the speed and course of b.

Solution. Let $\mathbf{i}$ be a unit vector pointing east and $\mathbf{j}$ a unit vector pointing north. Let o be a point fixed in the ocean. We then have from the information given

$$\mathbf{v}(oa) = 15\mathbf{i} \tag{19}$$

$$\mathbf{v}(ob) = A\mathbf{i} + B\mathbf{j} \tag{20}$$

where A and B are constants. From Eqs. (19) and (20) it follows that

$$\mathbf{v}(ab) \equiv \mathbf{v}(ao) + \mathbf{v}(ob) \equiv -\mathbf{v}(oa) + \mathbf{v}(ob) = (A - 15)\mathbf{i} + B\mathbf{j} \tag{21}$$

hence

$$\mathbf{r}(ab) = \left[(A - 15)t + C\right]\mathbf{i} + \left[Bt + D\right]\mathbf{j} \tag{22}$$

where C and D are constants. From the information given

$$\left[\mathbf{r}(ab)\right]_{t=0} = 10(\sin 30° \mathbf{i} + \cos 30° \mathbf{j}) \tag{23}$$

$$\left[\mathbf{r}(ab)\right]_{t=1/6} = G(\sin 45° \mathbf{i} + \cos 45° \mathbf{j}) \tag{24}$$

$$\left[\mathbf{r}(ab)\right]_{t=1/4} = H(\sin 60° \mathbf{i} + \cos 60° \mathbf{j}) \tag{25}$$

where G and H are constants. Substituting Eq. (22) in Eqs. (23)–(25) and equating components, we obtain six equations in the six unknowns A, B, C, D, G, H, which when solved give

$$A = \frac{35 - 15\sqrt{3}}{\sqrt{3} - 1} = 12.32 \tag{26}$$

$$B = -2A = -24.64 \tag{27}$$

$$C = 5 \tag{28}$$

$$D = 5\sqrt{3} = 8.66 \tag{29}$$

$$G = \frac{20\sqrt{2}}{6(\sqrt{3} - 1)} = 6.44 \tag{30}$$

$$H = 5 \tag{31}$$

Thus the velocity of the ship b is

$$\mathbf{v}(ob) = 12.32\mathbf{i} - 24.64\mathbf{j} \tag{32}$$

and the position of the ship b relative to a at the last observation is

$$[\mathbf{r}(ab)]_{t=1/4} = 5(\sin 60° \mathbf{i} + \cos 60° \mathbf{j}) \tag{33}$$

Hence the ship b is moving with speed 27.55 km/hr in a direction 63.43° south of east, and is 5 km from ship a at the last observation.

PROBLEMS

1. A particle moves with constant speed v along the cardioid $r = a(1 + \cos\theta)$. Show that the radial component of the acceleration is constant, and that both $d\theta/dt$ and the magnitude of the resultant acceleration are proportional to $r^{-1/2}$.

2. An airplane of speed v has a range (out and back) of R miles in calm weather. There is a wind of speed w from the direction $\alpha°$ east of north. Show that to fly in a direction whose true bearing is $\phi°$ east of north, the apparent bearing must make an angle $\beta°$ with the true one where $v\sin\beta = w\sin(\phi - \alpha)$. Prove also that the range in miles for a true bearing $\phi°$ east of north is $[R(v^2 - w^2)]/\{v[v^2 - w^2\sin^2(\phi - \alpha)]^{1/2}\}$.

3. A ship A is moving due east with constant speed u, while a second ship B is moving due north with constant speed $2u$, and third ship C is moving northeast with constant speed $2\sqrt{2}\,u$. When A is at a point O, it is observed that B and C cross the track of A simultaneously at distances a and $2a$, respectively, ahead of A. Prove that when $OA = x$ the line joining A to the midpoint of BC is rotating in space about the vertical through A with angular speed $12au/(9a^2 + 16x^2)$.

4. A flying target moves with uniform velocity v along the line $y = h$ in the x-y plane, where the x axis is horizontal and the y axis vertical. At time $t = 0$, the target is located at the point (o, h), and a guided missile P starts from the origin and moves with constant speed $2v$ in such a way that its velocity is always directed toward the target. Show that if at time t the coordinates of P are (x, y), and if $dx/dy \equiv q$, then $q(h - y) = vt - x$ and $1 + q^2 = 4v^2(dt/dy)^2$. By differentiating the first of these equations with respect to y and eliminating dt/dy, derive a differential equation relating q and y. Deduce that $[q + (1 + q^2)^{1/2}]^2 = h/(h - y)$.

5. A cyclist observes that while cycling due north at 25 km/hr the wind appears to come from due east, but when cycling due east at the same speed the wind appears to come from 22.5° south of east. Determine the velocity of the wind.

6. A steamship is traveling at the rate of 15 km/hr due south. A man on the ship notices that smoke pours out of the funnel in a direction that is due east with respect to the ship. If the speed of the ship is increased to 20 km/hr, he observes that the initial direction of the smoke is northeast. Find the velocity of the wind. Assume that as the smoke pours from the funnel it immediately takes up the velocity of the wind.

6

Relative Motion of Frames of Reference

INTRODUCTION

In the analysis of the motion of a particle or a system of particles it is sometimes useful or necessary to transform results from one coordinate system or frame to a second coordinate system or frame moving with respect to the first coordinate system. In this chapter we consider the problems involved in such transformations.

NOTATION

Let us consider two arbitrary Cartesian coordinate systems or frames S and S'.

Let O be the origin and $O1$, $O2$, and $O3$ the coordinate axes of the S frame; and O' the origin and $O'1'$, $O'2'$, and $O'3'$ the coordinate axes of the S' frame. The unit vectors in the $O1$, $O2$, and $O3$ directions are designated e_1, e_2, and e_3, respectively, and the unit vectors in the $O'1'$, $O'2'$, and $O'3'$ directions are designated $e_{1'}$, $e_{2'}$, and $e_{3'}$, respectively. The components of an arbitrary vector $\mathbf{A}$ with respect to the S frame are designated A_1, A_2, and A_3, respectively; and with respect to the S' frame they are designated $A_{1'}$, $A_{2'}$, and $A_{3'}$, respectively.

TRANSLATIONAL MOTION

The frame S' is said to be undergoing a translation with respect to the frame S if the position of the origin O' with respect to the origin O is changing in time. The translational motion of the frame S' with respect to the frame S is completely described by the vector $\mathbf{v}(OO')$, the velocity of the point O' with respect to the point O.

ROTATIONAL MOTION

The frame S' is said to be undergoing a rotation with respect to the frame S if the orientation of S' with respect to S is changing in time. Our first object in this chapter is to show that just as there is a vector $\mathbf{v}(OO')$ that describes the relative translational motion of the two frames, there is also a vector that describes the relative rotational motion of the two frames.

ANGULAR VELOCITY

The time rate of change of a quantity with respect to the frame S is not necessarily the same as the time rate of change of the same quantity with respect to the frame S'. For example, the time rate of change of a vector such as $\mathbf{e}_{1'}$, which is fixed in frame S', is zero with respect to S', but if S' is rotating with respect to S, it will not in general be zero with respect to S. We therefore distinguish between the time rate of change of an arbitrary quantity A with respect to S, which we designate $\dot{A}$, and the time rate of change of A with respect to S', which we designate $\dot{A}'$. We now prove two important facts.

Theorem 1. Let A be an arbitrary scalar, $\dot{A}$ the time rate of change of A with respect to a frame S, and $\dot{A}'$ the time rate of change of A with respect to a frame S'. Then

$$\dot{A} = \dot{A}' \tag{1}$$

PROOF: Since the value of a scalar does not depend on the position or orientation of the frame with respect to which it is measured, its time rate of change will be the same with respect to both frames.

Theorem 2. Let $\mathbf{A}$ be an arbitrary vector, $\dot{\mathbf{A}}$ the time rate of change of $\mathbf{A}$ with respect to a frame S, and $\dot{\mathbf{A}}'$ the time rate of change of $\mathbf{A}$ with respect to a frame S'. Then there exists a vector $\boldsymbol{\omega}(SS')$, called the angular velocity of frame S' with respect to S, such that

$$\dot{\mathbf{A}} = \dot{\mathbf{A}}' + \boldsymbol{\omega}(SS') \times \mathbf{A} \tag{2}$$

If $\bar{S}$ is a frame whose orientation coincides with the orientation of S but whose origin O' is the same as the origin of S', then at each instant of time there is a line L passing through O' whose points are instantaneously at rest with respect to both $\bar{S}$ and S'. The angular velocity vector $\boldsymbol{\omega}(SS')$ is parallel to L, points in the same direction S' would move if it were a right handed screw turning about L, and has a magnitude equal to the angular rate of rotation of S' about L.

PROOF: The vector **A** in terms of its components in the S' frame is given by

$$\mathbf{A} = \sum_{i'} \mathbf{e}_{i'} A_{i'} \tag{3}$$

Taking the time derivative of Eq. (3) with respect to frame S we obtain

$$\dot{\mathbf{A}} = \sum_{i'} \mathbf{e}_{i'} \dot{A}_{i'} + \sum_{i'} \dot{\mathbf{e}}_{i'} A_{i'} \tag{4}$$

Taking the time derivative of Eq. (3) with respect to frame S' we obtain

$$\dot{\mathbf{A}}' = \sum_{i'} \mathbf{e}_{i'} \dot{A}'_{i'} + \sum_{i'} \dot{\mathbf{e}}'_{i'} A_{i'} \tag{5}$$

But $A_{i'}$ is a scalar and therefore $\dot{A}'_{i'} = \dot{A}_{i'}$; and $\mathbf{e}_{i'}$ is fixed in the frame S' and therefore $\dot{\mathbf{e}}'_{i'} = 0$. Thus Eq. (5) can be written

$$\dot{\mathbf{A}}' = \sum_{i'} \mathbf{e}_{i'} \dot{A}_{i'} \tag{6}$$

If we substitute Eq. (6) in Eq. (4) we obtain

$$\dot{\mathbf{A}} = \dot{\mathbf{A}}' + \sum_{i'} \dot{\mathbf{e}}_{i'} A_{i'} \tag{7}$$

In terms of their components in the S' frame, the vectors $\dot{\mathbf{e}}_{i'}$ are given by

$$\dot{\mathbf{e}}_{i'} = \sum_{j'} (\dot{\mathbf{e}}_{i'} \cdot \mathbf{e}_{j'}) \mathbf{e}_{j'} \tag{8}$$

Of the nine terms $\dot{\mathbf{e}}_{i'} \cdot \mathbf{e}_{j'}$ only three are independent. To show this we note first that the unit vectors $\mathbf{e}_{i'}$ satisfy the relation

$$\mathbf{e}_{i'} \cdot \mathbf{e}_{j'} = \delta_{i'j'} \tag{9}$$

Where $\delta_{i'j'}$ is the Kronecker delta, which has the value $+1$ if $i' = j'$ and is zero otherwise. If we take the time derivative of Eq. (9) with respect to frame S we obtain

$$\dot{\mathbf{e}}_{i'} \cdot \mathbf{e}_{j'} + \mathbf{e}_{i'} \cdot \dot{\mathbf{e}}_{j'} = 0 \tag{10}$$

from which it follows that

$$\dot{\mathbf{e}}_{i'} \cdot \mathbf{e}_{j'} = -\dot{\mathbf{e}}_{j'} \cdot \mathbf{e}_{i'} \tag{11}$$

From Eq. (11) it follows that

$$\dot{\mathbf{e}}_{1'} \cdot \mathbf{e}_{1'} = \dot{\mathbf{e}}_{2'} \cdot \mathbf{e}_{2'} = \dot{\mathbf{e}}_{3'} \cdot \mathbf{e}_{3'} = 0 \tag{12}$$

Of the remaining six terms $\dot{\mathbf{e}}_{1'} \cdot \mathbf{e}_{2'}$, $\dot{\mathbf{e}}_{2'} \cdot \mathbf{e}_{3'}$, $\dot{\mathbf{e}}_{3'} \cdot \mathbf{e}_{1'}$, $\dot{\mathbf{e}}_{2'} \cdot \mathbf{e}_{1'}$, $\dot{\mathbf{e}}_{1'} \cdot \mathbf{e}_{3'}$, and $\dot{\mathbf{e}}_{3'} \cdot \mathbf{e}_{2'}$, only three are distinct. We use these three distinct terms to define a vector

$\boldsymbol{\omega}(SS')$ as follows:

$$\omega_{1'}(SS') \equiv \dot{\mathbf{e}}_{2'} \cdot \mathbf{e}_{3'} = -\dot{\mathbf{e}}_{3'} \cdot \mathbf{e}_{2'} \tag{13}$$

$$\omega_{2'}(SS') \equiv \dot{\mathbf{e}}_{3'} \cdot \mathbf{e}_{1'} = -\dot{\mathbf{e}}_{1'} \cdot \mathbf{e}_{3'} \tag{14}$$

$$\omega_{3'}(SS') \equiv \dot{\mathbf{e}}_{1'} \cdot \mathbf{e}_{2'} = -\dot{\mathbf{e}}_{2'} \cdot \mathbf{e}_{1'} \tag{15}$$

Eqs. (13)–(15) can be written more compactly as follows:

$$\dot{\mathbf{e}}_{i'} \cdot \mathbf{e}_{j'} = \sum_{k'} \epsilon_{i'j'k'} \omega_{k'}(SS') \tag{16}$$

where $\epsilon_{i'j'k'}$ is called a permutation symbol (see Appendix 4) and has the value $+1$ if ijk is an even permutation of the sequence 123, has the value -1 if ijk is an odd permutation of the sequence 123, and is zero otherwise. If we substitute Eq. (16) in Eq. (8) and the result in Eq. (7) we obtain

$$\dot{\mathbf{A}} = \dot{\mathbf{A}}' + \sum_{i'} \sum_{j'} \sum_{k'} \epsilon_{i'j'k'} \omega_{k'}(SS') A_{i'} \mathbf{e}_{j'} \tag{17}$$

The last term in Eq. (17) is just the cross product of the vector $\boldsymbol{\omega}(SS')$ and the vector $\mathbf{A}$, as can seen by writing out all the terms (see Appendix 4). We can thus rewrite Eq. (17)

$$\dot{\mathbf{A}} = \dot{\mathbf{A}}' + \boldsymbol{\omega}(SS') \times \mathbf{A} \tag{18}$$

This completes the first part of the theorem. In the second part of the theorem we let P' be a point that is fixed in the frame S' and at time t lies along a line that passes through O' and is parallel to $\boldsymbol{\omega}(SS')$. Let $\mathbf{r}(O'P')$ be the vector that joins O' and P'. Then $\dot{\mathbf{r}}'(O'P') = 0$, and $\boldsymbol{\omega}(SS') \times \mathbf{r}(O'P') = 0$, and therefore from Eq. (18)

$$\dot{\mathbf{r}}(O'P') = \dot{\mathbf{r}}'(O'P') + \boldsymbol{\omega}(SS') \times \mathbf{r}(O'P') = 0 \tag{19}$$

It follows that the set of points P', satisfying the condition above, forms a line L that passes through O' and whose orientation is fixed with respect to both frame S and frame S' or equivalently is at rest with respect to $\bar{S}$ and S'.

To determine the sense and magnitude of $\boldsymbol{\omega}(SS')$ let us consider the velocity of a point Q' that is fixed in S' and at time t lies along a line that passes through O' and is perpendicular to the line L. Using Eq. (18) with $\mathbf{A} = \mathbf{r}(O'Q')$ we obtain

$$\dot{\mathbf{r}}(O'Q') = \dot{\mathbf{r}}'(O'Q') + \boldsymbol{\omega}(SS') \times \mathbf{r}(O'Q')$$
$$= 0 + |\boldsymbol{\omega}(SS')| |\mathbf{r}(O'Q')| \mathbf{n} \tag{20}$$

where $|\boldsymbol{\omega}(SS')|$ is the magnitude of $\boldsymbol{\omega}(SS')$, $|\mathbf{r}(O'Q')|$ is the magnitude of $\mathbf{r}(O'Q')$, and $\mathbf{n}$ is a unit vector that is perpendicular to $\boldsymbol{\omega}(SS')$ and $\mathbf{r}(O'Q')$ and whose sense is such that $\boldsymbol{\omega}(SS')$, $\mathbf{r}(O'Q')$, and $\mathbf{n}$ in that order form a right handed system. But from a geometrical consideration, the magnitude of

$\dot{\mathbf{r}}(O'Q')$ is given by

$$|\dot{\mathbf{r}}(O'Q')| = |\mathbf{r}(O'Q')|\dot{\theta} \qquad (21)$$

where $\dot{\theta}$ is the angular rate of rotation of S' about the instantaneous axis of rotation. Comparing Eqs. (20) and (21) we see that

$$|\omega(SS')| = \dot{\theta} \qquad (22)$$

Finally we note that the direction of $\dot{\mathbf{r}}(O'Q')$ obtained using Eq. (20) will agree with the direction obtained using geometrical considerations only if $\omega(SS')$ is parallel to the axis of rotation in a right handed sense. Thus the sense must be right handed. This completes the proof of the theorem.

The vector $\omega(SS')$ provides a complete description of the rotational motion of frame S' with respect to S, for if we know $\omega(SS')$ we can determine the rate of change with respect to S of the orientation of all the unit vectors $\mathbf{e}_{i'}$, by using the relations

$$\dot{\mathbf{e}}_{i'} = \omega(SS') \times \mathbf{e}_{i'} \qquad (23)$$

Note that the orientation of the frame S' with respect to the frame S remains fixed if and only if $\omega(SS') = 0$. Thus $\omega(SS')$ determines the rotation only, not the translation.

THE INSTANTANEOUS AXIS OF ROTATION

The angular velocity ω of a rigid body with respect to a frame S is defined as the angular velocity of any frame $\bar{S}$ fixed in the rigid body with respect to S. If the frame S is not specified it is generally assumed to be an inertial frame. In the following remarks we assume that all quantities are measured with respect to a given frame S, but for the sake of economy of language we suppress mention of this frame.

If a rigid body is rotating about a fixed point O, then at every instant there is a line L of points in the body or in an extension of the body that are instantaneously at rest. The line L is called the instantaneous axis of rotation of the body. It can be shown that the line L passes through O and is parallel to ω. It follows that at a given instant the motion of a rigid body with one point fixed can be considered to be a rotation about the line L.

If no point in the rigid body is fixed, there is in general no line of points that are instantaneously at rest, but there is a line L of points that are all instantaneously moving along the line L, that is, a line L along which there is no motion perpendicular to L. The line L in this case is also called the instantaneous axis of rotation and can be shown to be parallel to ω. It follows that at a given instant the motion of a rigid body with no point fixed can be considered to consist of a rotation about L and a translation parallel to L.

The instantaneous axis of rotation of a rigid body in general changes with time and is not made up of a set of fixed points in the rigid body. For example, if a cone is rolling on a perfectly rough stationary horizontal surface, the instantaneous axis of rotation will be the line of contact of the cone and the surface.

SOME PROPERTIES OF THE ANGULAR VELOCITY VECTOR

In this section we derive a number of useful properties of the angular velocity vector.

Theorem 3. Let $\omega(SS')$ be the angular velocity of a frame S' with respect to a frame S, and $\omega(S'S)$ the angular velocity of frame S with respect to frame S'. Then

$$\omega(S'S) = -\omega(SS') \qquad (24)$$

PROOF: Let **A** be an arbitrary vector. Then from Theorem 2

$$\dot{\mathbf{A}} = \dot{\mathbf{A}}' + \omega(SS') \times \mathbf{A} \qquad (25)$$

and

$$\dot{\mathbf{A}}' = \dot{\mathbf{A}} + \omega(S'S) \times \mathbf{A} \qquad (26)$$

Adding Eqs. (25) and (26) we obtain

$$[\omega(SS') + \omega(S'S)] \times \mathbf{A} = 0 \qquad (27)$$

Equation (27) can be true for arbitrary **A** only if

$$\omega(SS') + \omega(S'S) = 0 \qquad (28)$$

Equation (24) follows immediately from Eq. (28), and the theorem is proved.

Theorem 4. Let $\omega(SS')$ be the angular velocity of a frame S' with respect to a frame S, $\omega(S'S'')$ the angular velocity of a frame S'' with respect to S', and $\omega(SS'')$ the angular velocity of frame S'' with respect to S. Then

$$\omega(SS'') = \omega(SS') + \omega(S'S'') \qquad (29)$$

PROOF: Let **A** be an arbitrary vector. Then from Theorem 2

$$\dot{\mathbf{A}} = \dot{\mathbf{A}}' + \omega(SS') \times \mathbf{A} \qquad (30)$$

$$\dot{\mathbf{A}}' = \dot{\mathbf{A}}'' + \omega(S'S'') \times \mathbf{A} \qquad (31)$$

$$\dot{\mathbf{A}}'' = \dot{\mathbf{A}} + \omega(S''S) \times \mathbf{A} \qquad (32)$$

Adding Eqs. (30)–(32) we obtain

$$[\omega(SS') + \omega(S'S'') + \omega(S''S)] \times \mathbf{A} = 0 \qquad (33)$$

hence
$$\omega(SS') + \omega(S'S'') + \omega(S''S) = 0 \qquad (34)$$

If we move the last term in Eq. (34) to the right hand side of the equation and make use of Theorem 3, we obtain Eq. (29), which is the desired result.

Theorem 5. Let $\omega(SS')$ be the angular velocity of a frame S' with respect to a frame S, $\dot{\omega}(SS')$ the time rate of change of $\omega(SS')$ with respect to an observer fixed in S, and $\dot{\omega}'(SS')$ the time rate of change of $\omega(SS')$ with respect to an observer fixed in S'. Then

$$\dot{\omega}(SS') = \dot{\omega}'(SS') \qquad (35)$$

PROOF: From Theorem 2

$$\dot{\omega}(SS') = \dot{\omega}'(SS') + \omega(SS') \times \omega(SS') \qquad (36)$$

The last term in Eq. (36) is identically zero; hence Eq. (35) follows immediately from Eq. (36).

TRANSFORMATION EQUATIONS FOR THE VELOCITY AND ACCELERATION OF A PARTICLE

If we know the position, velocity, and acceleration of a particle in one frame and wish to find its position, velocity, and acceleration in a second frame we can use the following theorem.

Theorem 6. Let S and S' be two frames with origins O and O', respectively; P be a particle; **R**, **V**, and **A** be the position, velocity, and acceleration in S of O' with respect to O; **R'**, **V'** and **A'** be the position, velocity, and acceleration in S' of O with respect to O'; **r**, **v**, and **a** be the position, velocity, and acceleration in S of P with respect to O; **r'**, **v'**, and **a'** be the position, velocity, and acceleration in S' of P with respect to O'; ω and α be the angular velocity and acceleration in S of S' with respect to S; and ω' and α' the angular velocity and acceleration in S' of S with respect to S'. Then

$$\mathbf{r} = \mathbf{R} + \mathbf{r}' \qquad (37)$$

$$\mathbf{v} = \mathbf{V} + \mathbf{v}' + \omega \times \mathbf{r}' \qquad (38)$$

$$\mathbf{a} = \mathbf{A} + \mathbf{a}' + \alpha \times \mathbf{r}' + 2\omega \times \mathbf{v}' + \omega \times (\omega \times \mathbf{r}') \qquad (39)$$

PROOF: We start with the relation

$$\mathbf{r}(OP) = \mathbf{r}(OO') + \mathbf{r}(O'P) \qquad (40)$$

where $\mathbf{r}(AB)$ designates the position of a point B with respect to a point A.

Taking the time derivative of Eq. (40) with respect to the frame S we obtain
$$\dot{\mathbf{r}}(OP) = \dot{\mathbf{r}}(OO') + \dot{\mathbf{r}}(O'P) \tag{41}$$
From Theorem 2
$$\dot{\mathbf{r}}(O'P) = \dot{\mathbf{r}}'(O'P) + \omega(SS') \times \mathbf{r}(O'P) \tag{42}$$
Substituting Eq. (42) in Eq. (41) we obtain
$$\dot{\mathbf{r}}(OP) = \dot{\mathbf{r}}(OO') + \dot{\mathbf{r}}'(O'P) + \omega(SS') \times \mathbf{r}(O'P) \tag{43}$$
Taking the time derivative of Eq. (43) with respect to the frame S we obtain
$$\ddot{\mathbf{r}}(OP) = \ddot{\mathbf{r}}(OO') + \overline{\dot{\mathbf{r}}'(O'P) + \omega(SS') \times \mathbf{r}(O'P)} \tag{44}$$
From Theorem 2
$$\overline{\dot{\mathbf{r}}'(O'P) + \omega(SS') \times \mathbf{r}(O'P)}$$
$$= \overline{\dot{\mathbf{r}}'(O'P) + \omega(SS') \times \mathbf{r}(O'P)}^{\,\cdot\,'}$$
$$+ \omega(SS') \times \left[\dot{\mathbf{r}}'(O'P) + \omega(SS') \times \mathbf{r}(O'P)\right]$$
$$= \ddot{\mathbf{r}}'(O'P) + \dot{\omega}'(SS') \times \mathbf{r}(O'P) + 2\omega(SS') \times \dot{\mathbf{r}}'(O'P)$$
$$+ \omega(SS') \times \left[\omega(SS') \times \mathbf{r}(O'P)\right] \tag{45}$$
Substituting Eq. (45) in Eq. (44), and noting from Theorem 5 that $\dot{\omega}'(SS') = \dot{\omega}(SS')$, we obtain
$$\ddot{\mathbf{r}}(OP) = \ddot{\mathbf{r}}(OO') + \ddot{\mathbf{r}}'(O'P) + \dot{\omega}(SS') \times \mathbf{r}(O'P) + 2\omega(SS') \times \dot{\mathbf{r}}'(O'P)$$
$$+ \omega(SS') \times \left[\omega(SS') \times \mathbf{r}(O'P)\right] \tag{46}$$
If we make use of the notation introduced at the beginning of the theorem, Eqs. (40), (43), and (46) reduce to Eqs. (37), (38), and (39). This completes the proof of the theorem.

EXAMPLES

Example 1. A wheel of radius a rolls on a horizontal plane and makes a constant angle β with the plane, the center of the wheel describing a circle of radius λa with speed V. Determine the angular velocity of the wheel.

Solution. Let C be the center of the wheel and O the center of the circle described by C. Let S be a frame with origin at O, 3 axes vertically up, and 1 axis fixed with respect to the plane; S' a frame with origin at O, $3'$ axis vertically up and $1'$ axis in the direction OC; and S'' a frame with origin at C, $1''$ axis normal to the wheel, and $2''$ axis fixed in the wheel. The situation is represented in Fig. 1 at a time when the 1-3 plane, the $1'$-$3'$ plane, and the $1''$-$3''$ planes coincide. The angular velocities $\omega(SS')$ and $\omega(S'S'')$ are of the

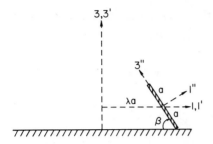

Fig. 1

form

$$\omega(SS') = A\mathbf{e}_3 = A\mathbf{e}_{3'} \tag{47}$$

$$\omega(S'S'') = B\mathbf{e}_{1''} = B\sin\beta\,\mathbf{e}_{1'} + B\cos\beta\,\mathbf{e}_{3'} \tag{48}$$

The values of A and B can be determined from our knowledge of the speed V of the point C, and the fact that the point on the wheel that is in contact with the plane must be momentarily at rest. The velocity of the point C is given by

$$\begin{aligned}\dot{\mathbf{r}}(OC) &= \dot{\mathbf{r}}'(OC) + \omega(SS') \times \mathbf{r}(OC) \\ &= 0 + A\mathbf{e}_{3'} \times (\lambda a \mathbf{e}_{1'}) \\ &= A\lambda a \mathbf{e}_{2'} \end{aligned} \tag{49}$$

But from the information given in the problem

$$\dot{\mathbf{r}}(OC) = V\mathbf{e}_{2'} \tag{50}$$

Combining Eqs. (49) and (50) we obtain

$$A = \frac{V}{\lambda a} \tag{51}$$

The velocity of P, the point on the wheel that is in contact with the plane, is given by

$$\begin{aligned}\dot{\mathbf{r}}(OP) &= \dot{\mathbf{r}}(OC) + \dot{\mathbf{r}}(CP) = \dot{\mathbf{r}}(OC) + \dot{\mathbf{r}}''(CP) + \omega(SS'') \times \mathbf{r}(CP) \\ &= \dot{\mathbf{r}}(OC) + 0 + \left[\omega(SS') + \omega(S'S'')\right] \times \mathbf{r}(CP) \\ &= V\mathbf{e}_{2'} \\ &\quad + \left(\frac{V}{\lambda a}\mathbf{e}_{3'} + B\sin\beta\,\mathbf{e}_{1'} + B\cos\beta\,\mathbf{e}_{3'}\right) \times (a\cos\beta\,\mathbf{e}_{1'} - a\sin\beta\,\mathbf{e}_{3'}) \\ &= \left(V + \frac{V}{\lambda}\cos\beta + Ba\right)\mathbf{e}_{2'} \end{aligned} \tag{52}$$

Since the wheel is not slipping,

$$\dot{\mathbf{r}}(OP) = 0 \tag{53}$$

Combining Eqs. (52) and (53) we obtain

$$B = -\frac{V}{\lambda a}(\lambda + \cos\beta) \tag{54}$$

Combining the results above we obtain

$$\omega(SS'') = \omega(SS') + \omega(S'S'')$$
$$= B \sin \beta \, \mathbf{e}_{1'} + (A + B \cos \beta) \mathbf{e}_{3'} \qquad (55)$$

where A and B are given by Eqs. (51) and (54).

PROBLEMS

1. A particle is fixed in a frame S at the point $(3, 1, 2)$. Find the velocity of the particle as noted by an observer fixed in a frame S' that has the same origin as S but is rotating with an angular velocity $\omega(SS') = 5\mathbf{e}_1 - 4\mathbf{e}_2 - 10\mathbf{e}_3$ with respect to S.

2. Two frames S and S' have a common origin, and the frame S' is rotating with respect to S with an angular velocity $\omega(SS') = 2\mathbf{e}_{1'} - 3\mathbf{e}_{2'} + 5\mathbf{e}_{3'}$. Let $\mathbf{A}$ be a vector defined by the equation $\mathbf{A} = \sin t \, \mathbf{e}_{1'} - \cos t \, \mathbf{e}_{2'} + \exp t \, \mathbf{e}_{3'}$. Determine $\dot{\mathbf{A}}$, $\mathbf{A}'$, $\ddot{\mathbf{A}}$, and $\mathbf{A}''$ in terms of their components with respect to the frame S'.

3. Two frames S and S' have a common origin and the frame S' is rotating with respect to S with an angular velocity $\omega(SS') = \mathbf{e}_3 \omega$, where ω is a constant. At time $t = 0$ the frames are coincident. Let $\mathbf{A}$ be a vector defined by the equation $\mathbf{A} = \sin t \, \mathbf{e}_{1'} - \cos t \, \mathbf{e}_{2'} + \exp t \, \mathbf{e}_{3'}$. Determine the following quantities in terms of their components with respect to the frames indicated:
   a. $\dot{\mathbf{A}}$ with respect to S.
   b. $\dot{\mathbf{A}}$ with respect to S'.
   c. $\mathbf{A}'$ with respect to S.
   d. $\mathbf{A}'$ with respect to S'.

4. A circle of radius a rotates in its own plane with uniform angular velocity ω about a fixed point O of the circumference. A point moves around the circumference with uniform speed V relative to the circle and in the same sense as ω. Prove that the locus of the point with respect to the plane in which the motion takes place may be written $r = 2a \sin[V\theta/(V + 2a\omega)]$, where r and θ are polar coordinates with O as origin.

5. A circular disk of radius r is pivoted about an axle through its center O perpendicular to its plane. To a point A of its rim a rod AB of length $4r$ is freely jointed; and to B is freely jointed a piston rod BC that is constrained to move in a straight line through O. The disk rotates with constant angular velocity ω. Initially $OABC$ (in order) is a straight line. Prove that after a time t the angular velocity is $\omega \cos \omega t / (16 - \sin^2 \omega t)^{1/2}$

and that the velocity of B is then $r\omega \sin \omega t \{1 + [\cos \omega t/(16 - \sin^2 \omega t)^{1/2}]\}$. Prove also that the acceleration of BC at the ends of its path are in the ratio $5:3$.

6. A merry-go-round is rotating with an angular velocity Ω about a vertical axis A. A child on the merry-go-round is sitting on a horse that is a distance a from the axis A. The horse is moving up and down, and its distance h above the floor of the merry-go-round is given by $h = b + c \sin \omega t$. Determine as a function of time the acceleration of the child relative to the ground.

7. A cylinder of radius b rolls on the inside of a larger cylinder of radius a. If the outer cylinder is rotating about its own axis, which is fixed, with angular velocity Ω, and the plane containing the axes of the two cylinders is turning about the fixed axes with angular velocity ω, find the angular velocity of the small cylinder. Assume that both ω and Ω are positive.

8. A cylinder A of radius c and center C is rolling without slipping on a horizontal plane. A second cylinder B of radius d and center D, and with axis parallel to the axis of the first cylinder, is rolling without slipping on the first cylinder. The point C is moving with a constant speed u with respect to the horizontal plane. The point D is moving with a constant speed v with respect to a frame fixed in cylinder A. Let P be a point on the cylinder B, and assume that at $t = 0$ the points C, D, and P lie in the same vertical plane. Find the velocity and acceleration of the point P with respect to the horizontal plane.

9. A right circular cone of vertical angle 2α rolls without slipping on the outside of another fixed right circular cone of vertical angle 2β, the vertices of the cones being always in contact. The axis of the moving cone makes n complete revolutions about the axis of the fixed cone every second. Find the angular velocity of the moving cone.

10. Two conic shells of semivertical angles β and $\beta + 2\alpha$, respectively, have a common axis and vertex. Between them is pressed a solid cone of semivertical angle α. The shells are made to rotate in opposite senses with angular speed ω about their common axes. Show that if there is no slipping, the speed of a point on the axis of the solid cone at a distance r from the vertex is $\omega r \cos(\alpha + \beta) \tan \alpha$ and that the component of the angular velocity of the solid cone in the direction of its axis is $\omega \sin(\alpha + \beta) \cot \alpha$. Find the instantaneous axis of rotation of the solid cone.

11. A rigid body is free to rotate about a fixed point O. Show that if $\mathbf{r}(A)$, $\mathbf{r}(B)$, $\mathbf{v}(A)$, and $\mathbf{v}(B)$ are the positions and velocities, respectively, of two points A and B fixed in the rigid body, then the angular velocity of the rigid body is given by $\boldsymbol{\omega} = [\mathbf{v}(A) \times \mathbf{v}(B)]/[\mathbf{v}(A) \cdot \mathbf{r}(B)] = [\mathbf{v}(B) \times \mathbf{v}(A)]/[\mathbf{v}(B) \cdot \mathbf{r}(A)]$.

7

The Description of Motion Using Orthogonal Curvilinear Coordinates

INTRODUCTION

Up till now the position, velocity, and acceleration vectors of a particle generally have been expressed in terms of their components with respect to a particular set of orthogonal unit vectors. It is sometimes convenient however to use different sets of unit vectors for different positions of the particle whose motion is being considered. If, for example, the motion of a particle is best described by spherical coordinates r, θ, ϕ, it is probably convenient to express the position, velocity, and acceleration vectors of the particle, when it is located at a particular point, in terms of a set of unit vectors whose respective directions are in the direction of increasing r, the direction of increasing θ, and the direction of increasing ϕ at the given point. In this chapter we consider cases in which the directions of the sets of unit vectors used differ from point to point.

ORTHOGONAL CURVILINEAR COORDINATES

The nature and properties of orthogonal curvilinear coordinates and the associated sets of unit vectors are discussed in Appendix 7. We shall let $\mathbf{q} \equiv q_1, q_2, q_3$ represent a set of orthogonal curvilinear coordinates and $\mathbf{e}_1(\mathbf{q})$, $\mathbf{e}_2(\mathbf{q})$, and $\mathbf{e}_3(\mathbf{q})$ be the set of unit vectors associated with the point $\mathbf{q}$.

TIME DERIVATIVES OF A VECTOR IN ORTHOGONAL CURVILINEAR COORDINATES

If **A** is a vector property of a particle that depends on the position **q** of the particle and the time t, then the time derivative of **A** is given by

$$\dot{\mathbf{A}} = \sum_i \frac{\partial \mathbf{A}(\mathbf{q},t)}{\partial q_i} \dot{q}_i + \frac{\partial \mathbf{A}(\mathbf{q},t)}{\partial t} \tag{1}$$

If we write the vector **A** in the form

$$\mathbf{A} = \sum_i A_i \mathbf{e}_i \tag{2}$$

then it follows that

$$\dot{\mathbf{A}} = \sum_i \dot{A}_i \mathbf{e}_i + \sum_i A_i \dot{\mathbf{e}}_i \tag{3}$$

where

$$\dot{\mathbf{e}}_i = \sum_j \frac{\partial \mathbf{e}_i(\mathbf{q})}{\partial q_j} \dot{q}_j \tag{4}$$

If we tabulate the quantities $\dot{\mathbf{e}}_i$ for a particular orthogonal curvilinear coordinate system we can readily obtain the time derivative of any vector in terms of the unit vectors $\mathbf{e}_i$. It follows that if we know the partial derivatives $\partial \mathbf{e}_i(\mathbf{q})/\partial q_j$ we can obtain the $\dot{\mathbf{e}}_i$, and if we know the $\dot{\mathbf{e}}_i$ we can obtain the time derivative of a vector function $\mathbf{A}(\mathbf{q}, t)$. Appendix 7 outlines the method for obtaining the partial derivatives $\partial \mathbf{e}_i(\mathbf{q})/\partial q_j$ and tabulates the results for cylindrical and spherical coordinates.

CYLINDRICAL COORDINATES

The cylindrical coordinates ρ, ϕ, z are defined by the equations

$$x = \rho \cos \phi \tag{5a}$$
$$y = \rho \sin \phi \tag{5b}$$
$$z = z \tag{5c}$$

Using the results of Appendix 7 we obtain for the time derivatives of the unit vectors $\mathbf{e}_\rho$, $\mathbf{e}_\phi$, $\mathbf{e}_z$

$$\dot{\mathbf{e}}_\rho = \dot{\phi} \mathbf{e}_\phi \tag{6a}$$
$$\dot{\mathbf{e}}_\phi = -\dot{\phi} \mathbf{e}_\rho \tag{6b}$$
$$\dot{\mathbf{e}}_z = 0 \tag{6c}$$

Using the results above we find that the position, velocity, and acceleration vectors for a particle are:

$$\mathbf{r} = \rho \mathbf{e}_\rho + z \mathbf{e}_z \tag{7}$$

$$\dot{\mathbf{r}} = \dot{\rho} \mathbf{e}_\rho + \rho \dot{\phi} \mathbf{e}_\phi + \dot{z} \mathbf{e}_z \tag{8}$$

$$\ddot{\mathbf{r}} = (\ddot{\rho} - \rho \dot{\phi}^2) \mathbf{e}_\rho + (\rho \ddot{\phi} + 2 \dot{\rho} \dot{\phi}) \mathbf{e}_\phi + \ddot{z} \mathbf{e}_z \tag{9}$$

SPHERICAL COORDINATES

The spherical coordinates r, θ, ϕ are defined by the equations

$$x = r \sin \theta \cos \phi \tag{10a}$$
$$y = r \sin \theta \sin \phi \tag{10b}$$
$$z = r \cos \theta \tag{10c}$$

Using the results of Appendix 7 we obtain for the time derivatives of the unit vectors $\mathbf{e}_r$, $\mathbf{e}_\theta$, $\mathbf{e}_\phi$

$$\dot{\mathbf{e}}_r = \dot{\theta} \mathbf{e}_\theta + \sin \theta \, \dot{\phi} \mathbf{e}_\phi \tag{11a}$$

$$\dot{\mathbf{e}}_\theta = -\dot{\theta} \mathbf{e}_r + \cos \theta \, \dot{\phi} \mathbf{e}_\phi \tag{11b}$$

$$\dot{\mathbf{e}}_\phi = -\sin \theta \, \dot{\phi} \mathbf{e}_r - \cos \theta \, \dot{\phi} \mathbf{e}_\theta \tag{11c}$$

Using the results above we find that the position, velocity, and acceleration vectors for a particle are:

$$\mathbf{r} = r \mathbf{e}_r \tag{12}$$

$$\dot{\mathbf{r}} = \dot{r} \mathbf{e}_r + r \dot{\theta} \mathbf{e}_\theta + r \sin \theta \, \dot{\phi} \mathbf{e}_\phi \tag{13}$$

$$\ddot{\mathbf{r}} = (\ddot{r} - r \dot{\theta}^2 - r \sin^2 \theta \, \dot{\phi}^2) \mathbf{e}_r + (r \ddot{\theta} + 2 \dot{r} \dot{\theta} - r \sin \theta \cos \theta \, \dot{\phi}^2) \mathbf{e}_\theta$$
$$+ (r \sin \theta \, \ddot{\phi} + 2 \sin \theta \, \dot{r} \dot{\phi} + 2 r \cos \theta \, \dot{\theta} \dot{\phi}) \mathbf{e}_\phi \tag{14}$$

CURVILINEAR COORDINATE SYSTEMS VIEWED AS MOVING COORDINATE SYSTEMS

We have assumed in this chapter that the use of an orthogonal curvilinear coordinate system is essentially equivalent to using a different set of orthogonal axes for each point in space. Alternatively we could have assumed that we had a single set of orthogonal axes, but these axes were moving.

From this point of view the use of cylindrical coordinates is equivalent to using a frame whose angular velocity is

$$\boldsymbol{\omega} = \dot{\phi} \mathbf{e}_z \tag{15}$$

and the use of spherical coordinates is equivalent to using a frame whose angular velocity is

$$\omega = \dot{\phi}\cos\theta\, \mathbf{e}_\rho - \dot{\phi}\sin\theta\, \mathbf{e}_\theta + \dot{\theta}\mathbf{e}_\phi \tag{16}$$

PROBLEMS

1. A particle moves with a constant speed v along the cardioid $r = k(1 + \cos\theta)$. Find:
   a. The radial component of the acceleration.
   b. The magnitude of the acceleration.
   c. The angular speed $\dot{\theta}$.

2. A particle moves on a spiral path such that the position in polar coordinates is given by $r = bt^2$, $\phi = ct$, where b and c are constants. Find the velocity and acceleration as functions of time in terms of the unit vectors $\mathbf{e}_r$ and $\mathbf{e}_\phi$.

3. A particle moves on a helical path with cylindrical coordinates varying with time as $\rho = a$, $\phi = bt$, $z = ct^2$, where a, b, and c are constants. Find the velocity and acceleration vectors as functions of t. Find the angle between the velocity and acceleration vectors at time $t = 1$.

4. Parabolic coordinates ξ, η, z are defined by the equations $x = \frac{1}{2}(\eta^2 - \xi^2)$, $y = \xi\eta$, $z = z$. Determine the position and velocity vectors of a particle in terms of the unit vectors $\mathbf{e}_\xi$, $\mathbf{e}_\eta$, and $\mathbf{e}_z$.

SECTION 3

The Basic Principles of Newtonian Dynamics

8

Mass and Momentum

INTRODUCTION

Up till now, we have been considering the motion of a single particle. In this chapter, we consider how the presence of a second particle influences the motion of a given particle.

THE INTERACTION OF PARTICLES

Consider two isolated particles, each of which, because it is isolated, is moving with a constant velocity with respect to an inertial frame. If these two particles come close to each other, the velocity of each one will be found to change, and we say that the particles interact with each other. Our first objective is to describe this interaction.

One of the simplest ways to imagine two particles to be interacting is to suppose that the one particle has some *thing* or property that it exchanges or passes on unchanged to the second particle. As a result of this exchange, each particle is modified, and the amount of whatever is exchanged is a measure of the interaction. In the remainder of this chapter we consider whether such a property can exist, and if it can, whether it does.

COLLISIONS

If two particles are approaching each other, and if the interaction between the particles has a finite range, or at least is negligible beyond a certain range, the motion can be broken down into three periods: an initial period during which the particles are moving toward each other with constant velocities, an intermediate period during which the particles interact, and a final period in which the original particles or some other set of particles are moving away with constant velocities. If these conditions are satisfied, we refer to the interaction as a collision. In the following section we assume that we can treat the interaction between two particles as a collision in which the interaction takes place instantaneously at a point, and we speak of the particles being free before and after the collision.

Since there are no preferred directions or points in an inertial frame, we can write down the following two theorems.

Theorem 1. For every collision that occurs, a collision in which the velocities of the incident and the outgoing particles are reflected in a plane is a possible collision.

Theorem 2. For every collision that occurs, a collision in which the velocities of the incident and the outgoing particles are rotated the same arbitrary amount is a possible collision.

QUANTITIES CONSERVED IN A COLLISION

If two particles a and b, with velocities $\mathbf{v}(a)$ and $\mathbf{v}(b)$, respectively, collide and two or more particles $c, d, \ldots$ emerge from the collision with velocities $\mathbf{v}(c), \mathbf{v}(d), \ldots$, respectively, and if there exists a quantity g, associated with a particle, that depends only on the intrinsic nature of the particle and its velocity and satisfies the condition

$$g[a,\mathbf{v}(a)] + g[b,\mathbf{v}(b)] = g[c,\mathbf{v}(c)] + g[d,\mathbf{v}(d)] + \cdots \qquad (1)$$

we say that the quantity g is conserved in the collision.

From the definition of conserved quantities, we can write down the following two theorems:

Theorem 3. If $g'(\mathbf{v})$ and $g''(\mathbf{v})$ are conserved quantities, then any linear combination of $g'(\mathbf{v})$ and $g''(\mathbf{v})$ is a conserved quantity, or equivalently

$$g(\mathbf{v}) = Ag'(\mathbf{v}) + Bg''(\mathbf{v}) \qquad (2)$$

where A and B are arbitrary constants, is a conserved quantity.

PROOF: The proof of this theorem follows immediately from the definition of a conserved quantity.

Theorem 4. If there exists a conserved quantity of the form

$$g(\mathbf{v}) = \alpha + \beta h(\mathbf{v}) \qquad (3)$$

where α and β are particle parameters, $h(\mathbf{v})$ is a universal function of the velocity other than a constant, and β is not identically zero, then the ratio of the value of β for one particle to the value of β for a second particle is unique for the given pair of particles.

PROOF: Consider a collision in which two particles a and b moving with velocities $\mathbf{v}(a)$ and $\mathbf{v}(b)$, respectively, collide and end up with velocities $\mathbf{v}^*(a)$ and $\mathbf{v}^*(b)$, respectively. If $\alpha + \beta h(\mathbf{v})$ is a conserved quantity, then

$$\alpha(a) + \beta(a)h[\mathbf{v}(a)] + \alpha(b) + \beta(b)h[\mathbf{v}(b)]$$
$$= \alpha(a) + \beta(a)h[\mathbf{v}^*(a)] + \alpha(b) + \beta(b)h[\mathbf{v}^*(b)] \tag{4}$$

If we let

$$\Delta h(a) \equiv h[\mathbf{v}^*(a)] - h[\mathbf{v}(a)] \tag{5}$$
$$\Delta h(b) \equiv h[\mathbf{v}^*(b)] - h[\mathbf{v}(b)] \tag{6}$$

we can rewrite Eq. (4) as

$$\beta(a)\Delta h(a) + \beta(b)\Delta h(b) = 0 \tag{7}$$

Now suppose there exists a second conserved quantity $\alpha' + \beta' h(\mathbf{v})$. Then by the same reasoning, we expect

$$\beta'(a)\Delta h(a) + \beta'(b)\Delta h(b) = 0 \tag{8}$$

Equations (7) and (8) constitute a set of simultaneous equations in the quantities $\Delta h(a)$ and $\Delta h(b)$ and will have a nontrivial solution if and only if

$$\frac{\beta(a)}{\beta'(a)} = \frac{\beta(b)}{\beta'(b)} \tag{9}$$

Since the particles a and b are arbitrary, we conclude that β/β' is a universal constant. The theorem follows immediately.

From Theorems 1 and 2 in the preceding section, we can derive the following two theorems.

Theorem 5. If $g(\mathbf{v})$ is a conserved quantity, then $g(-\mathbf{v})$ is a conserved quantity.

Theorem 6. If $g(v_i, v_j, v_k)$ is a conserved quantity where $i \neq j \neq k \neq i$, then $g(v_j, v_k, v_i)$ is a conserved quantity.

Theorem 7. If $g(v_i, v_j, v_k)$ is a conserved quantity where $i \neq j \neq k \neq i$, then $g(-v_i, v_j, v_k)$ is a conserved quantity.

Since all inertial frames are equivalent, we can write down the following theorem.

Theorem 8. If a quantity is conserved in one inertial frame, it is conserved in all inertial frames.

52 Mass and Momentum

If the appropriate transformation between one inertial frame and another is the Galilean transformation, then Theorem 8 is equivalent to the following theorem.

Theorem 9. If $g(\mathbf{v})$ is a conserved quantity, then $g(\mathbf{v}+\mathbf{V})$ is a conserved quantity for arbitrary constant $\mathbf{V}$.

POSSIBLE CONSERVED QUANTITIES

Consider a collision between two identical particles initially moving along a line with the same speed, but in opposite directions. If there exists a conserved quantity, then the possible products of the collision and the possible values of the velocities of the products will in some way be restricted. Conversely, if we know that certain products and velocities are either possible or impossible, this information tells us something about the possible conserved quantities. In this section we demonstrate that if certain fairly plausible products and velocities are assumed to be possible, we can put very severe restrictions on the possible conserved quantities. Whether our assumptions are correct or incorrect is, of course, a matter of experiment.

We assume first that a collision is possible in which the only effect is to change the directions of the two identical colliding particles. Secondly, if the two particles are approaching the point of collision along the same line, then from symmetry we expect that they will have to recede from the point of collision along a common line. We do not assume that this is necessarily so, merely that it is possible. Finally, and this is the least plausible assumption we make, we assume that the line of recession may assume any direction. With these very plausible assumptions, we show in Theorems 10 and 11 that the possible conserved quantities are extremely limited.

Theorem 10. If one assumes that when two identical particles moving along a line with the same speed but in opposite directions collide, it is possible for the same particles to emerge moving with their original speeds and in opposite directions but along a different line, and the directions of the line of approach and the line of recession may assume any value; then any quantity, depending only on the nature and velocity of a particle, which is conserved must, in classical mechanics, be of the form

$$g = \lambda + \sum_i \mu_i v_i + \tfrac{1}{2} \nu v^2 \qquad (10)$$

where λ, μ_i, and ν are particle parameters, that is, quantities that are constant for a given particle, but whose values vary from particle to particle.

PROOF: Since the colliding particles and the products are all identical in the collision described, the functions g in Eq. (1) will all correspond to the same particle, which we indicate as simply $g(\mathbf{v})$, rather than $g[a,\mathbf{v}(a)]$. Since $g(\mathbf{v})$ must be conserved in the given collision, it follows that

$$g(v\mathbf{n}) + g(-v\mathbf{n}) = g(v\mathbf{n}^*) + g(-v\mathbf{n}^*) \tag{11}$$

where v is the speed of each particle, $\mathbf{n}$ is a unit vector parallel to the directions of the precollision velocities, and $\mathbf{n}^*$ a unit vector parallel to the directions of the post collision velocities. If there exists a quantity $g(\mathbf{v})$ that is conserved in one inertial frame, it will be conserved in all inertial frames. If S' is a frame that is moving with a velocity $-\mathbf{V}$ with respect to S, then a particle that is moving with a velocity $\mathbf{v}$ with respect to S will be moving with a velocity

$$\mathbf{v}' = \mathbf{V} + \mathbf{v} \tag{12}$$

with respect to S'. Converting the precollision and postcollision velocities in Eq. (11) to the values they would have in frame S', and assuming that g is still a conserved quantity, we obtain

$$g(\mathbf{V} + v\mathbf{n}) + g(\mathbf{V} - v\mathbf{n}) = g(\mathbf{V} + v\mathbf{n}^*) + g(\mathbf{V} - v\mathbf{n}^*) \tag{13}$$

We expect Eq. (13) to be valid for arbitrary values of v, $\mathbf{n}$, $\mathbf{n}^*$, and $\mathbf{V}$. This requirement puts severe restrictions on the form of the function g. We will now determine what these are. If we take the second derivative of Eq. (13) with respect to v and set $v = 0$, we obtain

$$\sum_i \sum_j n_i n_j g_{ij}(\mathbf{V}) = \sum_i \sum_j n_i^* n_j^* g_{ij}(\mathbf{V}) \tag{14}$$

where

$$g_{ij}(\mathbf{V}) \equiv \frac{\partial^2 g(\mathbf{V})}{\partial V_i \partial V_j} \tag{15}$$

Since $\mathbf{n}$ and $\mathbf{n}^*$ in Eq. (14) are arbitrary unit vectors, it follows that the function

$$f(\mathbf{n}) \equiv \sum_i \sum_j n_i n_j g_{ij} \tag{16}$$

must be independent of $\mathbf{n}$. Only two of the three components n_1, n_2, and n_3 are independent, since they must satisfy the condition

$$\sum_i n_i^2 = 1 \tag{17}$$

If we make use of the method of Lagrange multipliers (Appendix 15), we can show that the function $f(\mathbf{n})$ will be independent of our choice of any two of the three components n_1, n_2, n_3 if and only if there exists a constant h such that the function

$$F(\mathbf{n}) \equiv \sum_i \sum_j n_i n_j g_{ij} - h\left(\sum_i n_i^2 - 1\right) \tag{18}$$

is independent of our choice of any of the quantities n_1, n_2, and n_3. If $F(\mathbf{n})$ is to be independent of our choice of n_1, n_2, and n_3, the partial derivatives of F with respect to the n_i must vanish. In particular

$$\frac{\partial^2 F(\mathbf{n})}{\partial n_i \, \partial n_j} = 0 \tag{19}$$

Substituting Eq. (18) in Eq. (19) we obtain

$$g_{ij} = h\delta_{ij} \tag{20}$$

Recalling that g_{ij} is a function of $\mathbf{V}$ and noting that h, though not a function of $\mathbf{n}$, may be a function of $\mathbf{V}$, we can rewrite Eq. (20) as follows.

$$g_{ij}(\mathbf{V}) = h(\mathbf{V})\delta_{ij} \tag{21}$$

To determine $h(\mathbf{V})$ we note first that

$$g_{ijk} = h_k \delta_{ij} \tag{22}$$

where

$$h_k \equiv \frac{\partial h(\mathbf{V})}{\partial V_k} \tag{23}$$

The derivative g_{ijk} is equal to the derivative g_{kji}, hence

$$h_i \delta_{kj} = h_k \delta_{ij} \tag{24}$$

Choosing $i \neq j = k$ we obtain

$$h_i = 0 \tag{25}$$

Since all three partial derivatives of h vanish, it follows that

$$h = \nu \tag{26}$$

where ν is a constant. Substituting Eq. (26) in Eq. (21) we obtain

$$g_{ij}(\mathbf{V}) = \nu \delta_{ij} \tag{27}$$

Equation (27) provides us with all the second derivatives of g. If we solve for g we obtain

$$g(\mathbf{V}) = \lambda + \sum_i \mu_i V_i + \tfrac{1}{2}\nu V^2 \tag{28}$$

where λ, μ_1, μ_2, μ_3, and ν are constant. We have shown that if Eq. (13) is true for arbitrary values of v, $\mathbf{V}$, $\mathbf{n}$, and $\mathbf{n}^*$, then g must be of the form given by Eq. (28). Conversely, it can be shown by direct substitution that if g is of the form given by Eq. (28), then Eq. (13) is satisfied for all values of v, $\mathbf{V}$, $\mathbf{n}$, and $\mathbf{n}^*$.

Theorem 11. If there exists a quantity

$$g = \lambda + \sum_i \mu_i v_i + \tfrac{1}{2}\nu v^2 \tag{29}$$

which is conserved in a collision, then

a. There exists a particle parameter μ and a set of universal constants B_1, B_2, B_3, and D such that

$$\mu_i = B_i \mu \qquad (30)$$

$$\nu = D\mu \qquad (31)$$

If one of the parameters μ_1, μ_2, μ_3, or ν is not zero, the ratio of the value of μ for one particle to the value of μ for a second particle is unique for the given pair of particles.

b. The quantity

$$G = A\mu + \sum_i B_i \mu v_i + C(\lambda + \tfrac{1}{2} D\mu v^2) \qquad (32)$$

where $D \equiv \nu/\mu$, is a conserved quantity not only for the particular values of A, B_i, and C, for which G reduces to g, but for any other choice of the constants A, B_i, and C.

PROOF: Using Theorems 3, 5–7 and 9, we can show that if there exists a conserved quantity

$$g = \lambda + \sum_i \mu_i v_i + \tfrac{1}{2} \nu v^2 \qquad (33)$$

then the following quantities are also conserved quantities:

$$g_1 = \tfrac{1}{2}[g(\mathbf{v}) + g(-\mathbf{v})] = \lambda + \tfrac{1}{2}\nu v^2 \qquad (34)$$

$$g_2 = \tfrac{1}{2}[g(\mathbf{v}) - g(-\mathbf{v})] = \sum_i \mu_i v_i \qquad (35)$$

$$g_3 = \tfrac{1}{2}[g_2(v_i, v_j, v_k) - g_2(-v_i, v_j, v_k)] = \mu_i v_i \qquad (36)$$

$$g_4 = g_3(v_j, v_k, v_i) = \mu_i v_j \qquad (37)$$

$$g_5 = V_i^{-1}[g_3(\mathbf{v} + \mathbf{V}) - g_3(\mathbf{v})] = \mu_i \qquad (38)$$

$$g_6 = g_1(\mathbf{v} + \mathbf{V}) - g_1(\mathbf{v}) = \nu \sum_i v_i V_i + \tfrac{1}{2}\nu V^2 \qquad (39)$$

$$g_7 = V^{-2}[g_6(\mathbf{v}) + g_6(-\mathbf{v})] = \nu \qquad (40)$$

$$g_8 = \tfrac{1}{2}[g_6(\mathbf{v}) - g_6(-\mathbf{v})] = \nu \sum_i v_i V_i \qquad (41)$$

$$g_9 = \frac{1}{2V_i}[g_8(v_i, v_j, v_k) - g_8(-v_i, v_j, v_k)] = \nu v_i \qquad (42)$$

Since $\mu_1 v_i$, $\mu_2 v_i$, $\mu_3 v_i$, and νv_i are conserved quantities, by virtue of Theorem 4 there exists a set of constants B_1, B_2, B_3, and D and a particle parameter μ

such that

$$\mu_i = B_i \mu \tag{43}$$

$$\nu = D\mu \tag{44}$$

Finally, from Theorem 3, any linear combination of the quantities above is a conserved quantity; hence

$$G = A\mu + \sum_i B_i \mu v_i + C[\lambda + \tfrac{1}{2} D\mu v^2] \tag{45}$$

is a conserved quantity for any choice of the constants A, B_i, and C.

THE BASIC POSTULATE OF CLASSICAL DYNAMICS

From the preceding results, it follows that if there is any velocity dependent quantity at all that is conserved in a collision, there exists a quantity μv that is conserved. Experiment reveals that there is a quantity μv that is conserved. Furthermore, it is found that when the interaction between the two particles cannot be treated as a collision, the quantity μv is still conserved. This result is summarized in the following postulate.

Postulate IV. It is possible to associate with each particle i in the universe a parameter $\mu(i)$ having the following property: if two or more particles interact with one another, the quantity $\sum_i \mu(i) \mathbf{v}(i)$, where $\mathbf{v}(i)$ is the velocity of particle i, has the same value at the end of the interaction and at each instant during the interaction as it had at the beginning of the interaction.

NOTE: Implicit in Postulate IV is the assumption that the interaction between two or more particles occurs instantaneously without the intervention of a medium. If there were an upper limit to the speed of a particle with respect to an inertial frame, the interaction could not occur instantaneously and Postulate IV would have to be modified.

MASS

Given two particles i and j, the ratio $\mu(i)/\mu(j)$ is unique, but the values of $\mu(i)$ and $\mu(j)$ are not. We are free to assign any value to one particle in the universe. Once this value is assigned, however, the value of μ for all other particles is fixed. It is customary to pick some particle o as a reference and assign it unit value, that is, let $\mu(o) = 1$; then we refer to the quantity $\mu(i)/\mu(o) \equiv m(i)$ as the mass of particle i. The mass of any particle i can then, in principle, be determined by allowing the particle to interact with the reference particle o and noting that

$$\mu(i)\Delta\mathbf{v}(i) + \mu(o)\Delta\mathbf{v}(o) = 0 \tag{46}$$

where $\Delta\mathbf{v}(i)$ and $\Delta\mathbf{v}(o)$ are the changes in the velocities of particles i and o, respectively.

This fact is summarized in the following definition.

Definition 1. If we choose a particle o as a reference particle, the mass of any particle i is defined as the negative of the ratio of the change in the velocity of o to the change in the velocity of i when the two are allowed to interact and the velocities are measured with respect to an inertial frame, that is,

$$m(i) = -\frac{\Delta\mathbf{v}(o)}{\Delta\mathbf{v}(i)} \tag{47}$$

From Postulate IV and Theorem 11, we can derive the following theorems concerning the mass of a particle.

Theorem 12. If two or more particles interact, the sum of their masses remains constant.

Corollary 12a. If two or more particles unite to form a single particle, the mass of the resultant particle is equal to the sum of the masses of the component particles.

LINEAR MOMENTUM

The quantity $m\mathbf{v}$ is of great importance in classical dynamics and is given a special name, linear momentum.

Definition 2. If, in an inertial reference frame, a particle of mass m is moving with a velocity $\mathbf{v}$, its linear momentum $\mathbf{p}$ is defined by the relation

$$\mathbf{p} = m\mathbf{v} \tag{48}$$

CONSERVATION OF LINEAR MOMENTUM

From Postulate IV and Definitions 1 and 2, we obtain the following very important theorem.

Theorem (Conservation of Linear Momentum). If two or more otherwise isolated particles interact with each other, the total linear momentum of the system remains constant.

Mass and Momentum

CONCLUSION

The introduction of the concept of momentum provides us with a means of quantitatively describing the interaction between two particles. In the chapter that follows we pursue this subject further.

PROBLEMS

1. Two particles a and b are tied together by an elastic filament, then pulled apart and released in a nonresisting medium. With respect to an inertial frame of reference a and b are found to have accelerations of 10 and -5 cm/s^2, respectively. What is the mass of b relative to a? The particle b is now replaced by another particle c and the accelerations become: for c, -10 cm/s^2, and for a, 8 cm/s^2. Find the mass of c relative to a and the mass of c relative to b.

2. A particle of unit mass collides head on with a second particle of unknown mass that is initially at rest. The two particles are then observed to recoil from each other with equal speeds but in opposite directions, the speed of each being half the initial speed of the unit mass. What is the value of the unknown mass?

3. The position of a certain star of mass m relative to the background of fixed stars is given by $\mathbf{r}(t) = \mathbf{A} + \mathbf{B}t + C(\cos \omega t\, \mathbf{i} + \sin \omega t\, \mathbf{j})$, where $\mathbf{A}$, $\mathbf{B}$, C, and ω are constants and $\mathbf{i}$ and $\mathbf{j}$ are unit vectors in the x and y directions, respectively. It is believed that the star experiences its acceleration because of the presence of another star that is not visible, called a dark companion. Determine as completely as possible the position $\mathbf{R}(t)$ of the dark companion.

4. The two components of a double star are observed to move in circles of radii a and b, respectively. What is the ratio of their masses?

5. Newton defined the mass of a body as the product of its density and its volume. This definition has frequently been criticized as circular because the density of an object is defined as its mass per unit volume. Is this criticism entirely valid?

9

Force

INTRODUCTION

In analyzing the interaction between a given particle and one or more other particles, we are frequently interested in the behavior of the given particle only, not in the associated changes of the other particles. The concept of force, which we introduce in this chapter, is useful in enabling us to perform such analyses.

FORCE

If a system of particles, which we shall refer to as an agent, acts on a particle and causes the momentum of the particle to change, we say that the agent has exerted a force on the particle. More precisely, a force is defined as follows.

Definition 3. If in an inertial frame the momentum of a particle is changing with time, the particle is said to be acted on by a force $\mathbf{f}$ and the force is defined quantitatively by the relation

$$\mathbf{f} = \dot{\mathbf{p}} \qquad (1)$$

where $\dot{\mathbf{p}}$ is the time rate of change of the momentum $\mathbf{p}$ of the particle, or equivalently is defined by the relation

$$\mathbf{f} = m\mathbf{a} \qquad (2)$$

where m is the mass of the particle and $\mathbf{a}$ its acceleration.

NEWTON'S LAW

The equation

$$\mathbf{f} = \dot{\mathbf{p}} \qquad (3)$$

or equivalently,

$$\mathbf{f} = m\mathbf{a} \qquad (4)$$

was originally introduced by Newton as the second of three fundamental laws of motion and is called Newton's equation of motion or Newton's second law of motion. We will refer to it simply as Newton's equation, or Newton's law. Strictly it is not a law but a definition. Nevertheless it remains a significant contribution to mechanics. In the first place, the introduction of the concept of force enables us to consider the changes in the motion of one particle without detailed consideration of the agent producing these changes. In the second place, this particular choice of the definition of force, from among the infinite number of possible choices, tremendously simplifies the analysis of the motion of particles. For example, if we use the definition above to analyze the motion of the bodies making up our solar system, we find that their motion can be simply explained to a high degree of accuracy by assuming that each one of the bodies attracts each one of the other bodies with a force that is directed along the line joining the two bodies, and whose magnitude is inversely proportional to the square of the distance between the two bodies and directly proportional to the product of their masses. No other choice of definition of force would have yielded such a simple result.

THE LAW OF ACTION AND REACTION

If one particle is being acted on by a second particle, the force on the first particle due to the presence of the second particle is given by

$$\mathbf{f}(21) = \frac{d\mathbf{p}(1)}{dt} \qquad (5)$$

where $\mathbf{p}(1)$ is the momentum of the first particle. Conversely, the first particle is acting on the second particle and the force the first particle exerts on the second particle is given by

$$\mathbf{f}(12) = \frac{d\mathbf{p}(2)}{dt} \qquad (6)$$

where $\mathbf{p}(2)$ is the momentum of the second particle. Adding Eqs. (5) and (6) we obtain

$$\mathbf{f}(21) + \mathbf{f}(12) = \frac{d}{dt}\left[\mathbf{p}(1) + \mathbf{p}(2)\right] \qquad (7)$$

If there are no other forces acting on the particles, then from the law of conservation of linear momentum, the right hand side of Eq. (7) must be zero, hence

$$\mathbf{f}(12) = -\mathbf{f}(21) \qquad (8)$$

The result above is summarized in the following theorem.

Theorem (The Law of Action and Reaction). If two otherwise isolated particles interact with each other, the force the first particle exerts on the second is equal in magnitude but opposite in direction to the force the second particle exerts on the first.

The theorem above, generally referred to as the law of action and reaction, was introduced by Newton as the last of his three fundamental laws of motion.

CLASSIFICATION OF FORCES

In Newtonian dynamics, the forces that act on a particle fall into two general categories: contact forces and action at a distance forces. A *contact force* is a force exerted by an agent that is in direct contact with the particle. The force that a table exerts on an object resting on it is an example of a contact force. An *action at a distance force* is a force that is exerted by an agent that is at a distance from the particle without the intervention of a medium. The force that the sun exerts on the earth is an example of an action at a distance force.

NOTE 1: The foregoing categorization of forces into contact forces and action at a distance forces is strictly Newtonian. On the one hand, on a microscopic level the Newtonian forces turn out to be very complex forces resulting from the electromagnetic interactions of the atoms and electrons that are in contact, interactions that could loosely be categorized as interactions at a distance. On the other hand, if we consider the electromagnetic or gravitational field between two particles at a distance from each other to be a genuine dynamical entity, not simply a convenient mathematical construction as is done in Newtonian dynamics, then the interaction between particles at a distance from each other is a complicated interaction involving the interactions of the particles with the field, and the field with the particles–interactions that could loosely be classified as contact interactions. Even greater objections to the classification above arise from the fact that the nice clean-cut boundaries of a Newtonian object no longer exist when the object is viewed quantum mechanically, and thus the very meaning of being in contact or being at a distance breaks down.

NOTE 2: Since the action of particle 1 and the reaction of particle 2 must, according to Postulate IV, occur simultaneously, the transmission of the action of particle 1 to particle 2 and the transmission of the reaction of particle 2 back to particle 1 must occur instantaneously. Therefore all action at a distance forces in Newtonian dynamics are *instantaneous action at a distance* forces.

NOTE 3: The gravitational and electromagnetic interactions between two particles at a distance from each other are actually not transmitted instantaneously but are transmitted with the speed of light. The effect of this finite speed of transmission cannot be neglected if we are dealing with particles whose relative speed is comparable to the speed of light.

PROBLEMS

1. A particle of mass m is initially at rest. A force $\mathbf{f} = \mathbf{c}t^2$, where $\mathbf{c}$ is a constant is applied to the particle. Find the position and velocity of the particle as a function of the time.

2. The force acting on a particle of mass m is given by $f = -\mathbf{i}a^2 x$, where $\mathbf{i}$ is a unit vector in the x direction and a is a constant. At time $t = 0$ the particle is located at the origin and is moving with a velocity $\mathbf{v}_0 = \mathbf{i}v_0$. Find the position of the particle as a function of the time.

3. A particle of mass m is initially at rest. A constant force $\mathbf{f}_0$ acts on the particle for a time t_0. The force is then doubled to the value $2\mathbf{f}_0$ and remains constant at this value. Find the total distance the particle travels in a time $2t_0$.

4. A particle of mass m is moving along the x axis of an inertial frame of reference. Its position is given by the equation $x = c \sin pt$, where c and p are constants. Express the force acting on the particle as a function of x.

5. A particle of mass m is acted on by a force $\mathbf{f} = ax_1 \mathbf{e}_1 + (b\dot{x}_1 + c\dot{x}_2)\mathbf{e}_2 + (dx_3 + et)\mathbf{e}_3$, where a, b, c, d, and e are constants. At time t_0 the particle is on the positive x_1 axis, a distance s from the origin, and is moving along the x_1 axis with a speed v. What is the acceleration of the particle at time t_0?

6. At time $t = 0$ a particle of mass 50 kg is located at the center of a horizontal circular platform. The particle is made to move outward along the radius that at time $t = 0$ is pointing due east, at a constant speed of 2 m/s relative to the platform. If the platform rotates at a constant rate of 1 rpm, calculate the applied force acting on the particle when it is 10 m from the center of the platform.

7. Let $\mathbf{f}(1)$ and $\mathbf{f}(2)$ be two forces and let $\mathbf{f}(1 + 2)$ be the single force that when applied to a particle produces the same effect as the simultaneous application of forces $\mathbf{f}(1)$ and $\mathbf{f}(2)$. Show that $\mathbf{f}(1 + 2)$ is the vector sum of the forces $\mathbf{f}(1)$ and $\mathbf{f}(2)$.

8. A given force that is acting on a particle may be a function of the position of the particle, the velocity of the particle, and the time. Some authors have attempted to construct a theory of greater generality by

allowing the force to be a function of the acceleration as well. Show that this idea is inconsistent with the fundamental postulates of Newtonian dynamics.

9. Indicate the fallacy in the following argument. A boy pulls a wagon. But since to every action there is an equal and opposite reaction, the wagon pulls the boy backward with a force equal to the pull of the boy on the wagon. Therefore there can be no motion.

10. A particle is attracted toward a fixed line by a force whose direction is perpendicular to the line and whose magnitude varies as the distance from the line. Show that its path is a curve traced on an elliptical cylinder.

11. Discuss the following derivation of the law of action and reaction. Suppose that two bodies 1 and 2 attract each other. If 1 attracts 2 with a greater force than 2 attracts 1, it would follow that if an obstacle is interposed that prevents the meeting of the two bodies, the whole system would move in the direction from 2 to 1 and consequently would not remain in equilibrium but would accelerate indefinitely. A similar result would follow if 2 attracts 1 with a greater force than 1 attracts 2. Since this is nonsensical, it follows that body 1 attracts 2 with the same force that 2 attracts 1.

12. Discuss the following derivation of the law of action and reaction. Suppose that two particles 1 and 2 attract each other. Let $\mathbf{f}(21)$ be the force that 2 exerts on 1 and $\mathbf{f}(12)$ the force that 1 exerts on 2. Now suppose a force $\mathbf{f}$ is applied to particle 1; then $\mathbf{f} + \mathbf{f}(21) = m(1)\mathbf{a}(1)$ and $\mathbf{f}(12) = m(2)\mathbf{a}(2)$. Now suppose we adjust $\mathbf{f}$ so that the distance between particle 1 and particle 2 remains constant. It then follows that $\mathbf{a}(1) = \mathbf{a}(2) \equiv \mathbf{a}$. Thus (1) $\mathbf{f} + \mathbf{f}(21) = m(1)\mathbf{a}$; (2) $\mathbf{f}(12) = m(2)\mathbf{a}$; and since the whole system can be treated as a particle of mass $m(1) + m(2)$, acted on by the force $\mathbf{f}$ and undergoing the acceleration $\mathbf{a}$, we have (3) $\mathbf{f} = [m(1) + m(2)]\mathbf{a}$. Combining (1), (2), and (3) we obtain $\mathbf{f}(12) = -\mathbf{f}(21)$.

13. Discuss the following derivation of the law of action and reaction. Suppose that two particles 1 and 2 attract each other. Let $\mathbf{f}(12)$ be the force that 1 exerts on 2 and $\mathbf{f}(21)$ the force that 2 exerts on 1. Now suppose we consider two other particles 1' and 1" that are identical with particle 1 and are far removed from particles 1 and 2. Let $\mathbf{f}(1'1'')$ be the force that particle 1' exerts on particle 1" and $\mathbf{f}(1''1')$ be the force that particle 1" exerts on particle 1'. From symmetry we have: (1) $\mathbf{f}(1'1'') = -\mathbf{f}(1''1')$. We now adjust the distance between 1' and 1" until: (2) $\mathbf{f}(1'1'') = \mathbf{f}(12)$. Since 1' and 1 are identical particles, it follows that they will experience the same reaction when the forces they exert are equal. Thus from Eq. (2) above we obtain: (3) $\mathbf{f}(1''1') = \mathbf{f}(21)$. Combining (1), (2), and (3) we obtain $\mathbf{f}(12) = -\mathbf{f}(21)$.

14. A block of mass m rests on a block of mass M that lies on a frictionless table. The coefficient of static friction between the blocks is μ. What is the maximum horizontal force that can be applied to the blocks for them to accelerate without slipping on each other if the force is applied **(a)** to the block of mass m, and **(b)** to the block of mass M?

15. A string of N freight cars each of mass M is pulled with a force F by a locomotive. Find the force pulling the last n cars.

10

Center of Mass

INTRODUCTION

The postulates we introduced in the preceding chapters apply to the motion of a point particle. However most of the macroscopic objects of physical interest are not point particles. In this chapter we show that even when the parts of an extended object are all moving relative to one another, there is one point in the object that moves in the same way that a point particle of the same mass as the object acted on by the same forces would move. This point is called the center of mass of the object.

DYNAMICAL SYSTEMS

The analysis of the behavior of an extended object is greatly simplified if instead of viewing the object as a continuum we view it as a collection of a large number of point particles. We shall assume that such a procedure is legitimate and refer to the collection of particles as a *dynamical system*.

FORCE

The forces acting on a dynamical system can be divided into two categories, internal forces and external forces. An *internal force* is a force exerted on one particle in a dynamical system by another particle in the same system. An *external force* is a force exerted on a particle in a dynamical system by an agent that is external to the given system, or more formally it is defined as follows.

Definition 3′. If in an inertial frame the net momentum of a system of particles is changing with time, the system is said to be acted on by an

external force **F** and the force is defined quantitatively by the relation
$$\mathbf{F} = \dot{\mathbf{P}}$$
where **P** is the net momentum of the system of particles.

In the light of this generalization of the definition of force, the law of action and reaction can be extended as follows.

Theorem (The Law of Action and Reaction). If two otherwise isolated dynamical systems interact with each other, the force which the first system exerts on the second is equal in magnitude but opposite in direction to the force the second system exerts on the first.

THE CENTER OF MASS

The moment of the mass of a particle with respect to a given point is the product of the mass of the particle and the vector position of the particle with respect to the point. The *center of mass* of a dynamical system is the point with respect to which the sum of all the moments of the masses in the system vanishes. If we number the particles in a dynamical system from 1 to N and let $m(i)$ be the mass of particle i and $\mathbf{r}(ai)$ the vector position of particle i with respect to some point a, then the moment of the mass of particle i with respect to point a is $m(i)\mathbf{r}(ai)$, and the center of mass is the point c for which

$$\sum_i m(i)\mathbf{r}(ci) = 0 \tag{1}$$

Theorem 1. The center of mass of a dynamical system at a given instant of time is a unique point.

PROOF: Suppose there are two points c and c' that satisfy condition (1), that is,

$$\sum_i m(i)\mathbf{r}(ci) = 0 \tag{2}$$

$$\sum_i m(i)\mathbf{r}(c'i) = 0 \tag{3}$$

If we subtract Eq. (3) from Eq. (2), we obtain

$$\sum_i m(i)\left[\mathbf{r}(ci) - \mathbf{r}(c'i)\right] = 0 \tag{4}$$

But

$$\mathbf{r}(ci) - \mathbf{r}(c'i) = \mathbf{r}(cc') \tag{5}$$

where $\mathbf{r}(cc')$ is the position of point c' with respect to point c. If we substitute

Eq. (5) in Eq. (4) we obtain

$$\mathbf{r}(cc')\sum_i m(i) = 0 \qquad (6)$$

And thus

$$\mathbf{r}(cc') = 0 \qquad (7)$$

It follows that the points c and c' are the same point. Therefore the center of mass is a unique point.

Theorem 2. Consider a dynamical system. If $m(i)$ is the mass of particle i and $\mathbf{r}(ai)$ the vector position of particle i with respect to a point a, then the vector position of c, the center of mass of the system with respect to the point a, is given by

$$\mathbf{r}(ac) = \frac{\sum_i m(i)\mathbf{r}(ai)}{\sum_i m(i)} \qquad (8)$$

PROOF: The vector position of particle i with respect to the center of mass c can be written

$$\mathbf{r}(ci) = \mathbf{r}(ca) + \mathbf{r}(ai) \qquad (9)$$

If we substitute Eq. (9) in Eq. (1) we obtain

$$\sum_i m(i)[\mathbf{r}(ca) + \mathbf{r}(ai)] = 0 \qquad (10)$$

Solving for $\mathbf{r}(ca)$ and noting that $\mathbf{r}(ca) = -\mathbf{r}(ac)$, we obtain Eq. (8), which is the desired result.

THE MOTION OF THE CENTER OF MASS

Theorem 3. The motion of the center of mass of a dynamical system is governed by the equation

$$\mathbf{F} = M\mathbf{A} \qquad (11)$$

where $\mathbf{F}$ is the net external force acting on the system, M is the total mass of the system, and $\mathbf{A}$ is the acceleration of the center of mass with respect to an inertial frame of reference. It follows that the motion of the center of mass of a dynamical system of mass M that is acted on by a set of external forces is identical with the motion of a particle of mass M that is acted on by the same set of forces.

PROOF: From Definition 3'

$$\mathbf{F} = \dot{\mathbf{P}} \qquad (12)$$

Center of Mass

If o is a point fixed in an inertial frame, then

$$\mathbf{P} = \frac{d}{dt}\sum_i m(i)\mathbf{v}(oi) = \frac{d^2}{dt^2}\sum_i m(i)\mathbf{r}(oi) \tag{13}$$

But from Theorem 2

$$m(i)\mathbf{r}(oi) = \left[\sum_i m(i)\right]\mathbf{r}(oc) = M\mathbf{R} \tag{14}$$

Combining Eqs. (12)–(14), we obtain Eq. (11).

CONCLUSION

Our initial interest in this text is in the motion of particles. From the considerations in this chapter it is apparent that our results also are applicable to the center of mass motion of extended objects. It follows that the term "particle" can be interpreted more broadly than we have done up till now. We do not take advantage of this extra breadth in the theoretical developments, but we use it in the examples, problems, and applications.

PROBLEMS

1. A system consists of four particles of masses m, $2m$, $3m$, and $4m$, respectively, located at the points $(-1, -2, 2)$, $(3, 2, -1)$, $(1, -2, 4)$, and $(3, 1, 2)$. Find the coordinates of the center of mass.

2. The coordinates at an arbitrary time t of three particles of masses $2m$, m, and $3m$, respectively, are $5t\mathbf{i} - 2t^2\mathbf{j} + (3t - 2)\mathbf{k}$, $(2t - 3)\mathbf{i} + (12 - 5t^2)\mathbf{j} + (4 + 6t - 3t^2)\mathbf{k}$, and $(2t - 1)\mathbf{i} + (t^2 + 2)\mathbf{j} - t^3\mathbf{k}$. Find the velocity of the center of mass at time $t = 1$.

3. Three equal masses are located at the vertices of a triangle. Prove that the center of mass is located at the intersection of the medians of the triangle.

4. Show that the center of mass of three particles of equal mass that are located at points A, B, and C, respectively, coincides with the center of mass of a homogeneous lamina in the shape of a triangle with vertices at A, B, and C.

5. A uniform plate has the shape of the region bounded by the parabola $y = x^2$ and the line $y = a$. Find the center of mass.

6. The parallel sides of a trapezoid have the lengths of a and b, respectively. Their midpoints are connected by a line AB. Show that the center of mass C of the trapezoid lies on the line AB, and find the ratio of the segment AC to the segment CB.

7. Find the center of mass of a uniform right circular cone of radius a and height h.

8. Find the center of mass of a homogeneous hemisphere of radius a.

9. Determine the center of mass of a homogeneous solid system consisting of a cone and hemisphere placed base to base. The common radius of the bases is a, and the height of the cone is h.

10. Find the center of mass of a uniform solid bounded by the planes $4x + 2y + z = 8$, $x = 0$, $y = 0$, and $z = 0$.

11. A uniform solid is bounded by the paraboloid of revolution $x^2 + y^2 = cz$ and the plane $z = a$. Find the center of mass.

12. A right circular cone is divided into two equal parts by a plane through the axis. Find the center of mass of either half.

13. A rod of radius a is bent into the form of a semicircle of mean radius b. Determine the center of mass. Assume that the curved rod is homogeneous.

14. A thin circular lamina of radius a has a small circle of radius $a/2$ cut from it in a manner such that the periphery of the smaller circle passes through the center of the larger circle. Find the center of mass of the remainder.

15. A right circular cone of semivertical angle α is scooped out of a solid homogeneous sphere of radius a, the vertex of the cone being on the surface of the sphere and its axis being a diameter of the sphere. Find the center of mass of the remainder.

16. A hemispherical shell has inner radius a and outer radius b. Show that the distance of its center of mass from its geometrical center is $[3(a + b)(a^2 + b^2)]/[8(a^2 + ab + b^2)]$.

17. Find the center of mass of a semicircular filament of radius r.

18. Find the center of mass of a rod of length a whose density is proportional to the distance from one end.

19. The density at a point P of a solid hemisphere of radius a is directly proportional to the square of the distance of the point P from the center of the base. Find the center of mass.

20. A flat plate is shaped in the form of a quadrant of a circle of radius a. The density of the plate at an arbitrary distance r from the center of curvature is given by $m_0(r^2 + a^2)/a^2$, where m_0 is the density at $r = 0$. Find the center of mass of the plate.

SECTION 4

Elementary Applications of Newton's Law

11

Some Basic Forces

INTRODUCTION

In this chapter we introduce a few basic forces.

NEWTON'S LAW OF GRAVITY

If a particle p is in the presence of a second particle P, the particle p experiences a force $\mathbf{f}$ called a gravitational force whose direction is toward P and whose magnitude f is directly proportional to the mass m of particle p and the mass M of particle P and inversely proportional to the square of the distance r between p and P, that is,

$$f = \frac{GMm}{r^2} \tag{1}$$

where G is a proportionality constant that in MKS units is equal to 6.673×10^{-11} Nm^2/kg^2.

If a particle p of mass m is in the presence of a system S of particles then the resultant of all the gravitational forces the particles in S exert on the particle p will be a force $\mathbf{f}$ whose magnitude is proportional to m and whose magnitude and direction depend on the mass distribution in S, that is,

$$\mathbf{f} = m\mathbf{g} \tag{2}$$

where $\mathbf{g}$ is a function that depends on the mass distribution in S. If the particle is subjected to no forces other than the gravitational force $m\mathbf{g}$ the particle will undergo an acceleration $\mathbf{g}$; hence the quantity $\mathbf{g}$ is called the acceleration due to gravity.

In the particular case in which the mass distribution in S is spherically symmetric, the vector $\mathbf{g}$ is directed toward the center of mass of S, and if the particle p is a distance r from the center of mass c of S, the magnitude of $\mathbf{g}$ is equal to G/r^2 times the mass contained within the sphere whose center is at c

and whose radius is r. It should be carefully noted that this is a special case. In the case of an arbitrary mass distribution in S the vector $\mathbf{g}$ does not in general point toward the center of mass of S nor even toward the same point for different positions of p. Thus for the earth, which is not a spherically symmetric mass distribution, there is no fixed point toward which the vector $\mathbf{g}$ always points, nor is the magnitude of $\mathbf{g}$ inversely proportional to the square of the distance from some fixed point. The mass distribution in the earth is however *approximately* spherically symmetric and the magnitude of the acceleration due to gravity does not differ from the value 9.8 m/s^2 by more than 0.5% over the entire surface of the earth.

HOOKE'S LAW

Consider a spring AB. If the end A is attached to a fixed support, and the spring is extended or compressed by an agent that applies a force directed along the line of the spring to the free end B, the spring will exert a force $\mathbf{f}$ on the agent. If the spring is held in an extended or compressed position by the agent, the direction of $\mathbf{f}$ will be parallel to AB and its magnitude will be a function of the distance of B from its initial position. If we let l_0 be the unstretched length of the spring, the magnitude f is a function of $|l - l_0|$, that is,

$$f = \phi(|l - l_0|) \tag{3}$$

where $\phi(|l - l_0|)$ represents some function of $|l - l_0|$. If we expand $\phi(|l - l_0|)$ in a Taylor series in terms of the quantity $|l - l_0|$ and note that f must vanish when $l = l_0$, we obtain

$$f = a|l - l_0| + b|l - l_0|^2 + c|l - l_0|^3 + \cdots \tag{4}$$

where $a, b, c, \ldots$ are constants. If a is not equal to zero then for small displacements of the spring

$$f \approx a|l - l_0| \tag{5}$$

Thus for small displacements of the spring the magnitude of the force is approximately directly proportional to $|l - l_0|$. A linear spring is one in which the magnitude f is directly proportional to the quantity $|l - l_0|$ for all values of the displacement, that is,

$$f = k|l - l_0| \tag{6}$$

where k is a constant called the *spring constant*. For sufficiently small displacements most springs approximate the behavior of a linear spring, and it is customary in working problems involving springs to assume that the springs are linear springs if nothing to the contrary is stated.

Rather than using the quantity k to describe the elastic properties of a particular spring, one frequently uses a quantity λ, defined as follows

$$\lambda \equiv k l_0 \tag{7}$$

The quantity λ is called the *modulus of elasticity* of the spring. In terms of the modulus of elasticity

$$f = \frac{\lambda |l - l_0|}{l_0} \tag{8}$$

If a spring is cut in two, the moduli of elasticity of the resulting two springs will be the same as that of the original spring, whereas this is not true for the spring constants. The modulus of elasticity is thus a more useful quantity to use when we are tabulating the properties of elastic materials, since its value does not depend on the length of the sample.

The force law, Eq. (6), for a linear spring applies only if the spring is at rest and there are no gravitational forces acting on the spring. If the spring is moving or there are gravitational forces, then the force exerted by the spring at the end B will not be given in general by Eq. (6) but will be a complicated function of the mass distribution in the spring and the motion of the spring. In the limit as the mass of the spring approaches zero, Eq. (6) is valid even when the spring is moving or is in a gravitational field. It is customary in problems involving springs to assume that the mass of the spring is negligible unless we are told otherwise. Hence we ordinarily assume in problems that Eq. (6) is valid even when the spring is moving or gravitational forces are present.

AMONTON'S LAWS OF FRICTION

If two rigid flat surfaces A and B are in contact with each other, each will exert a force on the other and the forces will be equal in magnitude but opposite in direction. We let F be the magnitude of the tangential component of either force and N the magnitude of the normal component of either force.

Let us consider first the case in which the surfaces are at rest relative to each other. As long as the surfaces remain rigid there is no limit to the value of N. However since it is possible for the surfaces to slide relative to each other, there is an upper limit to the value of F. It is found experimentally that the upper limit F_m of F is primarily dependent on the nature of the surfaces in contact and on the value of N and is relatively independent of the area of contact. It is further found that for a given pair of surfaces the ratio of F_m to N is approximately constant over a wide range of values of N. The ratio F_m/N is called the coefficient of static friction for the contact and is usually designated by the letter μ, that is,

$$\mu = \frac{F_m}{N} \tag{9}$$

Since F must be less than or equal to F_m, it follows that

$$F \leq \mu N \tag{10}$$

Two things must be carefully kept in mind when using the law above. First the law is only approximately valid over a limited range of values of N. Second F

can have any value between 0 and μN and has the value μN only when the surfaces are on the verge of slipping.

If the two surfaces are in contact but are sliding relative to each other it is found experimentally that the ratio F/N remains approximately constant over a wide range of N. In this case the constant ratio F/N is called the coefficient of kinetic friction and is designated by $\bar{\mu}$. Thus for the case of sliding

$$F = \bar{\mu} N \qquad (11)$$

where $\bar{\mu}$ is an experimental constant that depends on the nature of the surfaces in contact.

If two surfaces in contact are incapable of exerting a tangential force F on each other, we say that the contact between the surfaces is *perfectly smooth*. If the contact between two surfaces is perfectly smooth, the coefficients of static and kinetic friction for the contact are zero.

If two surfaces in contact are incapable of sliding relative to each other, we say that the contact between the two surfaces is *perfectly rough*. If the contact between two surfaces is perfectly rough, the coefficient of static friction for the contact is infinite.

An ideal surface that is incapable of exerting a tangential force F on any other surface is said to be a *perfectly smooth surface*. An ideal surface on which it is impossible for any other surface to slide is said to be a *perfectly rough surface*.

Instead of using μ to describe the nature of the contact between two surfaces one sometimes uses a quantity ϵ, called the *angle of friction*, defined by the relation

$$\mu \equiv \tan \epsilon \qquad (12)$$

The angle of friction can be shown to be the maximum angle that the force exerted by the one surface on the other surface can make with the normal to the surface without slipping occurring. Another way of considering the angle of friction is to note that if a block rests on an inclined plane, and if the angle of friction for the contact between the block and the plane is ϵ, the block will be on the verge of slipping when the angle that the plane makes with the horizontal is equal to ϵ.

STRINGS

A string or a rope is frequently used to apply a force to a body. If a force is applied to one end of a string, the other end of which is attached to a body, the relation between the force applied to the string and the force transmitted to the body will in general be a complicated function depending on a variety of parameters including the mass distribution in the string, the motion of the string, and the flexibility of the string. We assume however unless stated otherwise that the word "string" means a string whose behavior approximates that of a perfectly flexible and massless string. If one end of such a string is connected to a body and a force is applied to the other end, the string will align itself with the direction of the force and the force that is applied to the body by the string will be equal in magnitude and direction to the force applied to the string. If the string is not directly connected to the body but passes over a series of smooth surfaces before it connects with the body, the force applied to the body will be equal in magnitude to the force applied to the string but its direction will be along the direction of the string at the point where the string joins the body.

A string may be inextensible or extensible. If it is inextensible its length remains constant. If it is extensible its length will depend on the force acting on the string. It is customary to assume unless stated otherwise that an extensible string behaves under tension like a linear spring, but differs from a linear spring in that it cannot support a compressive force.

THE SPECIFICATION OF A FORCE

The effect of a given force on a system consisting of more than one particle depends on the particle in the system on which the force acts or equivalently on the point of application of the force. Therefore the complete specification of a force consists in giving its magnitude, its direction, and its point of application.

If we are interested only in the motion of the center of mass of a system, it is sufficient to know the magnitude and direction of the force. In this chapter and in Chapters 12 and 13 we are interested only in the motion of the center of mass of a system; therefore we do not pay particular attention to the point of application of a force. Furthermore when we are dealing with a set of forces distributed over a surface as in the case of frictional forces, or distributed over a volume as in the case of the gravitational forces on an extended object, we replace these forces when possible by a single force whose magnitude and direction are obtained by taking the vector sum of all the forces in the set. It should be carefully kept in mind however that this force is not necessarily equivalent to the original set of forces, but is simply equivalent for the purposes we have in mind. Some conditions under which forces acting at different points can be combined are discussed in Chapter 35.

DETERMINATION OF THE FORCES ACTING ON AN OBJECT

One of the first tasks in ascertaining the theoretical motion of a given dynamical system is to determine the forces that are acting on the system. If we are interested only in the motion of the center of mass of a given system, we do not need to know the internal forces acting between the particles in the system, but only the external forces acting on these particles. Furthermore if we are interested only in the motion of the center of mass of the system, we do not need to know the point of application of a given external force but only its magnitude and direction.

The external forces acting on a system will be either action at a distance forces or contact forces. The action at a distance forces usually are either gravitational or electromagnetic, and to determine whether there are any action at a distance forces it is usually sufficient simply to ask whether the system is in a gravitational field, or whether there are any electromagnetic forces operative. Every system is subjected to a wide variety of gravitational forces, since every particle in the universe outside the system will exert a gravitational force on the system. In most problems however the effect of all but one or two mass distributions is negligible. For example on the surface of the earth the earth's gravitational attraction is usually but not always the only gravitational force that has to be taken into account. The contact forces will occur wherever the system is in contact with some other system. If the system can be isolated from the rest of the universe by a closed mathematical surface, then to determine the contact forces we simply find those points at which the surface is in contact with an outside agent.

GEOMETRICAL REPRESENTATION OF A FORCE

We frequently represent a force by a directed line segment or arrow. In such cases the number or letter adjacent to the arrow is assumed to be the algebraic magnitude of the force. If the algebraic magnitude is positive the force will be in the direction of the arrow. If the algebraic magnitude is negative the force will be in the direction opposite to that of the arrow. According to this convention the two forces represented in Fig. 1 have the same magnitude and direction.

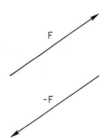

Fig. 1

EXAMPLES

Example 1. Determine the gravitational force that a homogeneous disk of radius a and mass σ per unit area exerts on a particle p of mass m that is located at a point on the axis of the disk and at a distance h from the disk.

Solution. Let us choose a Cartesian coordinate system with origin at the center of the disk and z axis along the axis of symmetry of the disk and pointing toward p. It then follows that p is located at the point $(0,0,h)$. Now consider the infinitesimal element of the disk located at $(x, y, 0)$ and of area $dx\,dy$. The mass of the element is $\sigma\,dx\,dy$. The distance of the element from p is $(x^2 + y^2 + h^2)^{1/2}$ and the unit vector pointing from p to the element is $(\mathbf{i}x + \mathbf{j}y - \mathbf{k}h)/(x^2 + y^2 + h^2)^{1/2}$; hence the force this element exerts on the particle p is

$$d\mathbf{f} = \frac{Gm(\sigma\,dx\,dy)}{x^2 + y^2 + h^2}\left[\frac{\mathbf{i}x + \mathbf{j}y - \mathbf{k}h}{(x^2 + y^2 + h^2)^{1/2}}\right] \tag{13}$$

The net gravitational attraction $\mathbf{f}$ that the disk exerts on p can be obtained by integrating $d\mathbf{f}$ over the whole disk. From symmetry it is apparent that the x and y components of $\mathbf{f}$ will vanish. The force $\mathbf{f}$ is thus directed along the z axis and has the algebraic magnitude

$$f = -Gm\sigma h \int_{-a}^{a}\int_{-(a^2-y^2)^{1/2}}^{(a^2-y^2)^{1/2}} (x^2 + y^2 + h^2)^{-3/2}\,dx\,dy \tag{14}$$

Switching to polar coordinates r, θ we obtain

$$f = -Gm\sigma h \int_{0}^{2\pi}\int_{0}^{a} \frac{r\,dr\,d\theta}{(r^2 + h^2)^{3/2}}$$

$$= -2\pi Gm\sigma\left[1 - \frac{h}{(a^2 + h^2)^{1/2}}\right] \tag{15}$$

Note that in the limit as the radius of the disk approaches infinity the force approaches a value that is independent of h.

Example 2. In Fig. 2 a block M of mass M is at rest on a horizontal perfectly smooth surface S. The block is attached by a spring of spring constant k and unstretched length l_0 to a vertical wall. A second block m of mass m rests on top of the first block. The coefficients of static and kinetic friction for the contact between m and M are μ and $\bar{\mu}$, respectively. The system is set in motion in such a way that M slides back and forth on S while m remains at rest with respect to M. Describe as completely as possible the forces acting on m and M, making use of the law of action and reaction but not Newton's law.

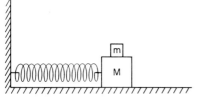

Fig. 2

80 Some Basic Forces

Fig. 3

Solution. The forces acting on m and M are represented geometrically in Fig. 3. The forces acting on m consist of the gravitational force mg and the contact force exerted by M, which we have decomposed into a normal component A and a tangential component B. The forces acting on M consist of the gravitational force Mg acting down, the contact force exerted by m (which by virtue of the law of action and reaction has a normal component A and a tangential component B in the directions shown), the contact force exerted by S (which because of the smoothness of the surface has only a normal component C), and finally the force D of the spring. The absolute value of B is less than or equal to μA, that is, $|B| \leqslant \mu A$. The value of D is given by $-k(l - l_0)$. If $l > l_0$ then $D < 0$, which indicates that the spring force in this case is directed to the left.

Example 3. In Fig. 4 a string $ABCD$ passes over a rough cylinder of radius a. The segment BC that is in contact with the cylinder forms an arc of ϕ radians. The tension in the string at A is T_0 and the tension in the string at D is T_1, where $T_1 > T_0$. Show that the maximum value T_1 can have without slipping taking place is

$$T_1 = T_0 \exp(\mu\phi) \qquad (16)$$

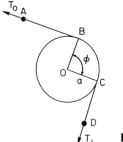

Fig. 4

Solution. Let us consider a small segment of the string as illustrated in Fig. 5. If we assume that the segment can be approximated by a straight line

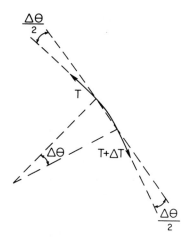

Fig. 5

joining its end points, the tangential component of the force on the segment is

$$F = (T + \Delta T)\cos\left(\frac{\Delta\theta}{2}\right) - T\cos\left(\frac{\Delta\theta}{2}\right) \approx \Delta T \tag{17}$$

and the normal component of the force on the segment is

$$N = (T + \Delta T)\sin\left(\frac{\Delta\theta}{2}\right) + T\sin\left(\frac{\Delta\theta}{2}\right) \approx T\Delta\theta \tag{18}$$

Setting $F = \mu N$ and proceeding to the limit of infinitesimal segments, we obtain

$$\frac{dT}{T} = \mu \, d\theta \tag{19}$$

If we integrate this equation between $\theta = \theta_0$ and $\theta = \theta_1$ and assume that $T = T_0$ when $\theta = \theta_0$ and $T = T_1$ when $\theta = \theta_1$, we obtain

$$T_1 = T_0 \exp\left[\mu(\theta_1 - \theta_0)\right] \tag{20}$$

Setting $\theta_1 - \theta_0 = \phi$ we obtain Eq. (16).

PROBLEMS

1. Three particles of masses m_1, m_2, and m_3, respectively, are placed at the vertices of an equilateral triangle each of whose sides is of length a. Show that the resultant attraction of any two of the particles on the third lies on a line that passes through the center of mass C of the three particles. Show that this would still be true if the force of attraction between two particles varied inversely as the nth power of the distance between them where n is different from 2.

2. Determine the gravitational force exerted by a uniform rectangular plate on a particle located at a point on the line that passes through the center

of the plate and is perpendicular to the plate. Let $2a$ and $2b$ be the lengths of the sides, m the mass of the particle, σ the mass per unit area of the plate, and z the distance of the particle from the plate.

3. Determine the gravitational force exerted by a homogeneous sphere of mass M on a particle of mass m located a distance r from the center of the sphere.

4. A small hole of negligible diameter is bored along a radius into a sphere whose density ρ at a given point is a function of r, the distance of the point from the center of the sphere. It is found that the gravitational force exerted on a particle in the hole is independent of the distance of the particle from the center of the sphere. Determine the dependence of ρ on r.

5. Determine the gravitational force exerted by a homogeneous cylinder of length $2b$, radius a, and density ρ on a particle of unit mass located at a point on the axis of the cylinder and at a distance z from the center of the cylinder.

6. Determine the gravitational force exerted by a homogeneous rod of length a and mass σ per unit length on a particle of unit mass located at a distance b from one end and c from the other end of the rod.

7. Matter is distributed on the curved part of the surface of a cone of height h and semivertical angle α. The mass σ per unit area at a point on the surface of the cone is inversely proportional to the distance of the point from the apex. Determine the gravitational attraction exerted by the mass distribution on a particle that is on the axis of the cone at a distance s from the apex.

8. The center of a uniform spherical shell of mass M and radius a is at a distance b $(b > a)$ from an infinite thin sheet having a mass σ per unit area. Find the force on the sheet due to the sphere.

9. A uniform rod of length b and mass α per unit length lies on the axis of symmetry of a circular loop of wire of mass β per unit length. The radius of the loop is a and the nearer end of the rod is located a distance h from the center of the loop. Find the net force the loop exerts on the rod.

10. A horse can exert a force of 7500 newtons on a rope, and a man can hold 500 newtons. How many times will it be necessary to wrap the rope around a snubbing post for the man to be able to hold the horse? The coefficient of static friction for the contact is 0.25.

12

Statics of a Particle

INTRODUCTION

A system of particles is said to be in *static equilibrium* if it is possible to find an inertial reference system with respect to which all the particles in the system remain at rest. In this chapter we consider systems in static equilibrium.

CONDITIONS OF STATIC EQUILIBRIUM FOR A PARTICLE

From Newton's law, it follows that for a particle to be in static equilibrium it is both necessary and sufficient that the sum of the forces acting on the particle vanish, that is,

$$\sum \mathbf{f} = 0 \qquad (1)$$

If we express the forces acting on the particle in terms of their components with respect to the x, y, and z axes of an inertial frame, the equilibrium condition becomes

$$\sum f_x = 0 \qquad (2a)$$

$$\sum f_y = 0 \qquad (2b)$$

$$\sum f_z = 0 \qquad (2c)$$

These equations can frequently be used to complete the determination of the forces acting on a particle when we know only some of the forces acting on the particle.

CONDITIONS OF STATIC EQUILIBRIUM FOR A SYSTEM OF PARTICLES

For a system of particles to be in static equilibrium it is a necessary but not a sufficient condition that the sum of the external forces vanish, that is,

$$\sum \mathbf{F} = 0 \qquad (3)$$

84 Statics of a Particle

or in component form

$$\sum F_x = 0 \qquad (4a)$$

$$\sum F_y = 0 \qquad (4b)$$

$$\sum F_z = 0 \qquad (4c)$$

The conditions above are necessary because the center of mass of the system would be accelerating if the sum of the external forces did not vanish. These conditions are not however sufficient. Consider for example a rigid body that is acted on by two forces that are equal in magnitude and opposite in direction but do not act along the same line. The net external force will be zero and thus the center of mass of the rigid body will remain at rest, but, since the forces do not have the same line of action, the body will begin to rotate about its center of mass.

The conditions above, though not sufficient, will suffice for the problems considered in this chapter. In later chapters we discuss in detail further conditions of equilibrium for systems of particles, and in particular for rigid bodies.

HELPFUL HINTS FOR SOLVING STATICS PROBLEMS

In applying the laws of dynamics to specific problems, a great deal of time and effort can be saved if one develops a systematic routine for attacking problems. We outline below a routine for statics problems that may help in the avoidance of some common pitfalls, or at least call attention to some of these pitfalls.

1. *Determine the system of interest.* You should be able to enclose the chosen system within a mathematical surface. If the problem involves a single object, that object is your system. If the problem involves N objects, you will generally have to consider N different systems, one at a time. These N systems may be the N objects, or any N independent subsets of these objects. For example if we have three objects A, B, and C, our set of three independent systems could be any three independent systems chosen from the set $A, B, C, A + B, A + C, B + C, A + B + C$.

2. *Draw an isolated diagram of the system and the external forces acting on the system.* In representing a force, draw an arrow to represent its direction, and put the magnitude beside it. If the direction is unknown, draw the arrow in an arbitrary direction and designate with a letter the angle that it makes with some known direction. If the magnitude is unknown, simply designate it with a letter. If the line of action of a force is known but not the direction along this line, choose either direction for the arrow. If your choice was wrong, the magnitude will when determined simply come out negative. In determining whether you have included all the forces, check

first to see whether any action at a distance forces have been excluded and then whether any contact forces have been excluded. Action at a distance forces generally consist of either the force of gravity or an electromagnetic force. In most cases in this book the only action at a distance force is gravity. A contact force occurs at any point of contact of the system with something outside it. Therefore to see whether a contact force has been excluded, check each point at which the system is in contact with something. If two different systems A and B are in contact with each other, be sure to make the force B exerts on A equal in magnitude and opposite in direction to the force A exerts on B.

3. *Choose a convenient set of Cartesian axes.* It is not necessary to use the same set of axes for each system.
4. *Apply the conditions for static equilibrium.* In this chapter this rule involves simply setting the sum of the x components of the forces equal to zero, the sum of the y components of the forces equal to zero, and, if the problem is three dimensional, the sum of the z components of the forces equal to zero.
5. *Check the completeness of the equations.* If there are N unknown parameters, there should be N equations. If you have too few equations for the number of unknowns, find out why.
6. *Solve for the unknown parameters.*

EXAMPLES

Example 1. In Fig. 1 a block A of mass M is at rest on an inclined plane C of angle α. The block A is connected by a string that passes over a smooth pulley to a block B, which is hanging free. The coefficient of static friction for the contact between A and C is μ, where $\mu < \tan \alpha$. Determine the range of possible values for the mass m of the block B consistent with static equilibrium.

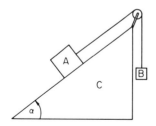

Fig. 1

Solution. There are two objects of interest, block A and block B. We choose as our two systems A and B, respectively. The forces on A and on B are represented in Fig. 2, where T is the tension in the string, and N and F are

86 Statics of a Particle

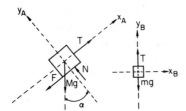

Fig. 2

the normal and tangential components, respectively, of the force exerted by C on A. A different set of coordinate axes has been chosen for each system. Applying the conditions of equilibrium we obtain

$$T - F - Mg \sin \alpha = 0 \tag{5}$$

$$N - Mg \cos \alpha = 0 \tag{6}$$

$$T - mg = 0 \tag{7}$$

We have three equations in the four unknowns T, F, N, and m. To obtain the fourth equation we note that for the block not to slip, the condition

$$|F| < \mu N \tag{8}$$

must be satisfied. Combining Eqs. (5)–(8) we obtain

$$|mg - Mg \sin \alpha| < \mu Mg \cos \alpha \tag{9}$$

from which it follows that

$$-\mu M \cos \alpha < m - M \sin \alpha < \mu M \cos \alpha \tag{10}$$

or equivalently

$$M(\sin \alpha - \mu \cos \alpha) < m < M(\sin \alpha + \mu \cos \alpha) \tag{11}$$

Since $\mu < \tan \alpha$, the lower limit is positive, as it must be. If μ were greater than $\tan \alpha$, the lower limit of m would simply be zero, since m cannot be negative.

PROBLEMS

1. Prove that if a particle is in equilibrium under the action of three forces **A**, **B**, **C** and if α is the angle between **B** and **C**, β the angle between **A** and **C**, and γ the angle between **A** and **B**, then $A/\sin \alpha = B/\sin \beta = C/\sin \gamma$.

2. A mass of 50 kg is suspended by two strings, each of which makes an angle of 60° with the vertical. What is the tension in each string?

3. A particle of mass m is suspended by three inelastic strings, each of length l, from the vertices of a horizontal equilateral triangle having sides of length s. What is the tension in each string?

4. How high can a bug crawl up the inside of a hemispherical bowl if the coefficient of friction for the contact between the bug and the bowl is 0.25.

5. Two rings each of mass m are free to slide on a rough horizontal pole. The coefficient of static friction for the contact between the rings and the pole is μ. The rings are connected by a string of length a that supports a smooth ring of mass M. Find the maximum distance between the rings of mass m consistent with static equilibrium.

6. A block of mass m rests on an inclined plane of angle α. The coefficient of static friction for the contact between the block and the plane is μ. What is the direction and magnitude of the smallest force that will start the body into motion down the plane?

7. A weight W is held on a smooth inclined plane by three forces in the same vertical plane, each of magnitude $W/3$. One force is directed vertically up, the second is directed up the plane, and the third is horizontal. What is the angle of inclination of the plane?

8. A smooth pulley is placed at the edge of a horizontal table. A block of mass M rests on the table and has attached to it a string that passes over the pulley to a second block of mass m, which hangs freely. The system is initially in equilibrium and the angle of friction for the contact between the mass M and the table is ϵ. Through what angle α can the table be tipped before the mass M begins to slip?

9. A weight is tied to two strings of lengths a and b and the other ends of the strings are attached to two points not necessarily at the same level. If the weights hang in equilibrium, the horizontal components of the two tensions have the same value H. Prove that if the lengths of the strings are varied in such a way that H remains constant, the weight describes a parabola that passes through the two points of suspension and has its axis vertical.

10. A small ring of weight W is free to slide on a vertical circular wire hoop of radius a. The angle of friction for the contact between the ring and the hoop is 10°. The ring is attached to an elastic string that is fastened to the highest point of the hoop. The natural length of the string is a and its modulus of elasticity is $7W$. Determine the values of the angle θ that the radius to the ring makes with the radius to the highest point that are consistent with static equilibrium.

11. A particle of mass m is at rest on an inclined plane of angle α. One end of a string of modulus of elasticity λ and natural length a is attached to the particle, and the other end to a point on the plane. Determine the region of the plane within which the particle can rest.

13

Dynamics Problems

INTRODUCTION

In this chapter we consider some elementary problems involving the application of Newton's law to the motion of a particle or a system consisting of a small number of particles.

A METHOD

If we have a system consisting of a small number of particles, and we are given sufficient information about the system to assure us that the forces acting on the particles and the resultant accelerations of the particles are determinate, any further information about the forces or the motion of the particles usually can be determined by simply applying Newton's laws to each of the particles in the system.

In applying Newton's law to the motion of a system of a small number of particles, it is helpful to develop some kind of systematic approach, such as that outlined in the following set of rules.

1. Determine the system of particles of interest. If you are interested only in the motion of the center of mass of an extended object, the object can be treated as a particle. A rigid body that is undergoing translational motion only can always be treated as a particle, since if you know how the center of mass is moving, you know how the body is moving.
2. Choose an inertial reference frame. Particular choices of origin o and orientation of axes x, y, and z frequently result in considerable simplifications in the later calculations. Sometimes it is helpful to vary the choices for the different particles.
3. Consider the system at some arbitrary instant during its motion. For each particle choose a convenient set of coordinates that uniquely specify the

position of the particle. Write down the relation between these coordinates and the Cartesian coordinates of the particle with respect to the x, y, and z axes.

4. For each particle in the system describe analytically or geometrically each of the forces acting on the particle, and introduce a parameter to tag any unknown property of these forces. Make use of the law of action and reaction where applicable.
5. Apply Newton's law by setting the sum of the components of the forces in the x direction equal to the mass of the particle times its acceleration in the x direction, the sum of the forces in the y direction equal to the mass of the particle times its acceleration in the y direction, and the sum of the forces in the z direction equal to the mass of the particle times its acceleration in the z direction.
6. If there are any constraints on the motion, express them mathematically.
7. Gather together the equations obtained. Count the number of independent equations and the number of unknowns. If you do not have a sufficient number of equations, check to see whether you have forgotten some constraint condition or failed to take into account some property of a particular force. Note that in a differential equation the time is not considered to be an unknown parameter; and different derivatives of the same quantity all correspond to the same unknown.
8. Solve the resulting equations for the unknown forces and the unknown motions of the system, introducing whatever boundary conditions apply. If the set of equations seems to be unnecessarily complex, perhaps some simplification could have been made at an earlier stage. Make sure that the answer you obtain is physically reasonable. Check limiting cases.

The procedure above is simply one possible set of rules that may help in avoiding some of the common errors one makes in solving dynamics problems; it is meant to aid the student in attaining an understanding of Newton's law, not to replace this understanding. With experience each student will develop individual methods that will implicitly or explicitly involve some such set of rules.

EXAMPLES

Example 1. A rock of mass m is thrown from a cliff at a height h above a level plain. The initial speed of the rock is v_0 and the initial direction of motion is at an angle α above the horizontal. Neglecting air friction, find the distance R from the foot of the cliff at which the rock hits the plain.

Solution. The system consists of the rock, which we treat as a particle. We choose our inertial frame as shown in Fig. 1. The position of the rock at some

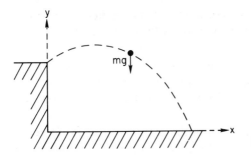

Fig. 1

arbitrary time can be specified by giving its x and y coordinates. There are no constraints acting on the rock, and the only force is the gravitational force, which has a magnitude mg and is in the negative y direction. Applying Newton's law we obtain

$$0 = m\ddot{x} \tag{1}$$

$$-mg = m\ddot{y} \tag{2}$$

We have two equations in the two unknowns x and y. Solving for x and y for the given initial conditions, we obtain

$$x = (v_0 \cos \alpha)t \tag{3}$$

$$y = -\tfrac{1}{2} gt^2 + (v_0 \sin \alpha)t + h \tag{4}$$

When the rock hits the plane, $x = R$ and $y = 0$. Setting $x = R$ and $y = 0$ in Eqs. (3) and (4) and solving for R we obtain

$$R = \frac{v_0 \cos \alpha}{g} \left[v_0 \sin \alpha + \left(v_0^2 \sin^2 \alpha + 2gh\right)^{1/2} \right] \tag{5}$$

Example 2. A wedge of mass M, height h, and angle θ rests on a horizontal plane. A block of mass m is placed at the top of the wedge and released. The contacts between the block and the wedge and between the wedge and the plane are perfectly smooth. Determine the motion of the block relative to the wedge up to the time it leaves the wedge, and also the force the block exerts on the wedge during this time.

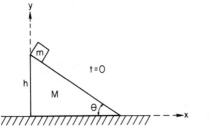

Fig. 2

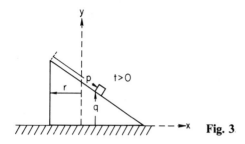

Fig. 3

Solution. Our system may be considered to consist of two particles, the block and the wedge. We choose our reference frame as shown in Fig. 2. The x axis lies along the horizontal surface, the y axis passes through the initial position of the block, and the xy plane is perpendicular to the top edge of the wedge. There will be no motion of the block or the wedge in the z direction, and no motion of the wedge in the y direction. We therefore drop consideration of these motions. At some arbitrary time t later, the system will be in a configuration such as that represented in Fig. 3. Let p, q, and r be the coordinates shown. These coordinates are related as follows to the x and y positions of m and M:

$$x_m = p \cos \theta - r + \text{constant} \qquad (6)$$

$$y_m = q + \text{constant} \qquad (7)$$

$$x_M = r + \text{constant} \qquad (8)$$

where x_m and y_m are the x and y coordinates of the center of mass of the block and x_M is the x coordinate of the center of mass of the wedge. The forces acting on the block and wedge are described by Fig. 4. Applying Newton's law to the x and y motion of m and the x motion of M we obtain

$$N \sin \theta = m\ddot{x}_m \qquad (9)$$

$$N \cos \theta - mg = m\ddot{y}_m \qquad (10)$$

$$-N \sin \theta = M\ddot{x}_M \qquad (11)$$

The block is constrained to remain on the wedge, therefore its velocity relative

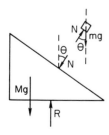

Fig. 4

to the wedge must be parallel to the top surface of the wedge, that is,

$$\frac{-\dot{y}_m}{\dot{x}_m - \dot{x}_M} = \tan\theta \tag{12}$$

Gathering results and eliminating x_m, y_m, and x_M we obtain

$$N\sin\theta = m(\ddot{p}\cos\theta - \ddot{r}) \tag{13}$$

$$N\cos\theta - mg = m\ddot{q} \tag{14}$$

$$N\sin\theta = M\ddot{r} \tag{15}$$

$$-\dot{q} = \dot{p}\sin\theta \tag{16}$$

We have four equations in the four unknowns p, q, r, and N. Solving for N we obtain

$$N = \frac{mMg\cos\theta}{m\sin^2\theta + M} \tag{17}$$

Note when $\theta = 0$, $N = mg$; when $\theta = \pi/2$, $N = 0$; when $M = \infty$, $N = mg\cos\theta$. Solving for $\ddot{p}$ we obtain

$$\ddot{p} = \frac{(M+m)g\sin\theta}{m\sin^2\theta + M} \tag{18}$$

Solving Eq. (18) for p, and noting that p and $\dot{p}$ are zero at $t = 0$, we obtain

$$p = \left[\frac{(M+m)g\sin\theta}{m\sin^2\theta + M}\right]\frac{t^2}{2} \tag{19}$$

Note when $\theta = 0$, $p = 0$; when $\theta = \pi/2$, $p = gt^2/2$; when $M = \infty$, $p = g\sin\theta\, t^2/2$.

Example 3. A particle of mass m is suspended from a fixed point O by a string of length a. A second particle of mass M is suspended from the first particle by a string of length b. If a horizontal velocity v is suddenly imparted

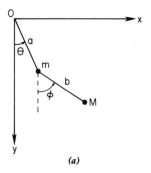
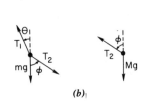

Fig. 5

to the mass m, show that the tensions in the strings are immediately increased by amounts that are in the ratio $\{1 + [mb/(Ma + Mb)]\} : 1$.

Solution. The system consisting of the two particles is shown in an arbitrary configuration in Fig. 5a. We use the angles θ and ϕ to describe the motion. The x and y components of the position, velocity, and acceleration of the two particles in terms of the angles θ and ϕ are

$$x(m) = a \sin \theta \tag{20}$$

$$y(m) = a \cos \theta \tag{21}$$

$$x(M) = a \sin \theta + b \sin \phi \tag{22}$$

$$y(M) = a \cos \theta + b \cos \phi \tag{23}$$

$$\dot{x}(m) = a \cos \theta \, \dot{\theta} \tag{24}$$

$$\dot{y}(m) = -a \sin \theta \, \dot{\theta} \tag{25}$$

$$\dot{x}(M) = a \cos \theta \, \dot{\theta} + b \cos \phi \, \dot{\phi} \tag{26}$$

$$\dot{y}(M) = -a \sin \theta \, \dot{\theta} - b \sin \phi \, \dot{\phi} \tag{27}$$

$$\ddot{x}(m) = a \cos \theta \, \ddot{\theta} - a \sin \theta \, \dot{\theta}^2 \tag{28}$$

$$\ddot{y}(m) = -a \sin \theta \, \ddot{\theta} - a \cos \theta \, \dot{\theta}^2 \tag{29}$$

$$\ddot{x}(M) = a \cos \theta \, \ddot{\theta} - a \sin \theta \, \dot{\theta}^2 + b \cos \phi \, \ddot{\phi} - b \sin \phi \, \dot{\phi}^2 \tag{30}$$

$$\ddot{y}(M) = -a \sin \theta \, \ddot{\theta} - a \cos \theta \, \dot{\theta}^2 - b \sin \phi \, \ddot{\phi} - b \cos \phi \, \dot{\phi}^2 \tag{31}$$

The force diagrams for the two particles are shown in Fig. 5b. If we apply Newton's law to each of the particles, we obtain

$$-T_1 \sin \theta + T_2 \sin \phi = m(a \cos \theta \, \ddot{\theta} - a \sin \theta \, \dot{\theta}^2) \tag{32}$$

$$-T_1 \cos \theta + T_2 \cos \phi + mg = m(-a \sin \theta \, \ddot{\theta} - a \cos \theta \, \dot{\theta}^2) \tag{33}$$

$$-T_2 \sin \phi = M(a \cos \theta \, \ddot{\theta} - a \sin \theta \, \dot{\theta}^2 + b \cos \phi \, \ddot{\phi} - b \sin \phi \, \dot{\phi}^2) \tag{34}$$

$$-T_2 \cos \phi + Mg = M(-a \sin \theta \, \ddot{\theta} - a \cos \theta \, \dot{\theta}^2 - b \sin \phi \, \ddot{\phi} - b \cos \phi \, \dot{\phi}^2) \tag{35}$$

For $t < 0, \theta = \phi = \dot{\theta} = \dot{\phi} = \ddot{\theta} = \ddot{\phi} = 0$. Substituting these values in Eqs. (33) and (35), and solving for T_1 and T_2, we obtain for $t < 0$,

$$T_1 = mg + Mg \tag{36}$$

$$T_2 = Mg \tag{37}$$

At $t = 0$: $x(m) = x(M) = 0$, $y(m) = a$, $y(M) = a + b$, $\dot{x}(m) = v$, $\dot{y}(m) = 0$, $\dot{x}(M) = 0$, $\dot{y}(M) = 0$. Substituting these values in Eqs. (20)–(27) we obtain $\theta = \phi = 0$, $\dot{\theta} = v/a$, $\dot{\phi} = -v/b$. Substituting these values in Eqs. (32)–(35) we obtain for $t = 0$

$$0 = ma\ddot{\theta} \tag{38}$$

$$-T_1 + T_2 + mg = -\frac{mv^2}{a} \tag{39}$$

$$0 = Ma\ddot{\theta} + Mb\ddot{\phi} \tag{40}$$

$$-T_2 + Mg = \frac{-Mv^2}{a} - \frac{Mv^2}{b} \tag{41}$$

Solving for T_1 and T_2 we obtain for time $t = 0$

$$T_1 = mg + Mg + \frac{mv^2}{a} + \frac{Mv^2}{a} + \frac{Mv^2}{b} \tag{42}$$

$$T_2 = Mg + \frac{Mv^2}{a} + \frac{Mv^2}{b} \tag{43}$$

It follows that the changes in T_1 and T_2 are

$$\Delta T_1 = \frac{mv^2}{a} + \frac{Mv^2}{a} + \frac{Mv^2}{b} \tag{44}$$

$$\Delta T_2 = \frac{Mv^2}{a} + \frac{Mv^2}{b} \tag{45}$$

and thus

$$\frac{\Delta T_1}{\Delta T_2} = 1 + \frac{mb}{Ma + Mb} \tag{46}$$

PROBLEMS

1. A man of mass m is lowered a distance L in an elevator. During the first third of the distance the downward acceleration is constant; during the middle third of the distance the velocity is constant; and during the final third of the distance the elevator experiences a constant retarding force. The elevator begins at rest and ends at rest. If the time of descent is equal to that taken by a particle in falling freely a distance $4L$, prove that the force exerted by the man on the elevator during the first third of the distance was $23\ mg/48$.

2. One end of a flexible rope is wrapped around a rough cylinder of radius a that is rotating with a constant angular velocity ω about a horizontal axis. A particle of mass m is suspended from the other end. Assuming that there is no horizontal motion of the mass, find:
   a. The tension in the rope as the mass ascends.
   b. The tension in the rope if the cylinder is rotating with a constant angular acceleration α.

3. A light spring AB is constrained to be in a vertical line and is fixed at its lower end B. At its upper end A the spring carries a cup of mass m. In the equilibrium state the cup compresses the spring, which obeys Hooke's law, by the amount a. Show that if a particle of mass m is gently placed in the cup, the ensuing motion is governed by the equation $2\ddot{x} + gx/a = 2g$, where x is the compression of the spring.

4. A particle is projected vertically upward in a constant gravitational field with an initial velocity v_0. Show that if there is a retarding force proportional to the square of the instantaneous velocity, the velocity of the particle when it returns to its initial position is $v_0 v_t/(v_0^2 + v_t^2)^{1/2}$, where v_t is the terminal velocity.

5. A particle of mass m and velocity v in the y direction is projected onto a horizontal belt that is moving with a uniform velocity V in the x direction. There is a coefficient of sliding friction $\bar{\mu}$ between the particle and the belt. Assuming that the particle first touches the belt at the origin of the fixed xy coordinate system and remains on the belt, find the coordinates (x, y) of the point where sliding stops.

6. A projectile is fired from a height h above a level plane. The initial speed of the projectile is V, and the initial direction of the projectile makes an angle α with the horizontal. Neglecting air friction show that for a given value of V the greatest horizontal range will occur when $\operatorname{cosec}^2\alpha = 2[1 + (gh/V^2)]$.

7. Show that if a projectile is thrown over a double inclined plane from one end of the horizontal base to the other and just grazes the summit in its flight, its angle of projection is $\tan^{-1}(\tan\theta + \tan\phi)$, where θ and ϕ are the slopes of the faces. Assume that the motion is in a vertical plane through the line of greatest slope.

8. A boy sitting on the back of a pickup truck traveling at a speed of 20 km/hr throws a snowball at 20 m/s just as he passes his friend. In what direction (azimuth and elevation) must he throw the snowball to hit the friend?

9. An airplane flying with constant speed u in a horizontal straight line at a height h passes directly over a gun on the ground. The gun is trained directly on the airplane, and when the angle of elevation of the gun reaches α, a shell is fired with muzzle speed v. Show that the plane will be hit if $gh\cot^2\alpha = 2u(v\cos\alpha - u)$.

10. A heavy particle rests on top of a smooth fixed sphere. Find the angular distance from the top at which it leaves the surface if it is slightly displaced.

11. A particle moves in a smooth circular tube of radius a, which rotates about a fixed vertical diameter with uniform angular speed ω. Prove that if θ is the angular distance of the particle from the lowest point and if initially it is at rest relative to the tube, with the value α for θ, where $\omega\cos(\alpha/2) = (g/a)^{1/2}$, then at any subsequent time t, $\cot(\theta/2) = \cot(\alpha/2)\cosh[\omega t \sin(\alpha/2)]$.

12. A projectile travels in a vertical plane through a medium that resists its motion with a force varying as the square of the resultant velocity. Show that the equations of motion can be written in the form $\ddot{x} = -k\dot{x}\dot{s}$, $\ddot{y} = -g - k\dot{y}\dot{s}$, where k is a constant and $\dot{s}^2 = \dot{x}^2 + \dot{y}^2$. Integrate to find the differential equation of the path in the form $d^2y/dx^2 = -ge^{2ks}/(v_0^2\cos^2\alpha)$, where v_0 is the initial speed, α the angle of projection, and s is the distance along the path from the point of projection. Assume a flat trajectory, that is, $|dy/dx| \ll 1$, and determine the equation of the trajectory in the form of a correction to the standard parabola obtained by neglecting air resistance.

13. A particle moves under gravity in a medium the resistance of which is proportional to the velocity. Prove that the range on a horizontal plane is a maximum, for a given velocity of projection, when the angles of elevation at the beginning and end of the trajectory are complementary.

14. A light inextensible string of length a is attached at one end to a particle P, of mass m. To the other end of the string is attached a ring Q, of mass $3m/2$, which is threaded on a fixed rough horizontal wire. The ring is at rest and the particle hangs in equilibrium under gravity. The particle is then projected horizontally and parallel to the wire with speed $(2ga)^{1/2}$. If θ is the inclination of the string to the downward vertical in the subsequent motion, show that provided the string does not slip on the wire, the tension in the string is $3mg\cos\theta$. Show that the ring will not slip if the coefficient of friction between it and the wire exceeds $1/\sqrt{3}$.

15. A light inextensible string of length $a + b$ hangs freely under gravity from a fixed point O. A particle of mass m is attached to the string at a point A, a distance a from the point O, and a second particle of mass M is attached to the string at a point B, a distance $a + b$ from the point O. The system is displaced and released in such a way that the particles move in the same vertical plane. Obtain a set of equations that can be used to find the positions of the two masses and the tensions in the string as a function of time.

16. Two steel balls each 0.25 m in diameter are placed at rest with their centers 4 m apart. If they are acted upon by no forces but their own mutual attraction, how long will it be before they collide, and what is the relative speed of collision?

17. Particles whose masses are 300 and 200 g, respectively, are attached to opposite ends of a string passing over a light frictionless pulley. Find the acceleration of the system and the tension in the string.

18. Two particles of masses m and M, respectively, are connected by a light inextensible string that passes over a smooth peg. The particles hang with the strings vertical, and the string is clamped to the peg so that no motion can take place. If the string is suddenly unclamped, what change will there be in the force exerted on the peg?

19. A bead of mass M is free to move on a smooth horizontal wire. A particle of mass m is suspended from the bead by a massless inextensible string of length a. The bead and the particle are initially held at rest, with the string making an angle of θ_0 with the vertical and the string and wire in the same vertical plane. The bead and the particle are then released. Obtain a set of equations that can in principle be used to solve for the angle that the string makes with the vertical, and the tension in the string as functions of the time.

20. A block of mass m and a block of mass m' slide down a plane that is inclined at an angle α with the horizontal. The blocks are connected by a link of negligible mass. The coefficients of sliding friction are as follows: between the block of mass m and the plane, $\bar{\mu}$, and between the block of mass m' and the plane, $\bar{\mu}'$. The block of mass m is located directly above the block of mass m' on the inclined plane and $\bar{\mu} > \bar{\mu}'$. Find the force in the connecting link and the acceleration of the two blocks.

21. Two particles A and B of equal mass m are attached to the ends of a light spring that exerts a tension of amount s per unit extension. Initially the particles are at rest on a smooth horizontal table with the spring unstretched; then a constant force of magnitude sa is applied to particle B in the direction AB. Obtain the differential equations for the displacements, x and y of the particles A and B, respectively, at time t, and show that $y + x = a\omega^2 t^2/4$ and $y - x = a(1 - \cos \omega t)/2$, where $m\omega^2 = 2s$.

22. A system consists of two particles of equal mass m that slide on a rigid massless rod that rotates freely about a vertical axis at O. Solve for the radial interaction force between the particles that will cause their distance from the axis to vary according to the formula $r = r_0 + A \sin \beta t$, where $A < r_0$. The initial angular velocity is ω_0.

23. A smooth wedge of mass M and angle α is free to move on a smooth horizontal plane in a direction perpendicular to its edge. A particle of mass m is projected directly up the face of the wedge with velocity V. Prove that it returns to the point on the wedge from which it was projected after a time $[2V(M + m\sin^2\alpha)]/[(m + M)g\sin\alpha]$. Also find the force between the particle and the wedge at any time.

24. A light inextensible string of length $2a$ has equal particles, each of mass m, attached to its ends, and a third particle of mass M attached to its midpoint. The particles lie in a straight line on a smooth horizontal table with the string just taut, and M is projected along the table with velocity V perpendicular to the string. Show that, if the two particles at the end collide after a time T when the displacement of M from its initial position is x, then $(M + 2m)x = MVT + 2ma$. Show also that the tension in the string just before the collision is $mM^2V^2/[(M + 2m)^2 a]$.

25. A particle of mass m is connected to a fixed point P on a horizontal plane by a string of length a. The plane rotates with constant angular speed ω about a vertical axis through a point O of the plane, where $OP = b$. Find the equations of motion of the particle. Assume that the string remains taut.

26. Two small rings of masses m and M, respectively, are moving on a smooth circular wire. The wire is fixed in a vertical plane. The two rings are connected by a straight, massless, inextensible string. Prove that as long as the string remains tight, its tension is $2mMg \tan\alpha \cos\theta/(m + M)$, where 2α is the angle that the string when tight subtends at the center of the circle, θ is the angle that the string makes with the horizontal, and g is the acceleration due to gravity.

27. A loop of string is spinning in the form of a circle about a diameter of the loop with constant angular velocity. Neglecting gravity, prove that the mass per unit length of the string must be proportional to $\operatorname{cosec}^3\theta$, θ being measured from the diameter.

28. A system of n particles, each of mass m, moves in a plane, the force between any two of the particles being a repulsion along the line joining them of magnitude μm^2 times their distance apart. Initially the particles are held at rest and are symmetrically arranged around the circumference of a circle of radius a and center O. The particles are released simultaneously. Find the equations of motion of the ith particle referred to rectangular axes with origin at O, and prove that when the distance of this particle from O is $r(i)$, its speed is $\{\mu nm[r^2(i) - a^2]\}^{1/2}$.

SECTION 5

Auxiliary Principles of Newtonian Dynamics

14

Torque and Angular Momentum

INTRODUCTION

In Chapters 8–10 we introduced all the basic principles that are needed to study the motion of a particle. There are however many auxiliary concepts that are extremely helpful not only in solving particular problems but also in opening up theoretical vistas in Newtonian dynamics. In this and the following few chapters we introduce some of these concepts. Their value will become evident only after we have made use of them.

LINEAR MOMENTUM WITH RESPECT TO A MOVING POINT

In Chapter 8 the linear momentum of a particle with respect to a particular inertial frame was defined as the quantity $\mathbf{p} \equiv m\mathbf{v}$, where m is the mass of the particle and $\mathbf{v}$ is the velocity of the particle with respect to a point fixed in the frame.

It is sometimes convenient to generalize this definition, and to define a quantity $\mathbf{p}(a)$, called the *linear momentum of a particle with respect to a point a*, as follows:

$$\mathbf{p}(a) \equiv m\mathbf{v}(a) \tag{1}$$

where $\mathbf{v}(a)$ is the velocity of the particle with respect to the point a, and the point a may be a moving point.

If the point a with respect to which the momentum $\mathbf{p}(a)$ is measured is not indicated, it will be understood to be the origin o of our inertial frame, or equivalently any point that is fixed with respect to the origin, that is,

$$\mathbf{p} \equiv \mathbf{p}(o) \tag{2}$$

In terms of the quantity $\mathbf{p}(a)$ we can rephrase Newton's law as follows.

Newton's Law. If a particle is acted on by a force **f** and if a is a point that is fixed with respect to an inertial frame, or a point that is moving with a constant velocity with respect to an inertial frame, then

$$\mathbf{f} = \dot{\mathbf{p}}(a) \tag{3}$$

where $\mathbf{p}(a)$ is the linear momentum of the particle with respect to the point a.

ANGULAR MOMENTUM

The angular momentum of a particle with respect to an arbitrary point a is defined as

$$\mathbf{h}(a) \equiv \mathbf{r}(a) \times \mathbf{p}(a) \tag{4}$$

where $\mathbf{r}(a)$ is the position of the particle with respect to point a, and $\mathbf{p}(a)$ is the linear momentum of the particle with respect to the point a.

A word of caution is in order. The definition above is not the only definition of angular momentum that is in use, nor is it necessarily the most common. In many textbooks the angular momentum of a particle with respect to a point a is defined as $\mathbf{r}(a) \times \mathbf{p}(o)$, where o is the origin of our inertial frame. If the point a is a fixed point, the two definitions are equivalent, but if the point a is a moving point, the two definitions are not equivalent The reader should therefore be careful to take this into account when comparing results in different texts.

If the point a with respect to which the angular momentum $\mathbf{h}(a)$, is measured is not indicated, it will be understood that the point is the origin o, that is,

$$\mathbf{h} \equiv \mathbf{h}(o) \tag{5}$$

TORQUE

The *torque with respect to an arbitrary point a* that a force **f** exerts on a particle is defined as

$$\mathbf{g}(a) \equiv \mathbf{r}(a) \times \mathbf{f} \tag{6}$$

where $\mathbf{r}(a)$ is the position of the particle with respect to the point a.

If the point a with respect to which the torque $\mathbf{g}(a)$ is measured is not indicated, it is understood to be the origin o, that is,

$$\mathbf{g} \equiv \mathbf{g}(o) \tag{7}$$

From the foregoing definition of a torque it is not difficult to show that the magnitude of a torque $\mathbf{g}(a) \equiv \mathbf{r}(a) \times \mathbf{f}$ is equal to the product of the magnitude of the force and the distance between the point a and the line of action of the force; and the direction is perpendicular to the plane formed by the point and the line of action of the force, with the positive sense being determined by the direction a right handed screw would progress when operated on by the torque.

TORQUE AND ANGULAR MOMENTUM

Theorem 1. If a particle is acted on by a force $\mathbf{f}$ and if a is a point that is fixed in an inertial reference frame or a point that is moving with a constant velocity with respect to an inertial frame, then

$$\mathbf{g}(a) = \dot{\mathbf{h}}(a) \tag{8}$$

where $\mathbf{g}(a)$ is the torque exerted by the force $\mathbf{f}$ with respect to the point a, and $\mathbf{h}(a)$ is the angular momentum of the particle with respect to the point a.

PROOF: From Newton's law

$$\mathbf{f} = \dot{\mathbf{p}}(a) \tag{9}$$

If we cross multiply this equation by $\mathbf{r}(a)$ we obtain

$$\mathbf{r}(a) \times \mathbf{f} = \mathbf{r}(a) \times \dot{\mathbf{p}}(a) \tag{10}$$

But

$$\mathbf{r}(a) \times \mathbf{f} \equiv \mathbf{g}(a) \tag{11}$$

and

$$\mathbf{r}(a) \times \dot{\mathbf{p}}(a) = \overline{\mathbf{r}(a) \times \mathbf{p}(a)} - \dot{\mathbf{r}}(a) \times \mathbf{p}(a)$$
$$= \dot{\mathbf{h}} - \mathbf{v}(a) \times m\mathbf{v}(a)$$
$$= \dot{\mathbf{h}} \tag{12}$$

If we substitute Eqs. (11) and (12) in Eq. (10) we obtain Eq. (8).

EXAMPLES

Example 1. A particle of mass m slides without friction on the inner surface of a smooth cone of semivertical angle α whose axis is in the vertical direction with the tip pointing down. Use the equation $\mathbf{g} = \dot{\mathbf{h}}$ to obtain the equations of motion of the bead. Show that it is possible for the particle to move in such a way that it always remains a fixed distance a from the axis of the cone. What must the angular speed of the particle be for such motion?

Solution. The cone and the forces acting on the particle are shown in Fig. 1. If we choose cylindrical coordinates ρ, ϕ, z to specify the position of the particle and note that $z = \rho \cot \alpha$, we obtain for the position and velocity of

Torque and Angular Momentum

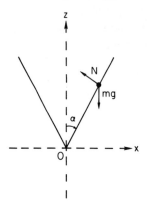

Fig. 1

the particle

$$\mathbf{r} = \rho \cos\phi \mathbf{i} + \rho \sin\phi \mathbf{j} + \rho \cot\alpha \mathbf{k} \tag{13}$$

$$\dot{\mathbf{r}} = (\dot\rho \cos\phi - \rho\dot\phi \sin\phi)\mathbf{i} + (\dot\rho \sin\phi + \rho\dot\phi \cos\phi)\mathbf{j} + (\dot\rho \cot\alpha)\mathbf{k} \tag{14}$$

The net force acting on the particle is

$$\mathbf{f} = -N\cos\alpha \cos\phi \mathbf{i} - N\cos\alpha \sin\phi \mathbf{j} + (N\sin\alpha - mg)\mathbf{k} \tag{15}$$

From Theorem 1

$$\mathbf{g}(o) = \dot{\mathbf{h}}(o) \tag{16}$$

From the definition of torque and angular momentum

$$\mathbf{g}(o) = \mathbf{r} \times \mathbf{f} = \rho \sin\phi \left(N\sin\alpha - mg + \frac{N\cos^2\alpha}{\sin\alpha} \right)\mathbf{i}$$
$$- \rho \cos\phi \left(N\sin\alpha - mg + \frac{N\cos^2\alpha}{\sin\alpha} \right)\mathbf{j} \tag{17}$$

$$\mathbf{h}(o) = \mathbf{r} \times m\dot{\mathbf{r}} = -m\cot\alpha \, \rho^2 \cos\phi \, \dot\phi \mathbf{i} - m\cot\alpha \, \rho^2 \sin\phi \, \dot\phi \mathbf{j} + m\rho^2 \dot\phi \mathbf{k} \tag{18}$$

Substituting Eqs. (17) and (18) in Eq. (16) and equating components we obtain

$$\rho \sin\phi \left(N\sin\alpha - mg + \frac{N\cos^2\alpha}{\sin\alpha} \right) = \frac{d}{dt}(-m\cot\alpha \, \rho^2 \cos\phi \, \dot\phi) \tag{19}$$

$$-\rho \cos\phi \left(N\sin\alpha - mg + \frac{N\cos^2\alpha}{\sin\alpha} \right) = \frac{d}{dt}(-m\cot\alpha \, \rho^2 \sin\phi \, \dot\phi) \tag{20}$$

$$0 = \frac{d}{dt}(m\rho^2 \dot\phi) \tag{21}$$

It would appear that we now have three equations in the three unknowns ρ, ϕ and N. However if we multiply Eq. (19) by $\cos\phi$ and Eq. (20) by $\sin\phi$ and add the resulting equations, we will obtain an equation that is equivalent to Eq.

(21). Hence the three equations are not independent. This will always be the case. The equation $\mathbf{g} = \dot{\mathbf{h}}$ applied at a single point is never sufficient to determine the motion of a particle, since it cannot taken into account the effect of forces whose direction is parallel to the line joining the point and the particle. To generate additional equations of motion we can either apply the equations $\mathbf{g} = \dot{\mathbf{h}}$ to some fixed point other than o or go back to Newton's law. The simplest approach in this problem is to use Newton's law. Setting the sum of the forces in the z direction equal to $m\ddot{z}$ we obtain

$$N \sin \alpha - mg = m \cot \alpha \, \ddot{\rho} \qquad (22)$$

Equations (19), (21), and (22) are independent, hence sufficient to determine ρ, ϕ and N. Gathering these equations together and carrying out the differentiations, we have

$$-m \cot \alpha (2\rho \cos \phi \, \dot{\rho}\dot{\phi} + \rho^2 \cos \phi \, \ddot{\phi} - \rho^2 \sin \phi \, \dot{\phi}^2)$$
$$= \rho \sin \phi \left(N \sin \alpha - mg + \frac{N \cos^2 \alpha}{\sin \alpha} \right) \qquad (23)$$

$$2\rho \dot{\rho} \dot{\phi} + \rho^2 \ddot{\phi} = 0 \qquad (24)$$

$$m \cot \alpha \, \ddot{\rho} = N \sin \alpha - mg \qquad (25)$$

If we assume a solution of the form $\rho = a$, $\dot{\phi} = \omega$, where a and ω are constants, then on substituting these values in Eqs. (23)–(25) we obtain

$$ma \cot \alpha \, \omega^2 = N \sin \alpha - mg + \frac{N \cos^2 \alpha}{\sin \alpha} \qquad (26)$$

$$0 = 0 \qquad (27)$$

$$0 = N \sin \alpha - mg \qquad (28)$$

These three equations will be satisfied if

$$N = \frac{mg}{\sin \alpha} \qquad (29)$$

and

$$\omega = \left(\frac{g \cot \alpha}{a} \right)^{1/2} \qquad (30)$$

PROBLEMS

1. A force with components $(-7, 4, -5)$ acts at the point $(2, 4, -3)$. Find:
   a. The torque with respect to the origin exerted by this force.
   b. The torque with respect to the point $(1, 1, 1)$.

2. A line L passes through the point (a, b, c) and its direction cosines are (l, m, n). Determine the component in the direction of L of the torque with

respect to the point (a, b, c) due to a force of unit magnitude pointing along the x axis.

3. A force with components (X, Y, Z) acts at the point (a, b, c). Determine the component in the direction of a line through the origin with direction cosines (l, m, n) of the torque with respect to the origin.

4. Determine the x, y, and z components of the angular momentum with respect to the origin of a particle of mass m in terms of the spherical coordinates r, θ, and ϕ.

5. A particle of mass 2 moves in a uniform time dependent force field $\mathbf{f} = 24t^2\mathbf{i} + (36t - 16)\mathbf{j} - 12t\mathbf{k}$. At $t = 0$ the particle is located at $3\mathbf{i} - \mathbf{j} + 4\mathbf{k}$ and has a velocity $6\mathbf{i} + 15\mathbf{j} - 8\mathbf{k}$. All the quantities above are in MKS units. Determine the torque and the angular momentum with respect to the origin for arbitrary time t.

6. Prove that the angular momentum with respect to a fixed point O of a system consisting of two moving particles is equal to the sum of the following parts: (1) the angular momentum with respect to O of a particle moving with the center of mass and having a mass equal to the total mass of the system, and (2) the angular momentum of the system with respect to the center of mass.

7. A long smooth straight wire is inclined at an angle α to the upward vertical and is rotated about the upward vertical through one end O with constant angular velocity ω. A small bead of unit mass is free to slide on the wire and is initially at rest relative to the wire and at a distance l from O. Use the equation $\mathbf{g} = \dot{\mathbf{h}}$ to obtain the equations of motion of the bead. Show that the bead will move continuously outward from the axis of rotation if $l > (g/\omega^2)\cot\alpha\,\mathrm{cosec}\,\alpha$. If $l = (2g/\omega^2)\cot\alpha\,\mathrm{cosec}\,\alpha$, show that at time t the component along the axis of rotation of the torque with respect to the point O acting on the particle is $(8g^2/\omega^2) \cdot \cot\alpha\cos\alpha\cosh^3(\tfrac{1}{2}\omega t \sin\alpha)\sinh(\tfrac{1}{2}\omega t \sin\alpha)$.

8. Show that the time rate of change of the angular momentum of a projectile fired from a point O is equal to the torque acting on the particle, where the angular momentum and torque are both with respect to the point O.

15

Work and Kinetic Energy

INTRODUCTION

In this chapter we introduce the concepts of work and kinetic energy. These concepts not only are useful in solving problems, they are fundamental to many of the reformulations of the laws of motion that are discussed in later chapters.

WORK

If a particle is acted on by a force $\mathbf{f}$ and undergoes an infinitesimal change in its position, then the *work* done on the particle by the force, with respect to an arbitrary point a, is defined as

$$dW(a) = \mathbf{f} \cdot d\mathbf{r}(a) \tag{1}$$

where $\mathbf{r}(a)$ is the position of the particle with respect to the point a.

If the point a with respect to which the work is measured is not indicated, it will be understood to be the origin o of our inertial frame, or equivalently a point that is fixed with respect to the origin, that is,

$$dW \equiv dW(o) \tag{2}$$

KINETIC ENERGY

The *kinetic energy* of a particle of mass m with respect to an arbitrary point a is defined as

$$T(a) \equiv \tfrac{1}{2} m \mathbf{v}(a) \cdot \mathbf{v}(a) \equiv \tfrac{1}{2} m v^2(a) \tag{3}$$

where $\mathbf{v}(a)$ is the velocity of the particle with respect to the point a.

If the point a, with respect to which the kinetic energy is measured, is not indicated, it will be understood to be the origin o of our inertial frame, or equivalently a point that is fixed with respect to the origin, that is,

$$T \equiv T(o) \tag{4}$$

WORK AND KINETIC ENERGY

Theorem 1. If a particle is acted on by a force $\mathbf{f}$ and if a is a point that is fixed in an inertial frame or a point that is moving with a *constant velocity* with respect to an inertial frame, then

$$dW(a) = dT(a) \tag{5}$$

where $dW(a)$ is the work done by the force $\mathbf{f}$ with respect to the point a and $T(a)$ is the kinetic energy of the particle with respect to the point a.

PROOF: From Newton's law

$$\mathbf{f} = \frac{d}{dt}[m\mathbf{v}(a)] \tag{6}$$

If we take the dot product of Eq. (6) and $d\mathbf{r}(a)$ we obtain

$$\mathbf{f} \cdot d\mathbf{r}(a) = \frac{d}{dt}[m\mathbf{v}(a)] \cdot d\mathbf{r}(a) \tag{7}$$

The left hand side of Eq. (7) can be rewritten

$$\mathbf{f} \cdot d\mathbf{r}(a) = dW(a) \tag{8}$$

Noting that $d\mathbf{r}(a) = \mathbf{v}(a)\,dt$, we can rewrite the right hand side of Eq. (7) as follows:

$$\begin{aligned}\frac{d}{dt}[m\mathbf{v}(a)] \cdot d\mathbf{r}(a) &= \frac{d}{dt}[m\mathbf{v}(a)] \cdot \mathbf{v}(a)\,dt \\ &= \frac{d}{dt}[\tfrac{1}{2}m\mathbf{v}(a) \cdot \mathbf{v}(a)]\,dt \\ &= dT(a)\end{aligned} \tag{9}$$

Substituting Eqs. (8) and (9) in Eq. (7) we obtain Eq. (5).

EXAMPLES

Example 1. A particle A of mass m is suspended from a fixed point O by an inextensible string of length a. A second particle B of mass M is suspended from particle A by an inextensible string of length b. Assuming that the system is free to move in a vertical plane, and letting θ be the angle that the string OA makes with the vertical and ϕ the angle that the string AB makes with the vertical, obtain an expression for the kinetic energy of the system as a function of θ, ϕ, $\dot{\theta}$, and $\dot{\phi}$.

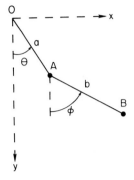

Fig. 1

Solution. Choosing the reference frame as shown in Fig. 1 and letting $x(OA)$, $y(OA)$ be the coordinates of particle A, and $x(OB)$, $y(OB)$ the coordinates of particle B, we obtain for the kinetic energy

$$T = \frac{1}{2}m\left\{\left[\frac{d}{dt}x(OA)\right]^2 + \left[\frac{d}{dt}y(OA)\right]^2\right\}$$
$$+ \frac{1}{2}M\left\{\left[\frac{d}{dt}x(OB)\right]^2 + \left[\frac{d}{dt}y(OB)\right]^2\right\}$$
$$= \frac{1}{2}m\left\{\left[\frac{d}{dt}(a\sin\theta)\right]^2 + \left[\frac{d}{dt}(a\cos\theta)\right]^2\right\}$$
$$+ \frac{1}{2}M\left\{\left[\frac{d}{dt}(a\sin\theta + b\sin\phi)\right]^2 + \left[\frac{d}{dt}(a\cos\theta + b\cos\phi)\right]^2\right\}$$
$$= \frac{1}{2}m\left\{[a\cos\theta\,\dot\theta]^2 + [-a\sin\theta\,\dot\theta]^2\right\}$$
$$+ \frac{1}{2}M\left\{[a\cos\theta\,\dot\theta + b\cos\phi\,\dot\phi]^2 + [-a\sin\theta\,\dot\theta - b\sin\phi\,\dot\phi]^2\right\}$$
$$= \frac{1}{2}m\{a^2\dot\theta^2\} + \frac{1}{2}M\{a^2\dot\theta^2 + b^2\dot\phi^2 + 2ab\cos(\theta-\phi)\dot\theta\dot\phi\}$$
$$= \frac{1}{2}(m+M)a^2\dot\theta^2 + \frac{1}{2}Mb^2\dot\phi^2 + Mab\cos(\theta-\phi)\dot\theta\dot\phi \qquad (10)$$

Example 2. A particle of mass m rests on a rough horizontal surface and is attached by an elastic string of modulus of elasticity λ and unstretched length a to a wall (Fig. 2). The coefficients of static and kinetic friction for the

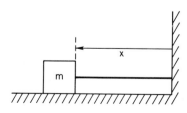

Fig. 2

contact between the particle and the surface are μ and $\bar{\mu}$, respectively. The particle is released from rest at a distance $a + b$ from the wall, slides along the surface, and hits the wall. What is the speed v of the particle at the instant just before it hits the wall?

Solution. The only forces doing work on the particle during its motion are the forces exerted by the string and the frictional force. The work done by the string on the particle is

$$W_s = -\int_{a+b}^{a} \lambda \frac{x-a}{a} dx = \frac{\lambda b^2}{2a} \tag{11}$$

The work done by the frictional force on the particle is

$$W_f = -\bar{\mu} m g (a + b) \tag{12}$$

The change in the kinetic energy of the particle is

$$\Delta T = \tfrac{1}{2} m v^2 \tag{13}$$

From the work-energy theorem

$$W_s + W_f = \Delta T \tag{14}$$

Substituting Eqs. (11)–(13) into Eq. (14) and solving for v we obtain

$$v = \left[\frac{\lambda b^2}{ma} - 2\bar{\mu} g (a + b) \right]^{1/2} \tag{15}$$

Note that the conditions of the problem require that $\lambda(a + b)/a > \mu m g$ and $\lambda b^2 / 2a > \bar{\mu} m g (a + b)$.

PROBLEMS

1. A particle is subjected to the force $\mathbf{f} = 2xz\mathbf{i} + 3z^2\mathbf{j} + y^2\mathbf{k}$. Determine the work done by the force when the particle moves along the straight line $x = 2y = 4z$ from the origin to the point $(4, 2, 1)$.

2. A particle is subjected to the force $\mathbf{f} = 4y\mathbf{i} + 2x\mathbf{j} + \mathbf{k}$. Determine the work done by the force when the particle moves along the helix $x = 4\cos\theta$, $y = 4\sin\theta$, $z = 2\theta$ from $\theta = 0$ to $\theta = 2\pi$.

3. A hole of radius a is cut from an infinite plane sheet. A uniform spherical shell of radius a is inside the hole. The centers of the hole and the spherical shell coincide. Both the sheet and the shell have a mass σ per unit area. How much work is required to move a particle of mass m from the center O of the hole to a point P, which is a distance h from O and lies on a line through O perpendicular to the sheet?

4. A uniform hemispherical shell of mass M and radius a is oriented so that it is concave upward, and the bottom of the shell is at a vertical distance h above an infinite thin sheet having a mass σ per unit area. The shell is

then shifted in orientation so that its axis of symmetry is parallel to the sheet and a height $h + a$ above it. How much work was done?

5. A spider of mass m is suspended from the ceiling on an elastic thread of negligible mass having a stretched length b and an unstretched length a. Calculate the total work the spider would have to do to climb to the ceiling. Compare this with the work required to climb an inextensible thread of length b.

6. A string that is looped over a smooth horizontal peg of radius a has a particle of mass m hanging from one side and a particle of mass M hanging from the other side. The mass M is larger than the mass m. The system is released from rest and moves under the action of gravity.
   a. Find the velocity of the particle of mass M after it has moved a distance x.
   b. If the peg were rough and the coefficient of kinetic friction for the contact between the string and the peg was $\bar{\mu}$, what would the velocity of the particle of mass M be after it had moved a distance x?

7. Two springs AB and BC of spring constants k and k' and unstretched lengths a and a', respectively, are joined in series to form a single spring AC. The spring AC is held in a vertical position with the bottom end fixed on a horizontal surface. A block of mass m is dropped from a height b above the top end C of the spring AC. When the block makes contact with the spring, the two become attached at the point of contact C.
   a. In the subsequent motion, how high above its original position does the point C rebound?
   b. What is the velocity of the point C when it is on the rebound, and a distance x above its original position?

8. A particle of mass m is sliding on a horizontal surface. The coefficient of sliding friction between the particle and the surface is $\bar{\mu}$. Use the concepts of work and energy to calculate the distance the particle will travel before it comes to rest, if it is initially moving with a speed v.

9. A particle of mass m is placed on an inclined plane of angle α. The coefficient of kinetic friction for the contact between the particle and the plane is $\bar{\mu}$. The particle slides down the plane. Use the work-energy theorem to determine how fast the particle is moving after it has traveled a distance s down the plane.

10. Two particles each of mass m are connected by an elastic string of modulus of elasticity λ and natural length a. The masses are pulled apart a distance $a + b$, placed on a rough horizontal surface, and released. The coefficient of kinetic friction for the contact between the particles and the surface is $\bar{\mu}$. For what values of b will the particles collide with each other?

Work and Kinetic Energy

11. Two particles of masses m and M, respectively, are connected by an elastic string of modulus of elasticity λ and natural length a. The masses are pulled apart a distance $a + b$, placed on a rough horizontal surface, and released. The coefficients of static and kinetic friction for the contact between the particles and the surface are μ and $\bar{\mu}$, respectively. For what values of b will the particles collide with each other?

12. A particle of mass m and velocity v in the y direction is projected onto a horizontal belt that is moving with a uniform velocity V in the x direction. The coefficient of kinetic friction for the contact between the particle and the belt is $\bar{\mu}$. Assuming that the particle first touches the belt at the origin of the fixed x-y coordinate system and remains on the belt, find the coordinates (x, y) of the point where sliding stops. [Hint: The inertial frame can be chosen in such a way as to appreciably simplify the solution.]

13. A particle moves under gravity on a rough vertical circle. It starts from rest at one end of the horizontal diameter and comes to rest at the lowest point of the circle. Find an equation that can be used to determine the coefficient of friction.

14. A particle of mass m is constrained to remain on a smooth horizontal table. A light string attached to the particle passes through a small hole in the table. The particle is originally spinning around the hole with a speed v_0 in a circle of radius a. The string is pulled through the hole until the radius of the circle is $a/2$. How much work was done on the particle?

16

Potential Energy

INTRODUCTION

Out of all the possible forces that can act on a particle, certain very important and frequently encountered classes of forces have properties that make them particularly tractable to analysis. In this chapter we consider some of these classes.

NOTATION

We will find it convenient in what follows to use the notation **x** to represent the set of Cartesian coordinates x_1, x_2, x_3. Thus we will say that a particle is located at the point **x** rather than at the point (x_1, x_2, x_3). Similarly if a quantity f is a function of the variables x_1, x_2, and x_3 we will write $f(\mathbf{x})$ rather than $f(x_1, x_2, x_3)$.

IRROTATIONAL FORCE FIELDS

If the force **f** that a particle would experience in a region R is a function of the position **x** of the particle, and possibly of the time t, we say that there exists a *force field* $\mathbf{f}(\mathbf{x}, t)$ in the region R.

Definition 1. A force field $\mathbf{f}(\mathbf{x}, t)$, defined in a region R, is an *irrotational force field* if and only if

$$\frac{\partial f_i}{\partial x_j} = \frac{\partial f_j}{\partial x_i} \tag{1}$$

everywhere in the region R.

Theorem 1. A force field $\mathbf{f}(\mathbf{x}, t)$ defined in a region R is an irrotational force field if and only if there exists a function $U(\mathbf{x}, t)$, called a potential function, such that

$$f_i = -\frac{\partial U(\mathbf{x}, t)}{\partial x_i} \qquad (2)$$

everywhere in the region R.

PROOF: The proof of this theorem is given in Appendix 11.

CONSERVATIVE FORCE FIELDS

Definition 2. A force field defined in a region R is said to be a *conservative force field* if and only if it is time independent and is such that when the particle on which the force acts moves from a point $\mathbf{x} = \mathbf{a}$ to a point $\mathbf{x} = \mathbf{b}$, the work done by the force does not depend on the path the particle follows but only on the initial point $\mathbf{a}$ and the final point $\mathbf{b}$.

Theorem 2. A force field f defined in a region R is a conservative force field if and only if it satisfies each of the following conditions:
 a. It is time independent.
 b. It is irrotational; that is, either $\partial f_i/\partial x_j = \partial f_j/\partial x_i$ everywhere in R, or equivalently there exists a potential function $U(\mathbf{x})$ such that $f_i = -\partial U(\mathbf{x})/\partial x_i$ everywhere in R.
 c. The potential $U(\mathbf{x})$ is single valued, or the region R is simple connected.

NOTE: If condition b is satisfied and the region R is simply connected, the potential will be single valued; but if condition b is satisfied and the potential is single valued, the region R is not necessarily simply connected.

PROOF: The proof of this theorem can be found in Appendix 11.

Since the potential function $U(\mathbf{x})$ associated with a conservative force field is single valued and since we know all its derivatives at all points in the region over which it is defined, we can determine the function within an arbitrary additive constant. The value of the additive constant has no physical significance, since the quantity that is observed is the force $\mathbf{f}$, not the potential function U, and the value of $\mathbf{f}$ does not depend on the value of the constant. We are therefore free to assign to the constant whatever value we choose. We refer to the potential function containing the arbitrary additive constant as the *general potential function*, and to a potential function that is obtained by assigning a particular value to the arbitrary constant as a *particular potential function*.

THE POTENTIAL ENERGY

Definition 3. If a particle is in a conservative force field $\mathbf{f}(\mathbf{x})$, the potential energy of the particle in the position $\mathbf{x}$ relative to a reference point $\mathbf{a}$ is defined as the work done by the force when the particle moves from the point $\mathbf{x}$ to the point $\mathbf{a}$. The potential energy is designated as $V(\mathbf{x})$.

NOTE: When the function $V(\mathbf{x})$ is known, the value of the reference point $\mathbf{a}$ can be found by determining the point at which the potential energy vanishes.

Theorem 3. If a particle is in a conservative force field, the potential energy $V(\mathbf{x})$ with respect to the point $\mathbf{a}$ is identical to the particular potential function $U(\mathbf{x})$ that is obtained by choosing the arbitrary additive constant in the general potential function so as to make $U(\mathbf{a})$ vanish.

PROOF: By definition the potential energy $V(\mathbf{x})$ with respect to the point $\mathbf{a}$ is given by

$$V(\mathbf{x}) = \int_{\mathbf{x}}^{\mathbf{a}} \mathbf{f} \cdot d\mathbf{r} = -\int_{\mathbf{a}}^{\mathbf{x}} \mathbf{f} \cdot d\mathbf{r} = -\int_{\mathbf{a}}^{\mathbf{x}} \sum_{i} f_i \, dx_i \qquad (3)$$

and the general potential function $U(\mathbf{x})$ is the most general function that satisfies the condition

$$-\frac{\partial U(\mathbf{x})}{\partial x_i} = f_i \qquad (4)$$

If we substitute Eq. (4) into Eq. (3) we obtain

$$V(\mathbf{x}) = \int_{\mathbf{a}}^{\mathbf{x}} \sum_{i} \frac{\partial U(\mathbf{x})}{\partial x_i} \, dx_i = \int_{\mathbf{a}}^{\mathbf{x}} dU(\mathbf{x}) = U(\mathbf{x}) - U(\mathbf{a}) \qquad (5)$$

If we choose the arbitrary additive constant in the general potential function $U(\mathbf{x})$ in such a way that $U(\mathbf{a}) = 0$, then

$$V(\mathbf{x}) = U(\mathbf{x}) \qquad (6)$$

This completes the proof of the theorem.

Theorem 4. If a particle is in a conservative force field, then the work done by the force when the particle moves from one point to another is equal to the negative of the change in the potential energy, that is,

$$dW = -dV \qquad (7)$$

NOTE: The theorem is true irrespective of our choice of the reference point for the potential energy.

PROOF: The proof of this theorem follows immediately from the definition of the potential energy.

EXAMPLES

Example 1. Show that the force field $\mathbf{f} = e^y\mathbf{i} + (z + xe^y)\mathbf{j} + (1+y)\mathbf{k}$ is a conservative force field and find the potential energy relative to the origin.

Solution. The force $\mathbf{f}$ is time independent and satisfies the conditions

$$\frac{\partial}{\partial x}(z + xe^y) = \frac{\partial}{\partial y}(e^y) \tag{8}$$

$$\frac{\partial}{\partial x}(1 + y) = \frac{\partial}{\partial z}(e^y) \tag{9}$$

$$\frac{\partial}{\partial y}(1 + y) = \frac{\partial}{\partial z}(z + xe^y) \tag{10}$$

for all values of x, y, and z. It follows that the field is conservative, hence a potential energy $V(\mathbf{x})$ exists. The value of $V(\mathbf{x})$ relative to the origin is equal to the work done by the force when the particle on which the force is acting is moved from the point $\mathbf{x}$ to the origin along any path. If we choose the path made up of the straight lines joining the points (x, y, z), $(x, y, 0)$, $(x, 0, 0)$, and $(0, 0, 0)$ we obtain

$$V(\mathbf{x}) = \int_{x,y,z}^{x,y,0} \left[f_x(x, y, z) \, dx + f_y(x, y, z,) \, dy + f_z(x, y, z) \, dz \right]$$

$$+ \int_{x,y,0}^{x,0,0} \left[f_x(x, y, z) \, dx + f_y(x, y, z) \, dy + f_z(x, y, z) \, dz \right]$$

$$+ \int_{x,0,0}^{0,0,0} \left[f_x(x, y, z) \, dx + f_y(x, y, z) \, dy + f_z(x, y, z) \, dz \right]$$

$$= \int_z^0 f_z(x, y, z) \, dz + \int_y^0 f_y(x, y, 0) \, dy + \int_x^0 f_x(x, 0, 0) \, dx$$

$$= \int_z^0 (1 + y) \, dz + \int_y^0 xe^y \, dy + \int_x^0 dx$$

$$= -(1 + y)z + x - xe^y - x = -(1 + y)z - xe^y \tag{11}$$

Alternatively we could have obtained $V(\mathbf{x})$ by solving the set of equations

$$\frac{\partial V}{\partial x} = -e^y \tag{12}$$

$$\frac{\partial V}{\partial y} = -(z + xe^y) \tag{13}$$

$$\frac{\partial V}{\partial z} = -(1 + y) \tag{14}$$

$$V(O) = 0 \tag{15}$$

Integrating Eq. (12) we obtain

$$V = -xe^y + \phi(y, z) \tag{16}$$

where $\phi(y, z)$ is an unknown function of y and z. Substituting Eq. (16) in Eqs.

(13) and (14) we obtain

$$\frac{\partial \phi}{\partial y} = -z \tag{17}$$

$$\frac{\partial \phi}{\partial z} = -(1+y) \tag{18}$$

Integrating Eq. (17) we obtain

$$\phi = -yz + \psi(z) \tag{19}$$

where ψ is an unknown function of z. Substituting Eq. (19) in Eq. (18) we obtain

$$\frac{d\psi}{dz} = -1 \tag{20}$$

Solving, we obtain

$$\psi = -z + C \tag{21}$$

where C is a constant. Combining Eqs. (16), (19), and (21) we obtain

$$V(\mathbf{x}) = -xe^y - z(y+1) + C \tag{22}$$

Applying Eq. (15) we obtain

$$C = 0 \tag{23}$$

Substituting Eq. (23) in Eq. (22) we again obtain Eq. (11).

Example 2. Consider a fixed particle of mass M. Show that the gravitational force exerted by this particle on a second particle of mass m is a conservative force and determine the potential energy.

NOTE: It is customary in problems involving the gravitational potential energy of a particle in the presence of another particle or a mass distribution to assume the reference point for the particle to be at infinity. Unless otherwise stated this assumption is always made.

Solution. Let us choose a Cartesian coordinate system with the origin at the position of the particle of mass M. The force exerted on the particle of mass m is given by

$$\mathbf{f} = -GMm \left[\frac{\mathbf{i}x + \mathbf{j}y + \mathbf{k}z}{(x^2 + y^2 + z^2)^{3/2}} \right] \tag{24}$$

This force can be shown to satisfy the criteria for a conservative force and thus a potential energy $V(\mathbf{x})$ exists. To determine $V(\mathbf{x})$ we calculate the work done by the force when the particle is moved to infinity along a radial line. Doing this we obtain

$$V = \int_r^\infty \frac{-GMm}{r^2} dr = -\frac{GMm}{r} \tag{25}$$

where $r = (x^2 + y^2 + z^2)^{1/2}$.

Example 3. Consider a uniform circular disk of radius a and mass σ per unit area. Determine the gravitational potential energy of a particle of mass m located at a point p on the axis of the disk at a distance h from the disk.

Solution. Let us choose a Cartesian coordinate system with origin at the center of the disk and z axis along the axis of symmetry of the disk and pointing toward p. It then follows that p is located at the point $(0, 0, h)$. Now consider the infinitesimal element of the disk located at $(x, y, 0)$ and of an area $dx\,dy$. The mass of the element is $\sigma\,dx\,dy$. The distance of the element from p is $(x^2 + y^2 + h^2)^{1/2}$. The work done by the gravitational force that the element exerts on the particle when the particle is moved from $(0,0,h)$ to infinity is

$$dV = -\frac{Gm\sigma\,dx\,dy}{(x^2 + y^2 + h^2)^{1/2}} \tag{26}$$

The net work done by the gravitational force that the disk exerts on the particle when the particle is moved from $(0, 0, h)$ to infinity is obtained by integrating dV over the whole disk. Doing this we obtain

$$V = -Gm\sigma \int_{-a}^{a} \int_{-(a^2-y^2)^{1/2}}^{+(a^2-y^2)^{1/2}} (x^2 + y^2 + h^2)^{-1/2} dx\,dy$$

$$= -Gm\sigma \int_{0}^{2\pi} \int_{0}^{a} (r^2 + h^2)^{-1/2} r\,dr\,d\theta$$

$$= -2\pi Gm\sigma \left[(a^2 + h^2)^{1/2} - h \right] \tag{27}$$

PROBLEMS

1. Evaluate the force corresponding to the potential function $V = cz/r^3$, where c is a constant. Write your answer in vector form.

2. Find the force fields for which the following are potential functions:
   a. $U = \frac{1}{2}\ln(x^2 + y^2 + z^2)$
   b. $U(r, \theta, \phi) = a\cos\theta / r^2$
   c. $U = \frac{1}{2}(k_1 x^2 + k_2 y^2 + k_3 z^2)$
   d. $U(r, \theta, \phi) = e^{-kr}/r$

3. Consider a rectangular plate of dimensions $2a \times 2b$ and of mass σ per unit area. Determine the gravitational potential energy of a particle of mass m that is located a distance z from the plate on the perpendicular line through the center of the plate.

4. Consider a uniform spherical shell of density ρ, inner radius a, and outer radius b. Determine the gravitational potential energy of a particle of mass m a distance r from the center of the shell.

5. Consider a homogeneous rod of mass σ per unit length and of length $2a$. Determine the gravitational potential energy of a particle of mass m that

is a distance r_1 from one end of the rod and a distance r_2 from the other end.

6. Show that the force field $\mathbf{f} = (y^2z^3 - 6xz^2)\mathbf{i} + 2xyz^3\mathbf{j} + (3xy^2z^2 - 6x^2z)\mathbf{k}$ is a conservative force field and find the potential energy relative to the origin of coordinates.

7. Show that the force $\mathbf{f} = e^y\mathbf{i} + xe^y\mathbf{j} + \mathbf{k}$ is a conservative force and find the potential energy with reference to the origin of coordinates.

8. a. The components of the force acting on a particle are $f_x = Ax$ and $f_y = Bx + Cy^2$. Is this a conservative force? Calculate the work done by the force on the particle when it moves along the path consisting of straight lines successively joining the points $(0,0)$, $(a,0)$, $(0,b)$, and $(0,0)$.

   b. The components of a force acting on a particle are $f_x = Ax + By$ and $f_y = Bx + Cy^2$. Determine whether this force is conservative, and calculate the work done, as in part a.

9. Determine whether each of the following forces is conservative and if so find the general potential function.
   a. $\mathbf{f} = [y/(x^2+y^2)^{1/2}]\mathbf{i} - [x/(x^2+y^2)^{1/2}]\mathbf{j}$
   b. $\mathbf{f} = x\phi(r)\mathbf{i} + y\phi(r)\mathbf{j} + z\phi(r)\mathbf{k}$, where $\phi(r)$ is an arbitrary function of the distance from the origin
   c. $\mathbf{f} = [y/(x^2+y^2)]\mathbf{i} - [x/(x^2+y^2)]\mathbf{j}$
   d. $\mathbf{f} = 2xy^3z^4\mathbf{i} + 3x^2y^2z^4\mathbf{j} + 4x^2y^3z^3\mathbf{k}$

10. Determine whether each of the following forces is conservative and if so find the general potential function.
    a. $\mathbf{f} = \phi_1(x)\mathbf{i} + \phi_2(y)\mathbf{j} + \phi_3(z)\mathbf{k}$
    b. $\mathbf{f} = [a/(x^2+y^2)]\mathbf{i} + [b/(x^2+y^2)]\mathbf{j}$
    c. $\mathbf{f} = (xy^2z\mathbf{i} + x^2yz\mathbf{j} + \frac{1}{2}x^2y^2\mathbf{k})$

11. Determine whether each of the following forces is conservative and if so find the general potential function:
    a. $\mathbf{f} = (ax + by^2)\mathbf{i} + (az + 2bxy)\mathbf{j} + (ay + bz^2)\mathbf{k}$
    b. $\mathbf{f} = \mathbf{a} \times \mathbf{r}$
    c. $\mathbf{f} = a\mathbf{r}$
    d. $\mathbf{f} = \mathbf{a}(\mathbf{a} \cdot \mathbf{r})$

12. Prove that a force field in which the force at any point P is directed along a line joining P and a fixed point O and the magnitude of the force is a function of the distance of P from O is a conservative force field.

13. Find the potential energy function of a particle that is attracted toward a fixed point by a force of magnitude k/r^n where r is the distance from the fixed point and k and n are constants. Let the reference point be at $r = \infty$.

17

Constants of the Motion

CONSTANTS OF THE MOTION

The dynamical state of a particle is completely determined at a given time t, if we know its position $\mathbf{x}$, and its velocity $\dot{\mathbf{x}}$. If we know the dynamical state of a particle at some time t_0, and if the force $\mathbf{f}$ acting on the particle is a known function of $\mathbf{x}$, $\dot{\mathbf{x}}$, and t, we can determine the state of the particle at any later time t from the equations of motion

$$f_i = m\ddot{x}_i \qquad i = 1, 2, 3 \tag{1}$$

This set of equations consists of three second order differential equations in the three variables x_1, x_2, and x_3. The solution of this set of equations for $\mathbf{x}$ and $\dot{\mathbf{x}}$ can be written in the form

$$x_i = x_i(\mathbf{c}, t) \qquad i = 1, 2, 3 \tag{2}$$

$$\dot{x}_i = \dot{x}_i(\mathbf{c}, t) \qquad i = 1, 2, 3 \tag{3}$$

where $\mathbf{c} \equiv c_1, \ldots, c_6$ is a set of six independent and arbitrary constants. If we know the values of $\mathbf{x}$ and $\dot{\mathbf{x}}$ at some instant of time t, the values of the six constants will be uniquely determined. Conversely if the six constants are known, the values of $\mathbf{x}$ and $\dot{\mathbf{x}}$ at any arbitrary time t are determined.

The set of six equations given by Eqs. (2) and (3) can be solved for the set of six constants c_i. If we do this we obtain six equations,

$$\phi_i(\mathbf{x}, \dot{\mathbf{x}}, t) = c_i \qquad i = 1, \ldots, 6 \tag{4}$$

From Eq. (4) it follows that the value of any one of the functions ϕ_i remains constant during the motion of the particle. These functions are called constants of the motion.

More generally any function $\phi(\mathbf{x}, \dot{\mathbf{x}}, t)$ whose value remains constant during the motion is called a *constant of the motion*.

Theorem 1. If a particle is acted on by a force $\mathbf{f}$, which is a known function of $\mathbf{x}$, $\dot{\mathbf{x}}$, and t, there will be exactly six independent constants of the motion.

Constants of the Motion 121

PROOF: We note first that the six constants of the motion $\phi_1, \ldots, \phi_6$ given by Eq. (4) are independent constants of the motion, since the six constants c_i can be independently and arbitrarily varied by varying the initial conditions. It follows that there are at least six independent constants of the motion. To prove that there are no more than six independent constants of the motion, let us suppose that we have seven arbitrary constants of the motion $\phi_1, \ldots, \phi_7$. Since by definition a constant of motion is a function whose value remains constant during the motion of the system, it then follows that

$$\frac{d\phi_i}{dt} \equiv \sum_j \frac{\partial \phi_i}{\partial x_j} \dot{x}_j + \sum_j \frac{\partial \phi_i}{\partial \dot{x}_j} \ddot{x}_j + \frac{\partial \phi_i}{\partial t} \equiv 0 \qquad i = 1, \ldots, 7 \qquad (5)$$

This set of seven equations may be considered to be a set of seven simultaneous equations in the six unknowns $\dot{x}_1, \dot{x}_2, \dot{x}_3, \ddot{x}_1, \ddot{x}_2$, and $\ddot{x}_3$. The set will be consistent only if the following condition is satisfied (see Appendix 13):

$$\begin{vmatrix} \frac{\partial \phi_1}{\partial x_1} & \cdots & \frac{\partial \phi_1}{\partial t} \\ \cdots & \cdots & \cdots \\ \frac{\partial \phi_7}{\partial x_1} & \cdots & \frac{\partial \phi_7}{\partial t} \end{vmatrix} \equiv 0$$

But from the results of Appendix 14 this condition implies that there exists a functional relationship between the seven constants of motion $\phi_1, \ldots, \phi_7$. It follows that no more than six of the constants of motion can be independent.

Theorem 2. If a particle is acted on by a force $\mathbf{f}$, which is a function of $\mathbf{x}$ and $\dot{\mathbf{x}}$ but not of t, there are five independent constants of the motion that are not explicit functions of the time t.

PROOF: If $\mathbf{f}$ is not a function of time then the equations of motion are of the form

$$f_i(\mathbf{x}, \dot{\mathbf{x}}) = m\ddot{x}_i \qquad i = 1, 2, 3 \qquad (7)$$

If the set of functions $\mathbf{x}(t) \equiv x_1(t), x_2(t), x_3(t)$ is a solution to the set of Eqs. (7) then the equations

$$f_i[\mathbf{x}(t), \dot{\mathbf{x}}(t)] = m\ddot{x}_i(t) \qquad i = 1, 2, 3 \qquad (8)$$

are identities in the variable t. We now consider what happens in the equations above if the variable t is replaced by $t + t_0$. Since the derivatives $\dot{\mathbf{x}}$ and $\ddot{\mathbf{x}}$ are derivatives with respect to the variable t, it will be to our advantage to spell out this fact by expressing these derivatives as $d\mathbf{x}/dt$ and $d^2\mathbf{x}/dt^2$ rather than as $\dot{\mathbf{x}}$ and $\ddot{\mathbf{x}}$. With this change in notation Eq. (8) becomes

$$f_i\left[\mathbf{x}(t), \frac{d\mathbf{x}(t)}{dt}\right] = m\frac{d^2\mathbf{x}(t)}{dt^2} \qquad i = 1, 2, 3 \qquad (9)$$

If we now replace the variable t by $t + t_0$ and note that

$$\frac{d^n x_i(t + t_0)}{d(t + t_0)^n} = \frac{d^n x_i(t + t_0)}{dt^n} \qquad (10)$$

we obtain

$$f_i\left[\mathbf{x}(t + t_0), \frac{d\mathbf{x}(t + t_0)}{dt}\right] = m\frac{d^2\mathbf{x}(t + t_0)}{dt^2} \qquad i = 1, 2, 3 \qquad (11)$$

Comparing Eqs. (9) and (11), we see that if $\mathbf{x}(t)$ is a solution to the set of Eq. (7), then $\mathbf{x}(t + t_0)$ is also a solution. It follows that if $\mathbf{f}$ is not a function of time, the solution to the equations of motion for $\mathbf{x}$ and $\dot{\mathbf{x}}$ can be written in the form

$$x_i = x_i(\mathbf{c}, t + t_0) \qquad i = 1, 2, 3 \qquad (12)$$

$$\dot{x}_i = \dot{x}_i(\mathbf{c}, t + t_0) \qquad i = 1, 2, 3 \qquad (13)$$

where the constants $\mathbf{c} \equiv c_1, c_2, \ldots, c_5$ and t_0 are determined from the initial conditions. If we solve one of these equations for $t + t_0$ in terms of $\mathbf{x}$, $\dot{\mathbf{x}}$, and $\mathbf{c}$ and substitute the result in the remaining equations we obtain five equations involving $\mathbf{x}$, $\dot{\mathbf{x}}$, and $\mathbf{c}$. If we solve this set of equations for the c_i, we obtain five equations of the form

$$\phi_i(\mathbf{x}, \dot{\mathbf{x}}) = c_i \qquad i = 1, \ldots, 5 \qquad (14)$$

We have thus obtained five independent constants of the motion that do not depend on the time.

THE AIM OF DYNAMICS

If a particle is acted on by a force $\mathbf{f}$, which is a known function of $\mathbf{x}$, $\dot{\mathbf{x}}$, and t, then if we can determine any six independent constants of the motion, we can use the six constants of the motion to determine $\mathbf{x}(\mathbf{c}, t)$. Thus the determination of the motion of a particle may be considered to be equivalent to the determination of the independent constants of the motion. The main aim of dynamics may therefore be said to be to find the independent constants of the motion.

From the considerations above it follows that if we can find even one constant of the motion, we have made a step toward solving for the motion. In the remainder of this chapter we enumerate some constants of the motion that can be immediately identified.

CONSERVATION OF LINEAR MOMENTUM

Theorem 3. If the ith component of the force $\mathbf{f}$ acting on a particle is zero, then the ith component of the momentum $\mathbf{p}$ is a constant of the motion.

PROOF: From Newton's law $f_i = \dot{p}_i$. If $f_i = 0$ then $\dot{p}_i = 0$, and if $\dot{p}_i = 0$, then p_i remains constant during the motion of the system, hence p_i is a constant of the motion.

CONSERVATION OF ANGULAR MOMENTUM

Theorem 4. Let o be a point that is at rest or is moving with a constant velocity with respect to an inertial frame of reference. If the ith component of the torque $\mathbf{g}(o)$ is zero then the ith component of the angular momentum $\mathbf{h}(o)$ is a constant of the motion.

PROOF: The proof follows immediately from Theorem 1 in Chapter 14.

CONSERVATION OF ENERGY

Theorem 5. If the force $\mathbf{f}$ acting on a particle is a conservative force, the sum of the kinetic energy T and the potential energy V is a constant of the motion.

PROOF: From Theorem 1 in Chapter 15:
$$dW = dT \tag{15}$$
But from Theorem 4 in Chapter 16,
$$dW = -dV \tag{16}$$
Combining Eqs. (15) and (16) we obtain
$$d(T + V) = 0 \tag{17}$$
It follows that $T + V$ is a constant of the motion.

EXAMPLES

Example 1. A particle is free to move on a smooth surface of revolution, the equation of which, in cylindrical coordinates with the z axis pointing vertically up, is $\rho = f(z)$. Determine the equations of motion.

Solution. There are two forces acting on the particle, the gravitational force $m\mathbf{g}$, which acts in the negative z direction, and the surface force $\mathbf{N}$, which because of the smoothness of the surface is normal to the surface. The force $\mathbf{N}$ does no work during the motion. The force $m\mathbf{g}$ is conservative and the potential energy function V relative to the origin is mgz. The sum of the kinetic energy T and the potential energy V is a constant of motion, that is,
$$\tfrac{1}{2}m[\dot{\rho}^2 + \rho^2\dot{\phi}^2 + \dot{z}^2] + mgz = E \tag{18}$$
where E is a constant. The z components of the torque with respect to the

origin due to **N** and $m\mathbf{g}$ are zero. Hence the z component of the angular momentum is a constant of the motion, that is,

$$m\rho^2\dot{\phi} = h \tag{19}$$

where h is a constant. These two equations together with the equation

$$\rho = f(z) \tag{20}$$

provide us with three equations in the three unknowns ρ, ϕ, and z. Note that not only are Eqs. (18) and (19) simpler to obtain than the equations of motion that would be obtained by a direct application of Newton's law, but they are also first order rather than second order differential equations.

PROBLEMS

1. A heavy particle moves on a smooth surface. Show that its speed is the same whenever its path cuts a given horizontal curve on the surface.

2. A particle moves under gravity on a smooth surface of revolution with axis vertical. The equation of the surface in cylindrical coordinates is given in the form $\rho = f(z)$. If the velocity is horizontal and of magnitude v_1 at a height z_1 and again horizontal and of magnitude v_2 at a height z_2, determine v_1 and v_2 in terms of z_1 and z_2.

3. A smooth narrow tube has the form of the circular helix $x = c\cos\alpha\cos\theta$, $y = c\cos\alpha\sin\theta$, $z = c\theta\sin\alpha$, where c, and α are constants and the z axis is vertically downward. A particle of mass m inside the tube is released from rest at the point for which $\theta = 0$, and it slides down under gravity. Prove that for arbitrary value of θ the force the particle exerts on the tube is of magnitude $mg(\theta^2\sin^2 2\alpha + \cos^2\alpha)^{1/2}$.

4. A particle is moving in the x-y plane under the action of a constant gravitational force acting in the negative y direction. Find four independent constants of the motion for the two dimensional motion. Find three independent constants of the motion that do not explicitly depend on the time.

5. One end of an elastic string of unstretched length a and modulus of elasticity λ is attached to a fixed point O on a smooth horizontal surface. A particle of mass m is attached to the other end and is initially at rest on the surface. The particle is suddenly given a blow that imparts to it a velocity of magnitude v in a direction normal to the string. In the subsequent motion the string is stretched to a maximum length $3a$. What was the initial speed v of the particle?

6. A particle of mass m moves under the influence of gravity on the inner surface of a smooth paraboloid of revolution. The equation of the paraboloid in cylindrical coordinates ρ, ϕ, z is $\rho^2 = az$, and the z axis is

pointing vertically up. The particle is initially projected from the point $\rho = a$, $\phi = 0$, $z = a$ with a speed V in the horizontal direction. Determine the speed of the particle as a function of z. Determine the maximum and minimum values of z.

7. A particle of mass m is located at the end B of a massless rod AB of length a. The end A of the rod is free to move on a smooth horizontal wire. The rod is held parallel to the wire, then released. Determine the horizontal and vertical components of the velocity of the particle as a function of the angle that the rod makes with the wire.

8. A particle of unit mass moves on the inner surface of a paraboloid of revolution whose equation in cylindrical coordinates is $\rho^2 = 4az$. The particle is subjected to a repulsive force of magnitude $\mu\rho$ directed away from the z axis. Show that if the particle is projected along the surface in a direction perpendicular to the z axis with velocity $2a\mu^{1/2}$ it will describe a parabola.

9. A particle of mass m is suspended by a massless inextensible string of length $3\pi a$ from the end of a horizontal diameter of a cylinder of radius a whose axis is horizontal. The particle is projected with speed V horizontally so that it moves in a plane perpendicular to the axis of the cylinder and winds the string around the cylinder. Prove that the string will not be slack at the top of the path if $V^2 \geqslant 12\pi ga$. If $V^2 = 12\pi ga$ find the tension in the string when the particle is first moving vertically upward.

18

Impulse

INTRODUCTION

If a particle strikes a hard surface and bounces off, the force the surface exerts on the particle will last for a very short period of time, and the magnitude of the force will undergo very rapid changes during this time. A detailed analysis of the behavior of the particle during the collision would require a detailed knowledge of the force that the surface exerts on the particle. In most cases however we are not interested in the behavior of the particle during the collision; we simply want to know what final effect the collision has on the motion of the particle. In this chapter we consider the problem of determining the net effect on the particle of such a collision.

IMPULSE

Definition 1. If a particle is acted on by a force $\mathbf{f}$ then the impulse imparted to the particle between the times t_1 and t_2 is defined as

$$\hat{\mathbf{f}}(t_1, t_2) \equiv \int_{t_1}^{t_2} \mathbf{f}\, dt \tag{1}$$

Theorem 1. If a particle is acted on by a force $\mathbf{f}$ then the change in the linear momentum $\mathbf{p}$ of the particle between the times t_1 and t_2 is equal to the impulse imparted by the force between the times t_1 and t_2, that is,

$$\mathbf{p}(t_2) - \mathbf{p}(t_1) = \hat{\mathbf{f}}(t_1, t_2) \tag{2}$$

PROOF: From Newton's law

$$\mathbf{f} = \frac{d\mathbf{p}}{dt} \tag{3}$$

Integrating Eq. (3) between times t_1 and t_2 we obtain

$$\int_{t_1}^{t_2} \mathbf{f}\, dt = \int_{t_1}^{t_2} \frac{d\mathbf{p}}{dt}\, dt \tag{4}$$

The right hand side of Eq. (4) can be rewritten

$$\int_{t_1}^{t_2} \frac{d\mathbf{p}}{dt}\, dt = \int_{t_1}^{t_2} d\mathbf{p}(t) = \mathbf{p}(t_2) - \mathbf{p}(t_1) \tag{5}$$

Combining Eqs. (1), (4), and (5) we obtain Eq. (2).

Definition 2. If a particle is acted on by a torque $\mathbf{g}$ then the angular impulse imparted to the particle between the times t_1 and t_2 is defined as

$$\hat{\mathbf{g}}(t_1, t_2) \equiv \int_{t_1}^{t_2} \mathbf{g}\, dt \tag{6}$$

Theorem 2. If a particle is acted on by a torque $\mathbf{g}$ then the change in the angular momentum $\mathbf{h}$ of the particle between the times t_1 and t_2 is equal to the angular impulse imparted by the torque between the times t_1 and t_2, that is,

$$\mathbf{h}(t_2) - \mathbf{h}(t_1) = \hat{\mathbf{g}}(t_1, t_2) \tag{7}$$

PROOF: If we integrate the equation $\mathbf{g} = \dot{\mathbf{h}}$ between t_1 and t_2 we will immediately obtain the desired result.

INSTANTANEOUS IMPULSE

If a particle that is moving with a velocity $\mathbf{v}(t)$ at time t is acted on by a constant force $\mathbf{f}$ for a time Δt, the velocity of the particle will be changed by an amount

$$\Delta \mathbf{v} = \frac{\mathbf{f}\, \Delta t}{m} \tag{8}$$

and the position of the particle will be changed by an amount

$$\Delta \mathbf{r} = \left[\mathbf{v}(t) + \frac{\mathbf{f}\, \Delta t}{2m} \right] \Delta t \tag{9}$$

For suitably large values of $\mathbf{f}$ and small values of Δt, it is possible to observe an appreciable change in the velocity of the particle while the change in the position of the particle is negligible. We are thus led to the following definition.

Definition 3. If a particle is acted on by a large force $\mathbf{f}$ between the times t and $t + \Delta t$, and if the duration Δt of the force is so short that the change in the position of the particle during the time the force is acting is negligible, we can assume without appreciable error that the impulse is applied instantaneously at the time t, producing a discontinuous change $\Delta \mathbf{p}$ in the momentum of the particle, while leaving the position of the particle unchanged. In this case we refer to the impulse as an instantaneous impulse and designate it as $\hat{\mathbf{f}}(t)$, or simply as $\hat{\mathbf{f}}$. From a mathematical point of view, if a particle is acted on by a large force $\mathbf{f}$ between the times t and $t + \Delta t$, where Δt is a short period of time, the instantaneous impulse $\hat{\mathbf{f}}(t)$ associated with the impulse $\hat{\mathbf{f}}(t, t + \Delta t)$ is the impulse obtained by letting the duration Δt of the force approach zero and the magnitude of the force approach infinity while maintaining the value of the impulse constant, that is,

$$\hat{\mathbf{f}}(t) \equiv \hat{L}\hat{\mathbf{f}}(t, t + \Delta t) \tag{10}$$

where $\hat{L} \equiv$ limit as $\Delta t \to 0$, $f \to \infty$, and $\hat{\mathbf{f}}(t, t + \Delta t)$ is held constant.

Theorem 3. If a particle is acted on by an instantaneous impulse $\hat{\mathbf{f}}$, then the position $\mathbf{r}$ of the particle will be unchanged, but the momentum $\mathbf{p}$ will undergo a finite discontinuous change $\Delta \mathbf{p}$ equal to the instantaneous impulse, that is,

$$\Delta \mathbf{r} = 0 \tag{11}$$

$$\Delta \mathbf{p} = \hat{\mathbf{f}} \tag{12}$$

PROOF: The change in the momentum at the time t as a result of the application of the instantaneous impulse $\hat{\mathbf{f}}(t)$ at time t is given by

$$\Delta \mathbf{p}(t) = \hat{L}[\mathbf{p}(t + \Delta t) - \mathbf{p}(t)]$$
$$= \hat{L}[\hat{\mathbf{f}}(t, t + \Delta t)]$$
$$= \hat{\mathbf{f}}(t) \tag{13}$$

The change in the position at the time t is given by

$$\Delta \mathbf{r}(t) = \hat{L}[\mathbf{r}(t + \Delta t) - \mathbf{r}(t)]$$
$$= \hat{L}\left[\int_t^{t+\Delta t} \mathbf{v}\, dt\right]$$
$$= \hat{L}\left[\frac{1}{m} \int_t^{t+\Delta t} \mathbf{p}\, dt\right] \tag{14}$$

Since $\mathbf{p}$ starts at a finite value and changes by a finite amount as a result of the impulse, the integrand in the integral above remains finite; hence in the limit as $\Delta t \to 0$ the integral vanishes. We thus obtain

$$\Delta \mathbf{r}(t) = 0 \tag{15}$$

This completes the proof of the theorem.

INSTANTANEOUS ANGULAR IMPULSE

If a particle is acted on by an instantaneous linear impulse, then with respect to a given point, there is an associated instantaneous angular impulse. It is possible to develop the theory of instantaneous angular impulses by starting with Theorem 2 and paralleling the method used in developing the theory of instantaneous linear impulses. However we can get to the heart of the matter much more quickly by starting directly with Theorem 3. Following this route we are led to the following definition and theorem.

Definition 4. If a particle is acted on by an instantaneous linear impulse $\hat{\mathbf{f}}$, then the associated instantaneous angular impulse with respect to a given point is defined as

$$\hat{\mathbf{g}} \equiv \mathbf{r} \times \hat{\mathbf{f}} \qquad (16)$$

where $\mathbf{r}$ is the position of the particle with respect to the given point.

NOTE: The instantaneous angular impulse $\hat{\mathbf{g}}$ could also have been defined as $\hat{L}[\hat{\mathbf{g}}(t, t + \Delta t)]$. However it can be shown, since $\Delta \mathbf{r} = 0$ in the impulse, that both definitions are equivalent.

Theorem 4. If a particle is acted on by an instantaneous angular impulse $\hat{\mathbf{g}}$, the angular momentum will undergo a finite discontinuous change $\Delta \mathbf{h}$ equal to the instantaneous angular impulse, that is,

$$\Delta \mathbf{h} = \hat{\mathbf{g}} \qquad (17)$$

PROOF: From Theorem 3

$$\Delta \mathbf{p} = \hat{\mathbf{f}} \qquad (18)$$

If we cross multiply Eq. (18) by $\mathbf{r}$ we obtain

$$\mathbf{r} \times \Delta \mathbf{p} = \mathbf{r} \times \hat{\mathbf{f}} \qquad (19)$$

From Theorem 3 $\Delta \mathbf{r} = 0$, hence

$$\mathbf{r} \times \Delta \mathbf{p} = \Delta(\mathbf{r} \times \mathbf{p}) \equiv \Delta \mathbf{h} \qquad (20)$$

Combining Eqs. (16), (19), and (20) we obtain Eq. (17), which is the desired result.

NOTE: In the preceding discussion very little was said about the point with respect to which the linear momentum, angular momentum, and instantaneous angular impulse are measured. In the absence of this information we are certainly safe in assuming that at least the origin of our inertial frame is a suitable point. However it can be shown that Theorems 3 and 4 are valid with respect to any point a, whether moving or at rest, provided the velocity of the point does not change discontinuously at the moment the instantaneous impulse is applied. To show this we note simply that if the velocity of the point a with respect to the origin does not change at the instant of impulse, the change in the velocity of the particle with respect to the origin is the same as the change in the velocity of the particle with respect to the point a. It follows that the change in the momentum will be the same whether measured with reference to the origin or to the point a, and consequently Theorem 3 is true with reference to the point a, as well as to the origin. In a similar manner we can show that Theorem 4 is also true with reference to the point a.

THE IMPULSIVE EQUATIONS OF MOTION

The techniques developed in the preceding sections have been introduced to simplify the analysis of dynamical problems—not, as it may seem, to complicate them. Thus if a particle collides with a hard surface and rebounds, it is simpler to assume that the momentum undergoes a discrete change at the surface due to an instantaneous impulse, rather than that the momentum changes continuously from its initial to its final value due to a rapidly changing force. The situation is analogous to the case in electrodynamics at the boundary between two different media, where one assumes that the change from the one medium to the other is a discontinuous change rather than a sharp but continuous change.

To stress the basic simplicity of the ideas just introduced, let us consider the following recapitulations of the results obtained previously for ordinary forces, and the results obtained in this chapter for instantaneous impulses.

1. The time rates of change of the linear momentum **p** and angular momentum **h** of a particle acted on by a force **f** can be obtained from the equations

$$\mathbf{f} = \dot{\mathbf{p}} \qquad (21)$$

$$\mathbf{g} = \dot{\mathbf{h}} \qquad (22)$$

where

$$\mathbf{g} = \mathbf{r} \times \mathbf{f} \qquad (23)$$

2. The change in the linear momentum **p** and the angular momentum **h** of a particle acted on by an instantaneous impulse $\hat{\mathbf{f}}$ can be obtained from the equations

$$\hat{\mathbf{f}} = \Delta \mathbf{p} \qquad (24)$$

$$\hat{\mathbf{g}} = \Delta \mathbf{h} \qquad (25)$$

where
$$\hat{\mathbf{g}} = \mathbf{r} \times \hat{\mathbf{f}} \tag{26}$$

To obtain statement 2 from statement 1, we simply replace "time rate of change" by "change" and "force" by "instantaneous impulse."

COLLISIONS

If one studies the collision of a particle with a surface it is found that the following hypothesis gives results in reasonable agreement with experiment.

Hypothesis 1. If a particle collides with a surface, the collision can be broken down into two stages. In the first stage, called the period of compression, the particle strikes the surface and the surface is compressed until the normal component of the velocity of the particle with respect to the surface is reduced to zero. In the second stage, called the period of restitution, the surface expands and the particle rebounds from the surface. The impulse imparted to the particle by the surface during the period of compression is equivalent to an instantaneous impulse $\hat{\mathbf{f}}'$ and the impulse imparted to the particle by the surface during the period of restitution is equivalent to an instantaneous impulse $\hat{\mathbf{f}}''$. The normal component of $\hat{\mathbf{f}}''$ is equal to e times the normal component of $\hat{\mathbf{f}}'$, where e is an experimental constant, called the *coefficient of restitution*, which depends only on the natures of the particle and the surface, and whose value lies between 0 and 1. If the surface is smooth, there are no tangential components to the impulses $\hat{\mathbf{f}}'$ and $\hat{\mathbf{f}}''$. If the surface is rough then there is little experimental evidence that would enable one to predict beforehand the value of the tangential components of $\hat{\mathbf{f}}'$ and $\hat{\mathbf{f}}''$. However in the absence of any other information it is customary to assume that the tangential components of $\hat{\mathbf{f}}'$ and $\hat{\mathbf{f}}''$ bear the same relation to the normal components as they would if they were ordinary forces and the ordinary laws of friction were in effect. If the surface is smooth and $e = 1$, we say that the collision is an *elastic collision*. If the surface is rough or $e \neq 1$, we say that the collision is an *inelastic collision*. If the surface is perfectly rough, or sufficiently rough to reduce the tangential velocity of the particle to zero, and $e = 0$, we say that the collision is a *perfectly inelastic collision*.

The following theorem is an immediate consequence of the hypothesis above.

Theorem 5. If a particle collides with a surface, which is at rest or moving, and if the velocity of the surface does not appreciably change as a result of the impact, then the normal component of the postcollision velocity of the particle relative to the surface is equal to $-e$ times the normal component of the precollision velocity of the particle relative to the surface, that is,

$$\mathbf{v}'' \cdot \mathbf{n} = -e\mathbf{v}' \cdot \mathbf{n} \tag{27}$$

where $\mathbf{v}'$, and $\mathbf{v}''$ are the pre- and postcollision velocities of the particle relative to the surface, and $\mathbf{n}$ is a unit vector normal to the surface at the point of impact.

PROOF: The normal component of the velocity of the particle with respect to the surface at the beginning of the period of compression is $\mathbf{v}' \cdot \mathbf{n}$ and at the end of the period of compression is zero. Equating the change in this component of the velocity to the normal component of the instantaneous impulse $\hat{\mathbf{f}}'$ imparted by the wall during the period of compression, we obtain

$$-\mathbf{v}' \cdot \mathbf{n} = \hat{\mathbf{f}}' \cdot \mathbf{n} \tag{28}$$

The normal component of the velocity of the particle with respect to the surface at the beginning of the period of restitution is zero and at the end is $\mathbf{v}'' \cdot \mathbf{n}$. Equating the change in this component of the velocity to the normal component of the instantaneous impulse $\hat{\mathbf{f}}''$ imparted by the wall during the period of restitution, we obtain

$$\mathbf{v}'' \cdot \mathbf{n} = \hat{\mathbf{f}}'' \cdot \mathbf{n} \tag{29}$$

But from Hypothesis 1 we have

$$\hat{\mathbf{f}}'' = e\hat{\mathbf{f}}' \tag{30}$$

Combining Eqs. (28)–(30) we obtain Eq. (27).

EXAMPLES

Example 1. A particle of mass m moving with a speed v strikes a rough wall at an angle θ with the normal to the wall. The coefficients of static and kinetic friction for the contact between the particle and the wall are μ and $\bar{\mu}$, respectively. The coefficient of restitution for the collision is e. Determine the magnitude and direction of the velocity of the particle after it rebounds.

Solution. We choose our reference frame with the origin at the point of impact, the x axis normal to the surface and pointing inward, and the x-y plane the plane of motion with the initial velocity in the first quadrant as shown in Fig. 1a. The impulses administered to the particle during the compressive stage (Fig. 1b) consist of a normal impulse of magnitude $\hat{N}$ in the negative x direction and a tangential impulse of magnitude $\hat{T}$ in the negative y direction. The magnitude of the tangential impulse cannot exceed $\mu \hat{N}$. The velocity of the particle at the end of the compressive stage is shown in Fig. 1c. The x component of the velocity is zero and the y component will have some value v_0 between 0 and $v \sin \theta$. Applying the impulse equation $\hat{\mathbf{f}} = \Delta \mathbf{p}$ to the

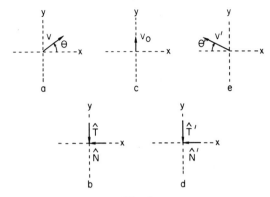

Fig. 1

compressive stage we obtain

$$-\hat{N} = 0 - mv\cos\theta \tag{31}$$

$$-\hat{T} = mv_0 - mv\sin\theta \tag{32}$$

If $v_0 = 0$ then $\hat{T}/\hat{N} = \tan\theta$. But $\hat{T}/\hat{N}$ cannot exceed μ, hence if $\tan\theta \leq \mu$ then $v_0 = 0$, but if $\tan\theta > \mu$ then slipping occurs and $\hat{T} = \bar{\mu}\hat{N}$. Using these results in Eqs. (31) and (32) we obtain

$$\hat{N} = mv\cos\theta \tag{33}$$

$$\begin{aligned}v_0 &= 0 & \text{if} \quad \tan\theta \leq \mu \\ v_0 &= v(\sin\theta - \bar{\mu}\cos\theta) & \text{if} \quad \tan\theta > \mu\end{aligned} \tag{34}$$

The impulses administered to the particle during the restitution stage (Fig. 1d) consist of a normal impulse of magnitude $\hat{N}'$ in the negative x direction and a tangential impulse $\hat{T}'$ in the negative y direction. The velocity of the particle at the end of the restitution stage is shown in Fig. 1e. Applying the impulse equation $\hat{\mathbf{f}} = \Delta\mathbf{p}$ to the restitution stage, we obtain

$$-\hat{N}' = -mv'\cos\theta' \tag{35}$$

$$-\hat{T}' = mv'\sin\theta' - mv_0 \tag{36}$$

From Hypothesis 1

$$\hat{N}' = e\hat{N} \tag{37}$$

The value of $\hat{T}'$ will be equal to $\bar{\mu}\hat{N}'$, unless this value is more than enough to stop the motion in the y direction, in which case it will be equal to mv_0. Thus

$$\begin{aligned}\hat{T}' &= \bar{\mu}\hat{N}' & \text{if} \quad \bar{\mu}\hat{N}' \leq mv_0 \\ \hat{T}' &= mv_0 & \text{if} \quad \bar{\mu}\hat{N}' > mv_0\end{aligned} \tag{38}$$

Combining Eqs. (33)–(38) we obtain

$$v' \cos \theta' = ev \cos \theta \tag{39}$$

$$\begin{aligned} v' \sin \theta' &= 0 & \text{if} \quad \tan \theta \leq \mu \text{ or } \bar{\mu}(1+e) \\ v' \sin \theta' &= v \sin \theta - \bar{\mu}(1+e)v \cos \theta & \text{if} \quad \tan \theta > \mu \text{ and } \bar{\mu}(1+e) \end{aligned} \tag{40}$$

Solving Eqs. (39) and (40) for v' and θ' we obtain the following two cases:

Case 1. $\tan \theta \leq \mu$ or $\bar{\mu}(1+e)$.

$$v' = ev \cos \theta \tag{41}$$

$$\theta' = 0 \tag{42}$$

Case 2. $\tan \theta > \mu$ and $\bar{\mu}(1+e)$.

$$v' = v\left\{ e^2 \cos^2 \theta + \left[\sin \theta - \bar{\mu}(1+e) \cos \theta\right]^2 \right\}^{1/2} \tag{43}$$

$$\theta' = \tan^{-1}\left\{ (\tan \theta / e) - \left[\bar{\mu}(1+e)/e\right] \right\} \tag{44}$$

NOTE: The solution to this problem could have been shortened by not breaking the collision into two stages. The two stage process is however the more fundamental approach and we have employed it for illustrative purposes. The reader should rework the problem using the condensed approach.

PROBLEMS

1. A tug boat of mass m is attached to a barge of mass M by a cable the mass of which may be neglected. The cable is slack. The tug moves and has acquired a speed v when the cable becomes taut and the barge is jerked into motion. Assuming that the cable has a coefficient of restitution $\frac{1}{2}$ and neglecting the impulsive resistance of the water, find (a) the speed imparted to the barge and (b) the average tension in the cable during the jerk, supposing this to last a time Δt.

2. A steel block of mass M initially at rest on a smooth horizontal surface is fired on by bullets each having a mass m very much less than M, a speed v, and a coefficient of restitution e. The bullets are fired at the rate of N per second and strike the surface of the block normally. Find the velocity of the block at any time t after the firing has begun.

3. A block whose coefficient of restitution is e slides without friction along a horizontal groove at a speed v_1. It strikes a slower moving block of equal mass whose speed is v_2 in the same direction as the first. Compute the magnitudes and directions of the velocities of the two blocks after the collision. How much kinetic energy is lost in the collision?

4. A ball is dropped on the floor from a height h. If the coefficient of restitution is e, find the height of the ball at the top of the nth rebound.

5. A ping-pong ball is dropped from a height a onto a hard, smooth, fixed tabletop. It bounces to a height b. Compute the coefficient of restitution of the ball and the kinetic energy lost after n bounces of the ball.

6. A ball of mass m moving with a speed v strikes a ball of equal mass which is at rest. The line of contact of the balls makes an angle α with the initial velocity of the first ball. The balls are smooth and the coefficient of restitution is e. Find the directions and the magnitudes of the velocities after impact.

7. A particle of mass m moving with a speed v strikes a rough wall at an angle θ with the normal to the wall. The coefficients of static and kinetic friction for the contact between the particle and the wall are μ and $\bar{\mu}$, respectively. The coefficient of restitution for the collision is e. Determine the magnitude and direction of the velocity of the particle after it rebounds.

8. A particle moving with a speed of 30 m/s in a direction making an angle of 60° with the horizontal strikes a smooth horizontal plane and rebounds, the coefficient of restitution being $\frac{1}{3}$. Find the speed and direction of the motion of the particle immediately after impact.

9. A particle of mass m and speed v collides head on with one of the atoms of a molecule consisting of two atoms, each of mass M, connected by a rigid linear bond. The molecule is initially at rest, the collision is elastic, and the velocity of the particle makes an angle θ with the bond line.
  a. Determine the final velocity of the particle.
  b. Determine the final velocity of the center of mass of the molecule.
  c. Determine the ratio of the rotational to the translational kinetic energy imparted to the molecule.

19

The Equations of Motion in Noninertial Reference Frames

INTRODUCTION

If we want to determine the motion of a particle with respect to an inertial frame of reference we use Newton's law. Once we have determined the motion of a particle with respect to one frame, we can determine its motion with respect to another frame by simply transforming our results from one frame to the other. Thus to determine the motion of a particle with respect to a noninertial frame, we might first determine its motion with respect to an inertial frame and then transform our results to the noninertial frame. However it is frequently more convenient to work right from the start with the noninertial frame. In this chapter we derive the equations that allow us to directly determine the motion of a particle with respect to a noninertial frame.

THE EQUATION OF MOTION IN A NONINERTIAL REFERENCE FRAME

Theorem 1. Let S be an inertial reference frame, S^* an arbitrary noninertial reference frame, O the origin of frame S, O^* the origin of frame S^*, P a particle of mass m, $\mathbf{r}(OP)$ the position of P with respect to O, $\mathbf{r} \equiv \mathbf{r}(O^*P)$ the position of P with respect to O^*, $\omega \equiv \omega(SS^*)$ the angular velocity of frame S^* with respect to frame S, $\mathbf{R} \equiv \mathbf{r}(OO^*)$ the position of O^* with respect to O, $\mathbf{f}$ the force acting on the particle P, $\dot{\mathbf{A}}$ the time rate of change of an arbitrary vector $\mathbf{A}$ with respect to an observer fixed in frame S, and $\overset{*}{\mathbf{A}}$ the time rate of change of an arbitrary vector $\mathbf{A}$ with respect to an observer fixed in frame S^*. Then the equation of motion of particle P can be written in the form

$$\mathbf{f} = m\overset{**}{\mathbf{r}} + m\ddot{\mathbf{R}} + m\overset{*}{\omega} \times \mathbf{r} + 2m\omega \times \overset{*}{\mathbf{r}} + m\omega \times (\omega \times \mathbf{r}) \tag{1}$$

PROOF: The equation of motion with respect to frame S is

$$\mathbf{f} = m\ddot{\mathbf{r}}(OP) \tag{2}$$

But
$$r(OP) = r(OO^*) + r(O^*P) \equiv \mathbf{R} + \mathbf{r} \qquad (3)$$

Substituting Eq. (3) in Eq. (2) we obtain

$$\mathbf{f} = m\ddot{\mathbf{R}} + m\ddot{\mathbf{r}} \qquad (4)$$

But from Theorem 2 in Chapter 6

$$\dot{\mathbf{r}} = \overset{*}{\mathbf{r}} + \omega \times \mathbf{r} \qquad (5)$$

$$\ddot{\mathbf{r}} = \overline{[\overset{*}{\mathbf{r}} + \omega \times \mathbf{r}]}^* + \omega \times [\overset{*}{\mathbf{r}} + \omega \times \mathbf{r}]$$

$$= [\overset{**}{\mathbf{r}} + \overset{*}{\omega} \times \mathbf{r} + \omega \times \overset{*}{\mathbf{r}}] + [\omega \times \overset{*}{\mathbf{r}} + \omega \times (\omega \times \mathbf{r})]$$

$$= \overset{**}{\mathbf{r}} + \overset{*}{\omega} \times \mathbf{r} + 2\omega \times \overset{*}{\mathbf{r}} + \omega \times (\omega \times \mathbf{r}) \qquad (6)$$

Substituting Eq. (6) in Eq. (4) we obtain Eq. (1), which is what we wanted to prove.

FICTITIOUS FORCES

The terms other than $m\overset{**}{\mathbf{r}}$ on the right hand side of Eq. (1) arise from the motion of the frame S^* with respect to the frame S. These terms make the equation of motion look quite unlike Newton's law. However if we rearrange Eq. (1) as follows:

$$\mathbf{f} - m\ddot{\mathbf{R}} - m\overset{*}{\omega} \times \mathbf{r} - 2m\omega \times \overset{*}{\mathbf{r}} - m\omega \times (\omega \times \mathbf{r}) = m\overset{**}{\mathbf{r}} \qquad (7)$$

and treat the terms other than $\mathbf{f}$ on the left hand side as if they were forces, our interpretation of the equations is appreciably simplified. From this point of view an observer in a noninertial frame will simply note the presence of forces that are not present in an inertial frame, but he will use the same equation of motion. Since these forces are not forces in the sense in which we have previously defined "forces," but are due to the motion of the frame S^* with respect to the frame S, we call them fictitious forces or inertial forces. The introduction of the concept of fictitious forces, and the consequent formal retention of Newton's law, allows us freely to make use of the techniques and ideas already developed for motion with respect to an inertial frame.

In the following sections we consider some of these fictitious forces in more detail.

Fictitious Gravitational Forces

The fictitious force $-m\ddot{\mathbf{R}}$ in Eq. (7) is present if and only if the frame S^* is undergoing a translational acceleration with respect to frame S. If we compare the force $-m\ddot{\mathbf{R}}$ with the force $m\mathbf{g}$ that is produced by gravity, we see that the force $-m\ddot{\mathbf{R}}$ is equivalent to the force that would be exerted by a gravitational field for which $\mathbf{g} = -\ddot{\mathbf{R}}$. Thus the presence of a fictitious gravitational force is the only unusual effect that would be noticed by an observer who is in a frame that is undergoing a uniform translational acceleration with respect to an inertial frame. Although this force is fictitious, it will produce exactly the same effects as a real gravitational force. We regularly experience this fictitious force when we are in an accelerating elevator or automobile.

If the frame of reference S^* is freely falling in a gravitational field $\mathbf{g}$, then $\ddot{\mathbf{R}} = \mathbf{g}$. In this case the fictitious force $-m\ddot{\mathbf{R}}$ will just cancel out the real force $m\mathbf{g}$. It follows that if we are in a frame that is freely falling in the earth's gravitational field, we will experience a condition of apparent weightlessness. An astronaut in a capsule that is freely orbiting the earth experiences this condition.

Centrifugal Force

The fictitious force $-m\boldsymbol{\omega} \times (\boldsymbol{\omega} \times \mathbf{r})$ is called the centrifugal force. Noting that $\mathbf{r}$ is the vector that joins the origin O^* of the frame S^* with the position P of the particle, and $\boldsymbol{\omega} \times (\boldsymbol{\omega} \times \mathbf{r})$ is perpendicular to both $\boldsymbol{\omega}$ and $\boldsymbol{\omega} \times \mathbf{r}$, we can readily show that the force $-m\boldsymbol{\omega} \times (\boldsymbol{\omega} \times \mathbf{r})$ is directed perpendicularly away from an axis through O^* and parallel to $\boldsymbol{\omega}$ and has a magnitude $m\omega^2 r \sin\theta$, where $r \sin\theta$ is the distance from the particle to the axis through O^* and parallel to $\boldsymbol{\omega}$.

We regularly experience the effect of this force when we round a corner in an automobile, train, or plane.

Coriolis Force

The fictitious force $-2m\boldsymbol{\omega} \times \overset{*}{\mathbf{r}}$ is called the Coriolis force. It is the only one of the fictitious forces that depends on the velocity of the particle with respect to the frame S^*.

We list below several examples around us of effects produced by the Coriolis force arising from the earth's rotation.

1. The Coriolis force plays a large role in the great mass movements of the oceans and the atmosphere. In particular it plays a significant part in the formation of cyclones and the trade winds.
2. Long-range rockets are considerably deflected by the Coriolis force. The deflection is to the right of the plane of firing in the Northern Hemisphere and to the left of the plane of firing in the Southern Hemisphere.
3. Rivers and railroad trains experience a slight force to their right in the Northern Hemisphere and to their left in the Southern Hemisphere. Whether there are any noticeable effects of this force however is questionable.
4. If a simple pendulum, consisting of a heavy bob suspended vertically by a very long string from an essentially frictionless pivot, is set in motion in a vertical plane, the plane of motion is observed to rotate gradually about a vertical axis. The rate of rotation of the plane of motion will depend on

the angular velocity of the earth and the latitude at which the pendulum is located. Such a pendulum is called a Foucault pendulum. The rotation of the plane of motion is due to the Coriolis force.

EXAMPLES

Example 1. The Foucault Pendulum. A heavy particle suspended by a long string at a latitude λ in the Northern Hemisphere is initially set in motion in a vertical plane. Show that the plane of motion of the particle will slowly precess about the upward vertical with an angular velocity $-\omega \sin \lambda$, where ω is the angular velocity of the earth.

Solution. We assume for the purposes of this problem that the earth can be approximated by a uniform sphere of radius R and mass M, which is rotating about a diameter fixed with respect to an inertial frame of reference. We then define S to be an inertial frame with origin O at the center of the earth, and S^* to be a frame fixed to the earth with its origin O^* on the surface of the earth. If a particle of mass m is hanging from a string and is at rest with respect to the frame S^*, and if **T** is the force that the string exerts on the particle, then from Eq. (7)

$$\mathbf{T} - \frac{GMm(\mathbf{R} + \mathbf{r})}{|\mathbf{R} + \mathbf{r}|^3} - m\ddot{\mathbf{R}} - m\omega \times (\omega \times r) = 0 \tag{8}$$

or equivalently

$$\mathbf{T} - \frac{GMm(\mathbf{R} + \mathbf{r})}{|\mathbf{R} + \mathbf{r}|^3} - m\omega \times [\omega \times (\mathbf{R} + \mathbf{r})] = 0 \tag{9}$$

The second term in Eq. (9) is the gravitational force the earth exerts on the particle and the third term, which is small compared to the second term, is the centrifugal force. It is customary to combine these two forces into a single effective gravitational force

$$m\mathbf{g} \equiv -\frac{GMm(\mathbf{R} + \mathbf{r})}{|\mathbf{R} + \mathbf{r}|^3} - m\omega \times [\omega \times (\mathbf{R} + \mathbf{r})] \tag{10}$$

The value of **g** does not vary greatly over short distances on the surface of the earth, and it is approximately radial in direction. For the purposes of this problem we therefore assume that **g** has a constant value and is directed toward the center of the earth. The equation of motion with respect to the frame S^* of a particle suspended from a string can then be approximated by the equation

$$\mathbf{T} + m\mathbf{g} - 2m\omega \times \overset{*}{\mathbf{r}} = m\overset{**}{\mathbf{r}} \tag{11}$$

where **T** is the tension in the string and **g** is a constant directed toward the center of the earth. We now choose the 3 axis of frame S to be in the direction of the angular velocity ω, and we choose the frame S^* as the frame fixed in the earth with origin O^* at the equilibrium position of the particle, 3* axis in

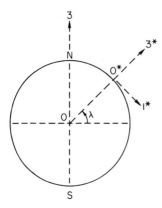

Fig. 1

the direction of $-\mathbf{g}$, 2* axis pointing east, and 1* axis pointing south as shown in Fig. 1. The latitude of the point O^* is the angle between the 3* axis and the equatorial plane as shown in Fig. 1. If we let $\mathbf{i}, \mathbf{j}, \mathbf{k}$ be the unit vectors in the 1*, 2*, and 3* directions, respectively, and let x, y, z be the 1*, 2*, and 3* coordinates of the particle, then

$$\mathbf{r} = x\mathbf{i} + y\mathbf{j} + z\mathbf{k} \tag{12}$$

$$\mathbf{g} = -g\mathbf{k} \tag{13}$$

$$\boldsymbol{\omega} = -\omega \cos\lambda \, \mathbf{i} + \omega \sin\lambda \, \mathbf{k} \tag{14}$$

If a is the length of the string then $\mathbf{T}$, which is in the direction of the string, is of the form

$$\mathbf{T} = A\big[-x\mathbf{i} - y\mathbf{j} + (a-z)\mathbf{k}\big] \tag{15}$$

where A is a positive quantity. Substituting Eqs. (12)–(15) in Eq. (11) we obtain

$$-Ax + 2m\omega\dot{y}\sin\lambda = m\ddot{x} \tag{16}$$

$$-Ay - 2m\omega(\dot{x}\sin\lambda + \dot{z}\cos\lambda) = m\ddot{y} \tag{17}$$

$$A(a-z) - mg + 2m\omega\dot{y}\sin\lambda = m\ddot{z} \tag{18}$$

If the length of the string is large compared to the displacement of the particle from its equilibrium position, the average value of z will be small compared to the average values of x and y, and $\omega\dot{y}$ will be very much smaller than g; hence the equations above can be approximated by the equations

$$-Ax + 2m\omega\dot{y}\sin\lambda = m\ddot{x} \tag{19}$$

$$-Ay - 2m\omega\dot{x}\sin\lambda = m\ddot{y} \tag{20}$$

$$Aa - mg = 0 \tag{21}$$

Eliminating A and rearranging terms we obtain

$$\ddot{x} + n^2 x = 2\Omega\dot{y} \tag{22}$$

$$\ddot{y} + n^2 y = -2\Omega\dot{x} \tag{23}$$

where

$$n^2 = \frac{g}{a} \tag{24}$$

$$\Omega = \omega \sin \lambda \tag{25}$$

If we now introduce a new frame S' whose origin and 3 axis coincide with those of S^* but which is rotating with an angular velocity $\omega(S^*S') = -\Omega \mathbf{k}$ with respect to S^*, and if we let x' and y' be the coordinates of the particle with respect to S', then

$$x = x' \cos \Omega t + y' \sin \Omega t \tag{26}$$

$$y = -x' \sin \Omega t + y' \cos \Omega t \tag{27}$$

Substituting Eqs. (26) and (27) into Eqs. (22) and (23) and neglecting terms in Ω^2 we obtain

$$\ddot{x}' + n^2 x' = 0 \tag{28}$$

$$\ddot{y}' + n^2 y' = 0 \tag{29}$$

which are the usual equations of motion of a simple pendulum. If the particle is initially swinging in a vertical plane in the frame S' it will continue to do so, hence in the frame S^* the plane of motion is precessing with an angular velocity $-\omega \sin \lambda \mathbf{k}$. Alternatively we could have solved the problem by switching to the frame S' at an earlier stage. In the frame S' the equation of motion is approximately

$$\mathbf{T} + m\mathbf{g} - 2m[\omega(SS') \times \dot{\mathbf{r}}] = m\ddot{\mathbf{r}}' \tag{30}$$

If we assume that the z' motion can be ignored and that the magnitude of $\mathbf{T}$ is approximately mg, and note that S' has been so chosen that $\omega(SS')$ has no component in the z' direction, then

$$\mathbf{T} \approx -\frac{mgx'}{a}\mathbf{i}' - \frac{mgy'}{a}\mathbf{j}' + T_z \mathbf{k}' \tag{31}$$

$$\mathbf{g} = -mg\mathbf{k}' \tag{32}$$

$$\omega(SS') = \omega_x \mathbf{i}' + \omega_y \mathbf{j}' \tag{33}$$

$$\mathbf{r} \approx x'\mathbf{i}' + y'\mathbf{j}' \tag{34}$$

Substituting Eqs. (31)–(34) in Eq. (30) we obtain Eqs. (28) and (29), which is the desired result.

PROBLEMS

1. The origin O^* of a frame S^* is moving through space with a constant velocity and the frame S^* is rotating about its 2^* axis with an angular velocity that is zero at $t = 0$ but is increasing at a constant rate b. Write the component equations of motion in the frame S^* of a particle that is acted on by no real forces.

2. Let S be an inertial frame and S^* a noninertial frame. The velocity of the origin O^* of S^* with respect to the origin O of S is $bt\mathbf{e}_1$, where b is a constant. The angular velocity of S^* with respect to S is $\omega \mathbf{e}_3$, where ω is a constant. At time $t = 0$ the two frames coincide. Write the component equations of motion in the frame S^* of a particle of mass m that is constrained to move in the 1*-2* plane, but is otherwise acted upon by no forces. If the particle is initially at rest with respect to S^* at the origin O^* find x_{1*} and x_{2*} as functions of the time.

3. A truck of mass M moves with a linear acceleration a along a horizontal road. On the truck a horizontal turntable rotates at angular velocity ω. A small object of mass m rests on the turntable at a distance r from its center of rotation. The coefficient of static friction between the object and the surface of the turntable is μ. What is the maximum value that ω may have before the object will start to move?

4. A particle slides on a perfectly smooth plane that is located at a latitude λ on the earth and is oriented so as to be perpendicular to the plumb line at that latitude. If the angular velocity of rotation of the earth has the magnitude ω, find the equation of the path of the particle as it appears to an observer sitting on the plane. Choose the z axis upward parallel to the plumb line, x to the south, and y to the east, and assume that when $t = 0$ then $x = 0$, $y = a$, $\dot{x} = u$, and $\dot{y} = 0$.

5. A particle slides on a smooth plane that is tangent to the earth's surface at a latitude λ. Determine the reaction, due to the earth's rotation, of the plane on the particle.

6. Water is flowing northward with a speed v along a channel of width b located at a latitude of λ. Show that the height of the water on the east side of the channel exceeds that on the west side by $2bv\omega \sin\lambda/g$, where ω is the angular velocity of the earth.

7. A particle of mass m is thrown vertically upward to a height h at a latitude λ in the Northern Hemisphere. Approximately where will it strike the ground?

8. Solve the equation of motion of a particle falling freely from a height h to second order in ω where ω is the angular velocity of the earth. Show that in the Northern Hemisphere there is a deviation to the south as well as that to the east. Calculate both components of the deviation for $h = 400$ m and a latitude $50°$N. How would an inertial observer interpret the southerly deviation?

9. A projectile at a latitude λ is fired eastward with a speed V and angle of elevation α. Show that the lateral deviation because of the earth's rotation is $4\omega V^3 \sin^2\alpha \cos\alpha \sin\lambda/g^2$, where ω is the angular velocity of the earth.

10. A projectile at a latitude λ is fired eastward with a speed V and angle of elevation α. Show that the change in the range of the particle due to the earth's rotation is $(2R^3/g)^{1/2}\omega\cos\lambda[\cot^{1/2}\alpha - (\tan^{3/2}\alpha/3)]$, where R is the range in the absence of the earth's rotation, and ω is the angular velocity of the earth.

11. A particle of mass m slides on a smooth rod that is rotating in a plane about an axis through one end of the rod and perpendicular to the plane. The axis is at rest in an inertial frame, and the angular velocity ω of the rod with respect to the inertial frame is constant. Determine the reaction R of the rod on the particle first by employing a nonrotating coordinate frame, and then by employing a rotating coordinate frame.

12. A particle of mass m is constrained to slide on a smooth straight wire, which in turn is forced to rotate in a horizontal plane about a point O on the wire at a constant angular velocity ω. Gravity is acting vertically downward, and there is a force of magnitude mk/r^2 acting on the particle and directed toward O, where r is the distance from O to the particle and k is a positive constant. Initially the particle is at rest a distance b from O. Show that the magnitude of the reaction of the wire on the particle is $(m^2g^2 + 4m^2\omega^2\{\omega^2(r^2 - b^2) + [2k(b-r)/rb]\})^{1/2}$.

13. A bead executes small oscillations about the lowest point of the smooth parabolic wire $x^2 = 4az$, where the z axis is vertically up. Show that the period of oscillation is the same as that of a simple pendulum of length $2a$. Show also that if the wire is constrained to rotate about the z axis with constant angular speed ω, the equivalent simple pendulum is of length $2ag/(g - a\omega^2)$, provided $g > 2a\omega^2$. If $\omega^2 > g/2a$, show that there is no position of stable equilibrium.

14. A bead moves on a smooth wire that is bent in the form of a vertical circle of radius a. The wire rotates about a fixed vertical diameter with a uniform angular velocity ω. Show that if θ is the angular distance from the lowest point, and the particle is initially at rest relative to the tube at $\theta = \alpha$, where $\omega\cos(\alpha/2) = (g/a)^{1/2}$, then $(\dot\theta)^2 = \omega^2(1 - \cos\theta)(\cos\theta - \cos\alpha)$.

15. A bead slides freely on a smooth circular wire of radius a that is constrained to rotate in a horizontal plane with constant angular speed ω about a fixed point O on the circumference. Show that if the speed of the bead is u when it passes through O, then its speed relative to the wire when it passes through the point diametrically opposite to O is $(u^2 + 4a^2\omega^2)^{1/2}$.

20

The Equations of Motion in Orthogonal Curvilinear Coordinate Systems

INTRODUCTION

In many problems the analysis is considerably simplified if one uses orthogonal curvilinear coordinates to describe the motion of a particle. In Chapter 7 we indicated how to handle particle kinematics using curvilinear coordinates. The extension of the results to particle dynamics is straightforward. Therefore in this chapter we simply state the result for cylindrical and spherical coordinates, and provide the student with an example and problems.

CYLINDRICAL COORDINATES

The motion of a particle is determined by Newton's law:

$$\mathbf{f} = m\ddot{\mathbf{r}} \tag{1}$$

Using cylindrical coordinates

$$\mathbf{f} = f_\rho \mathbf{e}_\rho + f_\phi \mathbf{e}_\phi + f_z \mathbf{e}_z \tag{2}$$

and

$$\ddot{\mathbf{r}} = (\ddot{\rho} - \rho\dot{\phi}^2)\mathbf{e}_\rho + (\rho\ddot{\phi} + 2\dot{\rho}\dot{\phi})\mathbf{e}_\phi + \ddot{z}\mathbf{e}_z \tag{3}$$

Substituting Eqs. (2) and (3) in Eq. (1) and equating components we obtain

$$m(\ddot{\rho} - \rho\dot{\phi}^2) = f_\rho \tag{4}$$

$$m(\rho\ddot{\phi} + 2\dot{\rho}\dot{\phi}) = f_\phi \tag{5}$$

$$m\ddot{z} = f_z \tag{6}$$

SPHERICAL COORDINATES

Using spherical coordinates

$$\mathbf{f} = f_r \mathbf{e}_r + f_\theta \mathbf{e}_\theta + f_\phi \mathbf{e}_\phi \tag{7}$$

and

$$\ddot{\mathbf{r}} = (\ddot{r} - r\dot{\theta}^2 - r\sin^2\theta\,\dot{\phi}^2)\mathbf{e}_r + (r\ddot{\theta} + 2\dot{r}\dot{\theta} - r\sin\theta\cos\theta\,\dot{\phi}^2)\mathbf{e}_\theta$$
$$+ (r\sin\theta\,\ddot{\phi} + 2\dot{r}\dot{\phi}\sin\theta + 2r\cos\theta\,\dot{\theta}\dot{\phi})\mathbf{e}_\phi \tag{8}$$

Substituting these expressions in Newton's law and equating components we obtain

$$m(\ddot{r} - r\dot{\theta}^2 - r\sin^2\theta\,\dot{\phi}^2) = f_r \tag{9}$$

$$m(r\ddot{\theta} + 2\dot{r}\dot{\theta} - r\sin\theta\cos\theta\,\dot{\phi}^2) = f_\theta \tag{10}$$

$$m(r\sin\theta\,\ddot{\phi} + 2\dot{r}\dot{\phi}\sin\theta + 2r\cos\theta\,\dot{\theta}\dot{\phi}) = f_\phi \tag{11}$$

EXAMPLES

Example 1. A bead of mass m is free to slide on a smooth helical wire, the equation of which in cylindrical coordinates is $\rho = a$, $z = b\phi$. Gravity is acting in the positive z direction. The particle is released from rest at the point $\rho = a$, $\phi = 0$, $z = 0$. Determine as a function of ϕ the force the wire exerts on the bead.

Solution. Letting $\mathbf{R}$ be the force the wire exerts on the bead, and $m\mathbf{g}$ the force of gravity, we obtain for the force acting on the particle

$$\mathbf{f} = \mathbf{R} + m\mathbf{g} \tag{12}$$

In cylindrical coordinates

$$\mathbf{R} = R_\rho \mathbf{e}_\rho + R_\phi \mathbf{e}_\phi + R_z \mathbf{e}_z \tag{13}$$

$$m\mathbf{g} = mg\mathbf{e}_z \tag{14}$$

Using these results in Eqs. (4)–(6) and letting $\rho = a$ and $z = b\phi$ we obtain

$$R_\rho = m(\ddot{\rho} - \rho\dot{\phi}^2) = -ma\dot{\phi}^2 \tag{15}$$

$$R_\phi = m(\rho\ddot{\phi} + 2\dot{\rho}\dot{\phi}) = ma\ddot{\phi} \tag{16}$$

$$R_z = m\ddot{z} - mg = mb\ddot{\phi} - mg \tag{17}$$

Equations (15)–(17) provide us with three equations in the four unknowns R_ρ, R_ϕ, R_z, and ϕ.

The additional equation required before we can solve for these unknowns is generated by noting that the wire is smooth and thus the force $\mathbf{R}$ cannot have a component in the direction of the wire. To express this fact mathematically we note that the velocity of the bead is along the direction of the wire and is given by

$$\dot{\mathbf{r}} = \dot{\rho}\mathbf{e}_\rho + \rho\dot{\phi}\mathbf{e}_\phi + \dot{z}\mathbf{e}_z = a\dot{\phi}\mathbf{e}_\phi + b\dot{\phi}\mathbf{e}_z \tag{18}$$

Hence at a given point on the wire the unit vector in the direction of the wire is

$$\mathbf{n} = \frac{a\mathbf{e}_\phi + b\mathbf{e}_z}{(a^2 + b^2)^{1/2}} \tag{19}$$

Setting $\mathbf{R} \cdot \mathbf{n} = 0$ we obtain

$$aR_\phi + bR_z = 0 \tag{20}$$

Equation (20) together with Eqs. (15)–(17) now provide us with four equations in the four unknowns R_ρ, R_ϕ, R_z, and ϕ. We could alternatively use the principle of conservation of energy to generate a fourth equation, since this principle implicitly includes the smoothness of the wire by ignoring the force R, which because of the smoothness does no work when the particle moves. Applying the conservation of energy principle we obtain

$$T + V = \tfrac{1}{2}m(\dot{\rho}^2 + \rho^2\dot{\phi}^2 + \dot{z}^2) + mgz = \text{constant} \tag{21}$$

Setting $\rho = a$ and $z = b\phi$, and making use of the initial conditions to determine the constant, we obtain

$$\dot{\phi} = \left(\frac{2gb}{a^2 + b^2}\right)^{1/2} \phi^{1/2} \tag{22}$$

From Eq. (22) we obtain

$$\ddot{\phi} = \frac{gb}{a^2 + b^2} \tag{23}$$

Substituting Eqs. (22) and (23) in Eqs. (15)–(17) we obtain

$$R_\rho = -\frac{2mgab\phi}{a^2 + b^2} \tag{24}$$

$$R_\phi = \frac{mgab}{a^2 + b^2} \tag{25}$$

$$R_z = -\frac{mga^2}{a^2 + b^2} \tag{26}$$

PROBLEMS

1. A particle of mass m is suspended from a fixed point by an elastic string of modulus of elasticity λ and unstretched length a. Write down the equations of motion for the particle in terms of the spherical coordinates r, θ, and ϕ. Choose the polar axis in the downward vertical direction.

2. A bead of mass m is free to slide along a rough rod that makes an angle θ with a vertical axis passing through the rod. The rod rotates about this axis with an angular velocity αt, where α is a positive constant and t is the time. The coefficients of static and kinetic friction are both equal to the same constant μ. Find the equation of motion of the particle along the rod.

3. Two small rings of masses m and M are moving on a smooth circular wire. The wire is fixed in a vertical plane. The two rings are connected by a straight massless inextensible string. Prove that as long as the string remains tight, its tension is $2mMg \tan\alpha \cos\theta/(m+M)$, where 2α is the angle that the string when tight subtends at the center of the circle, θ is the angle that the string makes with the horizontal, and g is the acceleration due to gravity.

4. A smooth surface of revolution has the equation $z\rho = c^2$ in cylindrical coordinates ρ, ϕ, and z. If the axis of z is vertically downward and a particle is projected horizontally along the inner surface at depth $z = a$, with a speed due to a free fall from rest through a distance na, prove that the particle rises initially if $n > \tfrac{1}{2}$ and that its coordinate ρ is bounded if $n < 1$. If $\tfrac{1}{2} < n < 1$ and the reaction R makes an acute angle α with the z axis when the particle is at the highest point of its path, show that R is then given by $R = mg \cos\alpha [1 + 2(1-n)\tan^2\alpha]$.

5. A particle A of mass m is held in contact with the inner surface of a smooth cone of vertical angle 2α whose axis is vertical and vertex O downward. The particle A is connected to a second particle B of mass $nm \cos\alpha$ by a light inextensible string of length a that passes through a small hole in the vertex to the second particle B, which hangs in equilibrium under gravity below O. If A is projected horizontally from a point whose vertical height above O is b, with speed $(2gb)^{1/2}$, show that B will start to rise or fall according to whether $n < 1$ or $n > 1$. If $n = 5$ prove that B oscillates through a vertical distance $b \sec\alpha/2$ and find the greatest reaction between A and the cone.

SECTION 6

Applications

21

One Dimensional Motion in an Arbitrary Potential

INTRODUCTION

In this chapter we consider the motion of a particle that is constrained to move along a straight line and is acted on by a force, directed along the line of motion, that is a function only of the position of the particle on the line.

THE EQUATION OF MOTION

If we choose our coordinate axes in such a way that the x axis lies along the line of motion, and let $f(x)$ be the force acting on the particle, the equation of motion of the particle is

$$m\ddot{x} = f(x) \tag{1}$$

The work done by the force $f(x)$, when the particle moves from an arbitrary point a to an arbitrary point b on the line, does not depend on the path but only on the initial point a and the final point b; hence the force field $f(x)$ is a conservative force field. We can therefore define a potential energy function with reference to a point a as

$$V(x) = \int_x^a f(x)\,dx \equiv -\int_a^x f(x)\,dx \tag{2}$$

Since the force field $f(x)$ is conservative, the total energy $T + V$ will be a constant of the motion, that is,

$$\tfrac{1}{2}m\dot{x}^2 + V(x) = E \tag{3}$$

where E is an arbitrary constant whose value is determined by the initial conditions. This result could also have been obtained directly from Eq. (1) by multiplying by $\dot{x}$ and integrating with respect to the time. Equation (3) can be rewritten

$$\frac{dx}{dt} = \pm \left[\frac{2}{m}(E - V)\right]^{1/2} \quad (4)$$

or equivalently

$$dt = \frac{\pm dx}{[(2/m)(E - V)]^{1/2}} \quad (5)$$

The plus sign is used when dx is positive and the minus sign when dx is negative. If we integrate Eq. (5) between (x_0, t_0) and (x, t) we obtain

$$t - t_0 = \int_{x_0}^{x} \frac{\pm dx}{[(2/m)(E - V)]^{1/2}} \quad (6)$$

where the plus sign is used when dx is positive and the minus sign when dx is negative. If we know $V(x)$ we can integrate Eq. (6) to obtain $t(x)$, and then invert the function $t(x)$ to obtain $x(t)$. Thus Eq. (6) is essentially the solution to the differential equation Eq. (1).

TURNING POINTS

Many characteristics of the motion of the particle under the action of the force $f(x)$ can be determined from the properties of the potential energy $V(x)$ without actually solving for x as a function of the time.

Suppose for example the potential energy $V(x)$ is the function shown in Fig. 1, and suppose that at $t = 0$ the particle is located at the point $x = x_0$ and is moving with a speed v_0 in the positive x direction.

The value of the total energy at the point x_0 is given by

$$\tfrac{1}{2}mv_0^2 + V(x_0) \equiv E_0 \quad (7)$$

Since the total energy is a constant of the motion, we can determine the speed v of the particle at any other point from the condition

$$\tfrac{1}{2}mv^2 + V(x) = E_0 \quad (8)$$

Since the kinetic energy $mv^2/2$ is always positive, it is impossible for the

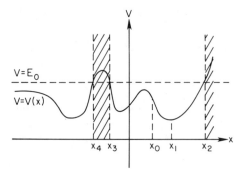

Fig. 1

particle to be in a region where $V > E_0$. If for example E_0 has the value shown by the dashed line in Fig. 1, then V will be greater than zero in the shaded regions, that is, the region between x_4 and x_3 and the region beyond x_2. Since the particle cannot enter these regions, it will be confined, for the given initial conditions, to the region between x_3 and x_2. If the particle is at a point where $V \leq E_0$, its speed will be $[(2/m)(E_0 - V)]^{1/2}$. Hence lower values of the potential energy correspond to higher values of the speed. If the particle is at the point where $V = E_0$ its speed is zero.

The force acting on the particle at an arbitrary point x can be determined from the potential energy by using the relation

$$f(x) = -\frac{dV(x)}{dx} \tag{9}$$

If the slope of the potential is positive the force will be in the negative x direction, and if the slope is negative the force will be in the positive x direction. It follows that the force on the particle in our example will be positive in the region between x_0 and x_1 and negative in the region between x_1 and x_2. If the particle starts at x_0 moving in the positive x direction, it will speed up between x_0 and x_1, slow down between x_1 and x_2, come to rest at x_2, and then proceed to move off in the negative x direction until it eventually comes to the point x_3, where it will again stop, and change its direction of motion. A point at which the particle stops and changes its direction is called a *turning point*. The turning points for a particular value E_0 of the total energy are the points of intersection of the curves $V = V(x)$ and $V = E_0$.

OSCILLATORY MOTION

If a particle is trapped between two turning points $x = a$ and $x = b$, it will oscillate back and forth between the two points. Some information about the oscillations can be obtained without explicitly solving for x as a function of the time.

If for example the particle starts from rest at the time $t = t_a$ at the point $x = a$, the time that elapses between the time $t = t_a$ and the time $t = t_b$ at which it arrives at the point $x = b$ can be found by integrating Eq. (5) between the points a and b. If we do this we obtain

$$t_b - t_a = \int_a^b \frac{dx}{\left[(2/m)(E - V)\right]^{1/2}} \tag{10}$$

where we have assumed that $b > a$, and the particle is moving in the positive x direction. Since the time it takes the particle to return from b to a is the same as the time it takes to get from a to b, it follows that the period τ of the motion is given by twice the integral above, that is,

$$\tau = (2m)^{1/2} \int_a^b \frac{dx}{(E - V)^{1/2}} \tag{11}$$

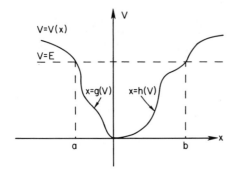

Fig. 2

The period of the motion can thus be obtained without solving explicitly for x as a function of the time.

For different values of the total energy E, the period generally differs. Experimentally it is often easier to determine the period τ of a given oscillatory motion as a function of E than to obtain the force as a function of x. It is therefore of interest to know whether we can determine $f(x)$ or equivalently $V(x)$ from a knowledge of $\tau(E)$.

Let us suppose that the potential energy has only one minimum in the range over which the oscillations occur, and let us for convenience choose our origin of coordinates and the reference point for the potential energy as the point where the potential energy has its minimum. Such a situation is illustrated in Fig. 2. Let $x = a$ and $x = b$ be the turning points for a given value E of the total energy. The period of the oscillation for the given value of E is then

$$\tau(E) = (2m)^{1/2} \int_a^b \frac{1}{(E-V)^{1/2}} dx \qquad (12)$$

If we break the integral down into two parts, one from a to 0, and the other from 0 to b, and if we let $x = g(V)$ represent the inverse of the function $V = V(x)$ in the region a to 0 and $x = h(V)$ the inverse of the function $V = V(x)$ in the region 0 to b, we can rewrite Eq. (12) as follows:

$$\tau(E) = (2m)^{1/2} \int_{g(E)}^{0} \frac{1}{[E-V(g)]^{1/2}} dg + (2m)^{1/2} \int_0^{h(E)} \frac{1}{[E-V(h)]^{1/2}} dh$$

$$= (2m)^{1/2} \int_E^0 \frac{1}{(E-V)^{1/2}} \frac{dg(V)}{dV} dV$$

$$+ (2m)^{1/2} \int_0^E \frac{1}{(E-V)^{1/2}} \frac{dh(V)}{dV} dV$$

$$= (2m)^{1/2} \int_0^E \frac{1}{(E-V)^{1/2}} \left(\frac{dh}{dV} - \frac{dg}{dV} \right) dV \qquad (13)$$

If we now divide Eq. (13) by the quantity $(z-E)^{1/2}$, where z is a dummy

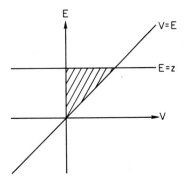

Fig. 3

variable introduced for analytical convenience, and then integrate with respect to E, between 0 and z, we obtain

$$\int_0^z \frac{\tau(E)}{(z-E)^{1/2}} dE = (2m)^{1/2} \int_0^z \int_0^E \frac{1}{(z-E)^{1/2}(E-V)^{1/2}} \left(\frac{dh}{dV} - \frac{dg}{dV} \right) dV\, dE \tag{14}$$

The range of integration for the double integration is the shaded region shown in Fig. 3. Interchanging the order of integration and noting from Fig. 3 that

$$\int_0^z \int_0^E (\ )\, dV\, dE = \int_0^z \int_V^z (\ )\, dE\, dV \tag{15}$$

we obtain

$$\int_0^z \frac{\tau(E)}{(z-E)^{1/2}} dE = (2m)^{1/2} \int_0^z \left(\frac{dh}{dV} - \frac{dg}{dV} \right) \int_V^z \frac{1}{(z-E)^{1/2}(E-V)^{1/2}} dE\, dV$$

$$= (2m)^{1/2} \pi \int_0^z \left(\frac{dh}{dV} - \frac{dg}{dV} \right) dV$$

$$= (2m)^{1/2} \pi [h(z) - g(z)] \tag{16}$$

and therefore

$$h(z) - g(z) = \frac{1}{(2m)^{1/2}\pi} \int_0^z \frac{\tau(E)}{(z-E)^{1/2}} dE \tag{17}$$

Replacing the dummy variable z by V we obtain

$$h(V) - g(V) = \frac{1}{(2m)^{1/2}\pi} \int_0^V \frac{\tau(E)}{(V-E)^{1/2}} dE \tag{18}$$

It follows that the function $\tau(E)$ can be used to determine the difference between the functions $g(V)$ and $h(V)$, but the functions themselves remain indeterminate. This means that there is not one but an infinity of curves $V(x)$ that give the prescribed dependence of period on energy. If however the potential is known to be symmetric about the V axis, then

$$h(V) = -g(V) = \frac{1}{2\pi(2m)^{1/2}} \int_0^V \frac{\tau(E)dE}{(V-E)^{1/2}} \tag{19}$$

It follows that knowledge of $\tau(E)$ is equivalent to knowledge of $V(x)$ for a symmetric potential.

PROBLEMS

1. A particle of mass m is moving in the potential $V(x) = A[\exp(-2\alpha x) - 2\exp(-\alpha x)]$, where A and α are positive constants. Such a potential is called a Morse potential.
   a. Determine $x(t)$.
   b. Determine the period for the case in which the motion is oscillatory.

2. A particle of mass m moves under a conservative force with potential energy $V = cx/(x^2 + a^2)$, where a and c are positive constants. Find the position of stable equilibrium and the period for small oscillations about this point. If the particle starts from this point with velocity v, find the values of v for which it (a) oscillates, (b) escapes to $-\infty$, and (c) escapes to $+\infty$.

3. Describe the motion of a particle in the following potentials, where A and α are positive constants:
   a. $V(x) = -A/\cosh^2(\alpha x)$
   b. $V(x) = A\tan^2(\alpha x)$

4. Solve for the motion of a particle moving in a field of potential $V = -x^{-1} + x^{-2}$. Show that for small total energies the motion is oscillatory, but for larger energies it is nonperiodic and extends to infinity. Find the energy that forms the dividing line between these two cases. Determine the period of the oscillatory motion as a function of the energy.

5. Describe the motion of a particle in the potential $V(x) = -Ax^4$ for the case in which its energy is equal to zero.

6. Describe the motion of a simple pendulum for an arbitrary value of the energy. (*Hint*: The time dependence of the angle that the pendulum makes with the vertical can be expressed in terms of elliptic functions.)

7. Show that the period of a simple pendulum consisting of a mass m suspended by a massless rod of length l is given by $\tau = 4(l/g)^{1/2} \cdot K[\sin(\phi_0/2)]$, where

$$K(k) \equiv \int_0^{\pi/2} (1 - k^2\sin^2 z)^{-1/2} dz$$

is the complete elliptic integral of the first kind and ϕ_0 is the maximum value of the angle between the rod and the vertical. Show that for small oscillations $\tau \approx 2\pi(l/g)^{1/2}[1 + (\phi_0^2/16)]$.

8. A particle of mass m is oscillating in the potential $V = m\omega_0^2(x^2 - bx^4)/2$. Show that the period for oscillation of amplitude a is

$$\tau = \frac{2}{\omega_0} \int_{-a}^{a} (a^2 - x^2)^{-1/2} \left[1 - b(a^2 + x^2)\right]^{-1/2} dx.$$

Using the binomial theorem to expand in powers of b, show that for small amplitude $\tau \approx (2\pi/\omega_0)[1 + (3ba^2/4)]$.

9. A bead is constrained to remain on a smooth vertical hoop of radius a. Describe its motion for the case in which its kinetic energy T at the lowest point of the hoop is equal to $2mga$. Show that if $0 < T - 2mga \ll 2mga$ then the period of rotation is given approximately by $(a/g)^{1/2}\ln[(\pi^2 mga/2)/(T - 2mga)]$.

10. A particle of mass m is oscillating in a symmetrical potential. The period of the oscillation is found to be independent of the energy of the particle. Determine the shape of the potential.

22

The Harmonic Oscillator

INTRODUCTION

Consider a particle that is constrained to move on a straight line and is confined to a small region on the line by a conservative force that attracts the particle toward a point on the line. The point toward which the particle is attracted will be a point of minimum potential energy; hence the particle is confined to a small region in the neighborhood of a potential energy minimum. For convenience we choose our axes in such a way that the line of motion lies along the x axis, and the minimum of the potential energy $V(x)$ occurs at $x = 0$. We furthermore choose the point $x = 0$ to be the reference point for the potential energy, so that $V(0) = 0$. For small displacements about the point $x = 0$ the potential may be approxmated by the first nonvanishing term in a Taylor series expansion. Expanding $V(x)$ in a Taylor series about the point $x = 0$, we obtain

$$V(x) = V(0) + V'(0)x + \frac{V''(0)x^2}{2!} + \frac{V'''(0)x^3}{3!} + \cdots \tag{1}$$

where $V'(x) \equiv dV(x)/dx$, $V''(x) \equiv d^2V(x)/dx^2$, and so forth. Since $x = 0$ is the reference point for the potential energy it follows that $V(0) = 0$, and since $x = 0$ is the point where the potential energy is a minimum, it follows that $V'(0) = 0$, and $V''(0) \geq 0$. If we assume that $V''(0) \neq 0$ then for small displacements from the point $x = 0$ we can write approximately

$$V(x) = \tfrac{1}{2}kx^2 \tag{2}$$

where

$$k \equiv V''(0) > 0 \tag{3}$$

The potential above is one of the most important potentials encountered in dynamics, since there are innumerable cases of a particle being confined to a region in the neighborhood of a potential energy minimum, and in such cases the potential can frequently be approximated by the potential above.

The force corresponding to the potential above is

$$f(x) = -kx \tag{4}$$

That is, it is a restoring force that varies directly as the displacement of the particle from the point $x = 0$.

SIMPLE HARMONIC MOTION

If a particle of mass m that is constrained to move on the x axis is acted on by the force $f(x) = -kx$, the equation of motion of the particle is

$$m\ddot{x} = -kx \tag{5}$$

or equivalently

$$\ddot{x} + \omega_0^2 x = 0 \tag{6}$$

where

$$\omega_0^2 \equiv \frac{k}{m} \tag{7}$$

Any system whose motion is governed by the differential equation (6) is called a *simple harmonic oscillator*, and the resulting motion is called *simple harmonic motion*.

The complete solution to the differential equation above has been obtained in Appendix 9 (Case 1, with $b = 0$) and can be written in any one of the following equivalent forms:

$$x = A_1 \cos\omega_0 t + A_2 \sin\omega_0 t \equiv A\cos(\omega_0 t - \phi) \equiv A\sin(\omega_0 t - \psi) \tag{8}$$

where the constants $A_1, A_2, A, \phi,$ and ψ are related as follows:

$$A = \left[A_1^2 + A_2^2\right]^{1/2} \tag{9}$$

$$\phi = \psi + \frac{\pi}{2} = \tan^{-1}\left(\frac{A_2}{A_1}\right) \tag{10}$$

and their values are determined by the initial conditions.

The motion is sinusoidal with an *amplitude A*, a *frequency*

$$\nu_0 = \frac{\omega_0}{2\pi} \tag{11}$$

and a *period*

$$\tau_0 = \frac{2\pi}{\omega_0} \tag{12}$$

The quantity

$$\omega_0 \equiv 2\pi\nu_0 \equiv \frac{2\pi}{\tau_0} \tag{13}$$

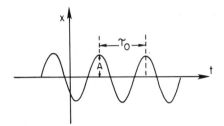

Fig. 1

is called the *angular frequency* or sometimes just the frequency. The motion is illustrated in Fig. 1.

THE DAMPED HARMONIC OSCILLATOR

If a particle is acted on by a retarding or damping force f_d whose value increases with increasing speed, then for small speeds the damping force can be approximated as

$$f_d = -\alpha \dot{x} \tag{14}$$

where α is a positive constant.

The equation of motion of a particle of mass m that is acted on by the linear restoring force $f = -kx$ and is acted on by a damping force $f_d = -\alpha \dot{x}$ is

$$m\ddot{x} = -kx - \alpha \dot{x} \tag{15}$$

or equivalently

$$\ddot{x} + 2\beta \dot{x} + \omega_0^2 x = 0 \tag{16}$$

where

$$\omega_0 \equiv \left(\frac{k}{m}\right)^{1/2} \tag{17}$$

and

$$\beta \equiv \frac{\alpha}{2m} \tag{18}$$

The complete solution to the differential equation above is obtained in Appendix 9. There are three possible cases that may arise.

Case 1. *Underdamped motion.* If $\beta^2 < \omega_0^2$ the solution can be written in any one of the following equivalent forms (see Case 1 in Appendix 9):

$$\begin{aligned} x &= \exp(-\beta t)[A_1 \cos \omega_1 t + A_2 \sin \omega_1 t] \\ &\equiv \exp(-\beta t)[A \cos(\omega_1 t - \phi)] \\ &\equiv \exp(-\beta t)[A \sin(\omega_1 t - \psi)] \end{aligned} \tag{19}$$

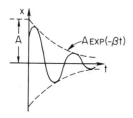

Fig. 2

where
$$\omega_1^2 \equiv \omega_0^2 - \beta^2 \tag{20}$$
and the constants A_1, A_2, A, ϕ, and ψ are related as follows:
$$A \equiv (A_1^2 + A_2^2)^{1/2} \tag{21}$$
$$\phi \equiv \psi + \frac{\pi}{2} \equiv \tan^{-1}\left(\frac{A_2}{A_1}\right) \tag{22}$$

The motion consists of an oscillatory motion in which the maximum displacement decreases with each successive swing as illustrated in Fig. 2. The quantities $A \exp(-\beta t)$, ω_1, and $\nu_1 \equiv \omega_1/2\pi$ are called the *amplitude*, *angular frequency*, and *frequency*, respectively, of the damped oscillator. The quantity $\tau_1 \equiv 2\pi/\omega_1$ is called the *period*, even though the motion is not strictly periodic because the motion does not repeat itself periodically. The period τ_1 is equal to the time between successive passages of the particle in the same direction through the point $x = 0$, or equivalently the time between successive maxima, and is larger than the period τ_0 for undamped motion.

A variety of parameters is used to specify the rate at which the oscillations are damped. The *decay modulus* or *relaxation time* is defined as the time required for the amplitude to decay to $(1/e)$th of its original value, and is equal to $1/\beta$. The smaller the decay modulus, the greater is the damping. The *decrement* is defined as the ratio of the displacement at a time t to the displacement at a time $t + \tau_1$ and is equal to $\exp(\beta\tau_1)$. The *logarithmic decrement*, the logarithm of the decrement, is the difference between the logarithm of the displacement at a time t and the logarithm of the displacement at a time $t + \tau_1$, and is equal to $\beta\tau_1$. The larger the decrement or logarithmic decrement, the greater is the damping.

Case 2. Overdamped motion. If $\beta^2 > \omega_0^2$ then the solution to the differential Eq. (16) can be written (see Case 2 in Appendix 9):
$$x = \exp(-\beta t)\left[A \exp(\omega_2 t) + B \exp(-\omega_2 t)\right] \tag{23}$$
where
$$\omega_2^2 \equiv \beta^2 - \omega_0^2 \tag{24}$$

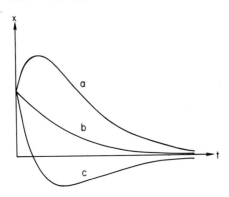

Fig. 3

The motion is nonoscillatory, and, after a preliminary stage that depends on the initial conditions, consists of an asymptotic approach to the zero position. The preliminary stage falls into one of three modes. Let x_0 and v_0 be the initial position and velocity, respectively, of the particle, and let us assume $x_0 > 0$. We then have the following three cases:

   a. If $v_0 > 0$, the particle will ascend to a maximum, then decay monotonically toward zero.
   b. If $-(\beta + \omega_2)x_0 \leq v_0 \leq 0$, the particle will decay monotonically toward zero.
   c. If $v_0 < -(\beta + \omega_2)x_0$, the particle will descend to a minimum value less than zero, then decay monotonically toward zero.

These three situations are illustrated in Fig. 3.

Case 3. Critically damped motion. If $\beta^2 = \omega_0^2$, then the solution to the differential Eq. (16) can be written (see Case 3 in Appendix 9)

$$x = (A + Bt)\exp(-\beta t) \tag{25}$$

The motion in this case is qualitatively similar to the case of overdamped motion. After a preliminary stage depending on the initial conditions the particle asymptotically approaches zero. If $x_0 > 0$, the motion of the particle will follow one of the modes illustrated in Fig. 3, the particular mode depending on whether (a) $v_0 > 0$, (b) $-\beta x_0 \leq v_0 \leq 0$, or (c) $v_0 < -\beta x_0$.

THE DRIVEN HARMONIC OSCILLATOR

If a particle of mass m is acted on by a linear restoring force $-kx$, a damping force $-\alpha \dot{x}$ and a sinusoidal driving force $f_0 \cos \omega t$, the equation of motion of the particle is

$$m\ddot{x} = -kx - \alpha \dot{x} + f_0 \cos \omega t \tag{26}$$

or equivalently

$$\ddot{x} + 2\beta \dot{x} + \omega_0^2 x = \frac{f_0}{m} \cos \omega t \tag{27}$$

where

$$\omega_0 \equiv \left(\frac{k}{m}\right)^{1/2} \tag{28}$$

$$\beta \equiv \frac{\alpha}{2m} \tag{29}$$

The complete solution to Eq. (27) is the sum of the complete solution to the homogeneous equation $\ddot{x} + 2\beta\dot{x} + \omega_0^2 x = 0$, which was obtained in the preceding section, and any particular solution of Eq. (27). The term consisting of the complete solution to the homogeneous equation will be damped out with time as long as $\beta \neq 0$ and is called the *transient term*. The remaining term in the solution persists in time and is called the *steady state solution*. This section considers the steady state solution. The steady state solution to Eq. (27) was obtained in Appendix 9 and is given by

$$x = A \cos(\omega t - \phi) \tag{30}$$

where

$$A \equiv \frac{f_0/m}{\left[(\omega_0^2 - \omega^2)^2 + 4\beta^2\omega^2\right]^{1/2}} \tag{31}$$

$$\tan \phi \equiv \frac{2\omega\beta}{\omega_0^2 - \omega^2} \tag{32}$$

The motion is sinusoidal. The angular frequency ω is the same as the angular frequency of the driving force. The phase $\omega t - \phi$ lags behind the phase ωt of the driving force by ϕ radians. The phase angle ϕ depends on the frequency of the driving force and ranges from zero to π as ω ranges from zero to infinity. The amplitude A, which depends on the magnitude and the frequency of the driving force, is considered in greater detail in the section on resonance.

ENERGY CONSIDERATIONS IN THE DRIVEN OSCILLATOR

In many problems we are interested in the amount of energy that is stored in the oscillator, in the form of kinetic energy or potential energy, and also in the amount of work that must be done by the driving force to maintain a given amount of energy in the oscillator.

The average potential energy and the average kinetic energy stored in the oscillator can be found by averaging the quantities $\frac{1}{2}kx^2$ and $\frac{1}{2}m\dot{x}^2$ over a single period $\tau \equiv 2\pi/\omega$. Doing this we obtain

$$\left\langle \frac{1}{2}kx^2 \right\rangle = \frac{1}{\tau} \int_t^{t+\tau} \frac{1}{2}k[A\cos(\omega t - \phi)]^2 \, dt = \frac{kA^2}{4} \equiv \frac{m\omega_0^2 A^2}{4} \tag{33}$$

$$\left\langle \frac{1}{2}m\dot{x}^2 \right\rangle = \frac{1}{\tau} \int_t^{t+\tau} \frac{1}{2}m[-A\omega\sin(\omega t - \phi)]^2 \, dt = \frac{m\omega^2 A^2}{4} \tag{34}$$

where A is given by Eq. (31).

In the steady state, the kinetic energy has the same value at the beginning and the end of a period, and the potential energy has the same value at the beginning and the end of a period. Hence the only work that must be done by the driving force over one period is the work that is necessary to supply the energy dissipated by the damping force. The energy dissipated by the damping force in one period, or equivalently the negative of the work done by the damping force in one period, is

$$-\Delta W = \int_{x(t)}^{x(t+\tau)} \alpha \dot{x} \, dx = \int_{t}^{t+\tau} (\alpha \dot{x}) \dot{x} \, dt = \beta m \omega^2 A^2 \tau \qquad (35)$$

A measure of the efficiency of a given oscillator for energy storage is given by the *Q-factor*, which is defined as follows:

$$Q = 2\pi \frac{\text{average stored energy}}{\text{energy dissipated per cycle}} \qquad (36)$$

If we use Eqs. (33)–(35) in Eq. (36) we obtain

$$Q = 2\pi \frac{m\omega_0^2 A^2/4 + m\omega^2 A^2/4}{\beta m \omega^2 A^2 \tau}$$

$$= \frac{\omega_0^2 + \omega^2}{4\beta \omega} \qquad (37)$$

RESONANCE

For a driven harmonic oscillator the displacement function $x(t)$ and the velocity function $\dot{x}(t)$ are sinusoidal functions with amplitudes A and A', respectively, where

$$A \equiv \frac{f_0/m}{\left[(\omega_0^2 - \omega^2)^2 + 4\beta^2 \omega^2\right]^{1/2}} \qquad (38)$$

$$A' \equiv \frac{(f_0/m)\omega}{\left[(\omega_0^2 - \omega^2)^2 + 4\beta^2 \omega^2\right]^{1/2}} \qquad (39)$$

If the magnitude f_0 of the driving force is held fixed and the frequency ω is varied, and if $\beta < \omega_0$, the quantities A and A' will each have a maximum value for a certain frequency. If $\beta \ll \omega_0$ the quantities A and A' will be sharply peaked around their maximum values as illustrated in Figs. 4a and 4b. The rapid change in the value of A or A' in the neighborhood of a certain frequency is called a *resonance*, and the frequency at which the resonance occurs is called the *resonant frequency*.

The frequency ω_R at which the amplitude A takes on its maximum value is called the *displacement resonance frequency* and is found by maximizing Eq. (38) with respect to the variable ω, or more simply by minimizing the square of the denominator of Eq. (38) with respect to the variable ω^2. If we do this we

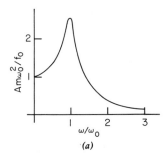 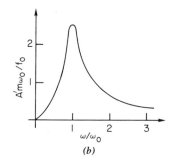

Fig. 4

obtain

$$\omega_R = \left(\omega_0^2 - 2\beta^2\right)^{1/2} \tag{40}$$

If $\beta \ll \omega_0$ then

$$\omega_R \approx \omega_0 \tag{41}$$

The frequency, ω_R', at which the amplitude A' takes on its maximum value, is called the *velocity resonance frequency* and is found by maximizing Eq. (39) with respect to the variable ω^2. If we do this we obtain

$$\omega_R' = \omega_0 \tag{42}$$

When $\beta \ll \omega_0$, the resonance will occur in both cases at the frequency ω_0. When the term "resonance frequency" is used without modification it generally means that the value is ω_0.

One is sometimes interested in the resonance curves for the average potential and kinetic energies. The average potential energy is given by $m\omega_0^2 A^2/4$ and the average kinetic energy is given by $m\omega^2 A^2/4 \equiv m(A')^2/4$. It follows that the resonance curve for the average potential energy is just $m\omega_0^2/4$ times the square of the resonance curve for the displacement, and the resonance curve for the kinetic energy is $m/4$ times the square of the resonance curve for the velocity. It follows that the average potential energy will have its maximum value at $\omega = \omega_R$ and the average kinetic energy will have its maximum at $\omega = \omega_R'$.

The sharpness of a resonance peak can be described by giving its full width at the point where the amplitude has dropped to half its maximum value. This quantity is called the *full width at half-maximum* and is designated as $\Delta\omega$. The value of $\Delta\omega$ for the velocity resonance curve, and also, when $\beta \ll \omega_0$, for the displacement resonance curve, is given by

$$\Delta\omega = 2\sqrt{3}\,\beta \tag{43}$$

The *half-width at half-maximum* is frequently used instead of the full width at half-maximum, and can be obtained by taking half the value above. The value of $\Delta\omega$ for the kinetic energy resonance curve and also, when $\beta \ll \omega_0$, for the potential energy resonance curve, is 2β. The important thing to notice about these results is that the width of the curve is a measure of the strength of the damping. The greater the damping, the wider the resonance curve.

When $\omega = \omega_0$ the Q-factor of the oscillator is given by

$$Q_0 = \frac{\omega_0}{2\beta} \qquad (44)$$

It follows that Q_0 is inversely proportional to the width of the resonance curve. The sharper the resonance, the larger the value of Q_0.

In the section on the damped harmonic oscillator we showed that the relaxation time of a damped harmonic oscillator that had been initially excited and then allowed to decay was equal to $1/\beta$. It follows that the relaxation time is inversely proportional to the width. The narrower the resonance, the longer the relaxation time.

From the analysis above we see that a great deal of information about a harmonic oscillator can be obtained from one of its resonance curves. The location of the resonance supplies us with the natural frequency ω_0 of the oscillator, and the width supplies us with the damping factor β. These two quantities can then be related to the Q-factor, the relaxation time, the restoring force, the damping force, and so forth. The fact that a resonance curve supplies us with this information is very important, since it frequently happens that we cannot directly measure these quantities, but we can obtain a resonance curve.

THE EFFECT OF A CONSTANT FORCE ON A HARMONIC OSCILLATOR

If a particle of mass m is acted on by a linear restoring force $-kx$, a damping force $-\alpha \dot{x}$, a driving force $f(t)$, and a constant force f_0, the equation of motion of the particle is

$$m\ddot{x} + \alpha\dot{x} + kx = f(t) + f_0 \qquad (45)$$

If we introduce a new variable

$$y = x - \frac{f_0}{k} \qquad (46)$$

the equation of motion can be written

$$m\ddot{y} + \alpha\dot{y} + ky = f(t) \qquad (47)$$

This is identical with the equation that would result if we had $f_0 = 0$, except that the variable x is replaced by $x - (f_0/k)$. It follows that the only effect of the constant force f_0 is to translate the point about which the oscillations take place from the point $x = 0$ to the point $x = f_0/k$, which is just the equilibrium position of the particle when it is at rest.

It follows, for example, that if we are analyzing the motion of a particle that is suspended from a linear spring in a gravitational field, we can neglect the effect of the gravitational field, provided we assume that the spring exerts a restoring force that is proportional to the displacement of the particle from its equilibrium position, rather than from the position corresponding to the unstretched length of the spring.

THE DRIVEN OSCILLATOR WITH ARBITRARY DRIVING FORCE

In the preceding sections we have shown how one can solve for the motion of a driven harmonic oscillator when the driving force is periodic. Now we solve for the motion of a driven harmonic oscillator when the driving force is either periodic or non periodic. We restrict our consideration to the case of the driven undamped oscillator. The result however can be generalized without too much trouble to the case of the driven damped oscillator. The method is summarized in the following theorem.

Theorem 1. If a time dependent force $F(t)$ is applied to an undamped simple harmonic oscillator whose natural frequency is ω_0, the general solution for the motion is

$$x(t) = a\sin(\omega_0 t + \phi) + \int_{-\infty}^{t} \frac{\sin[\omega_0(t - t_1)]}{m\omega_0} F(t_1)\, dt_1 \qquad (48)$$

where a and ϕ are constants whose value can be determined from the boundary conditions.

PROOF: Consider an oscillator whose instantaneous position and velocity for times $t < t_0$, are given by

$$x(t) = a\sin(\omega_0 t + \phi) \qquad (49)$$

$$\dot{x}(t) = a\omega_0 \cos(\omega_0 t + \phi) \qquad (50)$$

If at time $t = t_0$ the oscillator receives an impulse $\hat{f}$ then for $t > t_0$ Eqs. (49) and (50) are not valid. To determine the proper equations for $t > t_0$, let us first note that for $t > t_0$ we can write the new equations in the form

$$x(t) = a\sin(\omega_0 t + \phi) + A\sin(\omega_0 t + \Phi) \qquad (51)$$

$$\dot{x}(t) = a\omega_0 \cos(\omega_0 t + \phi) + A\omega_0 \cos(\omega_0 t + \Phi) \qquad (52)$$

where A and Φ are unknown constants. When the impulse is applied, the position of the oscillator remains unchanged and the velocity changes by an amount $\hat{f}/m$. It follows that

$$a\sin(\omega_0 t_0 + \phi) + A\sin(\omega_0 t_0 + \Phi) = a\sin(\omega_0 t_0 + \phi) \qquad (53)$$

$$a\omega_0 \cos(\omega_0 t_0 + \phi) + A\omega_0 \cos(\omega_0 t_0 + \Phi) = a\omega_0 \cos(\omega_0 t_0 + \phi) + \frac{\hat{f}}{m} \qquad (54)$$

Solving for A and Φ we obtain

$$\Phi = -\omega_0 t_0 \qquad (55)$$

$$A = \frac{\hat{f}}{m\omega_0} \qquad (56)$$

168 The Harmonic Oscillator

It follows that the motion of the oscillator is given by

$$x(t) = a\sin(\omega_0 t + \phi) \qquad t \leq t_0 \qquad (57a)$$

$$x(t) = a\sin(\omega_0 t + \phi) + \frac{\hat{f}}{m}\sin[\omega_0(t - t_0)] \qquad t \geq t_0 \qquad (57b)$$

Introducing the step function $\eta(t)$, defined by the conditions

$$\eta(t) = 0 \qquad t < 0 \qquad (58a)$$

$$\eta(t) = 1 \qquad t \geq 0 \qquad (58b)$$

we can express the solution in the form

$$x(t) = a\sin(\omega_0 t + \phi) + \frac{\hat{f}}{m\omega_0}\sin[\omega_0(t - t_0)]\eta(t - t_0) \qquad (59)$$

The effect of an arbitrary time dependent force $F(t)$ on the motion of an oscillator can be found by first determining the effect of a series of impulses applied at the times $\ldots, -3\Delta t, -2\Delta t, -\Delta t, 0, \Delta t, 2\Delta t, 3\Delta t, \ldots$, and of respective magnitudes $\ldots, F(-3\Delta t)\Delta t, F(-2\Delta t)\Delta t, F(-\Delta t)\Delta t, F(0)\Delta t, F(\Delta t)\Delta t, F(2\Delta t)\Delta t, F(3\Delta t)\Delta t, \ldots$, and then taking the limit as Δt approaches zero. Proceeding in this fashion we find that if the motion in the absence of the force $F(t)$ is given by $x(t) = a\sin(\omega_0 t + \phi)$, the motion in the presence of the force is given by

$$x(t) = a\sin(\omega_0 t + \phi) + \lim_{\Delta t \to 0} \sum_{n=-\infty}^{\infty} \frac{F(n\Delta t)\Delta t}{m\omega_0} \sin[\omega_0(t - n\Delta t)]\eta(t - n\Delta t)$$

$$\equiv a\sin(\omega_0 t + \phi) + \int_{-\infty}^{\infty} \frac{\sin[\omega_0(t - t_1)]}{m\omega_0} F(t_1)\eta(t - t_1)\,dt_1$$

$$\equiv a\sin(\omega_0 t + \phi) + \int_{-\infty}^{t} \frac{\sin[\omega_0(t - t_1)]}{m\omega_0} F(t_1)\,dt_1 \qquad (60)$$

This completes the proof of the theorem.

EXAMPLES

Example 1. An undamped simple harmonic oscillator of natural frequency ω_0 that is at rest at $x = 0$ is subjected, starting at time $t = 0$, to the force $F(t) = F_0[1 - \exp(-t/\tau)]$. Determine the resulting motion. Let $E(\tau)$ be the energy eventually received by the oscillator for a given value of τ over and above the energy required to change the point about which the oscillations occur. Show that $E(\tau)/E(0) = 1/(1 + \omega_0^2\tau^2)$. Note that $E(\tau)$ decreases as the force is imposed more slowly and becomes zero when $\tau = \infty$.

Solution. Making use of Theorem 1, and noting that when $t = 0$ then $x = 0$, and $\dot{x} = 0$, we obtain

$$x(t) = \frac{F_0}{m\omega_0}\int_0^t \sin[\omega_0(t - t_1)]\left[1 - \exp\left(\frac{-t_1}{\tau}\right)\right]dt_1 \qquad (61)$$

Introducing the variable $z = \omega_0(t - t_1)$ we can rewrite Eq. (61) as

$$x(t) = \frac{F_0}{m\omega_0^2} \int_0^{\omega_0 t} \sin z \left[1 - \exp\left(\frac{z}{\omega_0 \tau} - \frac{t}{\tau} \right) \right] dz$$

$$= A + B \sin \omega_0 t + C \cos \omega_0 t + D \exp\left(\frac{-t}{\tau} \right) \qquad (62)$$

where

$$A \equiv \frac{F_0}{m\omega_0^2} \qquad (63)$$

$$B \equiv \frac{F_0 \omega_0 \tau}{m\omega_0^2(\omega_0^2\tau^2 + 1)} \qquad (64)$$

$$C \equiv \frac{-F_0}{m\omega_0^2(\omega_0^2\tau^2 + 1)} \qquad (65)$$

$$D \equiv \frac{-F_0 \tau^2}{m(\omega_0^2\tau^2 + 1)} \qquad (66)$$

In the limit of large t the last term vanishes and the system oscillates about the point $x = A$ with frequency ω_0 and amplitude $(B^2 + C^2)^{1/2}$. The energy $E(\tau)$ is equal to the maximum value of the kinetic energy, and thus has the value

$$E(\tau) = \tfrac{1}{2} m\omega_0^2 [B^2 + C^2] \qquad (67)$$

It follows that

$$\frac{E(\tau)}{E(0)} = \frac{1}{\omega_0^2 \tau^2 + 1} \qquad (68)$$

PROBLEMS

1. The natural length of a spring is 10 cm. When a mass is suspended from the end of the spring, the equilibrium length is increased to 12 cm. The mass is given a blow that starts it moving with speed 4 cm/s in the downward direction.
   a. Find the period and amplitude of the resulting oscillations.
   b. Write an expression giving the distance of the mass below the spring support as a function of the time. Assume that the blow was delivered at time $t = 0$.

2. Find the frequency for small oscillations for particles moving in each of the following potentials:
   a. $V(x) = A \cos(\alpha x) - Bx$
   b. $V(x) = A[(\alpha x)^2 - \sin^2(\alpha x)]$

3. An endless light elastic string of unstretched length $2a$ passes around two small smooth pegs A and B, which are a distance a apart on the same

horizontal line. A particle P of mass m is attached to the string. In equilibrium the string forms an equilateral triangle. Show that if P is displaced vertically through a small distance and then released, it will oscillate with period $2\pi[2(3)^{1/2}a/7g]^{1/2}$.

4. A pendulum whose period in a vacuum is 2 s is placed in a resistive medium. Its amplitude on each swing is observed to be half that of the previous swing. What is its new period? If the pendulum is subjected in turn to sinusoidal forces of equal amplitude with periods of 1, 2, and 2.5 s, respectively, what is the ratio of the amplitudes of the resulting forced oscillations?

5. A plank of mass M rests on four equal massless springs, one at each corner. A mass m is placed symmetrically on the plank, and the whole rests in static equilibrium under gravity. The plank is raised a distance b above the equilibrium position and released. What is the force constant of each spring such that m will just not leave the plank during the subsequent motion?

6. A body of mass m is suspended from a fixed point by a light elastic string of natural length b and modulus of elasticity mg. The body is set in motion and makes vertical oscillations of amplitude a. The body is passing through the equilibrium position and moving in the upward direction when it picks up another body also of mass m. Show that the amplitude of the subsequent oscillations is $[b^2 + (a^2/2)]^{1/2}$.

7. A mass M is suspended in a medium whose resistance is $2kM \times$ (velocity) by a light spring of modulus λMl and natural length l from a point A, which moves up and down with a simple harmonic motion so that its height y above a fixed point O is $y = a(1 - \cos pt)$, M being at rest and in equilibrium at time $t = 0$. If $\lambda < k^2$, find the complete expression for the downward displacement of M at time t, and show that when t is large, the average rate of dissipation of energy is approximately $kMa^2\lambda^2p^2/[(\lambda - p^2)^2 + 4k^2p^2]$.

8. A straight light elastic string AB of natural length a and modulus of elasticity mg lies unstretched on a rough horizontal table. A particle of mass m is attached to the string at A, the coefficient of friction between the particle and the table being μ. The end B of the string is moved with uniform velocity u in the direction AB. Show that at time t after the end A begins to move, the length of the string is $(u/\omega)\sin \omega t + (\mu + 1)a$, where $g = a\omega^2$, and find the greatest tension in the string during the motion, assuming that $u < \mu a\omega$.

9. Two particles whose masses are m and $3m$, respectively, are attached to the ends of an inextensible string that hangs over a small smooth pulley. To the first of these another particle of mass $2m$ is attached by means of a light elastic string of natural length a and spring constant $k = 4mg/a$. The system is let go from rest, the springs being vertical where not in

contact with the pulley, in a position in which the elastic string is of length a. Prove that the particle of mass $3m$ performs simple harmonic oscillations of period $2\pi(a/3g)^{1/2}$ about a point a distance $a/6$ below its initial position.

10. Two particles A and B of masses km and m, respectively, are connected by a light elastic string of natural length a and modulus of elasticity λ. Initially they lie at rest with the string just taut and perpendicular to the edge of a smooth horizontal table and B close to the edge. If B is then gently pushed over the edge and released, show that A will still be on the table when the string becomes slack, provided λ is greater than $2\pi^2 kmg/(k+1)^2$.

11. A particle P of mass m is tied to one end of a light elastic string, and the other end is fastened to the center C of a smooth board of mass M. The modulus of elasticity of the string is Mg and its unstretched length is a. The board is placed on a horizontal plane, and the particle is placed on the board. The system is initially held with P at a distance $2a$ from C and is then released. Prove that the particle will return to its original position relative to the board after a time $2(\pi + 2)/n$, where $n^2 = g(M+m)/am$. Show that the distance the board has described relative to the plane when the particle first passes over C is $2ma/(M+m)$.

12. An undamped linear oscillator, originally at rest in its equilibrium position, is subjected to a forcing function $F(t)$, where $F(t) = 0$ for $t < 0$, $F(t) = mat/\tau$ for $0 < t < \tau$, and $F(t) = ma$ for $t > \tau$. Determine the subsequent motion.

13. Obtain the response of an undamped linear oscillator to the forcing function $F(t)$, where $F(t) = 0$ for $t < 0$, $F(t) = ma \sin \omega t$ for $0 < t < \pi/\omega$, and $F(t) = 0$ for $t > \pi/\omega$.

14. Prove that if a time dependent force $F(t)$ is applied to an underdamped simple harmonic oscillator, the general solution for the motion is

$$x = a\exp(-\beta t)\sin(\omega_1 t + \phi)$$
$$+ \frac{1}{m\omega_1} \int_{-\infty}^{t} \exp[-\beta(t-t_1)]\sin[\omega_1(t-t_1)]F(t_1)\,dt_1$$

where the quantities ω_1 and β are defined in the section on the damped harmonic oscillator, and a and ϕ are arbitrary constants whose values depend on the boundary conditions.

23

Central Force Motion

INTRODUCTION

If the force acting on a particle is always directed toward or away from some fixed point, the force is said to be a *central force*.

In this chapter we investigate the behavior of a particle under the action of a *conservative* central force. The following theorem provides us with some immediate properties of such forces.

Theorem. A force field is a *conservative central force* field if and only if either of the following equivalent conditions is satisfied:

a. The direction of the force is along the line joining the particle and a fixed point, and the magnitude of the force depends only on the distance of the particle from the point.
b. The force is derivable from a potential energy function that depends only on the distance of the particle from a fixed point.

PROOF: We first prove that conditions a and b are equivalent. We then prove that condition b is a necessary and sufficient condition for the force to be a conservative central force.

Let $\mathbf{f}$ be the force and $\mathbf{r}$ the position of the particle with respect to the fixed point; and let f and r be the respective magnitudes of $\mathbf{f}$ and $\mathbf{r}$. If the force is directed along the line joining the particle and the point, and its magnitude depends only on the distance of the particle from the point, we can express the force as follows:

$$\mathbf{f} = \frac{f(r)\mathbf{r}}{r} \qquad (1)$$

Hence the work done by the force in an infinitesimal displacement $d\mathbf{r}$ is given by

$$dW = \mathbf{f} \cdot d\mathbf{r} = \frac{f(r)\mathbf{r} \cdot d\mathbf{r}}{r} = \frac{f(r)d(\mathbf{r} \cdot \mathbf{r})}{2r}$$

$$= \frac{f(r)d(r^2)}{2r} = f(r)\,dr \tag{2}$$

If Eq. (2) is used to calculate the work done by the force when the particle moves from one point to another, it is apparent that the work done will depend only on the initial and final values of r, not on the path. The force is therefore a conservative force, and the potential energy at a given point—that is, the work done by the force when the particle moves from the given point to a chosen reference point—is a function of r only. Conversely if there exists a potential energy function that depends only on r, the force is given by

$$\mathbf{f} = -\nabla V = -\sum_i \mathbf{e}_i \frac{\partial V}{\partial x_i} = -\sum_i \mathbf{e}_i \frac{dV}{dr}\frac{\partial r}{\partial x_i}$$

$$= -\frac{dV}{dr}\sum_i \mathbf{e}_i \frac{\partial r}{\partial x_i} = -\frac{dV}{dr}\nabla r$$

$$= -\frac{dV}{dr}\frac{\mathbf{r}}{r} \tag{3}$$

Hence the force is directed along the line joining the particle and the fixed point, and since $dV(r)/dr$ is a function of r only, the magnitude of the force depends only on r. This completes the proof of the equivalence of conditions a and b.

Now suppose there exists a potential energy function $V(r)$. Since the potential energy function exists, the force is conservative; and since the potential energy function depends only on r, the force is central, as we have proved above. Conversely if the force is central and conservative, then

$$\mathbf{f} = f\frac{\mathbf{r}}{r} \tag{4}$$

and there exists a potential energy function V such that

$$\mathbf{f} = -\nabla V \tag{5}$$

At this stage nothing is known about whether f and V depend only on r. Equating the right hand sides of Eqs. (4) and (5) and taking the dot product of the resulting equation with $d\mathbf{r}$ we obtain

$$\nabla V \cdot d\mathbf{r} = -\frac{f\mathbf{r} \cdot d\mathbf{r}}{r} \tag{6}$$

The left hand side of Eq. (6) can be rewritten

$$\nabla V \cdot d\mathbf{r} = \sum_i \frac{\partial V}{\partial x_i} dx_i = dV \tag{7}$$

and by the same argument used in obtaining Eq. (2) the right hand side of Eq.

(6) can be rewritten

$$\frac{-f\mathbf{r}\cdot d\mathbf{r}}{r} = -f\,dr \tag{8}$$

Combining Eqs. (6)–(8) we obtain

$$dV = -f\,dr \tag{9}$$

It follows that V changes only when r changes, hence V is a function of r only. Thus if the force is conservative and central, there exists a potential energy function and the potential energy function depends only on r. We have thus shown that condition b is a necessary and sufficient condition for the force to be central and conservative. This completes the second half of the proof of the theorem.

CONSTANTS OF THE MOTION

Let us consider a particle that is moving under the action of a conservative central force, and let us for simplicity choose our coordinate axes such that the fixed point toward which or away from which the force is directed lies at the origin.

Since the line of action of the force passes through the origin, the force exerts no torque with respect to the origin; hence the angular momentum $\mathbf{h}$ of the particle with respect to the origin is a constant of the motion, that is,

$$\mathbf{r}\times\mathbf{p} \equiv \mathbf{h} = \text{constant} \tag{10}$$

Since the force is a conservative force, the total energy E is also a constant of the motion, that is,

$$\tfrac{1}{2}m\dot{\mathbf{r}}\cdot\dot{\mathbf{r}} + V = E = \text{constant} \tag{11}$$

Recognition of the two constants of the motion above enables us to simplify considerably the analysis of the motion of a particle in a conservative central force field. We note first from Eq. (10) that the position vector $\mathbf{r}$ of the particle is always perpendicular to the constant vector $\mathbf{h}$. Hence the orbit of the particle must lie in a plane that is perpendicular to $\mathbf{h}$ and passes through the origin. For convenience we choose our z axis in the direction of $\mathbf{h}$. The motion of the particle will then lie in the x-y plane. If furthermore we specify the position of the particle in the x-y plane by polar coordinates r and ϕ we can write

$$\mathbf{r} = \mathbf{i}r\cos\phi + \mathbf{j}r\sin\phi \tag{12}$$

$$\dot{\mathbf{r}} = \mathbf{i}[\dot{r}\cos\phi - r\dot{\phi}\sin\phi] + \mathbf{j}[\dot{r}\sin\phi + r\dot{\phi}\cos\phi] \tag{13}$$

Substituting Eqs. (12) and (13) into Eqs. (10) and (11) we obtain

$$mr^2\dot{\phi} = h \tag{14}$$

$$\tfrac{1}{2}m\dot{r}^2 + \tfrac{1}{2}mr^2\dot{\phi}^2 + V = E \tag{15}$$

Our problem is thus reduced to the solution of two simultaneous first order differential equations for the two unknowns r and ϕ. We will find it more convenient in what follows to express the result above in a slightly different form. Solving Eqs. (14) and (15) for $\dot{\phi}$ and $\dot{r}$ we obtain

$$\dot{\phi} = \frac{h}{mr^2} \tag{16}$$

$$\dot{r} = \pm \left[\frac{2}{m}\left(E - V - \frac{h^2}{2mr^2} \right) \right]^{1/2} \tag{17}$$

The plus sign is used when r is increasing and the minus sign when r is decreasing. Noting that $\dot{\phi} \equiv d\phi/dt$ and $\dot{r} \equiv dr/dt$, we can rewrite Eqs. (16) and (17) as follows:

$$dt = \frac{mr^2 \, d\phi}{h} \tag{18}$$

$$dt = \frac{\pm dr}{\{(2/m)[E - V - (h^2/2mr^2)]\}^{1/2}} \tag{19}$$

The plus sign is used when dr is positive and the minus sign when dr is negative. Equations (18) and (19) are the fundamental equations that are used to analyze the motion of a particle in a conservative central force field.

THE ORBIT EQUATION

A complete solution to Eqs. (18) and (19) would involve the determination of $r(t)$ and $\phi(t)$. Frequently one is interested simply in the orbit the particle describes; that is, $r(\phi)$. This section shows how one can obtain the orbit equation without first solving for $r(t)$ and $\phi(t)$.

Equating the right hand sides of Eqs. (18) and (19) and then multiplying by h/mr^2 we obtain

$$d\phi = \frac{\pm h \, dr}{r^2\{2m[E - V - (h^2/2mr^2)]\}^{1/2}} \tag{20}$$

The plus sign is used when dr is positive and the minus sign when dr is negative. Integrating between (r_0, ϕ_0) and (r, ϕ) we obtain

$$\phi - \phi_0 = \int_{r_0}^{r} \frac{\pm h \, dr}{r^2\{2m[E - V - (h^2/2mr^2)]\}^{1/2}} \tag{21}$$

where the plus sign is used when dr is positive and the minus sign when dr is negative. Equation (21) can be solved for $\phi(r)$, which can be inverted to obtain $r(\phi)$.

THE RADIAL MOTION

The radial motion of the particle, that is, $r(t)$, can be obtained from Eq. (19). Integrating Eq. (19) between (r_0, t_0) and (r, t) we obtain

$$t - t_0 = \int_{r_0}^{r} \frac{\pm dr}{\{(2/m)[E - V - (h^2/2mr^2)]\}^{1/2}} \tag{22}$$

The plus sign is used when dr is positive and the minus sign when dr is negative. Since V is a function of r only, Eq. (22) can in principle be solved to obtain $r(t)$.

If we define an effective potential

$$U \equiv V(r) + \frac{h^2}{2mr^2} \tag{23}$$

then Eq. (22) can be rewritten

$$t - t_0 = \int_{r_0}^{r} \frac{\pm dr}{[(2/m)(E - U)]^{1/2}} \tag{24}$$

where the plus sign is used when dr is positive and the minus sign when dr is negative. Equation (24) is formally identical to the equation we would obtain for the one dimensional motion of a particle in the potential U (see Chapter 21). Hence the radial motion of a particle whose angular momentum is h is equivalent to the one dimensional motion of a particle in the effective potential U. This is a very useful result because it enables us to apply the general analysis developed in Chapter 21 for one dimensional motion in a conservative field to the radial motion of a particle in a conservative central force field. Note carefully however that the effective potential U depends on the value of the angular momentum h, hence there is a different potential for each value of h.

THE ANGULAR MOTION

The angular motion of the particle, that is, $\phi(t)$, can be obtained in several ways.

If we know $r(\phi)$ then we can substitute $r(\phi)$ in Eq. (18) and integrate to obtain

$$t = \frac{m}{h} \int [r(\phi)]^2 d\phi + A \tag{25}$$

where A is an integration constant. Equation (25) gives us $t(\phi)$, which can be inverted to give $\phi(t)$.

If we know $r(t)$ then we can multiply Eq. (18) by h/mr^2, substitute $r(t)$ in the resulting equation, and integrate to obtain

$$\phi = \frac{h}{m} \int \frac{dt}{[r(t)]^2} + B \qquad (26)$$

where B is an integration constant. Equation (26) gives us $\phi(t)$.

As a particle moves under the action of a central force from a point **r** to a point **r** + d**r**, the radius vector, or equivalently the line segment that joins the force center and the particle, sweeps out an area

$$dA = \tfrac{1}{2} r(r\, d\phi) \qquad (27)$$

If we combine Eqs. (18) and (27) we obtain

$$\frac{dA}{dt} = \frac{h}{2m} = \text{constant} \qquad (28)$$

Thus the area swept out per unit time by the radius vector **r** is a constant. This fact is very useful in helping one to develop an intuitive grasp of the angular motion of a particle in a central force field.

APSIDES

An *apsis* (pl. apsides) or an *apse* (pl. apses) is a point on an orbit for which r has a maximum or a minimum value. At an apsis $dr/d\theta = 0$ and $\dot{r} = 0$.

The value of r at an apsis is called an *apsidal distance*. Since r has a minimum or maximum value at an apsis, an apsidal distance is a turning point for the radial motion. If a particle is moving in a conservative central force field, there may be many apsides, but there will be at most two apsidal distances, since there can be no more than two turning points for the radial motion in a given orbit. If there are two apsidal distances then the motion will be bounded; the orbit will lie between two concentric circles, the radii of which are the two apsidal distances, and will be tangent alternately to the one circle and then to the other at successive apsides.

In a conservative central force field, the radial velocity of a particle that is approaching an apsis is a function of r; and the radial velocity of the particle as it is receding from the apsis is the negative of the same function. The angular velocity is also a function of r, but in this case the same function applies to both stages of the motion. It follows that the orbit of the particle is symmetric with respect to a line drawn from the force center to the apsis. If we know the orbit between two consecutive apsides, we can use the symmetry property above to generate the entire orbit. If we do this in the case of a motion that is bounded, we will find that the angle $\Delta\phi$ subtended between consecutive apsidal radii is a constant, which we shall call the *apsidal angle*; and the radial position r of the particle is a periodic function of its angular position ϕ, and the period is $2\Delta\phi$. If there exists an integer n such that

Central Force Motion

$2n \, \Delta\phi = 2\pi$, then the orbit will close on itself at the end of one complete revolution; otherwise it will not.

DETERMINATION OF THE POTENTIAL

In the preceding sections we have assumed that the potential energy $V(r)$ is known and that we are interested in determining the motion of the particle. It may happen however that the motion of the particle is known and we wish to determine the force $f(r)$, or equivalently the potential $V(r)$, which is responsible for the motion.

If $r(\phi)$ is known, we can determine $V(r)$ from the relation

$$V = E - \frac{h^2}{2mr^2} - \frac{h^2}{2mr^4}\left(\frac{dr}{d\phi}\right)^2 \tag{29}$$

which is obtained by dividing Eq. (18) by Eq. (19) and solving for V. Note that to obtain $V(r)$, $dr/d\phi$ must be expressed as a function of r.

If $r(t)$ is known then we can determine $V(r)$ from the relation

$$V = E - \frac{h^2}{2mr^2} - \frac{m}{2}\left(\frac{dr}{dt}\right)^2 \tag{30}$$

which is obtained by solving Eq. (19) for V. Note that to obtain $V(r)$, dr/dt must be expressed as a function of r.

It should be noted that Eqs. (29) and (30) provide us with the dependence of V on r only over the range of values of r that is covered by the particle in its motion. Thus if the particle is moving in a circular orbit with its center at the origin, we can say nothing about the dependence of V on r.

THE INVERSE SQUARE FORCE FIELD

One of the most important central force fields is the inverse square force field, that is, the force field for which

$$f(r) = \pm \frac{k}{r^2} \tag{31}$$

where k is a positive integer, and the plus sign is used when the force is repulsive and the minus sign when the force is attractive. The potential energy in an inverse square field is ordinarily measured with respect to the reference point $r = \infty$. With this choice for the reference point the potential energy is given by

$$V(r) = \pm \frac{k}{r} \tag{32}$$

where the plus sign is used when the force is repulsive and the minus sign when the force is attractive.

In applying the techniques introduced in the preceding sections to the motion of a particle in an inverse square force field, the calculations and results can be appreciably simplified if position variables are expressed in units of h^2/mk, energy variables in units of mk^2/h^2, and time variables in units of h^3/mk^2. We indicate the value of a quantity in terms of these units by a bar over the quantity. Thus

$$r \equiv (h^2/mk)\bar{r} \tag{33}$$

$$E \equiv (mk^2/h^2)\bar{E} \tag{34}$$

$$V \equiv (mk^2/h^2)\bar{V} \tag{35}$$

$$U \equiv (mk^2/h^2)\bar{U} \tag{36}$$

$$t \equiv (h^3/mk^2)\bar{t} \tag{37}$$

Note that each of the barred quantities is dimensionless.

In terms of the quantities above the potential for an inverse square force field is

$$\bar{V} = \pm \frac{1}{\bar{r}} \tag{38}$$

the effective potential is

$$\bar{U} = \pm \frac{1}{\bar{r}} + \frac{1}{2\bar{r}^2} \tag{39}$$

and Eqs. (18) and (19) become

$$d\bar{t} = \bar{r}^2 \, d\phi \tag{40}$$

$$d\bar{t} = \frac{\pm d\bar{r}}{\left\{ 2\left[\bar{E} \mp (1/\bar{r}) - (1/2\bar{r}^2) \right] \right\}^{1/2}} \tag{41}$$

In the next two sections we consider separately motion in a repulsive inverse square force field and motion in an attractive inverse square force field.

MOTION IN A REPULSIVE INVERSE SQUARE FORCE FIELD

The effective potential for the radial motion in a repulsive inverse square force field is

$$\bar{U} = \frac{1}{\bar{r}} + \frac{1}{2\bar{r}^2} \tag{42}$$

which is plotted in Fig. 1. Since $\bar{U}$ is positive for all finite values of $\bar{r}$, the total energy $\bar{E}$ must also be positive. For a given value of $\bar{E}$ there will be one and only one turning point for the radial motion, and therefore if the particle is approaching the force center it will eventually arrive at a point where $\bar{E} = \bar{U}$, at which point the radial motion will be reversed and it will move away and

180 Central Force Motion

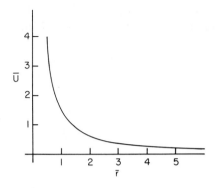

Fig. 1

never return. Setting $\bar{U} = \bar{E}$ and solving for $\bar{r}$ we obtain for the turning point

$$\bar{r}_0 = \frac{1 + (1 + 2\bar{E})^{1/2}}{2\bar{E}} \tag{43}$$

Combining Eqs. (40) and (41) for the repulsive case we obtain

$$d\phi = \frac{\pm d\bar{r}}{\bar{r}^2 \{2[\bar{E} - (1/\bar{r}) - (1/2\bar{r}^2)]\}^{1/2}} \tag{44}$$

If we choose our coordinates such that $\phi = 0$ when $\bar{r} = \bar{r}_0$, and integrate Eq. (44) from $(\bar{r}_0, 0)$ to $(\bar{r}, \phi)$ we obtain

$$\phi = \int_{\bar{r}_0}^{\bar{r}} \frac{d\bar{r}}{\bar{r}^2 \{2[\bar{E} - (1/\bar{r}) - (1/2\bar{r}^2)]\}^{1/2}} \tag{45}$$

The integration can be simplified if we introduce a new variable $\bar{u} \equiv 1/\bar{r}$. With this change

$$\phi = \int_{\bar{u}}^{\bar{u}_0} \frac{d\bar{u}}{(2\bar{E} - 2\bar{u} - \bar{u}^2)^{1/2}} \tag{46}$$

The value of $\bar{u}_0$, which can be determined from Eq. (43) or by setting the denominator in the integrand equal to zero, is given by

$$\bar{u}_0 = -1 + (1 + 2\bar{E})^{1/2} \tag{47}$$

Integrating Eq. (46) and then solving for $\bar{r}$ we obtain

$$\bar{r} = \frac{1}{\epsilon \cos \phi - 1} \tag{48}$$

where

$$\epsilon \equiv (1 + 2\bar{E})^{1/2} \geq 1 \tag{49}$$

Equation (48) is the equation of a hyperbola with focus at the origin. Figure 2 is a plot of $\bar{r}$ as a function of ϕ for the case $\epsilon = \frac{3}{2}$. Equation (48) is strictly

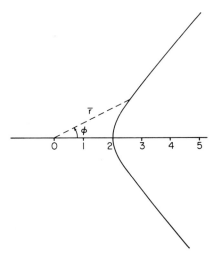

Fig. 2

valid only for values of ϕ ranging from 0 to $\cos^{-1}(1/\epsilon)$. Values of $\bar{r}$ for ϕ ranging from $-\cos^{-1}(1/\epsilon)$ to 0 can be obtained from the fact that the orbit must be symmetric about a line joining the force center and the apsis $(\bar{r}_0, 0)$. There is another branch to the hyperbola that is generated by allowing ϕ to range over the values we have excluded. This branch has no physical significance for the *repulsive* inverse square force, but as we show in the next section it is a possible orbit for the *attractive* inverse square force.

MOTION IN AN ATTRACTIVE INVERSE SQUARE FORCE FIELD

The effective potential for the radial motion in an attractive inverse square force field is

$$\bar{U} = -\frac{1}{\bar{r}} + \frac{1}{2\bar{r}^2} \qquad (50)$$

which is plotted in Fig. 3. The effective potential $\bar{U}$ has a minimum at $\bar{r} = 1$, at which point $\bar{U} = -\frac{1}{2}$. It follows that $\bar{E} \geq -\frac{1}{2}$. If $\bar{E} = -\frac{1}{2}$ then $\bar{r} = 1$ and the orbit is a circle. If $-\frac{1}{2} < \bar{E} < 0$, then there will be two turning points between which the radial motion will oscillate, and the orbit will thus be bounded. If $\bar{E} \geq 0$, then there will be one and only one turning point; therefore a particle that is approaching the force center will arrive at the turning point, where its radial motion will be reversed, and it will recede to infinity.

Combining Eqs. (40) and (41) for the attractive case we obtain

$$d\phi = \frac{\pm d\bar{r}}{\bar{r}^2 \left\{ 2\left[\bar{E} + (1/\bar{r}) - (1/2\bar{r}^2)\right]\right\}^{1/2}} \qquad (51)$$

If we choose our coordinates such that $\phi = 0$ when $\bar{r}$ has its minimum value,

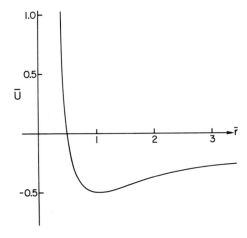

Fig. 3

which we designate $\bar{r}_0$, and integrate Eq. (51) from $(\bar{r}_0, 0)$ to $(\bar{r}, \phi)$ we obtain

$$\phi = \int_{\bar{r}_0}^{\bar{r}} \frac{d\bar{r}}{\bar{r}^2 \{ 2[\bar{E} + (1/\bar{r}) - (1/2\bar{r}^2)] \}^{1/2}} \tag{52}$$

Integrating Eq. (52) and solving for $\bar{r}$ we obtain

$$\bar{r} = \frac{1}{\epsilon \cos \phi + 1} \tag{53}$$

where

$$\epsilon \equiv (1 + 2\bar{E})^{1/2} \tag{54}$$

If $\bar{E} = -\frac{1}{2}$ then $\epsilon = 0$ and Eq. (53) is the equation of a circle. If $-\frac{1}{2} < \bar{E} < 0$ then $0 < \epsilon < 1$ and Eq. (53) is the equation of an ellipse with its focus at the force center. If $\bar{E} = 0$ then $\epsilon = 1$ and Eq. (53) is the equation of a parabola with its focus at the force center. If $\bar{E} > 0$ then $\epsilon > 1$ and Eq. (53) is the equation of a hyperbola with its focus at the force center. In Fig. 4 $\bar{r}$ is plotted as a function of ϕ for $\epsilon = 0$, $\epsilon = \frac{1}{2}$, $\epsilon = 1$, and $\epsilon = \frac{3}{2}$.

It should be remembered that Eq. (53) was obtained only for the portion of the orbit that lies between two successive apsides. The remainder of the orbit can however be obtained from symmetry. Failure to note this fact can lead at times to confusion with the mathematics. For example in the case in which $\epsilon > 1$, our solution encompasses only one branch of the hyperbola represented by Eq. (53).

One very interesting feature of the preceding results is the dependence of the determination of whether the orbit is a bound orbit (circle or ellipse) or an unbound orbit (parabola or hyperbola) on whether $\bar{E} < 0$ or $\bar{E} \geqslant 0$, hence, since $\bar{E} = (h^2/mk^2)E$, on whether $E < 0$ or $E \geqslant 0$. It follows that if a particle is projected from a point in an attractive inverse square field, there is a limiting speed of projection v_e called the escape speed, below which the particle cannot escape from the neighborhood of the force center, and above which the particle will escape to infinity irrespective of the direction of

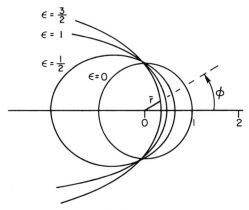

Fig. 4

propagation. The escape speed for a particle on the surface of the earth, considered to be a spherically symmetric mass distribution of radius R, can be shown to be $(2gR)^{1/2}$, where g is the acceleration due to gravity at the surface. Although the direction of propagation has no influence on whether the particle escapes, it does influence the value of h, hence does influence the character of the orbit.

If the orbit is an ellipse (or a circle) then the length of the semimajor axis can be found by adding the values of $\bar{r}$ at $\phi = 0$ and $\phi = \pi$, as obtained from Eq. (53), and dividing by 2. If we do this we obtain

$$\bar{a} = \frac{1}{1 - \epsilon^2} \tag{55}$$

The length of the semiminor axis, which we designate as $\bar{b}$, is equal to the maximum value of $\bar{r}\sin\phi$. If we multiply Eq. (53) by $\sin\phi$ and determine the maximum value of the resulting quantity, we obtain

$$\bar{b} = \frac{1}{(1 - \epsilon^2)^{1/2}} = (\bar{a})^{1/2} \tag{56}$$

The area $d\bar{A}$ swept out by the radius $\bar{r}$ in a change $d\phi$ is given by

$$d\bar{A} = \tfrac{1}{2}\bar{r}^2 d\phi = \tfrac{1}{2} d\bar{t} \tag{57}$$

Hence the time $\Delta \bar{t}$ between two successive positions is equal to twice the area swept out. It follows that the period $\bar{\tau}$ for an elliptical orbit is

$$\bar{\tau} = 2(\pi\bar{a}\bar{b}) = 2\pi\bar{a}^{3/2} \tag{58}$$

or equivalently

$$\tau = 2\pi(m/k)^{1/2} a^{3/2} \tag{59}$$

For an elliptical orbit there will be two apses. The closer apsis is in general called the *pericentron*, and the farther apsis is called the *apocentron*. For orbits around the sun the closer apsis is called the *perihelion*, and the farther apsis

184 Central Force Motion

the *aphelion*. For orbits around the earth the closer apsis is called the *perigee*, and the farther apsis the *apogee*.

EXAMPLES

Example 1. A particle of mass m moves in a circular orbit of center O and radius a under the action of a central force for which the potential energy is $V(r)$. Show that if the orbit is stable then

$$V''(a) + 3\frac{V'(a)}{a} > 0 \tag{60}$$

where $V'(r) \equiv dV(r)/dr$ and $V''(r) \equiv d^2V(r)/dr^2$. Show also that if it is stable and it is subjected to a small radial disturbance, the period of small oscillation about the original circular orbit is $2\pi/n$, where

$$n^2 = \frac{V''(a) + 3[V'(a)/a]}{m} \tag{61}$$

Solution. The radial motion of the particle is equivalent to the one dimensional motion of a particle in the potential

$$U = V(r) + \frac{h^2}{2mr^2} \tag{62}$$

If the orbit is a stable circular orbit of radius a, then U must have a minimum at a. The conditions for U to have a minimum at a are

$$U'(a) = V'(a) - \frac{h^2}{ma^3} = 0 \tag{63}$$

$$U''(a) = V''(a) + \frac{3h^2}{ma^4} > 0 \tag{64}$$

Combining Eqs. (63) and (64) we obtain Eq. (60). If the orbit is stable and the particle is subjected to a small radial displacement then to a first approximation the effective potential will be given by

$$U(r) = U(a) + \tfrac{1}{2}U''(a)(r-a)^2 \tag{65}$$

Hence the motion will be simple harmonic about $r = a$ with an angular frequency n, where

$$n^2 = \frac{U''(a)}{m} \tag{66}$$

Combining Eqs. (63), (64), and (66) we obtain Eq. (61).

PROBLEMS

1. A particle of mass m describes an ellipse under a force of magnitude $m\mu/r^2$ directed toward the focus. Show that if the particle is moving with a speed V when it is a distance c from the force center, the period of the motion is $(2\pi/\mu^{1/2})[(2/c)-(V^2/\mu)]^{-3/2}$.

2. A particle of mass m is describing a parabolic orbit of latus rectum $4a$ under an inverse square law of attraction to the focus. When the particle is at one end of the latus rectum it meets and coalesces with a particle of mass nm at rest. Show that the composite particle will trace out an elliptic orbit of eccentricity given by the equation $(n+1)^4(1-\epsilon^2) = 2n(n+2)$.

3. The minimum distance of a comet from the sun is observed to be half the radius of the earth's orbit (assumed to be circular), and its velocity at that point is twice the orbital velocity of the earth. Find the velocity of the comet when it crosses the earth's orbit, and the angle at which the orbits cross. Will the comet subsequently escape from the solar system?

4. A particle is moving in a central force field $f(r) = -kr^{-2} + Cr^{-3}$. Show that the equation of the orbit can be put in the form $r = a(1-\epsilon^2)/(1+\epsilon\cos\alpha\theta)$, which is an ellipse for $\alpha = 1$ but is a precessing ellipse for $\alpha \neq 1$. Derive an approximate expression for the rate of precession of the pericentron when α is close to unity in terms of the dimensionless quantity $\eta = C/ka$. The ratio η is a measure of the strength of the perturbing inverse cube term relative to the main inverse square term. What value of η would be necessary to account for the rate of precession of the perihelion of the planet Mercury? For Mercury: the rate of precession of the perihelion is $40''$ per century, the eccentricity of the orbit is 0.206, and the period is 0.24 year.

5. A particle P of mass m is describing an ellipse of major and minor axes $2a$ and $2b$, respectively, about a center of force at the center of the ellipse. When the particle reaches the end of the major axis, it strikes and coalesces with a particle of mass nm, which is at rest. The central attraction per unit mass is unchanged. Prove that the new orbit is an ellipse of major and minor axes $2a$ and $2b/(n+1)$, respectively.

6. A particle A is moving in a plane under an attractive force of magnitude $2/r^3$ per unit mass, which is directed toward a fixed point O in the plane. When $t = 0$ then $r = 2$ and the radial and transverse components of the velocity are $(3/2)^{1/2}$ and 1, respectively. Show that $\ddot{r} = 2/r^3$ and find r as a function of t.

7. A particle of mass m is acted on by an attractive force directed toward a point O and of magnitude $\mu m/r^3$, where r is the distance from the point O. The particle is projected from a point A at a distance a from O with a velocity of magnitude $\sqrt{m}/a$ in the direction AP, where the angle OAP is $45°$. Show that the particle moves in the orbit $r = a\exp\theta$.

8. A particle of mass m moves with initial speed v from infinity along a straight line that if continued would allow the particle to pass a distance a from a point P. The particle is attracted toward P by a force that

causes the particle to move in the trajectory $r = c \coth \phi$. What is the force law? Find ϕ as a function of the time.

9. A particle moves in the orbit $r = a \cos 2\phi$ under the action of a central force directed toward the origin. Show that the magnitude of the force is proportional to $(8a^2 - 3r^2)/r^5$.

10. Show that if a particle describes a circular orbit under the influence of an attractive central force directed toward a point on the circle, then the force varies as the inverse fifth power of the distance.

11. A particle P of unit mass is acted on by a force directed toward a point O and of magnitude $\omega^2 r + \omega^2 a^3 r^{-2}$, where r is the distance OP. If P is projected from a point that is a distance a from O with velocity $4a\omega/\sqrt{3}$ perpendicular to OP, show that subsequently the distance of P from O varies between $r = a$ and $r = 2a$, and find the speed of P when $r = 2a$.

12. The law of force in a central field is $(c/r^2)\exp(-kr)$, where k and c are constants, k being small. Show that the apsis line in a nearly circular orbit will advance in each revolution through the angle $\pi k a$ if a is the radius of the orbit.

13. A particle is moving under the action of a central force of magnitude $-k/r^3$. Show that it is possible to choose values of the total energy E and the angular momentum h such that the orbit is of the form $r = a \exp(b\phi)$.

14. Two particles A and B, each of mass m, are connected by an inextensible string of length $2a$, which passes through a small smooth hole in a smooth horizontal table. The particle A is free to slide on the table and B hangs freely. Initially OB is of length a and particle A is projected from rest with a speed $(8ga/3)^{1/2}$ in a direction perpendicular to OA. Show that in the subsequent motion particle B will just reach the hole.

24

The Differential Scattering Cross Section

INTRODUCTION

If a particle that is moving with some known velocity encounters a region of space within which there is a force field, the velocity of the particle will undergo a change. If the region within which the force acts is finite in extent, and the particle enters and leaves the region, we say that the particle has been scattered by the force field. If the region within which the force acts is infinite in extent, but the effect of the force on the particle is negligible beyond a certain range, we can still effectively treat the force field as if it were confined to a finite region of space. In what follows we therefore assume that the force field is confined to a finite region of space, even though we may apply the results to other cases, and we refer to the region containing the force field as a scattering region.

Frequently, when a particle with a given initial velocity is approaching a scattering region containing a known force field, we are interested not in the exact details of the encounters, but only in the probability that the particle will collide with the scattering region and have its final velocity in some definite range of velocities. On the other hand it may happen that the force field is unknown and we cannot measure it directly, but we are able to measure the initial and final velocities of particles scattered by the field. In this case we can use the particles to probe the field, and quite often we can obtain in this way a great deal of information about the field. In the analysis of either situation, the concept of the scattering cross section is extremely useful.

We initially restrict our attention to scattering by a spherically symmetric central force field. In a later chapter we allow for the possibility of other force fields. For a spherically symmetric central force field the trajectory of the particle lies in a single plane containing the force center.

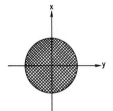

Fig. 1

THE TOTAL SCATTERING CROSS SECTION

Let us assume that the force center is located at the origin of a Cartesian coordinate frame and that the incident particle is moving with some known speed v in the positive z direction.

If we project the scattering region onto the x-y plane, as represented in Fig. 1, the area of the projection is defined as the *total scattering cross section* and is designated as $\sigma(v)$. If the initial values of the x and y coordinates of the incident particle lie within the range of values encompassed by the projected area, the particle will be scattered; otherwise it will miss the scattering region.

The total scattering cross section for a given speed v can be determined experimentally by directing a uniform beam of particles of speed v at the scattering region and measuring the fraction of the particles that are scattered. If the cross-sectional area A of the beam is large enough, the scattering region will be entirely encompassed by the beam, and the fraction F of the particles in the beam that are scattered will be $\sigma(v)/A$, and thus $\sigma(v) = FA$.

THE DIFFERENTIAL SCATTERING CROSS SECTION

We now consider the situation introduced in the preceding section in greater detail. To facilitate the discussion we use polar coordinates (s, ϕ) to specify the position of the initial line of motion of the particle, and spherical coordinates (θ, ϕ) to specify the direction of the final velocity, as shown in Fig. 2. The parameter s is called the *impact parameter*.

For a given force field and a given value of the initial speed v, the final velocity of the particle can in principle be determined if we know the values of s and ϕ. Thus for each value of s there is a unique value of θ. It follows that if we project the scattering region onto the x-y plane (Fig. 3a), then with each pair of values of s and ϕ in the projected area there is an associated pair of values of θ and ϕ (Fig. 3b); hence a given region R in the neighborhood of the point (s, ϕ) within the projected area will correspond to scattering into a definite solid angle Ω in the neighborhood of the direction (θ, ϕ). We now define $\sigma(v, \theta)\sin\theta\, d\theta\, d\phi$ to be the magnitude of the projected area that corresponds to scattering into the solid angle $\sin\theta\, d\theta\, d\phi$, or equivalently to

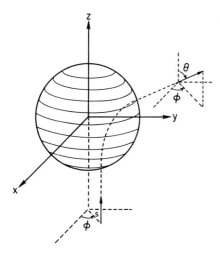

Fig. 2

scattering in which the value of θ lies between θ and $\theta + d\theta$, and the value of ϕ lies between ϕ and $\phi + d\phi$. The quantity $\sigma(v,\theta)$ is called the *differential scattering cross section*.

If we know the differential scattering cross section $\sigma(v,\theta)$ we can determine the total scattering cross section $\sigma(v)$ by integrating $\sigma(v,\theta)\sin\theta\, d\theta\, d\phi$ over the range of values of θ and ϕ covered by the scattered particles. It should be carefully noted that this range does not necessarily include all possible values of θ. There may for example be an upper limit to the scattering angle θ.

The differential scattering cross section $\sigma(v,\theta)$ of a given scattering region for a given speed v can be experimentally determined by directing a uniform beam of particles of speed v at the scattering region, and measuring the fraction of the particles that are scattered into each solid angle $d\Omega$. If the

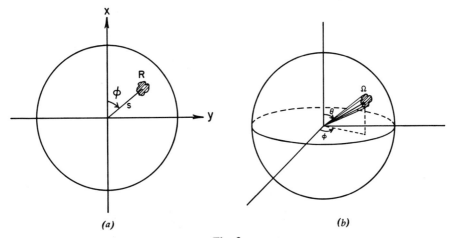

Fig. 3

cross-sectional area A of the beam is large enough, the scattering region will be entirely encompassed by the beam, and the fraction dF of the particles in the beam that are scattered into the solid angle $d\Omega$ centered about the direction (θ, ϕ) will be equal to $\sigma(v, \theta) d\Omega / A$, and thus $\sigma(v, \theta) = A dF / d\Omega$.

DETERMINATION OF THE DIFFERENTIAL SCATTERING CROSS SECTION FROM KNOWLEDGE OF THE IMPACT PARAMETER AS A FUNCTION OF THE SCATTERING ANGLE

The area of the region R in Fig. 3a is given by

$$\int\int_R s\, ds\, d\phi \tag{1}$$

and from the definition of the differential scattering cross section is also given by

$$\int\int_\Omega \sigma(v, \theta) \sin\theta\, d\theta\, d\phi \tag{2}$$

where Ω is the solid angle in Fig. 3b corresponding to the region R in Fig. 3a. Equating (1) and (2) we obtain

$$\int\int_R s\, ds\, d\phi = \int\int_\Omega \sigma(v, \theta) \sin\theta\, d\theta\, d\phi \tag{3}$$

If we change the variables of integration in the right hand integral from θ and ϕ to s and ϕ we obtain

$$\int\int_\Omega \sigma(v, \theta) \sin\theta\, d\theta\, d\phi = \int\int_R \sigma(v, \theta) \sin\theta \left|\frac{d\theta}{ds}\right| ds\, d\phi \tag{4}$$

Note that the absolute value of $d\theta/ds$ is used rather than simply $d\theta/ds$. By using the absolute value we eliminate the necessity of making a detailed conversion of the limits of integration. We simply choose the limits such that the area of R given by the integral $\int\int_R s\, ds\, d\phi$ comes out positive. Equating Eqs. (3) and (4) we obtain

$$\int\int_R s\, ds\, d\phi = \int\int_R \sigma(v, \theta) \sin\theta \left|\frac{d\theta}{ds}\right| ds\, d\phi \tag{5}$$

Since Eq. (5) must be true for arbitrary regions R, the integrands on both sides must be equal; hence

$$\sigma(v, \theta) = \frac{s}{\sin\theta} \left|\frac{ds}{d\theta}\right| \tag{6}$$

It follows that if we know s as a function of θ for the given value of v, we can determine $\sigma(v, \theta)$, from Eq. (6).

If $ds/d\theta$ changes sign in the region R then there will be more than one value of s for a given value of θ. In this case it will be necessary to break the region R down into subregions in each of which $ds/d\theta$ keeps a constant sign, and then to consider separately the contribution each region makes to the differential scattering cross section.

DETERMINATION OF THE DIFFERENTIAL SCATTERING CROSS SECTION ASSOCIATED WITH A GIVEN SCATTERING POTENTIAL

Consider a particle scattered by a potential $V(r)$ that vanishes at infinity. Let α be the angle that the final direction of the scattered particle makes with a line drawn from the scattering center to the point of closest approach. It then follows that the scattering angle θ is given by

$$\theta = |\pi - 2\alpha| \tag{7}$$

When the force is repulsive $\pi > 2\alpha$, hence $\theta = \pi - 2\alpha$. When the force is attractive $\pi < 2\alpha$, hence $\theta = 2\alpha - \pi$. From Eq. (21) in Chapter 23

$$\alpha = \int_{r_0}^{\infty} \frac{h \, dr}{r^2 \{2m[E - V - (h^2/2mr^2)]\}^{1/2}} \tag{8}$$

where r_0 is the distance of closest approach. The quantity h, the magnitude of the angular momentum, is given by

$$h = mvs \tag{9}$$

where v is the incident speed; and since the potential energy vanishes at infinity, the energy E is given by

$$E = \frac{mv^2}{2} \tag{10}$$

Using Eqs. (9) and (10) in Eq. (8) we obtain

$$\alpha = \int_{r_0}^{\infty} \frac{s \, dr}{r^2 [1 - (V/E) - (s^2/r^2)]^{1/2}} \tag{11}$$

If we define

$$z \equiv \frac{s}{r} \tag{12}$$

then Eq. (11) can be rewritten

$$\alpha = \int_0^{z_0} \frac{dz}{\{1 - [V(z)/E] - z^2\}^{1/2}} \tag{13}$$

where $z_0 \equiv s/r_0$. The value of r_0 is the minimum value of r, hence can be obtained by setting $dr/d\phi = 0$ in Eq. (20) in Chapter 23. Doing this and converting from the variable r to the variable z we find that z_0 must satisfy the equation

$$1 - \frac{V(z_0)}{E} - z_0^2 = 0 \tag{14}$$

Gathering results, we obtain for the scattering angle

$$\theta = \left| \pi - 2 \int_0^{z_0} \frac{dz}{\{1 - [V(z)/E] - z^2\}^{1/2}} \right| \tag{15}$$

where z is defined by Eq. (12) and z_0 obtained from Eq. (14). Equation (15) gives us θ as a function of s. Knowing $\theta(s)$ we can find $s(\theta)$ and then use Eq. (6) to find $\sigma(v,\theta)$.

EXAMPLES

Example 1. Determine the differential scattering cross section and the total scattering cross section for the scattering of a particle by a rigid elastic sphere of radius a.

Solution. A typical scattering event is illustrated in Fig. 4. From the figure it is apparent that

$$s = a \sin \frac{(\pi - \theta)}{2} \tag{16}$$

The same result could also have been obtained by substituting the hard sphere potential directly in Eq. (15). If we do this we obtain

$$\theta = \pi - 2 \int_0^{s/a} \frac{dz}{[1-z^2]^{1/2}} = \pi - 2\sin^{-1}\frac{s}{a} \tag{17}$$

which is the same result as Eq. (16). Using Eq. (16) in Eq. (6) we obtain for the

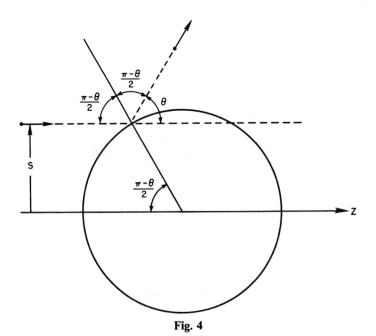

Fig. 4

differential scattering cross section

$$\sigma(v,\theta) = \frac{s}{\sin\theta}\left|\frac{ds}{d\theta}\right| = \frac{a^2}{4} \tag{18}$$

and for the total scattering cross section

$$\sigma(v) = \int_0^{2\pi}\int_0^{\pi}\sigma(v,\theta)\sin\theta\,d\theta\,d\phi = \pi a^2 \tag{19}$$

Example 2. Determine the differential scattering cross section for the scattering of a particle by an inverse square repulsive force of magnitude $+k/r^2$.

Solution. If we substitute $V = k/r \equiv kz/s$ in Eq. (15) we obtain

$$\theta = \pi - 2\int_0^{z_0}\frac{dz}{\left[1 - (kz/Es) - z^2\right]^{1/2}}$$

$$= \pi - 2\left[-\sin^{-1}\left\{\frac{-2z - (k/Es)}{\left[(k/Es)^2 + 4\right]^{1/2}}\right\}\right]_0^{z_0} \tag{20}$$

Letting $V(z_0) = kz_0/s$ in Eq. (14) we obtain

$$z_0 = \frac{-(k/Es) + \left[(k/Es)^2 + 4\right]^{1/2}}{2} \tag{21}$$

Substituting Eq. (21) in Eq. (20) we obtain

$$\theta = \pi - 2\left[-\sin^{-1}(-1) + \sin^{-1}\left\{\frac{-k/Es}{\left[(k/Es)^2 + 4\right]^{1/2}}\right\}\right]$$

$$= 2\sin^{-1}\left\{\frac{k/Es}{\left[(k/Es)^2 + 4\right]^{1/2}}\right\} \tag{22}$$

Solving Eq. (22) for s we obtain

$$s = \frac{k\cot(\theta/2)}{mv^2} \tag{23}$$

and thus

$$\sigma(v,\theta) = \frac{s}{\sin\theta}\left|\frac{ds}{d\theta}\right| = \left[\frac{k}{2mv^2\sin^2(\theta/2)}\right]^2 \tag{24}$$

The total scattering cross section for the force above will be infinite because the force field extends to infinity.

PROBLEMS

1. Consider the two dimensional scattering of a smooth elastic disk of radius r off a fixed smooth elastic disk of radius r.
   a. What is the total scattering cross section in two dimensions? (*Note*: The dimension of the cross section will be length.)
   b. What is the differential scattering cross section $\sigma(v,\theta)$, where $\sigma(v,\theta) \, d\theta$ is the effective cross-sectional length of the target for scattering between θ and $\theta + d\theta$?

2. Determine the differential scattering cross section for the scattering of a particle by an attractive force of magnitude k/r^2.

3. Determine the differential scattering cross section for the scattering of a particle by a repulsive central force of magnitude $f(r)$, where $f(r) = k/r^2$ for $r \leqslant a$ and $f(r) = 0$ for $r > a$.

4. Show that the differential scattering cross section for scattering from the repulsive central force $f = kr^{-3}$ is $\sigma(v,\theta) = k\pi^2(\pi - \theta)/mv^2\theta^2(2\pi - \theta)^2 \sin\theta$.

5. A beam of particles encounters the spherical potential
$$V(r) = -A \quad r \leqslant a$$
$$V(r) = 0 \quad r > a$$
where A is a positive constant. Show that the differential scattering cross section is given by
$$\sigma(v,\theta) = \left\{ \frac{a^2 n^2}{4 \cos(\theta/2)} \right\} \left\{ \frac{[n\cos(\theta/2) - 1][n - \cos(\theta/2)]}{[1 + n^2 - 2n\cos(\theta/2)]^2} \right\}$$
for $0 < \theta < 2\cos^{-1}(1/n)$ and is zero for $\theta > 2\cos^{-1}(1/n)$, where $n = [1 + (A/E)]^{1/2}$.

6. A beam of particles of mass m and energy E is incident normally on a thin foil of thickness d containing n fixed scattering centers per unit volume. The force between one of the particles and one of the scattering centers is an inverse square repulsive force of magnitude k/r^2. What fraction of the particles in the beam are scattered through an angle greater than some angle α?

7. A beam of particles of mass m and energy E is incident normally on a thin foil of thickness d containing n fixed scattering centers per unit volume. A particle detector of area A is placed a distance r, where $r^2 \gg A$, from the point O at which the beam passes through the foil. The line from the point O to the detector makes an angle α with the direction of the incident beam. If the number of particles per unit time that hit the foil is I, and the differential scattering cross section for an encounter between

one of the particles and one of the scattering centers is $\sigma(\theta)$, how many particles per unit time hit the detector?

8. A beam of particles encounters a target made up of many scattering centers. Let $I(x)$ be the intensity of the beam after it has penetrated a distance x into the target and σ be the total scattering cross section for an encounter of a particle in the beam with one of the scattering centers.
   a. Show that $I(x) = I(0)\exp(-\sigma n x)$.
   b. Show that the average distance traveled by a particle before it is scattered is $1/\sigma n$.

25

Two Particle Systems

INTRODUCTION

In a great number of situations in physics it is necessary to determine the motions of two particles that interact with each other but are otherwise effectively isolated from the rest of the universe.

COORDINATES

Consider two particles, which we label 1 and 2, respectively. The locations of the particles can be specified by giving the vector positions $\mathbf{r}(1)$ and $\mathbf{r}(2)$ of the particles with respect to the origin of an inertial frame of reference. A more useful pair of vectors for specifying the locations of the particles is the pair

$$\mathbf{r}(c) \equiv \frac{m(1)\mathbf{r}(1) + m(2)\mathbf{r}(2)}{m(1) + m(2)} \tag{1}$$

$$\mathbf{r}(21) \equiv -\mathbf{r}(12) \equiv \mathbf{r}(1) - \mathbf{r}(2) \tag{2}$$

The vector $\mathbf{r}(c)$ is the vector position of the center of mass c. The vector $\mathbf{r}(21)$ is the vector position of particle 1 with respect to particle 2. The inverse relations are obtained by solving Eqs. (1) and (2) for $\mathbf{r}(1)$ and $\mathbf{r}(2)$ and are given by

$$\mathbf{r}(1) = \mathbf{r}(c) + \left[\frac{m(2)}{m(1) + m(2)}\right]\mathbf{r}(21) \tag{3}$$

$$\mathbf{r}(2) = \mathbf{r}(c) - \left[\frac{m(1)}{m(1) + m(2)}\right]\mathbf{r}(21) \tag{4}$$

The relationship between $\mathbf{r}(1)$, $\mathbf{r}(2)$, $\mathbf{r}(c)$, and $\mathbf{r}(21)$ is illustrated in Fig. 1. The center of mass c lies between the particles 1 and 2 on the line joining them and

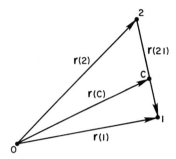

Fig. 1

is a distance $s(1)$ from particle 1 and $s(2)$ from particle 2 where $m(1)s(1) = m(2)s(2)$.

THE EQUATIONS OF MOTION

If the only forces acting on particles 1 and 2 are the mutual forces of interaction, and if we let $\mathbf{f}(21)$ be the force that particle 2 exerts on particle 1 and $\mathbf{f}(12)$ the force that particle 1 exerts on particle 2, and note, from the law of action and reaction (Chapter 9), that

$$\mathbf{f}(12) = -\mathbf{f}(21) \tag{5}$$

then the equations of motion for particles 1 and 2 are

$$\mathbf{f}(21) = m(1)\ddot{\mathbf{r}}(1) \tag{6}$$

$$-\mathbf{f}(21) = m(2)\ddot{\mathbf{r}}(2) \tag{7}$$

The equation of motion governing the vector $\mathbf{r}(c)$ can be obtained by adding Eqs. (6) and (7) and is given by

$$M\ddot{\mathbf{r}}(c) = 0 \tag{8}$$

where

$$M \equiv m(1) + m(2) \tag{9}$$

It follows that the velocity of the center of mass is constant.

The equation of motion governing the vector $\mathbf{r}(21)$ can be obtained by multiplying Eq. (6) by $m(2)/[m(1) + m(2)]$ and Eq. (7) by $m(1)/[m(1) + m(2)]$, and then taking the difference of the resulting equations. If we do this we obtain

$$\mathbf{f}(21) = \mu\ddot{\mathbf{r}}(21) \tag{10}$$

where

$$\mu \equiv \frac{m(1)m(2)}{m(1) + m(2)} \tag{11}$$

The parameter μ is called the *reduced mass*. Equation (10) is formally identical to Newton's law for a particle of mass μ, located at the point $\mathbf{r}(21)$ and acted on by the force $\mathbf{f}(21)$. It follows that particle 1 moves with respect to an observer fixed on particle 2 as if the observer's frame were an inertial frame and particle 1 a particle of mass μ acted on by the force $\mathbf{f}(21)$.

The motion of two particles moving under the action of their mutual forces of interaction can thus be reduced to a one particle problem.

KINETIC ENERGY

In dealing with a pair of particles, one is frequently interested in the kinetic energy of the pair. In terms of the velocities $\mathbf{v}(1) \equiv \dot{\mathbf{r}}(1)$ and $\mathbf{v}(2) \equiv \dot{\mathbf{r}}(2)$, the kinetic energy is given by

$$T = \tfrac{1}{2} m(1)[v(1)]^2 + \tfrac{1}{2} m(2)[v(2)]^2 \tag{12}$$

The kinetic energy can also be written in terms of the velocities $\mathbf{v}(c) \equiv \dot{\mathbf{r}}(c)$ and $\mathbf{v}(21) \equiv \dot{\mathbf{r}}(21)$, by simply substituting Eqs. (3) and (4) in Eq. (12). Doing this we obtain

$$T = \tfrac{1}{2} M[v(c)]^2 + \tfrac{1}{2} \mu [v(21)]^2 \tag{13}$$

where M and μ are defined by Eqs. (9) and (11). The first term on the right hand side of Eq. (13) is called the center of mass kinetic energy, and the second term the relative kinetic energy or kinetic energy with respect to the center of mass.

CENTRAL FORCE MOTION

If the force $\mathbf{f}(21)$ that particle 2 exerts on particle 1 is directed along the line joining the two particles and if its magnitude depends only on the distance between the two particles, the relative motion of particle 1 with respect to particle 2 is the same as the motion of a particle of mass μ moving in a conservative central force field $\mathbf{f}(21)$. It follows that the relative motion of particle 1 with respect to particle 2 can be obtained by appropriately modifying the results of Chapter 23, which apply to the motion of a particle in a stationary conservative force field. In converting the results of Chapter 23 we simply replace $\mathbf{r}$ by $\mathbf{r}(21)$, and m by μ everywhere, with one exception: namely, if m occurs in the potential $V(r)$ or equivalently in the force $f(r)$, as it would for example if the force of interaction were a gravitational force, we do not replace m by μ in that place, since this would have the undesired effect of replacing $\mathbf{f}(21)$, the force that particle 2 exerts on particle 1, by a different force.

MOTION WITH RESPECT TO THE CENTER OF MASS

In studying the motion of an isolated pair of particles one is frequently interested in the motion of the particles with respect to the center of mass rather than the relative motion of the particles. The vector position of particle 1 with respect to the center of mass can be obtained immediately from Eq. (3) and is given by

$$\mathbf{r}(c1) \equiv \mathbf{r}(1) - \mathbf{r}(c) = \left[\frac{m(2)}{m(1) + m(2)} \right] \mathbf{r}(21) \equiv \frac{\mu}{m(1)} \mathbf{r}(21) \quad (14)$$

The vectors $\mathbf{r}(c1)$ and $\mathbf{r}(21)$ are thus always in the same direction, and the ratio of their magnitudes is a positive constant dependent only on the masses of the particles. It follows that the trajectory of particle 1 with respect to the center of mass is simply a scaled-down version of the trajectory of particle 1 with respect to particle 2. This result could also have been obtained geometrically from Fig. 1. A similar result holds for the motion of particle 2 with respect to the center of mass.

THE EFFECT OF EXTERNAL FORCES ON THE RELATIVE MOTION

If in addition to the interaction forces $\mathbf{f}(12)$ and $\mathbf{f}(21)$, there is an external force $\mathbf{f}(1)$ acting on particle 1 and an external force $\mathbf{f}(2)$ acting on particle 2, then the equations of motion of particles 1 and 2 are

$$\mathbf{f}(1) + \mathbf{f}(21) = m(1)\ddot{\mathbf{r}}(1) \quad (15)$$

$$\mathbf{f}(2) - \mathbf{f}(21) = m(2)\ddot{\mathbf{r}}(2) \quad (16)$$

The equations of motion governing the vectors $\mathbf{r}(c)$ and $\mathbf{r}(21)$ can again be obtained as was done previously, and are given by

$$\mathbf{f}(1) + \mathbf{f}(2) = M\ddot{\mathbf{r}}(c) \quad (17)$$

$$\mu \left[\frac{\mathbf{f}(1)}{m(1)} - \frac{\mathbf{f}(2)}{m(2)} \right] + \mathbf{f}(21) = \mu \ddot{\mathbf{r}}(21) \quad (18)$$

Equation (17) is the equation of motion of a particle of mass M located at the point $\mathbf{r}(c)$ and acted on by the total external force $\mathbf{f}(1) + \mathbf{f}(2)$. The center of mass no longer moves with a constant velocity. The external forces also affect in a nontrivial manner the relative motion of the two particles, as can be seen from Eq. (18). However if the external force per unit mass exerted on particle 1 is equal to the external force per unit mass exerted on particle 2, that is, if

$$\frac{\mathbf{f}(1)}{m(1)} = \frac{\mathbf{f}(2)}{m(2)} \quad (19)$$

then the first term on the left hand side of Eq. (18) vanishes, and the equation reduces to the same equation obtained earlier. Thus in this case and this case only, the relative motion is unaffected by the presence of the external forces. In the case of two particles in a uniform gravitational field, the condition above is satisfied.

EXAMPLES

Example 1. Two particles of masses m and M, respectively, are connected by an elastic string of natural length a and spring constant k. Initially they are at rest on a smooth horizontal table at a distance a apart. The mass m is suddenly projected with a speed v in a direction perpendicular to the string. Prove that in the subsequent motion, and before the string becomes slack, the greatest distance r between the particles is a root of the equation

$$mMv^2(r + a) = k(m + M)r^2(r - a) \tag{20}$$

Solution. The motion of m with respect to M is the same as the motion of a particle of reduced mass $\mu = mM/(m + M)$ in a central force field whose potential is $V = k(r - a)^2/2$. We refer to this hypothetical particle as the particle μ. The energy and angular momentum of the particle μ are

$$E = \frac{\mu v^2}{2} \tag{21}$$

$$h = \mu v a \tag{22}$$

From the results of Chapter 23 the radial motion is the same as that of a particle moving in the one dimensional effective potential

$$U = \frac{k(r - a)^2}{2} + \frac{h^2}{2\mu r^2} \tag{23}$$

The turning points in the motion occur when $U = E$. Substituting Eq. (22) in Eq. (23) and setting $U = E$ we obtain

$$\frac{k(r - a)^2}{2} + \frac{\mu v^2 a^2}{2r^2} = \frac{\mu v^2}{2} \tag{24}$$

Setting $\mu = mM/(m + M)$ in Eq. (24) and rearranging terms we obtain Eq. (20).

PROBLEMS

1. A neutron of mass m and speed v collides with a stationary atom of mass M. Assuming that the collision is perfectly elastic, show that the maximum speed of the atom after collision is $V = 2mv/(M + m)$.

2. The parallax of a star (the angle subtended at the star by the radius of the earth's orbit) is p. The star's position is observed to oscillate with angular amplitude α and period τ. If this oscillation is attributed to the existence of a planet accompanying the star, show that its mass m is given by $m/M_0 = (\alpha/p)(M/M_0)^{2/3}(\tau/\tau_0)^{-2/3}$, where M is the total mass of the star and the planet, M_0 is the mass of the sun, and $\tau_0 = 1$ year. Evaluate the mass if $M = 0.25 M_0$, $\tau = 16$ years, $p = 0.5''$, and $\alpha = 0.01''$.

3. Show that if two planets of masses m_1 and m_2, respectively, describe ellipses relative to the sun with respective major axes $2a_1$ and $2a_2$, their periods satisfy the relation $T_1/T_2 = a_1^3(M + m_2)/a_2^3(M + m_1)$, where M is the mass of the sun.

4. Two identical charged particles each of mass m and charge e are initially far apart. One of the particles is at rest, and the other is moving with speed v and impact parameter $s = 2ke^2/mv^2$. Find the distance of closest approach of the particles, and the speed of each at the moment of closest approach.

5. Two particles of masses $4m$ and m, respectively, are free to move along the x axis. There is a force of attraction between the two particles of magnitude kr, where k is a constant and r is the distance between the two particles. At time $t = 0$ the particle of mass $4m$ is located at the point $x = 5a$ and the particle of mass m is located at $x = 10a$. Both particles are at rest at time $t = 0$.
   a. At what value of x do the two particles collide?
   b. What is the relative velocity of the two particles at the moment the collision occurs?
   c. If the two particles rebound elastically after the collision, how long will it be before they collide again?

6. Two particles A and B of masses m and M, respectively, describe orbits under their mutual attraction. Show that the total angular momentum of the system with respect to the center of mass is given by $M\mathbf{h}/(M + m)$, where $\mathbf{h}$ is the angular momentum of particle A with respect to the point B.

7. Two masses M and m, connected by a massless spring of force constant k, are at rest in equilibrium on a smooth horizontal table. A velocity of magnitude v and direction away from M is imparted to m. Find:
   a. The velocity of M relative to the table at the instant the string again becomes unstretched.
   b. The velocity of the center of mass.
   c. The period of the oscillatory motion.

8. Two particles of masses m and M, respectively, are attached to the ends of a light spring of unstretched length a and spring constant k. Initially the particles are at rest with m at a height a above M. At $t = 0$, m is projected vertically upward with speed v. Find the positions of the particles at any subsequent time. What is the largest value of v for which this solution applies?

9. Two particles, each of mass m, are connected by a spring of unstretched length a and spring constant k and are at rest on a smooth horizontal surface. A third particle of mass m, traveling with a speed v in a direction perpendicular to the spring, hits one of the particles and sticks to it. In the ensuing motion the spring stretches to a maximum length $3a$. Find the value of v.

26

Two Particle Collisions

INTRODUCTION

If two or more particles are approaching one another, and if the interaction between the particles has a finite range, or at least is negligible beyond a certain range, the motion can be broken down into three periods: an initial period during which the particles are moving toward one another with constant velocities, an intermediate period during which the particles interact, and a final period in which one or more particles, not necessarily the original particles, are moving away with constant velocities. If these conditions are satisfied, we refer to the interaction as a *collision*.

In this chapter we are interested in collisions satisfying the following description: (1) a particle of mass m collides with a particle of mass M that is initially at rest, (2) the interaction force between the particles is directed along the line joining the two particles, (3) there are no other forces besides the interaction force acting on the particles during the collision, and (4) the same two particles emerge from the collision. In a later chapter we consider more general collisions.

DESCRIPTION OF A COLLISION

To discuss the collision between the two particles m and M, we define the following *precollision* velocities:

$$\mathbf{v} \equiv \text{velocity of } m \qquad (1)$$

$$\mathbf{V} \equiv \text{velocity of } M \qquad (2)$$

$$\mathbf{w} \equiv \frac{m\mathbf{v} + M\mathbf{V}}{m + M} \equiv \text{velocity of the center of mass } C \text{ of the system} \qquad (3)$$

$$\mathbf{u} \equiv \mathbf{v} - \mathbf{V} \equiv \text{velocity of } m \text{ relative to } M \qquad (4)$$

$$\mathbf{s} \equiv \mathbf{v} - \mathbf{w} \equiv \frac{M}{M + m}\mathbf{u} \equiv \text{velocity of } m \text{ with respect to } C \qquad (5)$$

and the following *postcollision* velocities:

$$\mathbf{v}' \equiv \text{velocity of } m \tag{6}$$

$$\mathbf{V}' \equiv \text{velocity of } M \tag{7}$$

$$\mathbf{w}' \equiv \frac{m\mathbf{v}' + M\mathbf{V}'}{m + M} \equiv \text{velocity of the center of mass } C \text{ of the system} \tag{8}$$

$$\mathbf{u}' \equiv \mathbf{v}' - \mathbf{V}' \equiv \text{velocity of } m \text{ relative to } M \tag{9}$$

$$\mathbf{s}' \equiv \mathbf{v}' - \mathbf{w}' \equiv \frac{M}{M + m}\mathbf{u}' \equiv \text{velocity of } m \text{ with respect to } C \tag{10}$$

Since we have assumed that the particle of mass M is initially at rest, the precollision velocities can be simplified as follows:

$$\mathbf{V} = 0 \tag{11}$$

$$\mathbf{w} = \frac{m\mathbf{v}}{m + M} \tag{12}$$

$$\mathbf{u} = \mathbf{v} \tag{13}$$

$$\mathbf{s} = \frac{M}{M + m}\mathbf{v} \tag{14}$$

Since there are no external forces acting on the system during the collision, the center of mass velocity will remain constant during the collision. Thus

$$\mathbf{w}' = \mathbf{w} \tag{15}$$

Since the interaction force is along the line joining the two particles, and initially the motion of both particles lies in the plane defined by the initial position of M and the initial line of motion of m, it follows that the particles will not leave this plane. The problem of determining the final velocities is thus reduced from a three dimensional to a two dimensional problem.

If the collision is an elastic collision or if we know the change in the energy of the system as a result of the collision, we can determine still more about the postcollision velocities. Let us define $\Delta\epsilon$ as the change in the energy of the system, that is,

$$\Delta\epsilon \equiv \frac{1}{2}(M + m)(w')^2 + \frac{1}{2}\frac{Mm}{M + m}(u')^2$$

$$- \frac{1}{2}(M + m)w^2 - \frac{1}{2}\frac{Mm}{M + m}u^2 \tag{16}$$

Making use of Eq. (15) in Eq. (16), we obtain

$$\Delta\epsilon = \frac{1}{2}\frac{Mm}{M + m}(u')^2 - \frac{1}{2}\frac{Mm}{M + m}u^2 \tag{17}$$

Hence $\Delta\epsilon$ is also equal to the change in the kinetic energy with respect to the center of mass as well as the change in the total energy. If we let

$$\epsilon \equiv \frac{1}{2} \frac{Mm}{M+m} u^2 \qquad (18)$$

then Eq. (17) can be written

$$u' = \left(1 + \frac{\Delta\epsilon}{\epsilon}\right)^{1/2} u \qquad (19)$$

Although we can determine a great deal about the final velocities from knowledge of the initial velocities and $\Delta\epsilon$, we need more information to complete the description of the collision. In the section that follows we consider a number of quantities that are frequently used to describe the result of a collision.

THE LABORATORY AND CENTER OF MASS SCATTERING ANGLES

Let S be a Cartesian coordinate system with origin at the initial position of the mass M and 3 axis in the direction of the precollision velocity of m. We then define θ and ϕ as the spherical angles that determine the orientation of $\mathbf{u}'$ or $\mathbf{s}'$ with respect to the frame S, that is, θ and ϕ are the angles defined by the relation

$$\frac{\mathbf{u}'}{u'} \equiv \frac{\mathbf{s}'}{s'} \equiv \mathbf{e}_1 \sin\theta \cos\phi + \mathbf{e}_2 \sin\theta \sin\phi + \mathbf{e}_3 \cos\theta \qquad (20)$$

Similarly we define Θ and Φ as the spherical angles that determine the orientation of $\mathbf{v}'$ with respect to the frame S, that is, Θ and Φ are the angles defined by the relation

$$\frac{\mathbf{v}'}{v'} = \mathbf{e}_1 \sin\Theta \cos\Phi + \mathbf{e}_2 \sin\Theta \sin\Phi + \mathbf{e}_3 \cos\Theta \qquad (21)$$

The coordinates θ and ϕ are convenient coordinates to use when discussing the collision from a theoretical point of view and are called center of mass or relative coordinates. The coordinates Θ and Φ are convenient coordinates to use when the collision is being discussed from an experimental point of view and are called laboratory coordinates.

As we saw earlier, the motion of the particles is confined to the plane determined by the initial position of M which is located at the origin O of the frame S and the initial line of motion of m which is in the 3 direction. It follows that this plane contains the 3 axis, hence Φ and ϕ are equal (or possibly differ by π) and each is equal to the angle made by the projection onto the 1-2 plane of the position vector of m with the 1 axis. Hence the only angle needed to determine the direction of $\mathbf{u}'$ is the angle θ, and the only angle needed to determine the direction of $\mathbf{v}'$ is Θ. The angle θ is called the center of mass scattering angle. Note that since $\mathbf{u} = \mathbf{v}$, it is the angle between $\mathbf{u}'$ and $\mathbf{u}$ or between $\mathbf{s}'$ and $\mathbf{s}$. The angle Θ is called the laboratory scattering angle. It is the angle between $\mathbf{v}'$ and $\mathbf{v}$.

TRANSFORMATION BETWEEN THE LABORATORY SCATTERING ANGLE AND THE CENTER OF MASS SCATTERING ANGLE

The laboratory scattering angle Θ and the center of mass scattering angle are not independent. To obtain the relation between them we note first from the definition of $\mathbf{s}'$ that

$$\mathbf{v}' = \mathbf{w}' + \mathbf{s}' \tag{22}$$

It will be convenient to rewrite this in the form

$$\mathbf{v}' = s'(\boldsymbol{\gamma} + \mathbf{n}) \tag{23}$$

where

$$\mathbf{n} \equiv \frac{\mathbf{s}'}{s'} \tag{24}$$

$$\boldsymbol{\gamma} \equiv \frac{\mathbf{w}'}{s'} \tag{25}$$

Using Eqs. (15), (12), (10), (19), and (13) in Eq. (25) we obtain

$$\boldsymbol{\gamma} = \frac{\mathbf{w}'}{s'} = \frac{m\mathbf{v}/(M+m)}{[M/(M+m)]u'}$$

$$= \frac{m\mathbf{v}}{M[1+(\Delta\epsilon/\epsilon)]^{1/2}u} = \frac{\mathbf{e}_3 m v}{M[1+(\Delta\epsilon/\epsilon)]^{1/2}u} = \frac{m}{M}\left(\frac{\epsilon}{\epsilon+\Delta\epsilon}\right)\mathbf{e}_3 \tag{26}$$

Taking the dot product of Eq. (23) with $\mathbf{e}_3$ and making use of Eqs. (20), (21), and (26) we obtain

$$v'\cos\Theta = s'(\gamma + \cos\theta) \tag{27}$$

Taking the dot product of Eq. (23) with itself we obtain

$$(v')^2 = (s')^2[\gamma^2 + 2\gamma\cos\theta + 1] \tag{28}$$

Combining Eqs. (27) and (28) we obtain

$$\cos\Theta = \frac{\gamma + \cos\theta}{(\gamma^2 + 2\gamma\cos\theta + 1)^{1/2}} \tag{29}$$

or

$$\cos\theta = \gamma\cos^2\Theta \pm [\gamma^2\cos^2\Theta - \gamma^2 + 1]^{1/2}\cos\Theta - \gamma \tag{30}$$

where

$$\gamma = \frac{m}{M}\left(\frac{\epsilon}{\epsilon + \Delta\epsilon}\right) \tag{31}$$

If $\gamma^2 < 1$ then the plus sign is used in Eq. (30). If $\gamma^2 \geq 1$ then both signs are used. The ambiguity arises because it is possible for two values of θ to correspond to the same value of Θ. As an example suppose $m > M$ and $\Delta\epsilon = 0$; then for a head-on collision $\Theta = 0$ and $\theta = \pi$, while for a near miss $\Theta = 0$ and $\theta = 0$.

LABORATORY AND CENTER OF MASS DIFFERENTIAL CROSS SECTIONS

We are frequently interested not in the exact values of θ and ϕ or Θ and Φ but only in the probability that the collision will cause the values of these quantities to fall in some definite range. When this is the case the concept of a cross section is useful, just as it was in the case of a single particle that was scattered by a stationary force field.

The events leading up to a collision between the moving particle of mass m and the stationary particle of mass M are identical to the events leading up to a collision between a particle of mass m and a fixed scattering center. However the collision itself and the subsequent scattering of the particle of mass m differ because in the present case the particle of mass M can recoil.

Since there is a definite region surrounding M that m must enter if an interaction is to take place, we can as we did previously define the *total scattering cross section* as simply the area of the projection of this region onto the 1-2 plane. We designate the total scattering cross section as $q(v)$ or $Q(v)$.

The *center of mass differential cross section* $q(v,\theta)$ is defined by the condition that $q(v,\theta)\sin\theta\, d\theta\, d\phi$ is the magnitude of the projected area that corresponds to collisions in which the direction of $\mathbf{u}'$, the postcollision relative velocity, lies in the solid angle $\sin\theta\, d\theta\, d\phi$. The center of mass differential cross section can generally be found theoretically by first determining the differential cross section $\sigma(v,\theta)$ for a collision of a particle m, moving with a velocity $\mathbf{v}$, with particle M, held rigidly fixed; then the mass m is replaced by the reduced mass μ, except that if m occurs in the interparticle force, in that place it is not replaced by μ.

The *laboratory differential cross section* $Q(v,\Theta)$ is defined by the condition that $Q(v,\Theta)\sin\Theta\, d\Theta\, d\Phi$ is the magnitude of the projected area that corresponds to collisions in which the direction of $\mathbf{v}'$, the postcollision velocity of particle m, lies in the solid angle $\sin\Theta\, d\Theta\, d\Phi$. The laboratory differential cross section is the cross section one would normally measure experimentally.

Given the differential cross section $q(v,\theta)$ or $Q(v,\Theta)$, the total cross section is obtained from the relation

$$q(v) \equiv Q(v) = \int\int q(v,\theta)\sin\theta \, d\theta \, d\phi = \int\int Q(v,\Theta)\sin\Theta \, d\Theta \, d\Phi \quad (32)$$

The limits of integration are determined by the ranges of values of θ, ϕ and Θ, Φ resulting from the collision.

TRANSFORMATION BETWEEN LABORATORY AND CENTER OF MASS CROSS SECTIONS

The center of mass differential cross section $q(v,\theta)$ and the laboratory differential cross section $Q(v,\Theta)$ are not independent. To obtain the relationship between $q(v,\theta)$ and $Q(v,\Theta)$ we note first that a given region R on the projected area corresponds either to collisions in which the direction (θ,ϕ) lies in a certain solid angle ω, or alternatively to collisions in which the direction (Θ,Φ) lies in a certain solid angle Ω. From the definitions of $q(v,\theta)$ and $Q(v,\Theta)$ it follows that

$$\text{area of region } R = \int\int_\omega q(v,\theta)\sin\theta \, d\theta \, d\phi = \int\int_\Omega Q(v,\Theta)\sin\Theta \, d\Theta \, d\Phi \quad (33)$$

If we transform the variables of integration in the last integral from Θ and Φ to θ and ϕ we obtain

$$\int\int_\omega q(v,\theta)\sin\theta \, d\theta \, d\phi = \int\int_\omega Q(v,\Theta)\sin\Theta \left|\frac{d\Theta}{d\theta}\right| d\theta \, d\phi \quad (34)$$

Since this equation must be true for arbitrary regions R, or equivalently arbitrary solid angles ω, the integrands must be equal. Hence

$$q(v,\theta) = Q(v,\Theta)\frac{\sin\Theta}{\sin\theta}\left|\frac{d\Theta}{d\theta}\right|$$

$$= Q(v,\Theta)\left|\frac{d(\cos\Theta)}{d(\cos\theta)}\right| \quad (35)$$

If we make use of Eq. (29) in Eq. (35) we can further simplify the result, and obtain

$$q(v,\theta) = Q[v,\Theta(\theta)]\left|\frac{1+\gamma\cos\theta}{(1+\gamma^2+2\gamma\cos\theta)^{3/2}}\right| \quad (36)$$

or

$$Q(v,\Theta) = q[v,\theta(\Theta)]\left|\frac{\left[\gamma\cos\Theta \pm (1-\gamma^2+\gamma^2\cos^2\Theta)^{1/2}\right]^2}{(1-\gamma^2+\gamma^2\cos^2\Theta)^{1/2}}\right| \quad (37)$$

where as before

$$\gamma = \frac{m}{M}\left(\frac{\epsilon}{\epsilon + \Delta\epsilon}\right) \tag{38}$$

As long as $\gamma^2 < 1$ the plus sign is used. If $\gamma^2 \geq 1$ then both signs must be used.

PROBLEMS

1. Prove that the maximum deflection angle for a particle of mass m that is scattered off an initially stationary particle of mass M, where $m/M \geq 1$, is given by $\Theta = \sin^{-1}(M/m)$.

2. A theoretical physicist needs data on elastic proton-proton scattering for $\theta = \theta_0$ and $\epsilon = \epsilon_0$. What value of energy of the incident proton should an experimentalist use, and at what laboratory angle Θ should the detector be located?

3. Calculate the probability that a hard sphere of radius a that collides elastically with an identical hard sphere that is at rest will be scattered through an angle greater than 60°.

4. Show that the laboratory differential scattering cross section for protons of energy E incident on target protons at rest is given by

$$q(v, \Theta) = \left(\frac{k}{E}\right)^2 [\sin^{-4}\Theta + \cos^{-4}\Theta]\cos\Theta$$

for $\Theta \leq \pi/2$ and is zero for $\Theta > \pi/2$, where the force between two protons is given by k/r^2.

5. Show that if a particle of mass m is elastically scattered by an initially stationary particle of mass M, then the angle ψ between the final velocities of m and M is given by $\cot\psi = [(m - M)/(m + M)]\tan(\theta/2)$, where θ is the center of mass scattering angle.

6. A particle of mass m, charge q, and speed v is scattered through an angle of 90° by a particle of mass M and charge Q that is initially at rest. Calculate the distance of closest approach. Assume that all quantities are given in MKS units, and that the Coulomb force between the particles is the cause of the scattering.

7. A beam of particles of mass m, charge q, and speed v strikes a small thin target containing particles of mass M and charge Q that can be assumed to be initially at rest and free. The beam has a flux density J. The target is encompassed by the beam and has a total mass Z. A detector of cross-sectional area ΔA is placed at a distance d from the target, with the line from the target to the detector being at right angles to the beam

direction. Determine the rate at which the particles of mass m strike the detector. Determine the rate at which the particles of mass M strike the detector. Assume that all quantities are given in MKS units, and that the Coulomb force between the particles is the cause of the scattering.

8. Show that the laboratory scattering angle Θ is related to E, the energy of the incident particle before scattering, and E', the energy of the incident particle after scattering, according to the equation

$$\cos \Theta = \frac{M+m}{2m}\left(\frac{E'}{E}\right)^{1/2} - \frac{M-m}{2m}\left(\frac{E}{E'}\right)^{1/2} + \frac{M}{2m}\frac{\Delta\epsilon}{(EE')^{1/2}}$$

PART 2

THE NEWTONIAN MECHANICS OF SYSTEMS OF PARTICLES

SECTION 1

Basic Principles

27

Dynamical Systems

INTRODUCTION

The principles we have developed up to this point apply to the motion of a point particle, or to the motion of the center of mass of an extended object. In this and the chapters that follow we consider the motion of an extended object in greater detail.

DYNAMICAL SYSTEMS

The analysis of the behavior of an extended object is greatly simplified if instead of viewing the object as a continuum we view it as a collection of a large number of point particles. We assume that such a procedure is legitimate, and we refer to the collection of particles as a *dynamical system*. Unless otherwise stated, the systems we deal with are closed systems, that is, systems containing a definite and fixed number of particles. The total mass of a closed system is constant.

In discussing a dynamical system we assume that there are N particles in the system labeled from 1 to N, and we designate the mass of particle i as $m(i)$, and the position, velocity, acceleration, and linear momentum of particle i with respect to an arbitrary point a as $\mathbf{r}(ai)$, $\mathbf{v}(ai)$, $\mathbf{a}(ai)$, and $\mathbf{p}(ai)$, respectively. If the point a is not indicated, it will be understood to be the origin of an inertial frame of reference.

INTERNAL FORCES

The forces acting on a dynamical system can be divided into two categories, internal and external. An *internal force* is a force exerted on one particle in a dynamical system by another particle in the same system. An *external force* is a force exterted on a particle in a dynamical system by an agent that is external to the given system.

We designate the net external force acting on particle i as $\mathbf{f}(ei)$, the internal force exerted by a particular particle j on particle i as $\mathbf{f}(ji)$, and the net force acting on particle i, that is, $\mathbf{f}(ei) + \sum_j \mathbf{f}(ji)$, as $\mathbf{f}(i)$.

NEWTON'S LAW

The equation of motion of the ith particle in a dynamical system of N particles is

$$\mathbf{f}(i) = \dot{\mathbf{p}}(i) \tag{1}$$

or equivalently

$$\mathbf{f}(i) = m(i)\mathbf{a}(i) \tag{2}$$

The set of equations we obtain by letting i range from 1 to N is the basic set of equations governing the motion of a dynamical system. This set of equations is not, as we shall see, particularly practical in directly determining the motion of a system containing a large number of particles. However it serves as the starting point for our analysis.

INTERPARTICLE FORCES

If we knew all the internal and external forces acting on a dynamical system, we could use Newton's law to determine the motion of each particle in the system, hence obtaining, at least in principle, the equations of motion of the system. However, when the number of particles in the system is large, this is not practical. In the first place we usually cannot in this case determine the internal forces, and in the second place even if we succeeded, the number of equations of motion involved generally would be so large that their solution would be a practical impossibility. Nevertheless it is possible to determine the motion of the center of mass of a dynamical system on the basis of the limited knowledge of the interparticle forces supplied by the law of action and reaction, which can be stated as follows:

The Law of Action and Reaction. If one particle exerts a force on a second particle, the second particle simultaneously exerts a force on the first particle, and the two forces are equal in magnitude but opposite in direction.

In the following chapters we demonstrate that we are able to go even further in our analysis of the motion of a dynamical system if in addition to the conditions imposed by the law of action and reaction on the interparticle forces, we make the following postulate:

Postulate V. If the number of particles in a dynamical system is sufficiently large, the system will behave as if the interparticle forces between any two particles in the system are directed along the line joining the two particles.

NOTE 1: The only consequence of Postulate V that we will exploit is the fact that the sum of the torques with respect to any point exerted by the interparticle forces vanishes. It would therefore be possible to state this postulate in a weaker form. There is however no experimental evidence that disproves the stronger form.

NOTE 2: The number of particles must be large enough to ensure that any exceptions to Postulate V on the microscopic level will be averaged out on the macroscopic level.

THE CENTER OF MASS

In the following chapters we investigate a number of properties of dynamical systems. In most cases we find that properties whose values depend on a reference point can be simplified if we relate the values of the properties with respect to the reference point to their values with respect to the center of mass. We therefore in this section outline the general procedure to be followed, and introduce some notation that is used later.

If a property that is defined with respect to a reference point a can be expressed as a function of the position vectors $\mathbf{r}(ai)$ and the velocity vectors $\mathbf{v}(ai)$, then the relationship between the property with respect to the reference point a and the same property with respect to the center of mass c can generally be found by first decomposing the position vectors $\mathbf{r}(ai)$ and the velocity vectors $\mathbf{v}(ai)$ as follows:

$$\mathbf{r}(ai) \equiv \mathbf{r}(ac) + \mathbf{r}(ci) \tag{3}$$

$$\mathbf{v}(ai) \equiv \mathbf{v}(ac) + \mathbf{v}(ci) \tag{4}$$

and then noting from the definition of the center of mass that

$$\sum_i m(i)\mathbf{r}(ci) = 0 \tag{5}$$

$$\sum_i m(i)\mathbf{v}(ci) = 0 \tag{6}$$

The center of mass is such an important point, and one that occurs so often, that we use the abbreviated notation

$$\mathbf{R}(a) \equiv \mathbf{r}(ac) \quad \mathbf{V}(a) \equiv \mathbf{v}(ac) \quad \mathbf{A}(a) \equiv \mathbf{a}(ac) \tag{7}$$

to represent the position, velocity, and acceleration, respectively of the center of mass with respect to a point a. If the point a is not indicated, it is understood to be the origin. Thus

$$\mathbf{R} \equiv \mathbf{R}(o) \quad \mathbf{V} \equiv \mathbf{V}(o) \quad \mathbf{A} \equiv \mathbf{A}(o) \tag{8}$$

28

Force and Linear Momentum

INTRODUCTION

The linear momentum of a particle is one of the most if not the most important dynamical property of a particle. This chapter extends the concept of linear momentum to dynamical systems. Some of the ideas we discuss have been introduced in earlier chapters, but are repeated here for the sake of completeness.

LINEAR MOMENTUM

The *linear momentum of a particle i with respect to an arbitrary point a* is defined as

$$\mathbf{p}(ai) \equiv m(i)\mathbf{v}(ai) \qquad (1)$$

where $m(i)$ is the mass of particle i and $\mathbf{v}(ai)$ is the velocity of particle i with respect to the point a.

The *linear momentum of a dynamical system with respect to an arbitrary point a* is defined as

$$\mathbf{P}(a) \equiv \sum_i \mathbf{p}(ai) \qquad (2)$$

where $\mathbf{p}(ai)$ is the linear momentum of particle i with respect to the point a.

If the point a, with respect to which the momentum $\mathbf{P}(a)$ is measured, is not indicated, it will be understood that the point is the origin o or equivalently a point that is fixed with respect to the origin, that is,

$$\mathbf{P} \equiv \mathbf{P}(o) \qquad (3)$$

The relationship between the linear momentum of a dynamical system with respect to a point a, and its value with respect to the center of mass is contained in the following theorem.

Theorem 1. The linear momentum of a dynamical system with respect to an arbitrary point a can be written

$$\mathbf{P}(a) = M\mathbf{V}(a) \qquad (4)$$

where M is the total mass of the system and $\mathbf{V}(a)$ is the velocity of the center of mass with respect to the point a.

PROOF: From the definition of $\mathbf{P}(a)$ we have

$$\mathbf{P}(a) = \sum_i m(i)\mathbf{v}(ai) = \sum_i m(i)[\mathbf{v}(ac) + \mathbf{v}(ci)]$$

$$= \sum_i m(i)\mathbf{v}(ac) + \sum_i m(i)\mathbf{v}(ci) = M\mathbf{v}(ac) + 0$$

$$\equiv M\mathbf{V}(a) \qquad (5)$$

NOTE: From Theorem 1, it follows that the linear momentum of a dynamical system with respect to the center of mass is zero. Thus Theorem 1 could have been stated as follows: the linear momentum of a dynamical system with respect to a point a is equal to the linear momentum of the system with respect to the center of mass c plus the linear momentum of a particle of mass M located at the center of mass and moving with it.

FORCE

The net force acting on particle i in a dynamical system is

$$\mathbf{f}(i) = \mathbf{f}(ei) + \sum_j \mathbf{f}(ji) \qquad (6)$$

where $\mathbf{f}(ei)$ is the net external force acting on particle i and $\mathbf{f}(ji)$ is the force which particle j exerts on particle i.

The net force acting on the dynamical system will be designated $\mathbf{F}$ and is given by

$$\mathbf{F} = \sum_i \mathbf{f}(i) = \sum_i \mathbf{f}(ei) + \sum_i \sum_j \mathbf{f}(ji) \qquad (7)$$

Theorem 2. The net force $\mathbf{F}$ acting on a dynamical system is equal to the net external force acting on the system, that is,

$$\mathbf{F} \equiv \sum_i \mathbf{f}(i) = \sum_i \mathbf{f}(ei) \qquad (8)$$

where $\mathbf{f}(i)$ is the net force acting on particle i, and $\mathbf{f}(ei)$ is the net external force acting on particle i.

PROOF: From the law of action and reaction $\mathbf{f}(ij) = -\mathbf{f}(ji)$, hence

$$\sum_i \sum_j \mathbf{f}(ji) \equiv \sum_j \sum_i \mathbf{f}(ij) \equiv \frac{1}{2} \sum_i \sum_j [\mathbf{f}(ij) + \mathbf{f}(ji)] = 0 \qquad (9)$$

Substituting Eq. (9) in Eq. (7) we obtain Eq. (8).

FORCE AND LINEAR MOMENTUM

The relationship between the forces acting on a dynamical system and the momentum of the system is contained in the following theorem.

Theorem 3. The time rate of change of **P**, the linear momentum of a dynamical system with respect to the origin of an inertial frame of reference, is equal to **F**, the net external force acting on the system, that is,

$$\mathbf{F} = \dot{\mathbf{P}} \qquad (10)$$

PROOF: The equation of motion for the ith particle is

$$\mathbf{f}(i) = \dot{\mathbf{p}}(i) \qquad (11)$$

Summing over i we obtain

$$\sum_i \mathbf{f}(i) = \sum_i \dot{\mathbf{p}}(i) \qquad (12)$$

Substituting Eqs. (8) and (2) in Eq. (12) we obtain Eq. (10).

Theoerem 3 can also be written in terms of the motion of the center of mass as follows:

Theorem 4. The motion of the center of mass of a dynamical system is governed by the equation

$$\mathbf{F} = M\mathbf{A} \qquad (13)$$

where **F** is the net external force acting on the system, M is the total mass of the system, and **A** is the acceleration of the center of mass with respect to the origin of an inertial frame of reference.

PROOF: From Theorem 3, $\mathbf{F} = \dot{\mathbf{P}}$. From Theorem 1, $\dot{\mathbf{P}} = M\dot{\mathbf{V}}$. Combining results, we obtain Eq. (13).

EXAMPLES

Example 1. Show that the equation of motion of a rocket is given by

$$M \frac{d\mathbf{V}}{dt} = \mathbf{U} \frac{dM}{dt} + \mathbf{F} \qquad (14)$$

where M is the instantaneous mass of the rocket, $\mathbf{V}$ is the instantaneous velocity of the rocket, $\mathbf{F}$ is the instantaneous external force acting on the rocket, and $\mathbf{U}$ is the instantaneous velocity relative to the rocket at which the exhaust material is shot out of the rocket. The first term on the right is called the *thrust* of the rocket motor. Since dM/dt is negative, the thrust is opposite in direction to the exhaust velocity.

Solution. Consider the dynamical system consisting of the rocket including its fuel at time t. The momentum of this system at time t is given by

$$\mathbf{P}(t) = M(t)\mathbf{V}(t) \tag{15}$$

A short time Δt later, the particles that constitute the system are no longer all moving with the same velocity. During the time Δt a small portion of the system has been ejected from the remaining portion of the system. The momentum of the system at time $t + \Delta t$ is given by

$$\mathbf{P}(t + \Delta t) = M(t + \Delta t)\mathbf{V}(t + \Delta t) + [M(t) - M(t + \Delta t)][\mathbf{V}(\bar{t}) + \mathbf{U}(\bar{t})] \tag{16}$$

where $\bar{t}$ is some time between t and $t + \Delta t$. The first term is the momentum of the rocket minus the ejected fuel, and the second term is the momentum of the ejected fuel. It follows that the time rate of change of the momentum of the system is given by

$$\begin{aligned}
\dot{\mathbf{P}} &= \lim_{\Delta t \to 0} \left\{ \frac{\mathbf{P}(t + \Delta t) - \mathbf{P}(t)}{\Delta t} \right\} \\
&= \lim_{\Delta t \to 0} \left\{ \frac{M(t + \Delta t)\mathbf{V}(t + \Delta t) - M(t)\mathbf{V}(t)}{\Delta t} \right. \\
&\quad \left. - \frac{[M(t + \Delta t) - M(t)]}{\Delta t} [\mathbf{V}(\bar{t}) + \mathbf{U}(\bar{t})] \right\} \\
&= \frac{d}{dt}[M(t)\mathbf{V}(t)] - \frac{dM(t)}{dt}[\mathbf{V}(t) + \mathbf{U}(t)] \\
&= M(t)\frac{d\mathbf{V}(t)}{dt} - \mathbf{U}(t)\frac{dM(t)}{dt}
\end{aligned} \tag{17}$$

Substituting Eq. (17) into Eq. (10) we obtain Eq. (14).

Example 2. A uniform chain of length a and mass M is hanging vertically from its ends A and B, which are close together. At time $t = 0$ the end B is released. Determine the tension T at A when B has fallen a distance $x < a$.

Solution. Let the chain be the system of interest. The only forces acting on the system are the tension T at A, which is acting vertically up, and the gravitational force Mg, which is acting vertically down (Fig. 1). If we let x be the distance from A to B, and P be the component in the x direction of the

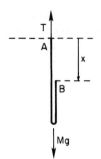

Fig. 1

momentum of the system, then equating the force acting on the system to the time rate of change of its momentum, we obtain

$$Mg - T = \dot{P} \qquad (18)$$

The end B of the chain is essentially free falling. Hence

$$x = \frac{gt^2}{2} \qquad (19)$$

Since the left portion of the chain as shown in Fig. 1 is at rest and the right portion is moving at the same speed as B, the component in the x direction of the momentum of the system is

$$P = \left(\frac{M}{a}\right)\left(\frac{a-x}{2}\right)\dot{x} \qquad (20)$$

Taking the time derivative of Eq. (20), making use of Eq. (19) to obtain $\dot{x}$ and $\ddot{x}$ in terms of x, and substituting the results in Eq. (18), we obtain

$$T = \frac{Mg}{2}\left(1 + \frac{3x}{a}\right) \qquad (21)$$

PROBLEMS

1. Every second n particles, each of mass m, strike the side of a box. Use the relation $\mathbf{F} = \dot{\mathbf{P}}$ to find the force required to hold the side of the box in place, assuming that each particle has the same speed v before and after hitting the side and that the particles move at right angles to the side. Explain precisely what dynamical system you use.

2. Show that if an airplane of mass M in horizontal flight drops a bomb of mass m, the airplane experiences an upward acceleration mg/M.

3. A spherical satellite of mass m and radius a moves with speed v through a tenuous atmosphere of density ρ. Find the frictional force on it, assuming that the speed of the air molecules can be neglected in

comparison with v and that each molecule that is struck becomes embedded in the skin of the satellite.

4. A rocket having a total mass of 50 kg contains 2 kg of propellant that is burned at a uniform rate in one second. The propellant has a specific impulse of 2000 N s/kg of propellant; that is, a thrust force of 2000 N is produced by burning one kilogram of propellant in one second.
   a. Assuming that the rocket moves horizontally with negligible frictional resisting forces, find the velocity of the rocket at the end of the burning time.
   b. If the propellant were burned in 2 seconds instead of one, would the velocity be different?

5. A rocket is to be fired vertically upward. The initial mass is M_0, the exhaust velocity $-u$ is constant, and the rate of exhaust $-dM/dt = A$ is constant. After a total mass ΔM has been exhausted, the rocket runs out of fuel.
   a. Neglecting air resistance and assuming that the acceleration g of gravity is constant, set up and solve the equation of motion.
   b. Show that if M_0, u, and ΔM are fixed, the larger the rate of exhaust A (i.e., the faster it uses up its fuel), the greater the maximum altitude reached by the rocket.

6. a. Determine the velocity at the end of 5 s and the maximum velocity of a rocket fired vertically upward. Aerodynamic forces may be neglected and the gravitational acceleration may be assumed to be constant. The following data apply: total loaded mass = 80 kg, empty rocket mass = 40 kg, jet velocity = 1200 m/s, time to burnout = 16 s.
   b. Calculate the altitude at burnout.

7. A lunar landing craft approaches the moon's surface. Assume that one-third of its mass is fuel, that the exhaust velocity from its rocket engine is 1500 m/s, and that the acceleration of gravity at the lunar surface is one-sixth of that at the earth's surface. How long can the craft hover over the moon's surface before it runs out of fuel?

8. A rocket, initially of total mass M, throws off every second a mass αM with constant velocity V relative to the rocket. Show that it cannot rise at once unless $\alpha V > g$ and that it cannot rise at all unless $\alpha V > \beta g$, where βM is the mass of the case of the rocket. Calculate the greatest height it can reach when the conditions are such that the rocket is just able to rise vertically at once.

9. A two stage rocket is to be built capable of accelerating a 100 kg payload to a velocity of 6000 m/s in free flight in empty space (no gravitational

field). (In a two stage rocket, the first stage is detached after exhausting its fuel, before the second stage is fired.) Assume that the fuel used can reach an exhaust velocity of 1500 m/s and that structural requirements imply that an empty rocket (without fuel or payload) will weigh 10% as much as the fuel it can carry. Find the optimum choice of masses for the two stages so that the total takeoff weight is a minimum. Show that it is impossible to build a single stage rocket that will do the job.

10. A jet of water of cross section A, density ρ, and absolute velocity v_0 moves horizontally and hits a block of mass m inelastically; that is, the water leaves with a zero horizontal component of velocity relative to the block. Find the terminal velocity of the block, assuming a coefficient of friction μ between the block and the horizontal plane on which it slides.

11. An open-topped freight car weighing 10^4 kg is coasting without friction along a level track. Rain is falling vertically. The initial velocity of the car is 2 m/s. There is a drain pipe in the floor so that the water runs out as fast as it comes in. What is the velocity of the car after it has traveled long enough for 2×10^3 kg of rainwater to run through it?

12. Water is poured into a barrel that is resting on the ground at the rate of 60 kg/min from a height of 5 m. The barrel has a mass of 12 kg. Find the force that the barrel is exerting on the ground after the water has been pouring into the barrel for 1 min.

13. Material is fed at a rate bg kg/s and with essentially zero absolute velocity from a hopper onto a belt of total mass m. It sticks to the belt until it reaches a second point a distance l from the first point, at which point it falls off. If the belt is inclined at 30° to the horizontal, and if the pulleys around which the belt slides are frictionless, find the steady state speed v of the belt motion.

14. A uniform chain of length $2a$ hangs over a smooth peg. If it is started from rest, prove that its velocity when it is leaving the peg is $\sqrt{ga}$.

15. A uniform flexible cord of length b and weight W hangs in the equilibrium configuration over a smooth peg. The equilibrium is disturbed. Find the force on the peg when the length of rope hanging on one side of the peg is x.

16. A uniform heavy chain of length a hangs initially with a part of length b hanging over the edge of a table. The remaining part, of length $a - b$, is coiled up at the edge of the table. Show that if the chain is released the speed of the chain when the last link leaves the end of the table is $[2g(a^3 - b^3)/3a^2]^{1/2}$.

17. A uniform heavy rope is coiled up on a smooth horizontal table. If one end is raised by hand with a uniform velocity v_0, show that when this

end is at a distance y above the table the force on the hand is equal to the weight of a length $y + v_0^2/g$ of the rope.

18. A flexible rope of mass ρ per unit length and total length l is suspended so that its upper end is at a height h above a horizontal floor, where $h > l$. Suddenly the rope is released, hitting the floor inelastically. Find the force on the floor as a function of time.

19. A ball of mass m and an attached flexible chain of linear density ρ and indefinite length originally rest on a smooth horizontal plane. If the ball is shot vertically into the air with initial velocity v_0, how high will it rise?

20. Find the differential equation of motion of a raindrop falling through a mist collecting mass as it falls. Assume that the drop remains spherical and that the rate of accretion is proportional to the cross-sectional area of the drop multiplied by the speed of fall. Show that if the drop starts from rest when it is infinitesimal, the acceleration is constant and equal to $g/7$.

29

Torque and Angular Momentum

INTRODUCTION

The purpose of this chapter is to extend the concept of angular momentum from particles to systems of particles.

ANGULAR MOMENTUM

The *angular momentum of a particle i with respect to an arbitrary point a* is defined as

$$\mathbf{h}(ai) \equiv \mathbf{r}(ai) \times \mathbf{p}(ai) \tag{1}$$

where $\mathbf{r}(ai)$ is the position of particle i with respect to point a, and $\mathbf{p}(ai)$ is the linear momentum of particle i with respect to the point a.

NOTE: As stated in Chapter 14 the definition above is not the only definition of angular momentum, nor is it necessarily the most common. In many textbooks the angular momentum of a particle i with respect to a point a is defined as $\mathbf{r}(ai) \times \mathbf{p}(oi)$, where o is the origin of an inertial frame. If the point a is a fixed point, then the two definitions are equivalent, but if the point a is a moving point then the two definitions are not equivalent. The reader should therefore be careful to take this into account when comparing results in different texts.

The *angular momentum of a dynamical system with respect to an arbitrary point a* is defined as

$$\mathbf{H}(a) \equiv \sum_i \mathbf{h}(ai) \tag{2}$$

where $\mathbf{h}(ai)$ is the angular momentum of particle i with respect to the point a.

If the point a, with respect to which the angular momentum $\mathbf{H}(a)$ is measured, is not designated, it is understood to be the origin o, that is,

$$\mathbf{H} \equiv \mathbf{H}(o) \tag{3}$$

The relationship between the angular momentum with respect to the point a and the angular momentum with respect to the center of mass c is contained in the following theorem.

Theorem 1. The angular momentum of a dynamical system with respect to an arbitrary point a is given by

$$\mathbf{H}(a) = \mathbf{H}(c) + \mathbf{R}(a) \times \mathbf{P}(a) \qquad (4)$$

where $\mathbf{H}(c)$ is the angular momentum with respect to the center of mass c, $\mathbf{R}(a)$ is the position of the center of mass with respect to the point a, and $\mathbf{P}(a)$ is the linear momentum of the dynamical system with respect to the point a.

PROOF: From the definition of $\mathbf{H}(a)$ we have

$$\begin{aligned}
\mathbf{H}(a) &= \sum_i \mathbf{r}(ai) \times m(i)\mathbf{v}(ai) \\
&= \sum_i m(i)[\mathbf{r}(ac) + \mathbf{r}(ci)] \times [\mathbf{v}(ac) + \mathbf{v}(ci)] \\
&= \left[\sum_i m(i)\right]\mathbf{r}(ac) \times \mathbf{v}(ac) + \mathbf{r}(ac) \times \left[\sum_i m(i)\mathbf{v}(ci)\right] \\
&\quad + \left[\sum_i m(i)\mathbf{r}(ci)\right] \times \mathbf{v}(ac) + \sum_i \mathbf{r}(ci) \times m(i)\mathbf{v}(ci) \\
&= M\mathbf{R}(a) \times \mathbf{V}(a) + 0 + 0 + \mathbf{H}(c) \\
&= \mathbf{R}(a) \times \mathbf{P}(a) + \mathbf{H}(c) \qquad (5)
\end{aligned}$$

TORQUE

The *torque with respect to an arbitrary point a* that the force $\mathbf{f}(i)$ exerts on particle i is defined as

$$\mathbf{g}(ai) \equiv \mathbf{r}(ai) \times \mathbf{f}(i) \qquad (6)$$

The *net torque with respect to a point a* that is exerted on the dynamical system is designated $\mathbf{G}(a)$ and is given by

$$\mathbf{G}(a) \equiv \sum_i \mathbf{g}(ai) \equiv \sum_i \mathbf{r}(ai) \times \mathbf{f}(i) \equiv \sum_i \mathbf{r}(ai) \times \left[\mathbf{f}(ei) + \sum_j \mathbf{f}(ji)\right] \qquad (7)$$

If the point a, with respect to which the torque $\mathbf{G}(a)$ is measured, is not indicated, it is understood to be the origin o, that is,

$$\mathbf{G} \equiv \mathbf{G}(o) \qquad (8)$$

Theorem 2. The net torque with respect to a point a, that is, $\mathbf{G}(a)$, which is exerted on a dynamical system, is equal to the net external torque with respect to the point a, which is exerted on the system, that is,

$$G(a) \equiv \sum_i r(ai) \times f(i) = \sum_i r(ai) \times f(ei) \tag{9}$$

where $r(ai)$ is the position of particle i with respect to point a, $f(i)$ is the net force exerted on particle i, and $f(ei)$ is the net external force exerted on particle i.

PROOF: The torque $G(a)$ is given by

$$G(a) \equiv \sum_i r(ai) \times f(i) \equiv \sum_i r(ai) \times \left[f(ei) + \sum_j f(ji) \right] \tag{10}$$

From the law of action and reaction $f(ij) = -f(ji)$, hence

$$\sum_i r(ai) \times \left[\sum_j f(ji) \right] \equiv \sum_i \sum_j r(ai) \times f(ji)$$

$$\equiv \sum_j \sum_i r(aj) \times f(ij)$$

$$= \frac{1}{2} \sum_i \sum_j \left[r(aj) - r(ai) \right] \times f(ij) \tag{11}$$

But $r(aj) - r(ai) \equiv r(ij)$ is the vector position of particle j with respect to particle i, and from Postulate V $f(ij)$ is parallel to $r(ij)$. Hence

$$\sum_i \sum_j \left[r(aj) - r(ai) \right] \times f(ij) \equiv \sum_i \sum_j r(ij) \times f(ij) = 0 \tag{12}$$

Combining Eqs. (10)–(12) we obtain Eq. (9).

The relationship between the torque with respect to the point a and the torque with respect to the center of mass is contained in the following theorem.

Theorem 3. The net external torque with respect to an arbitrary point a that the external forces exert on a dynamical system is given by

$$G(a) = G(c) + R(a) \times F \tag{13}$$

where $G(c)$ is the net external torque acting on the system with respect to the center of mass, $R(a)$ is the position of the center of mass with respect to the point a, and F is the net external force acting on the system.

PROOF: From the definition of $G(a)$ we have

$$G(a) = \sum_i r(ai) \times f(ei) = \sum_i \left[r(ac) + r(ci) \right] \times f(ei)$$

$$= r(ac) \times \sum_i f(ei) + \sum_i r(ci) \times f(ei)$$

$$= R(a) \times F + G(c) \tag{14}$$

TORQUE AND ANGULAR MOMENTUM

The relationship between the torque acting on a dynamical system and the angular momentum of the system is contained in the following theorem.

Theorem 4. The time rate of change of **H**, the angular momentum of a dynamical system with respect to the orgin of an inertial frame of reference, is equal to **G**, the net external torque with respect to the same point, that is,

$$\mathbf{G} = \dot{\mathbf{H}} \tag{15}$$

PROOF: From the results of Chapter 14

$$\mathbf{g}(i) = \dot{\mathbf{h}}(i) \tag{16}$$

Summing over i we obtain

$$\sum_i \mathbf{g}(i) = \sum_i \dot{\mathbf{h}}(i) \tag{17}$$

Substituting Eqs. (2) and (7) in (17) we obtain Eq. (15).

The equation $\mathbf{G} = \dot{\mathbf{H}}$ is true if the point with respect to which **G** and **H** are measured is the origin of an inertial frame. The equation $\mathbf{G} = \dot{\mathbf{H}}$ is also true if the point with respect to which **G** and **H** are measured is the center of mass, as is proved in the following theorem.

Theorem 5. The time rate of change of $\mathbf{H}(c)$, the angular momentum of a dynamical system with respect to the center of mass, is equal to $\mathbf{G}(c)$, the net external torque with respect to the center of mass, that is,

$$\mathbf{G}(c) = \dot{\mathbf{H}}(c) \tag{18}$$

PROOF: From Theorem 4

$$\mathbf{G} = \dot{\mathbf{H}} \tag{19}$$

But from Theorems 3 and 1

$$\mathbf{G} \equiv \mathbf{G}(o) = \mathbf{R}(o) \times \mathbf{F} + \mathbf{G}(c) \equiv \mathbf{R} \times \mathbf{F} + \mathbf{G}(c) \tag{20}$$

$$\mathbf{H} \equiv \mathbf{H}(o) = \mathbf{R}(o) \times \mathbf{P}(o) + \mathbf{H}(c) \equiv \mathbf{R} \times \mathbf{P} + \mathbf{H}(c) \tag{21}$$

If we substitute Eqs. (20) and (21) in Eq. (19) we obtain

$$\mathbf{R} \times \mathbf{F} + \mathbf{G}(c) = \dot{\mathbf{R}} \times \mathbf{P} + \mathbf{R} \times \dot{\mathbf{P}} + \dot{\mathbf{H}}(c) \tag{22}$$

The first term on the right hand side of Eq. (22) can be rewritten

$$\dot{\mathbf{R}} \times \mathbf{P} = \dot{\mathbf{R}} \times M\mathbf{V} = \mathbf{V} \times M\mathbf{V} = 0 \tag{23}$$

The second term on the right hand side of Eq. (22) can be rewritten

$$\mathbf{R} \times \dot{\mathbf{P}} = \mathbf{R} \times \mathbf{F} \tag{24}$$

If we substitute Eqs. (23) and (24) into Eq. (22) we obtain Eq. (18), which is the desired result.

There are other points besides the origin and the center of mass for which $\mathbf{G}(a)$ is equal to $\dot{\mathbf{H}}(a)$. It is obviously true for any point that is moving with constant velocity with respect to the origin, for such a point could itself be chosen as the origin of an inertial frame. But it is also true for an accelerating point if the acceleration of the point is directed toward the center of mass. This can readily be proved by solving Eq. (4) for $\mathbf{H}(c)$ and Eq. (13) for $\mathbf{G}(c)$, substituting the results in Eq. (18), and noting that $\mathbf{F} = \dot{\mathbf{P}} = \dot{\mathbf{P}}(o)$ and $\dot{\mathbf{P}}(o) - \dot{\mathbf{P}}(a) = M\mathbf{A}(o) - M\mathbf{A}(a) = M\mathbf{a}(oa)$, where $\mathbf{a}(oa)$ is the acceleration of the point a with respect to the origin. However the reader is advised to forget this special case. It is never necessary to use it, and the convenience that might be gained by retaining this case in one's repetoire is outweighed by the possibility of error it encourages.

EXAMPLES

Example 1. A uniform heavy rod of mass m and length a is smoothly pivoted at one end to a fixed support O. The rod rests in equilibrium, with the other end in contact with a smooth vertical wall; the distance of the fixed support from the wall is b, where $b < a$. The rod is gently displaced from the equilbrium position. Determine the position at which the rod will break contact with the wall.

Solution. Our system is the rod. Let us choose a Cartesian frame with origin at O, z axis normal to the wall, and x axis vertically up (Fig. 1). The torque with respect to the point O due to the gravitational force $-mg\mathbf{e}_x$ and the reaction force $-R\mathbf{e}_z$ of the wall is

$$\mathbf{G}(O) = \left(\frac{a}{2}\mathbf{e}_r\right) \times (-mg\mathbf{e}_x) + (a\mathbf{e}_r) \times (-R\mathbf{e}_z)$$

$$= \left(\frac{a}{2}\mathbf{e}_r\right) \times (-mg\sin\theta\cos\phi\,\mathbf{e}_r - mg\cos\theta\cos\phi\,\mathbf{e}_\theta + mg\sin\phi\,\mathbf{e}_\phi)$$

$$+ (a\mathbf{e}_r) \times (-R\cos\theta\,\mathbf{e}_r + R\sin\theta\,\mathbf{e}_\theta)$$

$$= -\left(\frac{mga}{2}\right)\sin\phi\,\mathbf{e}_\theta + \left[-\left(\frac{mga}{2}\right)\cos\theta\cos\phi + aR\sin\theta\right]\mathbf{e}_\phi \qquad (25)$$

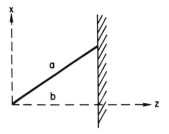

Fig. 1

The element dr of the rod, which is located at the point $r\mathbf{e}_r$, has a mass $(m/a)dr$ and is moving with a velocity $r\sin\theta\,\dot\phi\mathbf{e}_\phi$. It follows that the total angular momentum of the rod with respect to the point O is

$$\mathbf{H}(O) = \int_0^a (r\mathbf{e}_r) \times (r\sin\theta\,\dot\phi\mathbf{e}_\phi)\frac{m}{a}\,dr$$

$$= -\frac{ma^2\sin\theta\,\dot\phi}{3}\mathbf{e}_\theta \tag{26}$$

Taking the time derivative of $\mathbf{H}(O)$ and noting that $\dot{\mathbf{e}}_\theta = \cos\theta\,\dot\phi\mathbf{e}_\phi$ we obtain

$$\dot{\mathbf{H}}(O) = -\frac{ma^2\sin\theta\,\ddot\phi}{3}\mathbf{e}_\theta - \frac{ma^2\sin\theta\cos\theta\,\dot\phi^2}{3}\mathbf{e}_\phi \tag{27}$$

Making use of the equation $\mathbf{G}(O) = \dot{\mathbf{H}}(O)$ we obtain

$$\frac{mga}{2}\sin\phi = \frac{ma^2\sin\theta}{3}\ddot\phi \tag{28}$$

$$-\left(\frac{mga}{2}\right)\cos\theta\cos\phi + aR\sin\theta = -\frac{ma^2\sin\theta\cos\theta}{3}\dot\phi^2 \tag{29}$$

Solving Eq. (28) for $\dot\phi$ we obtain

$$\dot\phi^2 = \frac{3g}{a\sin\theta}(1 - \cos\phi) \tag{30}$$

The rod will break contact with the wall when $R = 0$. Substituting Eq. (30) in Eq. (29) and setting $R = 0$ we find that the value of ϕ at which the rod leaves the wall is given by ϕ_0 where

$$\cos\phi_0 = \tfrac{2}{3} \tag{31}$$

PROBLEMS

1. Show that if the net external force $\mathbf{F}$ acting on a system of particles is zero, the net external torque has the same value when taken about any point.

2. Explain how a man standing on a smooth sheet of ice can turn around by moving his arms.

3. Explain how a man standing on a swing can increase the amplitude of the oscillations by crouching and standing up at suitable times.

4. A system of particles moves in a plane. Prove that there exists at time t a straight line L such that the angular momentum about any point on L is zero. Furthermore, show that if no external forces act on the system, the line L is fixed for all values of t.

5. A system consists of three particles of mass 1, 2, and 3 kg, respectively. At time $t = 0$ the positions and velocities of the particles with respect to the origin of coordinates are $\mathbf{r}(1) = \mathbf{i} + \mathbf{j}$, $\mathbf{r}(2) = \mathbf{j} + \mathbf{k}$, $\mathbf{r}(3) = \mathbf{k}$, $\mathbf{v}(1) = 2\mathbf{i}$, $\mathbf{v}(2) = \mathbf{j}$, $\mathbf{v}(3) = \mathbf{i} + \mathbf{j} + \mathbf{k}$, respectively, where $\mathbf{i}$, $\mathbf{j}$, and $\mathbf{k}$ are unit vectors in the x, y, and z directions, respectively, and MKS units are used.

a. Find the linear momentum of the system with respect to the origin of coordinates.

b. Find the angular momentum with respect to the origin of coordinates.

c. Find the linear momentum at $t = 0$ with respect to a point a whose position as a function of time is given by $\mathbf{r}(a) = 4t^2\mathbf{i} + 5t\mathbf{j} + \mathbf{k}$.

d. Find the angular momentum at $t = 0$ with respect to the point a.

6. A homogeneous circular disk with a mass m and radius a rolls without slipping on a horizontal plane with an angular velocity of ω. The disk remains in a vertical plane. Determine the linear momentum and the angular momentum of the disk:

   a. With respect to a point that is on the horizontal surface and in the vertical plane.

   b. With respect to the center of the disk.

   c. With respect to a point on the circumference of the disk at the instant it is at its lowest point.

   d. With respect to the point of contact considered to be moving along with the disk.

7. A disk of mass M and radius R rotates about a perpendicular axis through its center with an angular velocity ω. Find the angular momentum of a ring in the disk between r and $r + dr$, then integrate to obtain the angular momentum of the whole disk.

8. Three particles of masses 2, 3, and 5 move under the influence of a force field so that their position vectors relative to a fixed coordinate system are given respectively by $\mathbf{r}(1) = 2t\mathbf{i} - 3\mathbf{j} + t^2\mathbf{k}$, $\mathbf{r}(2) = (t + 1)\mathbf{i} + 3t\mathbf{j} - 4\mathbf{k}$, and $\mathbf{r}(3) = t^2\mathbf{i} - t\mathbf{j} + (2t - 1)\mathbf{k}$, where t is the time, and $\mathbf{i}$, $\mathbf{j}$, and $\mathbf{k}$ are unit vectors in the x, y, and z directions, respectively. All units are MKS units. Find:

   a. The total angular momentum and the total external torque with respect to the origin.

   b. The total angular momentum and the total external torque about a point whose position is given by $\mathbf{r} = t\mathbf{i} - 2t\mathbf{j} + 3\mathbf{k}$.

9. A rod of mass m and length a is rotating with an angular speed ω about a vertical axis that passes through one end of the rod. The angle θ between the upward vertical and the rod is maintained constant. Choosing a frame with origin at the fixed end of the rod, and z axis vertically up, determine the angular momentum of the rod at the instant when the rod is in the x-z plane.

10. A uniform circular disk of radius a and mass m is free to turn about a horizontal axis through its center O and perpendicular to its plane. Over the rim of the disk hangs a light string that carries particles A and B of

masses m and $2m$, respectively, at its free ends. In addition to the torque exerted by the string on the disk there is a restoring torque of magnitude $mga\theta$ acting on the disk, where θ is the angular displacement of the disk. The system is released from rest at $\theta = 0$. Assuming that the string does not slip on the disk, find the acceleration of B when it has descended a distance x, and show that it describes a simple harmonic motion of period $\pi(14a/g)^{1/2}$. Find the ratio of the tensions in the two vertical parts of the string as a function of x.

11. A uniform circular disk of radius a is free to rotate in a vertical plane about a smooth horizontal axis through its center C. An insect A, whose mass is one-tenth that of the disk, is at the lowest point of the disk and the whole system is at rest. The insect suddenly starts to crawl along the rim with uniform speed V relative to the disk. Show that the initial spin of the disk is $V/6a$. If in the subsequent motion, θ is the inclination of CA to the downward vertical, prove that $6a\ddot{\theta} + g\sin\theta = 0$. Show that the component of the force exerted by the insect on the disk in the direction perpendicular to the radius is $5mg\sin\theta/6$, where m is the mass of the insect.

12. Two equal beads of mass m are free to slide on a frictionless and massless wire that is free to rotate in a horizontal plane about a vertical axis that passes through the wire at a point O. At time $t = 0$ the beads are at rest on opposite sides of O and each one is a distance a from O. A torque with respect to the point O is applied to the wire. The torque is in the vertical direction and of magnitude $b + ct$, where b and c are constants. Determine the angular speed and the angular acceleration of the wire as functions of time. Determine the distance of the beads from the point O as functions of the time.

13. A thin uniform rod of mass m and length a is free to rotate in a vertical plane about a smooth pin passing through one end of the rod. The rod is initially held at rest with its free end vertically up and released. Find the angular velocity of the rod, and the force the rod exerts on the pin as functions of the angle θ through which the rod has turned.

14. Show that the acceleration of a thin circular ring, rolling without sliding down a plane of inclination α to the horizontal, is $g\sin\alpha/2$ and that the least coefficient of friction necessary to prevent sliding is $\tan\alpha/2$.

15. A bar of length $2b$ is fitted at its middle point with a nut that moves without friction on a fixed vertical screw of pitch p; the bar remains horizontal and turns with the nut. Find the acceleration of the nut.

16. A uniform sphere rolls without slipping on a horizontal plane that is kept rotating with constant angular velocity Ω about a fixed vertical axis. Show that the center of the sphere describes a circle fixed in space, and that it makes a complete revolution in time $7\pi/\Omega$.

30

Work and Kinetic Energy

INTRODUCTION

The purpose of this chapter is to extend the concept of work and kinetic energy from particles to systems of particles.

KINETIC ENERGY

The *kinetic energy of a particle i with respect to an arbitrary point a* is defined as

$$T(ai) \equiv \tfrac{1}{2} m(i) \mathbf{v}(ai) \cdot \mathbf{v}(ai) \tag{1}$$

where $m(i)$ is the mass of particle i and $\mathbf{v}(ai)$ is the velocity of particle i with respect to the point a.

The *kinetic energy of a dynamical system with respect to an arbitrary point a* is defined as

$$T(a) \equiv \sum_i T(ai) \tag{2}$$

If the point a is not designated, it is understood to be the origin o, that is,

$$T \equiv T(o) \tag{3}$$

The relationship between the kinetic energy with respect to the point a and the kinetic energy with respect to the center of mass c is contained in the following theorem.

Theorem 1. The kinetic energy of a dynamical system with respect to an arbitrary point a is given by

$$T(a) = \tfrac{1}{2} M \mathbf{V}(a) \cdot \mathbf{V}(a) + T(c) \tag{4}$$

where $\mathbf{V}(a)$ is the velocity of the center of mass c with respect to the point a, and $T(c)$ is the kinetic energy with respect to the point c.

PROOF: From the definition of $T(a)$ we have

$$T(a) = \frac{1}{2} \sum_i m(i) \mathbf{v}(ai) \cdot \mathbf{v}(ai)$$

$$= \frac{1}{2} \sum_i m(i) [\mathbf{v}(ac) + \mathbf{v}(ci)] \cdot [\mathbf{v}(ac) + \mathbf{v}(ci)]$$

$$= \frac{1}{2} \left[\sum_i m(i) \right] \mathbf{v}(ac) \cdot \mathbf{v}(ac) + \left[\sum_i m(i) \mathbf{v}(ci) \right] \cdot \mathbf{v}(ac)$$

$$+ \frac{1}{2} \sum_i m(i) \mathbf{v}(ci) \cdot \mathbf{v}(ci)$$

$$= \frac{1}{2} M \mathbf{V}(a) \cdot \mathbf{V}(a) + T(c) \qquad (5)$$

WORK

If a particle i is acted on by a force $\mathbf{f}(i)$ and undergoes an infinitesimal change in its position, the work done on the particle by the force, with respect to an arbitrary point a, is defined as

$$dW(ai) = \mathbf{f}(i) \cdot d\mathbf{r}(ai) \qquad (6)$$

where $\mathbf{r}(ai)$ is the position of particle i with respect to the point a.

If a dynamical system is acted on by a set of forces $\mathbf{f}(i)$, where $\mathbf{f}(i)$ is the force on the ith particle, and if the dynamical system undergoes an infinitesimal change in its configuration, the work on the dynamical system by the set of forces $\mathbf{f}(i)$, with respect to an arbitrary point a, is defined as

$$dW(a) \equiv \sum_i dW(ai) \qquad (7)$$

If the point a is not indicated, it is understood to be the origin, that is,

$$dW \equiv dW(o) \qquad (8)$$

The relationship between the work done with respect to the point a and the work done with respect to the center of mass is contained in the following theorem.

Theorem 2. The work done by all the forces acting on a dynamical system, with respect to an arbitrary point a, when the system undergoes an infinitesimal change in its configuration, is given by

$$dW(a) = \mathbf{F} \cdot d\mathbf{R}(a) + dW(c) \qquad (9)$$

where $\mathbf{F}$ is the net external force acting on the system, $\mathbf{R}(a)$ is the position of the center of mass with respect to the point a, and $dW(c)$ is the work done by all the forces, with respect to the center of mass c.

PROOF: From the definition of $dW(a)$ we have

$$dW(a) = \sum_i \mathbf{f}(i) \cdot d\mathbf{r}(ai) = \sum_i \mathbf{f}(i) \cdot \left[d\mathbf{r}(ac) + d\mathbf{r}(ci) \right]$$

$$= \left[\sum_i \mathbf{f}(i) \right] \cdot d\mathbf{r}(ac) + \sum_i \mathbf{f}(i) \cdot d\mathbf{r}(ci)$$

$$= \mathbf{F} \cdot d\mathbf{R}(a) + dW(c) \tag{10}$$

WORK AND KINETIC ENERGY

The relationship between the work done on a dynamical system and the kinetic energy of the system is contained in the following theorem.

Theorem 3. If a dynamical system is acted on by a set of forces, then

$$dW = dT \tag{11}$$

where dW is the work done by the set of forces, with respect to the origin of an inertial frame of reference, and T is the kinetic energy of the system, with respect to the same point.

PROOF: From the results of Chapter 15 we have

$$dW(i) = dT(i) \tag{12}$$

Summing over i we obtain Eq. (11).

The equation $dW = dT$ is true if the point with respect to which dW and dT are measured is the origin of an inertial frame. The equation $dW = dT$ is also true if the point with respect to which dW and dT are measured is the center of mass, as is proved in the following theorem.

Theorem 4. If a dynamical system is acted on by a set of forces, then

$$dW(c) = dT(c) \tag{13}$$

where $dW(c)$ is the work done by the set of forces with respect to the center of mass, and $T(c)$ is the kinetic energy of the system with respect to the center of mass.

PROOF: From Theorem 3

$$dW = dT \tag{14}$$

But from Theorems 2 and 1 in this chapter

$$dW \equiv dW(o) = \mathbf{F} \cdot d\mathbf{R}(o) + dW(c) \equiv \mathbf{F} \cdot d\mathbf{R} + dW(c) \tag{15}$$

and

$$dT \equiv dT(o) = d\left[\tfrac{1}{2}M\mathbf{V}(o)\cdot\mathbf{V}(o) + T(c)\right]$$
$$\equiv d\left[\tfrac{1}{2}M\mathbf{V}\cdot\mathbf{V} + T(c)\right] = M\mathbf{A}\cdot\mathbf{V}\,dt + dT(c)$$
$$= \mathbf{F}\cdot d\mathbf{R} + dT(c) \tag{16}$$

If we substitute Eqs. (15) and (16) into Eq. (14) we obtain Eq. (13), which is the desired result.

EXAMPLES

Example 1. A uniform solid cylinder of mass M and radius a is free to rotate about its axis, which is fixed in a horizontal direction. One end of a uniform chain of mass m and length $b \equiv 2\pi an$, where n is an integer, is attached to the cylinder, and the chain is then wrapped around the cylinder, making n complete turns with a very small piece of chain hanging free. The system is released from rest and the chain unwraps itself. Find the angular velocity of the chain at the instant when the cylinder has undergone an angular displacement θ.

Solution. The system is shown in Fig. 1 at a time when the cylinder has undergone an angular displacement θ. The point P is the point on the cylinder that was originally at the top. The only work that has been done on the system has been done by the gravitational force acting on the chain. The mass of an element ds of the chain is $\rho\,ds$, where $\rho \equiv m/2\pi an$. If the loose end of the chain were rewrapped onto the cylinder, the element ds of the chain, which is located a distance s below the spot at which unwrapping occurs, would be located at a vertical distance $a\sin(s/a)$ below this level. The work required to move the element ds from the unwrapped state to the wrapped state is $(\rho\,ds)g[s - a\sin(s/a)] = (mg/2\pi an)[s - a\sin(s/a)]\,ds$. It follows that the

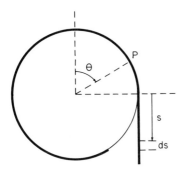

Fig. 1

work done on the chain is

$$\Delta W = \frac{mg}{2\pi a n} \int_0^{a\theta} \left(s - a\sin\frac{s}{a}\right) ds$$

$$= \frac{mga}{2\pi n}\left(\frac{\theta^2}{2} + \cos\theta - 1\right) \qquad (17)$$

When the cylinder is rotating with an angular velocity $\dot\theta$, each element of the chain is moving with a speed $a\dot\theta$, and thus the kinetic energy of the chain is

$$T(\text{chain}) = \frac{m(a\dot\theta)^2}{2} \qquad (18)$$

The segment of the cylinder lying in the region $r\,dr\,d\theta$ has a mass $(M/\pi a^2)\,r\,dr\,d\theta$, and when the cylinder is rotating with an angular velocity $\dot\theta$ is moving with a speed $r\dot\theta$, and thus the kinetic energy of the whole cylinder is

$$T(\text{cylinder}) = \frac{M}{\pi a^2}\int_0^{2\pi}\int_0^a \frac{(r\dot\theta)^2}{2}\,r\,dr\,d\theta$$

$$= \frac{Ma^2}{4}\dot\theta^2 \qquad (19)$$

From the work-energy theorem

$$\Delta W = T(\text{chain}) + T(\text{cylinder}) \qquad (20)$$

Substituting Eqs. (17)–(19) in Eq. (20) we obtain

$$\dot\theta^2 = \frac{mga(\theta^2 + 2\cos\theta - 2)}{\pi n(Ma^2 + 2ma^2)} \qquad (21)$$

PROBLEMS

1. A rod of length a and mass m is rotating in a horizontal plane with an angular speed ω about one end, which is fixed.
   a. What is the kinetic energy of the rod with respect to the fixed end?
   b. What is the kinetic energy of the rod with respect to the center of mass?

2. Two equal uniform rods AB and BC, each of mass m and length $2a$, are hinged freely at B and move freely on a smooth horizontal table with the center of mass of the system at rest. If 2θ is the angle between the rods and ϕ is the angle made by AC with a fixed direction in the plane, show that the kinetic energy of the system with respect to the center of mass is $T = (ma^2/3)[(1 + 3\cos^2\theta)\dot\theta^2 + (1 + 3\sin^2\theta)\dot\phi^2]$.

3. Show that the kinetic energy relative to the center of mass of a system of particles may be written in the form

$$\frac{1}{2} \sum_{i>j} \sum \frac{m(i)m(j)v^2(ij)}{M}$$

where $m(i)$ is the mass of particle i, $v(ij)$ is the speed of particle j with respect to particle i, and M is the total mass of the system.

4. Two particles each of mass m are connected by a rigid massless rod of length a. The system is initially at rest on a smooth horizontal surface. At time $t = 0$ a force **F** begins to act on one of the particles. The magnitude of the force remains constant, and the direction remains normal to the rod at all times. Determine by using the work-energy theorem the angular velocity of the rod at a time when the rod makes an angle θ with its original direction.

5. A block of mass m and a block of mass $2m$ are connected by an elastic string of unstretched length a and spring constant k. The system is on a rough horizontal table. The coefficient of kinetic friction for the contact between the blocks and the table is $\bar{\mu}$. The two blocks are initially held at rest a distance b apart ($b > a$) and then released. The blocks move toward each other and collide. What is the relative speed of the blocks just before the collision?

6. A uniform chain of length $2a$ hangs in equilibrium over a smooth peg. If it is started from rest, prove that its velocity when it is leaving the peg is $(ga)^{1/2}$.

7. A uniform rope of length $3a$ and mass $3m$ is hung on a rough circular peg in such a way that the length hanging on one side of the peg is a and the length hanging on the other side of the peg is $2a$. The rope immediately starts to slip off. If the coefficient of kinetic friction for the contact between the rope and the peg is $\bar{\mu}$, and the rope remains in contact with the peg until the end has slipped off, what will the speed of the rope be just as it leaves the peg?

31

Cartesian Configuration Space

INTRODUCTION

It is possible to treat the motion of a dynamical system in a manner very analogous to the motion of a single particle, which allows one easily to extend many of the results of single particle motion to the motion of a dynamical system. The purpose of this chapter is to demonstrate this possibility.

NEWTON'S LAW IN COMPONENT FORM

The motion of a dynamical system containing N particles can be determined in principle by the set of equations

$$\mathbf{f}(k) = m(k)\ddot{\mathbf{r}}(k) \qquad k = 1, 2, \ldots, N \tag{1}$$

where $\mathbf{f}(k)$ is the force on particle k, $m(k)$ is the mass of particle k, and $\mathbf{r}(k)$ is the position vector of particle k.

If we express Eq. (1) in terms of its components with respect to a Cartesian frame of reference, we obtain

$$f_j(k) = m(k)\ddot{x}_j(k) \qquad j = 1, 2, 3 \qquad k = 1, \ldots, N \tag{2}$$

Equation (2) can be formally simplified if we renumber the terms in such a way that a single number i between 1 and $3N$, rather than the two numbers j and k, suffices to indicate both the particle number k and the component number j. Explicitly if we let $i = 3k - 3 + j$, then a single value of i will correspond to a particular pair of values of j and k. For example if $i = 7$, then $k = 3$ and $j = 1$. Using the numbering scheme above, we define

$$f_1 \equiv f_1(1), f_2 \equiv f_2(1), \ldots, f_{3k-3+j} \equiv f_j(k), \ldots$$

$$x_1 \equiv x_1(1), x_2 \equiv x_2(1), \ldots, x_{3k-3+j} \equiv x_j(k), \ldots$$

$$m_1 \equiv m_2 \equiv m_3 \equiv m(1), \ldots, m_{3i-2} \equiv m_{3i-1} \equiv m_{3i} \equiv m(i), \ldots$$

With this notation Eq. (2) can be rewritten

$$f_i = m_i \ddot{x}_i \qquad i = 1, \ldots, 3N \tag{3}$$

In this form the equations of motion of a dynamical system are very analogous to the equations of motion of a point particle. We consider this analogy in greater detail in the sections that follow.

CARTESIAN CONFIGURATION SPACE

The position of a particle can be represented by a single point **x** in a three dimensional Cartesian space having coordinates x_1, x_2, and x_3. The geometrical state of a dynamical system containing N particles, which we refer to as the *configuration* of the system, can be represented by a set of N points **x**(1), **x**(2), ..., **x**(N), in the same space. Alternatively the configuration of the system can be represented by a single point **x** in a $3N$ dimensional space, the coordinates of which are the quantities $x_1, x_2, \ldots, x_{3N}$ defined in the preceding section. We refer to the latter space as a *Cartesian configuration space*.

THE EQUATIONS OF MOTION IN CONFIGURATION SPACE

The motion of a particle can be geometrically represented by the motion of a point **x** in the space (x_1, x_2, x_3). The set of equations governing the motion of the point is the set $f_i = m\ddot{x}_i$, where i ranges from 1 to 3.

The motion of a dynamical system containing N particles can be geometrically represented by the motion of a point **x** in the configuration space $(x_1, x_2, \ldots, x_{3N})$. The set of equations governing the motion of the point is the set $f_i = m_i \ddot{x}_i$, where i ranges from 1 to $3N$.

The only formal difference between the two sets of equations $f_i = m\ddot{x}_i$ and $f_i = m_i \ddot{x}_i$ other than their dimensionality is the fact that there are different masses m_i associated with different directions in the $3N$ dimensional configuration space. Thus the motion of a dynamical system when viewed in configuration space is, except for the number of dimensions involved, similar to the motion of a particle that has different masses for different directions of motion.

When we are using the concept of configuration space to analyze the motion of a dynamical system it is very convenient to carry over much of the terminology that is used in discussing the motion of a single particle. Consider a dynamical system consisting of N particles of masses $m(1), m(2), \ldots, m(N)$, respectively. The configuration of the system is determined by the positions **x**(1), **x**(2), ..., **x**(N) of the particles; the force acting on the system is determined by the forces **f**(1), **f**(2), ..., **f**(N) acting on the particles; and the motion of the system is determined by the velocities $\dot{\mathbf{x}}(1), \dot{\mathbf{x}}(2), \ldots, \dot{\mathbf{x}}(N)$ of the particles. From the configuration space point of view we simply say: the system is in the configuration $\mathbf{x} \equiv (x_1, x_2, \ldots, x_{3N})$; the force acting on the system is $\mathbf{f} \equiv (f_1, f_2, \ldots, f_{3N})$; and the velocity of the system is $\dot{\mathbf{x}} \equiv (\dot{x}_1, \dot{x}_2, \ldots, \dot{x}_{3N})$. In a similar vein we speak of the quantities $m_1, m_2, \ldots, m_{3N}$ as the components of the mass of the system.

CONCLUSION

As the following chapters indicate, the ideas introduced in this chapter can be used in many situations to simplify considerably the analysis of the motion of a dynamical system. There are however other situations in which we wish to emphasize, rather than play down, the individuality of the particles in a dynamical system. In these cases the analysis introduced in the earlier chapters in this section is more useful. We shall use whichever approach is most convenient for our purposes.

PROBLEMS

1. Two particles 1 and 2 of masses m and M, respectively, are connected by a rigid massless rod of length a. Initially particle 1 is at rest at the point $x = b$, $y = b$, $z = 0$ and particle 2 is at rest at the point $x = b$, $y = b$, $z = a$. Particle 1 is acted on by an external force $-kx\mathbf{e}_x - ky\mathbf{e}_y$ and particle 2 by an external force $-Mg\mathbf{e}_z$. Write down the equations of motion for the system in Cartesian configuration space, including any constraint equations. State the boundary conditions.

2. Two particles of masses m and M are constrained to remain on the inner surface of a smooth cylinder the equation of which is $x^2 + y^2 = a^2$. Gravity is acting in the negative z direction. The two particles are joined by a straight spring of spring constant k and unstretched length a. Write down the equations of motion for the system in Cartesian configuration space, including any constraint equations.

32

Potential Energy

INTRODUCTION

The purpose of this chapter is to extend the concept of potential energy from particles to systems of particles. If we use the concept of configuration space introduced in the preceding chapter, all the theorems and proofs introduced in Chapter 16, which considered the potential energy of a particle, can be immediataly generalized to dynamical systems. We therefore in this chapter simply state some theorems without proof.

IRROTATIONAL FORCE FIELDS

If the force **f** that a dynamical system would experience when its representative point lies in a region R of configuration space is a function of the configuration **x** of the system and possibly the time t, then we say that there exists a *force field* $\mathbf{f}(\mathbf{x}, t)$ in the region R of configuration space.

Definition 1. A force field $\mathbf{f}(\mathbf{x}, t)$, defined in a region R, is an *irrotational force field* if and only if

$$\frac{\partial f_i}{\partial x_j} = \frac{\partial f_j}{\partial x_i} \tag{1}$$

everywhere in the region R.

Theorem 1. A force field $\mathbf{f}(\mathbf{x}, t)$ defined in a region R is an irrotational force field if and only if there exists a function $U(\mathbf{x}, t)$, called a potential function, such that

$$f_i = -\frac{\partial U(\mathbf{x}, t)}{\partial x_i} \tag{2}$$

everywhere in the region R.

CONSERVATIVE FORCE FIELDS

Definition 2. A force field **f(x)** defined in a region R is said to be a *conservative force field* if and only if it is time independent and is such that when the system on which the force acts moves from a configuration **x = a** to a configuration **x = b**, the work done by the force does not depend on the path the system follows but only on the initial configuration **a** and the final configuration **b**.

Theorem 2. A force field **f** defined in a region R is a conservative force field if and only if it satisfies each of the following conditions:
  a. It is time independent.
  b. It is irrotational, that is, either $\partial f_i/\partial x_j = \partial f_j/\partial x_i$ everywhere in R, or equivalently there exists a potential function $U(\mathbf{x})$ such that $f_i = -\partial U(\mathbf{x})/\partial x_i$ everywhere in R.
  c. The potential $U(\mathbf{x})$ is single valued, or the region R is simply connected.

THE POTENTIAL ENERGY

Definition 3. If a dynamical system is in a conservative force field **f(x)**, the potential energy of the system in the configuration **x** relative to a reference configuration **a** is defined as the work done by the force when the system moves from the configuration **x** to the configuration **a**, and is designated as $V(\mathbf{x})$.

Theorem 3. If a dynamical system is in a conservative force field, the potential energy $V(\mathbf{x})$ with respect to the reference configuration **a** is identical with the particular potential $U(\mathbf{x})$, which is obtained by choosing the arbitrary additive constant in the general potential function so as to make $U(\mathbf{a})$ vanish.

Theorem 4. If a dynamical system is in a conservative force field, the work done by the force when the system moves from one configuration to another is equal to the negative of the change in the potential energy, that is,

$$dW = -dV \qquad (3)$$

EXAMPLES

Example. Two ladders AB and AC each of length $2a$ and mass M are freely hinged at A. The midpoints of the two ladders are joined by a massless elastic rope of spring constant k and unstretched length a. The bottom ends B and C of the ladders are constrained to remain on a smooth horizontal surface. Determine the potential energy of the system consisting of the two ladders as a function of the angle θ between the two ladders with respect to the reference configuration $\theta = \pi/3$.

Solution. As the system moves from one configuration to another, the only forces acting on the system that do work that is not negated by the work done by some other force are the gravitational force and the force exerted by the rope on the ladder. Both of these forces are conservative. The force exerted by the floor on the ladders does no work because the floor is smooth, hence the force is always perpendicular to the the displacement. The internal forces that maintain the rigidity of the ladders do no net work, because the work done by one force is always canceled out by the work done by the reaction force. Similarly the net work done at a smooth joint can be shown to be zero. The work done by the gravitational forces as the system moves from an arbitrary configuration θ to the reference configuration $\theta = \pi/3$ is $2Mg[a\cos(\theta/2) - a\cos(\pi/6)]$. The work done by the rope as the system moves from an arbitrary configuration θ to the reference configuration $\theta = \pi/3$ is $(k/2)[2a\sin(\theta/2) - a]^2$. Hence the potential energy of the system relative to the configuration $\theta = \pi/3$ is $V = Mga[2\cos(\theta/2) - \sqrt{3}] + (ka^2/2)[2\sin(\theta/2) - 1]^2$.

PROBLEMS

1. Show that if a rigid body of mass M is moved from one configuration to another in a uniform gravitational field, the change in the gravitational potential energy is $Mg\Delta h$, where Δh is the change in the vertical position of the center of mass.

2. A long uniform flexible chain of mass m_c and length l hangs over a pulley consisting of an axially symmetric disk of mass m_d and radius a, which can turn freely about a horizontal axis through its center. A particle of mass m_p is attached to the rim of the disk. The particle is at the lowest point of the disk when the lengths of the chain hanging on either side are equal. The angular displacement θ of the disk and the potential energy of the disk are both measured from this position.
   a. Determine the potential energy of the system as a function of θ.
   b. Find the values of θ for which the potential energy is a minimum or a maximum.

3. A uniform plank of length b and thickness d, which is initially resting in equilibrium in a horizontal position on top of a fixed rough cylinder of radius a, is slightly disturbed from its equilibrium position. Determine the potential energy of the plank with respect to its equilibrium configuration as a function of the angle that is made with the vertical by the plane containing the axis of the cylinder and the line of contact of the plank and the cylinder.

4. Three smooth circular cylinders, each of radius a and mass m, are held together along generators by a light elastic band of natural length $2\pi a$ and modulus of elasticity $n\pi mg$. Two of the cylinders rest on a horizontal table with the third above them. Determine the potential energy of the system as a function of the angle θ that the plane containing the axis of the upper cylinder and the axis of one of the lower cylinders makes with the horizontal. Assume that the potential energy is zero when $\theta = 60°$.

5. A uniform rod BC of length $2a$ and mass m is fastened at B to a light elastic string AB, which is fastened at A to a fixed point. The unstretched length of the string is a and its modulus of elasticity is nmg. The motion of the system is restricted to a vertical plane. Determine the potential energy of the system as a function of θ, the angle that the string AB makes with the vertical, ϕ, the angle that the rod BC makes with the vertical, and r, the length of the string. Take as reference configuration first the equilibrium configuration and then the configuration $\theta = 0$, $\phi = 0$, $r = a$.

33

Constants of the Motion

INTRODUCTION

The purpose of this chapter is to extend the discussion of constants of the motion introduced in Chapter 17 from particles to dynamical systems. If we use the concept of configuration space introduced in Chapter 31, the generalization is straightforward. We therefore simply state theorems without proof whenever the proof is an obvious generalization of our earlier results.

CONSTANTS OF THE MOTION

The dynamical state of a system of particles is completely determined at a given time t if we know its configuration $\mathbf{x}$ and its velocity $\dot{\mathbf{x}}$.

Any function $\phi(\mathbf{x}, \dot{\mathbf{x}}, t)$ whose value remains constant during the motion of a system is called a *constant of the motion*.

Theorem 1. If a dynamical system consisting of N particles is acted on by a force $\mathbf{f}$, which is a function of $\mathbf{x}$, $\dot{\mathbf{x}}$, and t, there will be exactly $6N$ independent constants of the motion.

Theorem 2. If a dynamical system consisting of N particles is acted on by a force that is a function of $\mathbf{x}$, and $\dot{\mathbf{x}}$ but not of t, then there are $6N - 1$ independent constants of the motion that are not explicit functions of the time.

THE AIM OF DYNAMICS

If we know the dynamical state of a system at one instant of time and the force $\mathbf{f}$ acting on the system, we can determine its state at a later instant of

248 Constants of the Motion

time from the equations of motion $f_i = m_i \ddot{x}_i$. The complete solution to this set of equations consists of $3N$ equations of the form $x_i = x_i(\mathbf{c}, t)$, where $\mathbf{c} \equiv c_1, c_2, \ldots, c_{6N}$ is a set of $6N$ independent arbitrary constants. The determination of the motion of the system therefore consists of finding the set of functions $x_i(\mathbf{c}, t)$ that is the solution to the set of equations $f_i = m_i \ddot{x}_i$. If we can determine $6N$ independent constants of the motion $\phi_i(\mathbf{x}, \dot{\mathbf{x}}, t)$ then the $6N$ equations $\phi_i(\mathbf{x}, \dot{\mathbf{x}}, t) = c_i$ can be used to find the $3N$ functions $x_i(\mathbf{c}, t)$. The determination of the motion of the system can therefore be said to consist of the determination of a complete set of independent constants of the motion. Each independent constant of the motion we find is thus one step toward the complete determination of the dynamical behavior of the system.

In the remainder of this chapter we enumerate some constants of the motion that can be immediately identified.

CONSERVATION OF LINEAR MOMENTUM

Theorem 3a. If the ith component of the net external force $\mathbf{F}$ acting on a dynamical system is zero, then the ith component of the momentum $\mathbf{P}$ is a constant of the motion.

PROOF: The proof follows immediately from Theorem 3 in Chapter 28.

Theorem 3b. If the ith component of the net external force $\mathbf{F}$ acting on a dynamical system is zero, then the ith component of the velocity $\mathbf{V}$ of the center of mass is a constant of the motion.

PROOF: The proof follows immediately from Theorem 4 in Chapter 28.

CONSERVATION OF ANGULAR MOMENTUM

Theorem 4a. Let o be a point that is at rest or moving with a constant velocity with respect to an inertial frame of reference. If the ith component of the net external torque $\mathbf{G}(o)$ acting on a dynamical system is zero, then the ith component of the total angular momentum $\mathbf{H}(o)$ is a constant of the motion.

PROOF: The proof follows immediately from Theorem 4 in Chapter 29.

Theorem 4b. Let c be the center of mass of a dynamical system. If the ith component of the net external torque $\mathbf{G}(c)$ acting on the system is zero, then the ith component of the total angular momentum $\mathbf{H}(c)$ is a constant of the motion.

PROOF: The proof follows immediately from Theorem 5 in Chapter 29.

CONSERVATION OF ENERGY

Theorem 5. If the force **f** acting on a dynamical system is a conservative force, then the sum of the kinetic energy T and the potential energy V is a constant of the motion.

PROOF: The proof is the same as the proof of Theorem 5 in Chapter 17.

PROBLEMS

1. A particle of mass $3m$ explodes into three equal pieces. Two of the pieces fly off at right angles to each other, one with a speed of $2v$ and the other with a speed $3v$. What is the magnitude and direction of the momentum of the third fragment?

2. Two particles of masses m and M are connected by a light elastic string of natural length a and modulus of elasticity λ, and are placed on a smooth horizontal table with the string at its natural length. The particle of mass M is projected with speed V along the table at right angles to the string. Show that the greatest length of the string during the subsequent motion is the value of r given by the equation $r^2(r - a) = [mMV^2a(r + a)]/[\lambda(m + M)]$.

3. Two equal particles A and B lie on a smooth horizontal table joined by a taut inextensible string. A third equal particle C moving on the table at right angles to the string hits it at its midpoint. Prove that in the subsequent motion C does not leave the string before A and B collide. Show also that at the time of this collision the speed of C is only one-third of its original speed.

4. Two massless rods AB and AC, each of length a, lie on a smooth horizontal plane. The end A of each of the rods is freely pinned to a fixed point A on the plane. At the end B of rod AB and at the end C of rod AC there is a particle of mass m. The ends B and C are connected by a spring of spring constant k and unstretched length $\sqrt{2}\, a$. Initially the rods are at rest and at right angles to one another, when a third particle also of mass m, moving on the plane with speed v in a direction perpendicular to rod AB and on the side away from the rod AC, strikes the mass at B and sticks. Obtain the equations of motion for the system in terms of the angle ϕ between the two rods, and the angle θ that the rod AB makes with its original position.

5. A bead A, of mass m, slides on a smooth horizontal rail, and a particle B, also of mass m, is attached to the bead by a light inelastic string of length $2a$. The system is released with the string taut and in the vertical plane containing the rail, and with AB making an acute angle α with the

downward vertical. Prove that when the inclination of the string to the vertical is θ then $\dot{\theta}^2/2 = (g/a)[(\cos\theta - \cos\alpha)]/(2 - \cos^2\theta)]$.

6. A light inextensible string of length $2a$ has equal particles, each of mass m, attached to its ends, and a third particle of mass M attached to its midpoint. The particles lie in a straight line on a smooth horizontal table with the string just taut, and M is projected along the table with velocity V perpendicular to the string. Show that if the two particles at the ends collide after a time τ when the displacement of M from its initial position is b, then $(M + 2m)b = MV\tau + 2ma$. Show also that the tension in the string just before the collision is $mM^2V^2/[(M + 2m)^2a]$.

7. A light rod of length $2a$ is free to rotate in a horizontal plane about its midpoint O, which is fixed. Two small smooth equal rings P and Q are free to slide on the rod, and are initially equidistant from O but on opposite sides of it. The rod is given an angular velocity Ω when P and Q are distant $3a/5$ from O and at rest relative to the rod. If the system is now left to itself, show that at the instant when the rings leave the rod, the rings are moving with speed $3a\Omega/5$ and the angular velocity of the rod is $9\Omega/25$.

8. Particles 1 and 2, each of mass m, are connected by an inextensible massless string of length $4a$. Initially each particle is sliding with velocity v along parallel paths separated by a distance $4a$ on a smooth horizontal floor. Suddenly a point on the string at a distance a from particle 1 strikes a fixed vertical nail of negligible diameter that projects from the floor. Assuming that the particles proceed to whirl in opposite directions without colliding, and that the string can slide freely on the nail, find for the ensuing motion:
   a. The maximum distance of particle 1 from the nail.
   b. The minimum tension in the string.

9. A uniform cylinder of mass M and radius R is free to rotate about its symmetry axis. Two small masses A' and A'', each of mass m, are connected by strings, each of length a, to hooks B' and B'' on opposite sides of the cylinder; the strings are wrapped in a clockwise direction around the cylinder, and the masses are connected directly to the cylinder by means of hooks C' and C''. The system is set in rotation with an arbitrary angular speed in a counterclockwise sense. As the system is spinning the hooks C' and C'' suddenly release the particles A' and A'', and the strings $A'B'$ and $A''B''$ begin to unwind. At the instant when the strings are completely unwound the hooks B' and B'' release the string and the particles fly off. Show that it is possible to choose the length a of the string such that the cylinder will be completely stopped regardless of its initial angular velocity, and find the length a. Such a mechanism could be used for despinning a satellite vehicle.

SECTION 2

Rigid Body Motion

34

Rigid Bodies

INTRODUCTION

One of the simplest and most important types of dynamical system is a rigid body. A *rigid body* can be defined as a system of particles bound together by internal forces that maintain the mutual distances between the particles fixed.

THE CONFIGURATION OF A RIGID BODY

If there are no constraints on the positions of the particles in a dynamical system, then three coordinates are required to specify the position of each particle; hence for a system containing N particles, $3N$ coordinates are required to specify the configuration of the system. In a rigid body the interparticle distances are fixed and are assumed to be known. It follows that the number of coordinates required to specify the configuration of a rigid body will be less than $3N$.

If a rigid body has one point fixed, and the particles in the rigid body are not all collinear, three coordinates are required to specify the configuration of the body. To see this note that it takes two coordinates to specify the direction of a line in a rigid body passing through the fixed point, and one additional coordinate to specify the orientation of the body with respect to this line.

If a rigid body has no point fixed, and the particles are not collinear, six coordinates are required to specify the configuration of the body: three to specify the position of a point in the body, and three to specify the orientation of the body.

In the special case of a rigid body in which the particles are collinear, one less coordinate is required to specify the configuration than is required for a rigid body in which the particles are not all collinear. Thus if one point is fixed, only two coordinates are required, and if no point is fixed, five coordinates are required. In discussing the properties of a rigid body we always assume unless stated otherwise that we are dealing with a rigid body in

Rigid Bodies

which the particles are not collinear. The results we obtain can easily be adapted to the special case of a rigid body in which the particles are collinear.

THE EQUATIONS OF MOTION FOR A RIGID BODY WITH ONE POINT FIXED

If a rigid body is free to rotate about a fixed point o, three coordinates are required to specify the orientation of the rigid body. It follows that if such a body is acted on by a known set of forces, three independent equations of motion are required to determine the motion. A sufficient set of equations is provided by the vector equation

$$\mathbf{G}(o) = \dot{\mathbf{H}}(o) \tag{1}$$

where $\mathbf{G}(o)$ is the net external torque with respect to the point o, $\mathbf{H}(o)$ is the angular momentum with respect to the point o, and the dot represents the time rate of change as noted by an observer fixed in an inertial frame.

THE EQUATIONS OF MOTION FOR A RIGID BODY WITH NO POINT FIXED

If no point in a rigid body is fixed, six coordinates are required to specify the configuration of the system, hence six independent equations of motion are required to determine the behavior of the system. In this case the motion can be broken down into the motion of the center of mass, and the motion with respect to the center of mass. The motion of the center of mass is governed by the equation

$$\mathbf{F} = M\ddot{\mathbf{R}} \tag{2}$$

where $\mathbf{F}$ is the net external force, M is the total mass, and $\mathbf{R}$ is the vector position of the center of mass with respect to a point fixed in an inertial frame. The motion with respect to the center of mass is governed by the equation

$$\mathbf{G}(c) = \dot{\mathbf{H}}(c) \tag{3}$$

where $\mathbf{G}(c)$ is the torque with respect to the center of mass, $\mathbf{H}(c)$ is the angular momentum with respect to the center of mass, and the dot represents the time rate of change as noted by an observer fixed in an inertial frame.

CONCLUSION

In the chapters that follow we consider in great detail the application of the preceding sets of equations. It is necessary however to do a considerable amount of preliminary work before we are in a position to fully exploit these equations.

35

Equivalent Systems of Forces

INTRODUCTION

The motion of a rigid body with no point fixed is determined by the equations

$$\mathbf{F} = M\ddot{\mathbf{R}} \tag{1}$$

$$\mathbf{G}(c) = \dot{\mathbf{H}}(c) \tag{2}$$

where $\mathbf{F}$ is the net external force, M the total mass, $\mathbf{R}$ the position of the center of mass in an inertial frame, $\mathbf{G}(c)$ the net external torque with respect to the center of mass, and $\mathbf{H}(c)$ the net angular momentum with respect to the center of mass. It follows that two systems of forces for which $\mathbf{F}$ and $\mathbf{G}(c)$ are the same will produce identical motions when they act on a given rigid body; hence for rigid body motion they can be considered to be equivalent. It should be carefully noted however that they are not absolutely equivalent but simply produce equivalent motions of a rigid body.

Consider for example two forces of equal magnitude that act in opposite directions and at different points along the same line. As far as rigid body motion is concerned this pair of forces can be shown to be equivalent to a zero force, since their sum is zero, and the net torque they exert with respect to any point is also zero. However if one of the forces is applied to one side of an object and the other to the opposite side of the object, the object will not experience the same effect as no force, even if the object remains approximately rigid; if the object does not remain rigid, the difference between the effect of the two forces and the effect of no force will be even more marked.

In this chapter we consider systems of forces that are equivalent for rigid body motion.

THE POINT OF APPLICATION OF A FORCE

The motion of the center of mass of an extended body acted on by a force $\mathbf{f}$ does not depend on the point of application of the force, but the motion with respect to the center of mass does depend on the point of application of the force. Hence in describing a force acting on an extended body it is necessary to specify not only the magnitude and direction of the force but also its point of application. We therefore in this chapter use the notation $\mathbf{f}(p)$ to designate a force $\mathbf{f}$ acting at a point p.

The torque with respect to a point o due to a force $\mathbf{f}(p)$ is given by $\mathbf{r}(op) \times \mathbf{f}(p)$, where $\mathbf{r}(op)$ is the vector position of point p with respect to point o. Since the effect of the point of application of the force $\mathbf{f}(p)$ on the value of the cross product $\mathbf{r}(op) \times \mathbf{f}(p)$ is included in the vector $\mathbf{r}(op)$, we write $\mathbf{r}(op) \times \mathbf{f}(p)$ as simply $\mathbf{r}(op) \times \mathbf{f}$.

THE LINE OF ACTION OF A FORCE

The line of action of a force is the line generated by moving the point of application of the force along the direction of the force.

Theorem 1. Two forces that have the same magnitude, direction, and line of action are equivalent for rigid body motion.

PROOF: Let $\mathbf{f}(p)$ be a force applied at the point p and $\mathbf{f}(q)$ a force equal in magnitude and direction but applied at a point q on the line of action of $\mathbf{f}(p)$. The torque with respect to a point o due to the force $\mathbf{f}(q)$ is $\mathbf{r}(oq) \times \mathbf{f}$ and the torque with respect to o due to the force $\mathbf{f}(p)$ is $\mathbf{r}(op) \times \mathbf{f}$. But $\mathbf{r}(oq) = \mathbf{r}(op) + \mathbf{r}(pq)$ and $\mathbf{r}(pq)$ is parallel to $\mathbf{f}$, and therefore

$$\mathbf{r}(oq) \times \mathbf{f} = [\mathbf{r}(op) + \mathbf{r}(pq)] \times \mathbf{f} = \mathbf{r}(op) \times \mathbf{f} \tag{3}$$

Since the two forces are equal and also produce the same torque with respect to an arbitrary point, they are necessarily equivalent for rigid body motion.

From the preceding theorem it follows that we can move a force along its line of action without altering its effect on the motion of a rigid body.

COUPLES

A couple is a pair of forces $\mathbf{f}(p)$ and $\bar{\mathbf{f}}(q)$, applied at different points, which are equal in magnitude and opposite in direction, that is, $\bar{\mathbf{f}} = -\mathbf{f}$. We designate a particular couple by the notation $\{\mathbf{f}(p), \bar{\mathbf{f}}(q)\}$. The plane determined by the lines of action of $\mathbf{f}(p)$ and $\bar{\mathbf{f}}(q)$ is called the *plane of action* of the couple.

Theorem 2. The torque $\mathbf{g}$ with respect to any point produced by a couple $\{\mathbf{f}(p), \bar{\mathbf{f}}(q)\}$ is given by

$$\mathbf{g} = \mathbf{r}(qp) \times \mathbf{f} \equiv \mathbf{r}(pq) \times \bar{\mathbf{f}} \tag{4}$$

where $\mathbf{r}(pq)$ is the vector position of point q with respect to point p. It follows that the magnitude of the torque produced by a couple is equal to the product of the magnitude of either of the forces and the distance between the lines of action of these two forces, and the direction of the torque is perpendicular to the plane of action of the couple, and its sense is in the direction in which a right handed screw would progress if acted on by the couple.

PROOF: The torque with respect to an arbitrary point o due to the two forces $\mathbf{f}(p)$ and $\bar{\mathbf{f}}(q)$ is given by

$$\begin{aligned}\mathbf{g}(o) &= \mathbf{r}(op) \times \mathbf{f} + \mathbf{r}(oq) \times \bar{\mathbf{f}} \\ &= \mathbf{r}(op) \times \mathbf{f} - \mathbf{r}(oq) \times \mathbf{f} \\ &= [\mathbf{r}(qo) + \mathbf{r}(op)] \times \mathbf{f} = \mathbf{r}(qp) \times \mathbf{f}\end{aligned} \qquad (5)$$

This proves the first part of the theorem. The remainder of the theorem follows directly from a geometrical analysis of the quantity $\mathbf{r}(qp) \times \mathbf{f}$ and is left to the reader.

Since the effect that a couple exerts on a rigid body depends only on the value of the torque produced by the couple, we can immediately write down the following theorem.

Theorem 3. The effect that a couple $\{\mathbf{f}(p), \bar{\mathbf{f}}(q)\}$ produces on the motion of a rigid body has the following properties:

  a. It does not depend on the choice of the point with respect to which the torque of the couple is measured.
  b. It is unaffected if the point of application of either force $\mathbf{f}$ or $\bar{\mathbf{f}}$ is translated an arbitrary amount along its line of action.
  c. It is unaffected if we translate the couple in its plane of action.
  d. It is unaffected if we rotate the couple in its plane of action.
  e. It is unaffected if we translate the plane of action of the couple an arbitrary amount parallel to itself.
  f. It is unaffected if we simultaneously change the magnitudes of $\mathbf{f}$, $\bar{\mathbf{f}}$, and $\mathbf{r}(pq)$ as long as we keep $\bar{\mathbf{f}} = -\mathbf{f}$ and the cross product of $\mathbf{r}(pq)$ and $\mathbf{f}(p)$ fixed.

PROOF: The effect that a couple produces on the motion of a rigid body depends only on the torque that it exerts on the rigid body. From the results of the preceding theorem one can show that none of the operations above alters the value of the torque.

Since two couples that produce the same torque are equivalent as far as rigid body motion is concerned, the nature of a couple is completely specified for rigid body motion by the value of the torque. We therefore frequently speak of a couple that produces a torque **g** as simply the couple **g**.

EQUIVALENT SYSTEMS OF FORCES

If a rigid body is acted on by a given system of forces, this system of forces can frequently be replaced by a simpler system of forces that will produce the same motion as the given system of forces. In this section we consider systems of forces that are equivalent for rigid body motion.

Theorem 4. A force system acting on a rigid body and consisting of a set of forces acting at a common point p may be reduced to a single force $\mathbf{F}(p)$, acting at the same point, where the value of $\mathbf{F}$ is equal to the vector sum of the forces in the original system.

PROOF: The net force and the net torque with respect to any point produced by both systems are the same, hence they are equivalent for rigid body motion.

Theorem 5. A force system acting on a rigid body and consisting of a set of couples may be reduced to a single couple $\mathbf{G}$, where the value of $\mathbf{G}$ is equal to the vector sum of the couples in the original system.

PROOF: Consider two couples $\mathbf{g}'$ and $\mathbf{g}''$. Since the torque produced by the couple $\mathbf{g}' + \mathbf{g}''$ is equivalent to the torque produced by $\mathbf{g}'$ and $\mathbf{g}''$, it follows that the couple $\mathbf{g}' + \mathbf{g}''$ is equivalent for rigid body motion to the pair of couples $\mathbf{g}'$ and $\mathbf{g}''$. This can be seen from a geometrical point of view by noting that if we move the forces that make up the couples $\mathbf{g}$ and $\mathbf{g}'$ along their respective lines of action until they all lie on the line formed by the intersection of the two planes of action, and adjust the points of application of the forces along this line until the pair of points of application for each couple are the same, we can add the forces acting at each point, and we will end up with a single couple $\mathbf{g}' + \mathbf{g}''$ whose torque is the vector sum of the torques of the component couples. From the considerations above it follows that we can successively add any number of couples in a system to form a resultant couple that is equivalent for rigid body motion to the original system of couples.

Theorem 6. An arbitrary force system acting on a rigid body may be reduced to a single force $\mathbf{F}(p)$ acting at an arbitrary point p and a couple $\mathbf{G}(p)$, where $\mathbf{F}$ is equal to the sum of the forces in the system and $\mathbf{G}(p)$ is equal to the sum of the torques with respect to the point p.

NOTE: With each point p there is associated a different couple; hence the notation $\mathbf{G}(p)$ rather than simply $\mathbf{G}$ is used.

PROOF: Any force $\mathbf{f}(q)$ is equivalent to the set of forces $\mathbf{f}(q), \mathbf{f}(p), -\mathbf{f}(p)$. But the pair of forces $\mathbf{f}(q)$ and $-\mathbf{f}(p)$ constitute a couple. Hence the force $\mathbf{f}(q)$ is equivalent to the force $\mathbf{f}(p)$ together with the couple $\{\mathbf{f}(q), -\mathbf{f}(p)\}$. If we have a system of forces, each force can be decomposed in this fashion, and we can then combine all the forces acting at the point p to obtain a single force $\mathbf{F}(p)$, and all the couples to obtain a single couple $\mathbf{G}(p)$. Since this system must be equivalent to the original system, it follows that the sum of the forces in the two systems must be equal, and the sum of the torques with respect to any point must also be equal. But the sum of the forces in the system $\mathbf{F}(p)$, $\mathbf{G}(p)$ is just $\mathbf{F}$, and the sum of the torques with respect to the point p is just $\mathbf{G}(p)$; hence $\mathbf{F}$ is equal to the sum of all the forces in the original system and $\mathbf{G}(p)$ is equal to the sum of all the torques with respect to the point p.

Corollary 7. A system of forces whose sum is zero is equivalent for rigid body motion to a single couple $\mathbf{G}$ whose value is equal to the net torque with respect to any point exerted by the system of forces.

The two lemmas that follow are very useful in the reduction of systems of forces.

Lemma 8. If the net force $\mathbf{F}$ associated with a system of forces does not vanish, the component in the direction of $\mathbf{F}$ of the net torque $\mathbf{G}$ is independent of the point with respect to which it is determined.

PROOF: From Theorem 3 in Chapter 29 we can show that the net torques with respect to two arbitrary points p and q are related as follows

$$\mathbf{G}(p) = \mathbf{G}(q) + \mathbf{r}(pq) \times \mathbf{F} \qquad (6)$$

where $\mathbf{r}(pq)$ is the vector position of q with respect to p. If we take the dot product of Eq. (6) with $\mathbf{F}$ and make use of the vector identity $(\mathbf{A} \times \mathbf{B}) \cdot \mathbf{B} = 0$ we obtain

$$\mathbf{G}(p) \cdot \mathbf{F} = \mathbf{G}(q) \cdot \mathbf{F} \qquad (7)$$

From Eq. (7) it follows that the component of $\mathbf{G}$ in the direction of $\mathbf{F}$ has the same value for all reference points. This completes the proof of the theorem.

Lemma 9. If the net force $\mathbf{F}$ associated with a system of forces does not vanish, there exists a line L with respect to whose points the net torque $\mathbf{G}$ has no component perpendicular to $\mathbf{F}$. The line L is parallel to $\mathbf{F}$ and passes through the point whose vector position with respect to an arbitrary point o is $\mathbf{F} \times \mathbf{G}(o)/F^2$, where $\mathbf{G}(o)$ is the net torque with respect to o.

260   Equivalent Systems of Forces

PROOF: If $G(o)$ is the net torque with respect to some point o, then the net torque with respect to a point p is given by

$$G(p) = G(o) + r(po) \times F \qquad (8)$$

If we assume that p is a point with respect to which the component perpendicular to F of the torque G vanishes, then

$$G(p) \times F = 0 \qquad (9)$$

Substituting Eq. (8) into Eq. (9) and making use of the vector identity $(A \times B) \times C = (C \cdot A)B - (C \cdot B)A$ we obtain

$$G(o) \times F + [F \cdot r(po)]F - F^2 r(po) = 0 \qquad (10)$$

Taking the cross product of Eq. (10) with F we obtain

$$[G(o) \times F - F^2 r(po)] \times F = 0 \qquad (11)$$

Equation (11) will be true if and only if the term in square brackets is in the direction of F, that is,

$$G(o) \times F - F^2 r(po) = \mu F \qquad (12)$$

or equivalently

$$r(op) = \lambda F + \left\{ \frac{F \times G(o)}{F^2} \right\} \qquad (13)$$

where $\lambda \equiv \mu/F^2$ is an arbitrary parameter. Conversely if p is a point satisfying Eq. (13) then Eq. (10) is also satisfied, hence the component of $G(p)$ perpendicular to F vanishes. It follows that any point p corresponding to a particular choice of λ in Eq. (13) is a point with respect to which the component of G perpendicular to F vanishes. If we let λ vary over arbitrary values then the point p traces out a straight line that is parallel to F and passes through the point whose position relative to q is given by $F \times G(o)/F^2$. This completes the proof of the lemma.

The following theorems, which we state without proof, follow immediately from the preceding lemmas.

Theorem 10. If a rigid body is acted on by a system of forces whose sum F does not vanish, and if the net torque G with respect to some point o is perpendicular to F, there exists a line L parallel to F with respect to whose points the torque vanishes. It follows that such a system of forces can be reduced to a single force F whose line of action is L. The line L passes through the point whose vector position with respect to o is given by $F \times G(o)/F^2$.

Corollary 11. If a rigid body is acted on by a system of coplanar forces whose sum F does not vanish, there is a line L in the plane and parallel to F

with respect to whose points the net torque **G** vanishes. It follows that a system of coplanar forces acting on a rigid body can be reduced to a single force **F** whose line of action is L. The line L passes through the point whose vector position with respect to an arbitrary point o is $\mathbf{F} \times \mathbf{G}(o)/F^2$.

Corollary 12. If a rigid body is acted on by a system of parallel forces whose sum **F** does not vanish, there is a line L parallel to **F** with respect to whose points the net torque **G** vanishes. It follows that a system of parallel forces acting on a rigid body can be reduced to a single force whose line of action is L. The line L passes through the point whose vector position with respect to an arbitrary point o is $\mathbf{F} \times \mathbf{G}(o)/F^2$.

A *wrench* is a system consisting of a force **f** acting at some point p, and a couple **g** which is parallel to **f**. The line of action of the force is called the *central axis* of the wrench, and the ratio **g/f** is called *the pitch* of the wrench. We will use the notation $\{\mathbf{f}(p)\|\mathbf{g}\}$ to represent a wrench.

The following theorem, which we state without proof, follows from the preceding definition and the lemmas proved earlier.

Theorem 13. If a rigid body is acted on by an arbitrary system of forces whose vector sum **F** does not vanish, then there is a line L parallel to **F** with respect to whose points the net torque **G** has no component perpendicular to **F**. It follows that the system of forces can be reduced to a wrench $\{\mathbf{F}(p) \| \mathbf{G}(p)\}$, where p is any point on L, and **F** and **G** are parallel to L. The line L passes through the point whose position relative to an arbitrary point o is given by $\mathbf{F} \times \mathbf{G}(o)/F^2$.

RIGID BODIES WITH ONE POINT FIXED

If a rigid body has one point o fixed, the results of this chapter are certainly applicable. However since the motion of a rigid body with one point o fixed is completely determined by the equation

$$\mathbf{G}(o) = \dot{\mathbf{H}}(o) \qquad (14)$$

it follows that any two systems of forces for which $\mathbf{G}(o)$ is the same will be equivalent for rigid body motion about the fixed point o. Hence a system of forces acting on a rigid body with one point o fixed can be reduced to a single couple **G** where the value of **G** is equal to the net torque with respect to the point o, or alternatively to a single force **F** acting at a point p, where p and **F** are chosen such that $\mathbf{r}(op) \times \mathbf{F}$ is equal to the net torque with respect to the point o.

EXAMPLES

Example 1. Show that if a rigid body of mass M is in a uniform gravitational field $\mathbf{g}$ then the gravitational force on the body is equivalent to a single force $M\mathbf{g}$ acting at the center of mass of the body.

Solution. Let us consider the rigid body to be made up of a set of particles of masses $m(1), m(2), \ldots$, respectively. Since the forces $m(1)\mathbf{g}, m(2)\mathbf{g}, \ldots$, acting on the particles are all parallel, it follows that there exists a line L with respect to whose points the net torque vanishes. The points p on the line L are defined by the condition

$$\sum_i \mathbf{r}(pi) \times m(i)\mathbf{g} = 0 \tag{15}$$

Since the center of mass c is defined by the condition

$$\sum_i m(i)\mathbf{r}(ci) = 0 \tag{16}$$

and since $\mathbf{g}$ is a constant, it follows that the point c lies on the line L. The proof of the problem now follows directly from Corollary 12. Note that if the rigid body were in a parallel but nonuniform gravitational field, that is, $\mathbf{g} = g(\mathbf{r})\mathbf{n}$, where $\mathbf{n}$ is a constant unit vector, the gravitational force would still be replaced by a single force; however the force would not be acting at the center of mass, but at some other point, which would be called the center of gravity.

PROBLEMS

1. A system of forces consists of three forces $R\mathbf{i}$, $2R\mathbf{j}$, and $3R\mathbf{k}$ acting at the origin, and a force $-R\mathbf{i} - R\mathbf{j} + R\mathbf{k}$ acting at the point $(a, a, 0)$. Find the equivalent wrench.

2. A thin wire is in the form of the arc of the helix $x = a\cos\theta$, $y = a\sin\theta$, $z = a\theta\tan\alpha$, obtained by varying θ from zero to $\pi/2$. Each line element ds of the wire is acted on by a force of magnitude $P\,ds$ that is directed away from the z axis in the direction of the perpendicular from the z axis to the element. Prove that the whole wire is acted on by a wrench whose axis is the line $x = y$, $z = (\pi a \tan\alpha)/4$, and find the pitch of this wrench.

3. Prove that a system of forces can be reduced to a force passing through a given point o and a force acting in a given plane that does not contain o.

4. A system consists of three forces $\mathbf{F}_1$, $\mathbf{F}_2$, and $\mathbf{F}_3$ defined as follows: $\mathbf{F}_1$ is of magnitude F and is directed along the positive x axis; $\mathbf{F}_2$ is of magnitude F and is directed in the positive y direction along the line $x = 0$, $z = -a$; $\mathbf{F}_3$ is of magnitude F and is directed along the line $x/\cos\phi = y/\sin\phi$

$= (z - a)/0$. If these three forces are reduced to two, one along the line $x = y = z$, show that the other will be perpendicular to the line $x = y = z$ if ϕ satisfies the equation $2\sin\phi - 2\sin^2\phi - 3\cos\phi = 0$.

5. A system consists of four forces whose lines of action form the edges of the tetrahedron $ABCD$ and whose values are given by $k\mathbf{r}(AB)$, $k\mathbf{r}(BC)$, $k\mathbf{r}(CD)$, and $k\mathbf{r}(DA)$, respectively, where k is a constant. Show that the system is equivalent to a couple of magnitude $6kV/p$, where V is the volume of the tetrehedron and p is the shortest distance between the edges AC and BD.

36

Statics of a Rigid Body

If a rigid body is in static equilibrium, then the sum of the external forces acting on the system must be equal to zero, and the sum of the external torques with respect to any point must also be equal to zero, that is,

$$\mathbf{F} = 0 \tag{1}$$

$$\mathbf{G}(o) = 0 \tag{2}$$

where o is any point. Since there is an infinite number of possible points o, there is an infinite number of equations that are necessary conditions for static equilibrium. However of these equations only six will be independent; hence once we have chosen six independent conditions of equilibrium, we cannot expect to extract any new information from the remaining conditions of equilibrium.

EXAMPLES

Example 1. A hollow cylinder c of radius b that is open at both ends is resting on a horizontal plane with its axis vertical. Two smooth spheres A and B, each of mass m and radius a, where $a > b/2$, are placed inside the cylinder. Prove that, in order for the cylinder not to be upset, its mass M must be at least $2m[(b - a)/b]$.

Solution. Since this problem involves three rigid bodies A, B, and C, we can choose any three independent systems for our consideration. We shall choose the systems A, $A + B$, and $A + B + C$. The forces acting on A are shown in Fig. 1b. The angle α is equal to $\cos^{-1}[(b - a)/a]$. The forces acting on $A + B$ are shown in Fig. 1c. The forces acting on $A + B + C$ are shown in Fig. 1d, where we have assumed that the cylinder is about to tip over. Setting the sum of the x and y components of the forces acting on A respectively

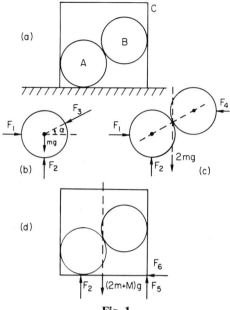

Fig. 1

equal to zero, we obtain

$$F_1 - F_3 \cos \alpha = 0 \qquad (3)$$

$$F_2 - mg - F_3 \sin \alpha = 0 \qquad (4)$$

Since the forces converge at the center of mass, this is the maximum number of independent equations of equilibrium we can write down. Setting the sum of the x and y components of the forces acting on $A + B$ respectively equal to zero, we obtain

$$F_1 - F_4 = 0 \qquad (5)$$

$$F_2 - 2mg = 0 \qquad (6)$$

Since the four equations (3)–(6) are sufficient to determine F_1, F_2, F_3, and F_4, the additional equation we would obtain by setting the torques about some point equal to zero would not give us any new information. Setting the sum of the x and y components of the forces acting on $A + B + C$ respectively equal to zero we obtain

$$-F_6 = 0 \qquad (7)$$

$$F_2 + F_5 - (2m + M)g = 0 \qquad (8)$$

Setting the torque with respect to the right hand corner equal to zero we obtain

$$(2m + M)gb - F_2(2b - a) = 0 \qquad (9)$$

We now have seven equations in the seven unknowns F_1, F_2, F_3, F_4, F_5, F_6, and M, and we can solve for any one of them. We are however interested only in M, which can be obtained directly from Eqs. (6) and (9) and is given by

$$M = 2m\frac{b-a}{b} \tag{10}$$

Note that if we had been clever we could have immediately spotted the two equations necessary to determine M, avoiding some work. However the purpose of this example is to demonstrate a method, not to obtain the answer by the shortest route. Note also that it would not have caused us any problems if we had not noticed that the torque equations for the systems A and $A + B$ did not lead to any new information. We would simply have obtained several equations that are already implicitly contained in the other equations.

PROBLEMS

1. A homogeneous hemisphere of mass m and radius a rests on a perfectly smooth horizontal table. One edge of the hemisphere is tied to a point on the table by means of a massless inextensible string of length b, where $b < a$. Find the tension in the string.

2. A uniform heavy ladder is in limiting equilibrium, standing on a rough floor and leaning against an equally rough wall. The angle of friction is ϵ. Find the angle that the ladder makes with the horizontal.

3. A uniform rod rests with one end on a rough horizontal plane and the other on a smooth plane inclined at an angle θ to the horizontal. If the coefficient of friction for the contact between the rod and the horizontal plane is μ, what is the least angle that the rod can make with the horizontal?

4. A uniform rod of mass m and length a is supported from a single point by means of two strings, each of length a, connected to the ends of the rod. An object of mass m is hung from one of the ends of the rod. Find the angle that the rod makes with the horizontal and the tension in each string.

5. Two smooth planes, inclined in opposite directions to the horizontal at angles α and β, have a common line of intersection with a horizontal plane. A uniform rod rests on the two planes. Show that its inclination θ to the horizontal can be expressed in the form $\tan\theta = \frac{1}{2}(\cot\alpha - \cot\beta)$.

6. A heavy circular cylinder of radius a rests with its curved surface in contact with a rough horizontal plane. A uniform bar AB, of length $4a$, rests in equilibrium, in a plane perpendicular to the axis of the cylinder, with the end A on the horizontal plane and touching the cylinder at P, where $AP = 3a$. By considering the equilibrium of the cylinder, find the

direction of the reaction between cylinder and bar. Show that the coefficient of friction at A must be at least $8/21$.

7. Two equal uniform ladders, AB, and BC, each of length $2a$ and mass M, are smoothly hinged together at B, while the ends A and C rest on a rough horizontal plane. The coefficient of friction is μ at A and C, and the angle ABC is 2α. Prove that a man of mass m can ascend a distance x given by $x = [2aM(2\mu - \tan\alpha)]/[m(\tan\alpha - \mu)]$, provided $x < 2a$ and $\mu < \tan\alpha < 2\mu$.

8. A uniform plank of negligible thickness having mass m and length b, rests in equilibrium horizontally across a rough fixed circular cylinder of radius a. The coefficient of friction between the plank and the cylinder is μ. What is the greatest mass that can be attached to one end of the plank without causing it to slip?

9. Two uniform circular cylinders of equal radii rest in contact on an inclined plane of inclination α. The mass of the lower cylinder is M and the mass of the upper cylinder is nM, where $n > 1$. If the coefficient of friction between any two surfaces in contact is μ, show that equilibrium is not possible if μ is less than $(n+1)/(n-1)$. Show further that if μ exceeds this value, equilibrium is possible for all values of α less than $\tan^{-1}[2n\mu/(\mu+1)(n+1)]$.

10. Two smooth planes intersect in a horizontal line L. They lie on opposite sides of the vertical plane through L, making angles θ, ϕ ($< \pi/2$) with this plane. Two smooth horizontal uniform circular cylinders, of equal mass and length but differing in radius, are in contact all along a generator and rest in equilibrium, each with a single generator in contact with one inclined plane. Show that the plane through the axes makes an angle ψ with the vertical where $2\cot\psi = \tan\theta - \tan\phi$.

11. A uniform bar of mass M and length $2a$ is suspended from two points in a horizontal plane by two equal strings of length b, which are originally vertical. The strings are connected to the ends of the bar. Show that the couple that must be applied to the bar in a horizontal plane to keep it at rest at right angles to its former direction is $Mga^2/(b^2 - 2a^2)^{1/2}$.

12. Two equal rough uniform circular cylinders of radius a and weight W rest on a rough horizontal plane with their axes $2a\sqrt{2}$ apart. A third cylinder, of radius a and weight $2W$, rests symmetrically on the first two. Show that for equilibrium the coefficient of friction for the contact between the cylinders must be greater than $\tan(\pi/8)$, and for the contact between the cylinders and the plane it must be greater than $[\tan(\pi/8)]/2$.

13. A uniform rod AB of length a and mass M rests with the end B on a rough wall and the end A on the ground, where it is smoothly pivoted at a point a distance b from the wall. The coefficient of static friction for

the contact between the rod and the wall is μ. Determine the minimum height above the floor that the end B can have without slipping occurring.

14. A uniform pole rests with one end A on a rough horizontal floor and the other end B against a smooth vertical wall, a distance d from A. Initially the pole is in a vertical plane perpendicular to the wall, with the end B at a height h above the ground. The end B is allowed to slide slowly down the wall, controlled by a light rope attached to the pole at B and passing through a ring in the wall at the initial position of B. If the end A slips when a length b of the rope has been paid out through the ring, show that the coefficient of friction between the pole and the floor is $[4h^2(b^2 + d^2) - b^4]^{1/2}/(4h^2 - b^2)$.

15. A uniform rod of length $2a$ and mass M rests with one end B on a smooth wall and the other end A on the ground, where it is smoothly pivoted. It is kept in equilibrium by a force $\mathbf{F}$ acting at B in the plane of the wall. The perpendicular from A to the wall meets it at M and the angle BAM is α. The inclination at MB to the vertical is θ. If the normal reaction at the wall is $\mathbf{R}$, find the direction of $\mathbf{F}$ that makes F/R a minimum.

16. The end A of a uniform heavy rod AB of length l can move freely on a smooth straight vertical wire, and the end B can move along a rough straight horizontal wire. The shortest distance apart of the wires is a, where $a < l$. The coefficient of static friction for the contact between the end B of the rod and the horizontal wire is μ. If $l > a(1 + \mu^2)^{1/2}$ and the end A is higher than the end B, show that equilibrium is possible only if $\sin^{-1}(a/l) \leq \theta \leq \cos^{-1}\{[l^2 - a^2(1 + \mu^2)]/[l^2(1 + 4\mu^2)]\}^{1/2}$, where θ is the angle AB makes with the downward vertical.

17. A tripod consists of three equal, uniform rods OA, OB, and OC of weight W and length $2a$, smoothly jointed at their common extremity O. When the feet are held at the vertices of an equilateral triangle ABC, a light spring attachment exerts on each rod a couple of moment $Wk\theta$ in the plane through the rod in question and the center G of the triangle ABC, where θ is the angle AOG and the tendency of the couple is to reduce θ. If $2/\pi < k/a < 1$, show that the tripod can stand in equilibrium on a smooth horizontal table with A, B, and C at the vertices of an equilateral triangle. If weights W_1, W_2, and W_3 are now hung at the midpoints of OA, OB, and OC and if A, B, and C are kept in the same positions by inextensible strings BC, CA, and AB, show that the tensions in the strings are equal and that their common value is $[(W_1 + W_2 + W_3)\tan\theta]/6\sqrt{3}$.

37

Uniplanar Motion of a Rigid Body

INTRODUCTION

In the chapters that follow we consider in detail the general motion of a rigid body. However before embarking on this complex project it will be helpful to consider the important special case of rigid body motion in which each particle in the system is constrained to move parallel to the same fixed plane. Such motion is called *uniplanar motion*. A lamina moving in its own plane and a rigid body rotating about a fixed axis are examples of uniplanar motion. If the particles are moving parallel to a given fixed plane, they are also moving parallel to any plane that is parallel to the given plane. For convenience we single out one of these planes and refer to it as the *fixed plane*. The instantaneous axis of rotation of a rigid body that is undergoing uniplanar motion always remains perpendicular to the fixed plane. We therefore refer to the direction perpendicular to the fixed plane in a given sense as the *axial direction*. Components in this direction of a vector are called axial components.

THE MOMENT OF INERTIA

Before writing down the equations of motion of a rigid body that is undergoing uniplanar motion we develop a number of auxiliary ideas that will simplify the final result.

If a body is acted on by a set of forces and we are interested only in the motion of the center of mass of the body, then all we need to know about the body is its total mass and the location of the center of mass. If on the other hand a rigid body is undergoing uniplanar motion under the action of a set of forces and we wish to describe its motion completely, we need to know more about the mass distribution in the body than simply the total mass and the location of the center of mass. However, as this chapter demonstrates, we still do not need a complete description of the mass distribution to determine the motion; we simply need to know, in addition to the total mass and the location of the center of mass, the moment of inertia about the axial direction at some point in the body.

The *moment of inertia* of a dynamical system about a given axis a is defined as follows:

$$I(a) \equiv \sum_i m(i)[\rho(ai)]^2 \qquad (1)$$

where $m(i)$ is the mass of particle i and $\rho(ai)$ is the distance of the ith particle from the axis a.

If we know the moment of inertia about one axis we can find the moment of inertia about any other axis that is parallel to the first axis by making use of the following theorem:

Theorem 1. (*Parallel Axis Theorem*). The moment of inertia of a dynamical system about an arbitrary axis a is given by

$$I(a) = I(c) + M[\rho(ac)]^2 \qquad (2)$$

where $I(c)$ is the moment of inertia of the system about an axis c that is parallel to the axis a and passes through the center of mass of the system, M is the total mass of the system, and $\rho(ac)$ is the distance between the axes a and c.

PROOF: If we let the axis i be the axis that passes through the particle i and is parallel to the axis a, and let $\rho(ai)$ be the vector position of the axis i with respect to the axis a, and $\rho(ac)$ the vector position of the axis c with respect to the axis a, the moment of inertia of the system with respect to the axis a can be written

$$\begin{aligned}
I(a) &= \sum_i m(i)[\rho(ai)]^2 \\
&= \sum_i m(i)\rho(ai) \cdot \rho(ai) \\
&= \sum_i m(i)[\rho(ac) + \rho(ci)] \cdot [\rho(ac) + \rho(ci)] \\
&= \left[\sum_i m(i)\right][\rho(ac)]^2 + \sum_i m(i)[\rho(ci)]^2 \\
&\quad + 2\rho(ac) \cdot \sum_i m(i)[\rho(ci)]
\end{aligned} \qquad (3)$$

But

$$\sum_i m(i) \equiv M \qquad (4)$$

$$\sum_i m(i)[\rho(ci)]^2 \equiv I(c) \qquad (5)$$

and from the definition of the center of mass

$$\sum_i m(i)\rho(ci) = 0 \tag{6}$$

Substituting Eqs. (4)–(6) in Eq. (3) we obtain Eq. (2), which is the desired result.

From the theorem above it follows that if we know the moment of inertia about the axis a we can find the moment of inertia about the axis c; and if we know the moment of inertia about the axis c, we can find the moment of inertia about any other parallel axis. Therefore if we know the moment of inertia about one axis we can use Theorem 1 to find the moment of inertia about any other parallel axis.

THE CONFIGURATION

To specify the configuration of a rigid body that is undergoing uniplanar motion we find it convenient to introduce two frames of reference, an inertial frame S and a frame $\bar{S}$ fixed in the rigid body. We let o be the origin and $x, y,$ and z the coordinate axes of the S frame; for the $\bar{S}$ frame, $\bar{o}$ is the origin and $\bar{x}, \bar{y},$ and $\bar{z}$ the coordinate axes. For convenience we choose the frames in such a way that the origins o and $\bar{o}$ are both in the fixed plane and the z and $\bar{z}$ axes are in the axial direction. The configuration of the rigid body can then be completely specified by the coordinates $x(\bar{o})$ and $y(\bar{o})$ of the point $\bar{o}$, and the angle θ that the $\bar{x}$ axis makes with the x axis. These coordinates are illustrated in Fig. 1.

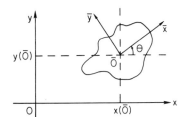

Fig. 1

ANGULAR MOMENTUM AND KINETIC ENERGY

In analyzing the uniplanar motion of a rigid body, the kinetic energy and the axial component of the angular momentum are useful dynamical properties.

Theorem 2. Consider a rigid body that is undergoing uniplanar motion. Let a be a point that is fixed in the rigid body. The axial component of the angular momentum with respect to the point a and the kinetic energy with respect to the point a are given, respectively, by

$$H(a) = I(a)\dot{\theta} \tag{7}$$

$$T(a) = \tfrac{1}{2}I(a)\dot{\theta}^2 \tag{8}$$

where $I(a)$ is the moment of inertia with respect to an axis through the point a and in the axial direction, θ is the angular displacement of the rigid body about the axis of rotation.

NOTE: The quantity $H(a)$, the axial component of the angular momentum $\mathbf{H}(a)$, should not be confused with the magnitude $H(a)$ of the angular momentum $\mathbf{H}(a)$. The two are different because the angular momentum of a rigid body is not necessarily in the same direction as the angular velocity; hence the angular momentum may have a component in a plane perpendicular to the angular velocity. However, as we shall see, when we are dealing with uniplanar motion of a rigid body, the analysis depends on the axial component only. Hence the duplication of notation should not cause any real trouble and is probably preferable to weighing down the notation with extra symbols.

PROOF: Let S be an inertial frame with the z axis in the axial direction, and $\overline{S}$ a frame fixed in the rigid body with the $\bar{z}$ axis also in the axial direction. The axial component of the angular momentum with respect to the point a and the kinetic energy with respect to the point a are given, respectively, by

$$H(a) = \sum_i \mathbf{k} \cdot \{\mathbf{r}(ai) \times [m(i)\dot{\mathbf{r}}(ai)]\} \tag{9}$$

$$T(a) = \tfrac{1}{2}\sum_i m(i)\dot{\mathbf{r}}(ai) \cdot \dot{\mathbf{r}}(ai) \tag{10}$$

where $m(i)$ is the mass of particle i, $\mathbf{k}$ is a unit vector in the z direction, $\mathbf{r}(ai)$ is the vector position of particle i with respect to the point a, and $\dot{\mathbf{r}}(ai)$ is the time rate of change of $\mathbf{r}(ai)$ as noted by an observer fixed in the frame S. If we let $\overline{\dot{\mathbf{r}}}(ai)$ be the corresponding time rate of change as noted by an observer fixed in $\overline{S}$, then from the results of Chapter 6

$$\dot{\mathbf{r}}(ai) = \overline{\dot{\mathbf{r}}}(ai) + \omega(S\overline{S}) \times \mathbf{r}(ai) \tag{11}$$

where $\omega(S\overline{S})$ is the angular velocity of frame $\overline{S}$ with respect to frame S. Since the points a and i are fixed in the rigid body, it follows that

$$\overline{\dot{\mathbf{r}}}(ai) = 0 \tag{12}$$

Since the $\bar{z}$ axis is in the direction of the angular velocity and $\dot{\theta}$ is the rate of rotation, it follows that

$$\omega(S\overline{S}) = \mathbf{k}\dot{\theta} \tag{13}$$

If we express **r**(*ai*) in terms of its components in the *x*, *y*, and *z* directions we have

$$\mathbf{r}(ai) = \mathbf{i}x(ai) + \mathbf{j}y(ai) + \mathbf{k}z(ai) \tag{14}$$

Combining Eqs. (9)–(14) we obtain

$$H(a) = \left\{ \sum_i m(i)\left[x^2(ai) + y^2(ai) \right] \right\} \dot{\theta} \tag{15}$$

$$T(a) = \tfrac{1}{2} \left\{ \sum_i m(i)\left[x^2(ai) + y^2(ai) \right] \right\} \dot{\theta}^2 \tag{16}$$

The quantity in the braces is just the moment of inertia of the body with respect to an axis in the *z* direction through the point *a*. We have thus obtained the desired result.

TORQUE

In analyzing the uniplanar motion of a rigid body it is necessary to determine the axial component of the external torque. The following theorem is helpful in doing this.

Theorem 3. If a force **F** is acting at a point *p* on a rigid body that is undergoing uniplanar motion, and if the vector position of point *p* with respect to some point *a* is **r**(*ap*), then *G*(*a*), the axial component of the torque **G**(*a*) ≡ **r**(*ap*) × **F** is the same as the axial component of the torque **G'**(*a*) ≡ **r'**(*ap*) × **F'**, where **r'**(*ap*) and **F'** are the components in the fixed plane of the vectors **r**(*ap*) and **F**.

NOTE 1: The quantity *G*(*a*), the axial component of the torque **G**(*a*), should not be confused with the magnitude *G*(*a*) of the torque **G**(*a*).

NOTE 2: It follows from the theorem above that we can ignore the axial components of the force **F** and the position vector **r**(*ap*) when we are calculating the axial component of the torque exerted by the force **F**.

PROOF: Let **k** be a unit vector in the axial direction. The axial component of the torque **G**(*a*) is then given by

$$G(a) = \mathbf{k} \cdot \left[\mathbf{r}(ap) \times \mathbf{F} \right] \tag{17}$$

If we let **r'**(*ap*) and **F'** be the components in the fixed plane of **r**(*ap*) and **F** and let **r''**(*ap*) and **F''** be the axial components of **r**(*ap*) and **F**, we can write Eq. (17) as follows:

$$G(a) = \mathbf{k} \cdot \left\{ \left[\mathbf{r'}(ap) + \mathbf{r''}(ap) \right] \times \left[\mathbf{F'} + \mathbf{F''} \right] \right\} \tag{18}$$

Of the four products **r'**(*ap*) × **F'**, **r'**(*ap*) × **F''**, **r''**(*ap*) × **F'**, and **r''**(*ap*) × **F''**, the latter three are all perpendicular to the axial direction, hence will vanish when dotted into the vector **k**. It follows that

$$G(a) = \mathbf{k} \cdot [\mathbf{r}'(ap) \times \mathbf{F}'] \tag{19}$$

which is the desired result.

THE EQUATIONS OF MOTION

If a rigid body is constrained to rotate about a fixed axis, the motion is uniplanar and can be determined by using the following theorem.

Theorem 4. Consider a rigid body that is constrained to rotate about a fixed axis. Let o be an arbitrary point on the axis. If the body is acted on by a set of external forces **F**, the rotational motion of the body is governed by the equation

$$G(o) = I(o)\ddot{\theta} \tag{20}$$

where $G(o)$ is the axial component of the torque with respect to the point o exerted by the external forces, $I(o)$ is the moment of inertia of the body with respect to an axis that passes through the point o and is in the axial direction, and θ is the angular displacement of the rigid body about the axis of rotation.

PROOF: From Theorem 4 in Chapter 29

$$\mathbf{G}(o) = \dot{\mathbf{H}}(o) \tag{21}$$

hence the axial component of $\mathbf{G}(o)$ is equal to the axial component of $\dot{\mathbf{H}}(o)$. It follows that

$$G(o) = \dot{H}(o) \tag{22}$$

But from Theorem 2 in this chapter

$$H(o) = I(o)\dot{\theta} \tag{23}$$

Substituting Eq. (23) in Eq. (22) and noting that $I(o)$ is a constant, we obtain Eq. (20), which is the desired result.

If a rigid body is undergoing uniplanar motion, and the instantaneous axis of rotation is not fixed, the behavior of the body can be determined by breaking the motion down into the motion of the center of mass and the motion with respect to the center of mass, as described in the following theorem.

Theorem 5. Consider a rigid body that is undergoing uniplanar motion and is acted on by a set of external forces **F**. Let S be an intertial frame with the z axis in the axial direction. The motion of the center of mass is governed by the equations

$$F_x = M\ddot{X} \quad (24)$$

$$F_y = M\ddot{Y} \quad (25)$$

where F_x and F_y are the x and y components of the force $\mathbf{F}$; M is the total mass of the body, and X and Y are the x and y position coordinates of the center of mass with respect to the origin of the frame S. The motion with respect to the center of mass is governed by the equation

$$G(c) = I(c)\ddot{\theta} \quad (26)$$

where $G(c)$ is the axial component of the torque with respect to the center of mass; $I(c)$ is the moment of inertia of the rigid body with respect to an axis that passes through the center of mass and is in the axial direction, and θ is the angular displacement of the rigid body.

PROOF: From Theorem 4 in Chapter 28

$$\mathbf{F} = M\mathbf{A} \quad (27)$$

Equations (24) and (25) follow directly from Eq. (27). From Theorem 5 in Chapter 29

$$\mathbf{G}(c) = \dot{\mathbf{H}}(c) \quad (28)$$

hence

$$G(c) = \dot{H}(c) \quad (29)$$

From Theorem 2 in the present chapter

$$H(c) = I(c)\dot{\theta} \quad (30)$$

Substituting Eq. (30) in Eq. (29) and noting that $I(c)$ is a constant, we obtain Eq. (26), which is the desired result.

EXAMPLES

Example 1. Determine the moment of inertia of a lamina in the shape of an isosceles triangle of mass m, base b, and height h, about an axis through the center of mass and parallel to the base.

Solution. Rather than directly calculating the desired moment of inertia, it is simpler to find the moment of inertia with respect to an axis parallel to the base and passing through the vertex and then to make use of the parallel axis theorem to obtain the desired moment of inertia. If we choose our x and y axis as shown in Fig. 2, the moment of inertia of the element $dx\,dy$ with respect to the x axis is $y^2\,dm \equiv y^2[m/(bh/2)]\,dx\,dy$. Summing over all the elements $dx\,dy$, we obtain for the moment of inertia of the triangular lamina

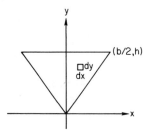

Fig. 2

about the x axis

$$I(o) = \int_0^h \int_{-by/2h}^{+by/2h} y^2 \frac{m}{bh/2} \, dx \, dy = \frac{mh^2}{2} \tag{31}$$

Using the parallel axis theorem, and noting that the center of mass is located at the point $(x, y) = (0, 2h/3)$ we obtain for the moment of inertia about an axis parallel to the x axis and through the center of mass,

$$I(c) = I(o) - m\left(\frac{2h}{3}\right)^2 = \frac{mh^2}{18} \tag{32}$$

Example 2. A circular disk of mass M, radius a, and center C is free to rotate in a vertical plane about a fixed point A on its circumference. Determine the angular velocity of the disk as a function of the angle θ, which the line AC makes with the vertical. Assume that the disk is released from rest with $\theta = \theta_0$.

Solution 1. Since the point A is fixed in space and in the rigid body, we can use the equation

$$G(A) = I(A)\ddot{\theta} \tag{33}$$

where θ is shown in Fig. 3a. The forces acting on the system at an arbitrary time t are shown in Fig. 3b, where R_x and R_y are the x and y components of the reaction force of the pin. Taking torques with respect to the point A, and noting that a torque that tends to rotate the system in the direction of increasing θ is positive and a torque that tends to rotate the system in the direction of decreasing θ is negative, we obtain

$$G(A) = -Mga \sin \theta \tag{34}$$

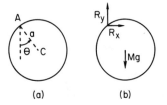

Fig. 3

By the parallel axis theorem

$$I(A) = I(C) + Ma^2 = \frac{Ma^2}{2} + Ma^2 = \frac{3Ma^2}{2} \tag{35}$$

Substituting Eqs. (34) and (35) in Eq. (33) we obtain

$$-Mga \sin\theta = \frac{3Ma^2}{2} \ddot{\theta} \tag{36}$$

If we multiply this equation by $\dot{\theta}$ and integrate with respect to the time, noting that $\ddot{\theta}\dot{\theta} = d(\dot{\theta}^2/2)/dt$, we can obtain $\dot{\theta}$ as a function of θ. If we do this, and make use of the initial conditions, we obtain

$$\dot{\theta} = \left(\frac{4g}{3a}\right)^{1/2} (\cos\theta - \cos\theta_0)^{1/2} \tag{37}$$

We could have obtained the same result by noting that the total energy is conserved. The kinetic and potential energies for arbitrary θ are given by

$$T = \frac{I(A)\dot{\theta}^2}{2} = \frac{3Ma^2}{4}\dot{\theta}^2 \tag{38}$$

$$V = -Mga\cos\theta + C \tag{39}$$

where C is a constant. The kinetic and potential energies for $t = 0$ are given by

$$T_0 = 0 \tag{40}$$

$$V_0 = -Mga\cos\theta_0 + C \tag{41}$$

Setting $T + V = T_0 + V_0$ we obtain Eq. (37).

Solution 2. It is instructive to consider the same problem in terms of the motion of the center of mass and the rotation with respect to the center of mass. In this case the basic equations are

$$F_x = M\ddot{X} \tag{42}$$

$$F_y = M\ddot{Y} \tag{43}$$

$$G(C) = I(C)\ddot{\theta} \tag{44}$$

Applying these three equations we obtain

$$R_x = M\frac{d^2}{dt^2}(a\sin\theta) \tag{45}$$

$$R_y - Mg = M\frac{d^2}{dt^2}(-a\cos\theta) \tag{46}$$

$$-R_x a\cos\theta - R_y a\sin\theta = \frac{Ma^2}{2}\ddot{\theta} \tag{47}$$

If we use Eqs. (45) and (46) to eliminate R_x and R_y from Eq. (47) we again obtain Eq. (37). The solution above is not as direct as the first solution presented. However it does yield expressions for the forces R_x and R_y and the first solution did not.

Example 3. A uniform rod of mass M and length $2a$ is placed like a ladder with one end against a smooth vertical wall and the other end on a smooth horizontal plane. It is released from rest at an angle θ_0 to the vertical. Determine the initial reactions of the wall and the floor on the rod. Determine whether the rod will lose contact with the floor or wall, and if so, the angle of inclination of the rod when this occurs.

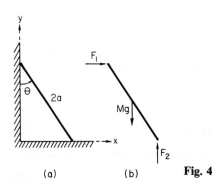

Fig. 4

Solution. We choose our inertial frame as shown in Fig. 4.a, and we let θ be the angle the rod makes with the vertical. The forces acting on the system are shown in Fig. 4b. Making use of Theorem 5 and noting that the coordinates of the center of mass of the rod are $X = a \sin \theta$, $Y = a \cos \theta$, and the moment of inertia of the rod about the center of mass is $Ma^2/3$, we obtain

$$F_1 = M \frac{d^2}{dt^2} (a \sin \theta) \tag{48}$$

$$F_2 - Mg = M \frac{d^2}{dt^2} (a \cos \theta) \tag{49}$$

$$-a \cos \theta \, F_1 + a \sin \theta \, F_2 = \frac{Ma^2}{3} \ddot{\theta} \tag{50}$$

The equations above provide us with three equations in the three unknowns F_1, F_2, and θ, and can in principle be solved to obtain $F_1(t)$, $F_2(t)$, and $\theta(t)$. Carrying out the differentiation in Eqs. (48) and (49) we obtain

$$F_1 = Ma\ddot{\theta} \cos \theta - Ma\dot{\theta}^2 \sin \theta \tag{51}$$

$$F_2 = -Ma\ddot{\theta} \sin \theta - Ma\dot{\theta}^2 \cos \theta + Mg \tag{52}$$

When $t = 0$ then $\theta = \theta_0$ and $\dot{\theta} = 0$. Substituting these values in Eqs. (50)–(52)

we obtain
$$F_1(0) = Ma\ddot{\theta}(0)\cos\theta_0 \tag{53}$$
$$F_2(0) = -Ma\ddot{\theta}(0)\sin\theta_0 + Mg \tag{54}$$
$$-a\cos\theta_0 F_1(0) + a\sin\theta_0 F_2(0) = \frac{Ma^2}{3}\ddot{\theta}(0)$$

Solving for $F_1(0)$ and $F_2(0)$ we obtain
$$F_1(0) = \frac{3Mg\sin\theta_0\cos\theta_0}{4} \tag{55}$$
$$F_2(0) = \frac{Mg(4 - 3\sin^2\theta_0)}{4} \tag{56}$$

This completes the first part of the problem.

To determine whether the rod will lose contact with the wall or floor we first assume that the rod is constrained to remain in contact with the wall and the floor, solve for the reaction forces that are necessary to maintain this constraint, and determine whether the surfaces can exert such forces. Thus if we solve Eqs. (48)–(50) for F_1 and F_2, assuming that the ends of the rod remain in contact with the wall and floor, and if at some time F_1 or F_2 becomes negative, at that time the rod will lose contact with the corresponding surface, since these surfaces can only exert forces for which F_1 and F_2 are positive. The solutions of Eqs. (48)–(50) can be facilitated by noting that the energy is a constant of the motion. Hence

$$\frac{1}{2}M\left\{\left[\frac{d}{dt}(a\sin\theta)\right]^2 + \left[\frac{d}{dt}(a\cos\theta)\right]^2\right\} + \frac{1}{2}\frac{Ma^2}{3}\dot{\theta}^2 + Mga\cos\theta$$
$$= Mga\cos\theta_0 \tag{57}$$

Simplifying Eq. (57) we obtain
$$\dot{\theta}^2 + \frac{3g}{2a}(\cos\theta - \cos\theta_0) = 0 \tag{58}$$

from which it follows that
$$\dot{\theta} = \left(\frac{3g}{2a}\right)^{1/2}(\cos\theta_0 - \cos\theta)^{1/2} \tag{59}$$
$$\ddot{\theta} = \frac{3g}{4a}\sin\theta \tag{60}$$

Substituting Eqs. (59) and (60) into Eqs. (51) and (52) we obtain
$$F_1 = \frac{3Mg}{4}\sin\theta(3\cos\theta - 2\cos\theta_0) \tag{61}$$
$$F_2 = \frac{Mg}{4}(1 + 9\cos^2\theta - 6\cos\theta\cos\theta_0) \tag{62}$$

The force F_2 remains positive for values of θ between θ_0 and $\pi/2$. The force F_1 remains positive for values of θ between θ_0 and θ_m, where

$$\theta_m = \cos^{-1}\left(\frac{2\cos\theta_0}{3}\right) < \frac{\pi}{2} \tag{63}$$

but becomes negative when $\theta > \theta_m$. Hence the rod will leave the wall when $\theta = \theta_m$.

PROBLEMS

1. Determine the moment of inertia of a uniform rod of mass m and length a:
   a. About an axis perpendicular to the rod through the center of the rod.
   b. About an axis perpendicular to the rod through one end of the rod.

2. Given a rectangular lamina of mass m, and sides $2a$ and $2b$, respectively:
   a. Show that the moment of inertia about an axis in its plane and perpendicular to the side $2a$ at its center is $ma^2/3$.
   b. Show that the moment of inertia about an axis through its center and perpendicular to its plane is $m(a^2 + b^2)/3$.

3. Determine the moment of inertia of a uniform circular disk:
   a. About an axis perpendicular to the disk and through the center of the disk.
   b. About a diameter.

4. Given an elliptical disk of mass m, and semiaxes a and b, respectively, show that the moment of inertia of the disk about a diameter of length d is ma^2b^2/d^2.

5. Determine the moment of inertia of a homogeneous sphere of mass m and radius a about a diameter.

6. Prove that the moment of inertia of a uniform circular cylinder of radius a, height $2h$, and mass M about an axis through its center of mass and perpendicular to the axis of symmetry is $M[(a^2/4) + (h^2/3)]$.

7. Two equal homogeneous spheres, each of mass m and radius a, are joined by a rigid massless rod that lies along the line joining the centers of the two spheres. The distance between the centers of the two spheres is $2b$. The system is rotating with a constant angular velocity ω about an axis that passes through and is perpendicular to the rod joining the two spheres. Find the moment of inertia of the system for arbitrary position of the axis, and determine the position of the axis for maximum angular momentum.

Problems 281

8. The end points of a uniform rod AB, of mass m and length $2a$, are moving in a plane with velocities **u** and **v**, respectively. Prove that the kinetic energy of the rod is $T = (m/6)(u^2 + \mathbf{u} \cdot \mathbf{v} + v^2)$.

9. A uniform solid sphere of mass m and radius a moves on a horizontal plane so that its center O has speed V, and it has an angular velocity Ω about its horizontal diameter. Find the kinetic energy of the sphere. Find the angular momentum of the sphere:
   a. About the point on the sphere in contact with the plane.
   b. About the point on the surface in contact with the sphere.
   c. About the contact point considered as a point moving along the surface with speed V.
   d. About its center.
   e. About its highest point.

10. A uniform rod AB of mass m and length $2a$ lies at rest on a smooth horizontal plane when a constant horizontal force of magnitude P is applied to B in a direction making an angle α with BA where $\alpha < \pi/2$. Prove that when the rod makes an angle θ with the direction of P, then $ma\dot\theta^2 = 6P(\cos\alpha - \cos\theta)$. Prove also that the rod oscillates with amplitude $2(\pi - \alpha)$.

11. A homogeneous rod of length $2a$ that is initially standing vertically upright on a perfectly rough floor is gently tipped over. What is its angular velocity when it hits the floor?

12. A rod of length $2a$ and mass m is standing vertically upright on a perfectly rough floor with the end A in contact with the floor. The center of mass C of the rod is located at the center of the rod, and the moment of inertia of the rod about an axis through C and perpendicular to the rod is mk^2. The rod is slightly displaced. Show that if $k^2 < a^2/3$, then A will leave the plane before the rod becomes horizontal. Show further that if the rod is uniform, the floor is not perfectly rough, and the rod slips when the inclination to the vertical is $\tan^{-1}\frac{3}{4}$, the coefficient of friction must be $18/49$.

13. A uniform plank of mass m and length a is suspended horizontally by two vertical ropes, one at each end. If one is severed what is the tension, immediately after, in the remaining rope?

14. A yo-yo is made of two uniform disks each of mass $m/2$ and radius R, which are joined along a common axis through their centers by a smaller disk of radius r and negligible mass. A string is wrapped around the smaller disk and its other end is fixed. If the yo-yo is released, with the string initially in a vertical direction, what will be the acceleration of its center as the string unwinds without slipping?

15. A uniform pulley of mass M and radius a is free to turn about a frictionless axle through its center. Two masses m_1 and m_2, where $m_2 > m_1$, are suspended at opposite ends of a massless inextensible string passing over the pulley. Determine the downward acceleration of mass m_2, assuming the string does not slip on the pulley. Determine the tension in the string supporting m_1, and the tension in the string supporting m_2.

16. A spur gear A of radius a drives a spur gear B of radius b. Each gear is free to turn about an axis through its center. The moment of inertia of gear A about its axis is I_A, and of gear B about its axis is I_B. If gear A is driven by a torque G with respect to its center, what will be the angular accelerations of gears A and B?

17. A homogeneous solid hemisphere is held with its base against a smooth vertical wall and its lowest point on a smooth floor. The hemisphere is released. Find the initial reactions of the wall and the floor.

18. A uniform solid hemisphere, of radius a, is initially held so that its curved surface is in contact with a smooth vertical wall and with a smooth horizontal floor, the plane face being parallel to the wall. If the solid is released from rest, show that it remains in contact with the wall until the plane face is horizontal and find the velocity of the center of mass in this position. Show that when the angular velocity subsequently vanishes, the plane face makes an angle $\cos^{-1}(45/128)$ with the horizontal.

19. A uniform solid circular cylinder, of mass m and radius a, rolls without slipping on the inside of a fixed rough hollow circular cylinder of internal radius b that has its axis horizontal. The angular velocity of the moving cylinder in its lowest position is Ω. Show that it will roll completely around the inside of the hollow cylinder if $3a^2\Omega^2 \geq 11g(b-a)$. Find the force of friction between the two cylinders.

20. A homogeneous sphere of radius a, which is initially resting in equilibrium on top of a perfectly rough fixed sphere of radius $b > a$, is given a small push. Find the angle that the line joining the centers of the two spheres makes with the vertical when the two spheres separate.

21. A uniform sphere of radius a is placed on a fixed rough sphere also of radius a and is released from rest when the line of centers makes an angle α with the vertical. If the surfaces are sufficiently rough to prevent slipping, show that in the ensuing motion the moving sphere will leave the fixed sphere when the line of centers makes with the vertical an angle $\cos^{-1}\chi$ where $17\chi = 10\cos\alpha$, and that the corresponding angular velocity of the moving sphere is then $(2g\chi/a)^{1/2}$.

22. A homogeneous sphere of radius a is set spinning with an angular velocity Ω about a horizontal axis. It is then released, just touching the

surface of a horizontal table. If the coefficient of sliding friction is μ, how far will the ball go before pure rolling sets in?

23. A homogeneous sphere is projected, without rotation, up a rough inclined plane of inclination α and coefficient of friction μ. Show that the time during which the sphere ascends the plane is the same as if the plane were smooth and that the ratio of the time during which the sphere slides to the time during which it rolls is $(2\tan\alpha)/7\mu$.

24. A circular hoop of mass m and radius a is projected with velocity v down a rough inclined plane of inclination α. At the same time the hoop is given an angular velocity such as to tend to make it roll up the plane. Initially the hoop is at a vertical height h above the foot of the plane. The hoop is constrained to remain in a vertical plane and it just comes to rest at the foot of the plane. Find the coefficient of friction between the hoop and the inclined plane and the initial angular velocity of the hoop.

25. A uniform rod is supported against a smooth fixed sphere by a horizontal string fastened to its upper end and to the highest point of the sphere. Show that if the string is cut, the reaction between the sphere and the rod is changed instantaneously by a factor $(\cos^2\alpha)/(1 + 3\sin^4\alpha)$, where α is the inclination of the rod to the horizontal.

26. A rough uniform rod, of mass m and length $2a$, is freely pivoted at one end and is supported at the other end in a horizontal position. A particle of mass αm is placed on the rod at its midpoint. The coefficient of friction between the particle and the rod is μ. Show that if the support is removed:
   a. The reaction at the hinge is instantaneously reduced by a factor $2/(4 + 3\alpha)$;
   b. The particle begins to slide when the rod is inclined to the horizontal at an angle θ given by $\mu = (10 + 9\alpha)\tan\theta$.

27. Two equal uniform rods AB and BC are smoothly hinged together at B, and the end A is smoothly hinged to a fixed point. The rods are held in a vertical plane with A and C on the same level and the angle ABC a right angle. Show that when the rods are released the ratio of the initial angular accelerations is $3:4$.

28. A pair of wheels joined by an axle stands on a rough slope, of inclination α, with the axle horizontal. The radius of the axle is a and the radius of each wheel is b. The system is axially symmetric, its mass is m and the moment of inertia of the system about its axis is mk^2. A light string is wound around the axle and emerges from the underside. It then passes up the plane parallel to the line of greatest slope, runs over a smooth peg, and hangs vertically with a particle, also of mass m, suspended from its end. The system is released from rest. Show that if the wheels roll without slipping, they roll down or up the plane according as a is greater

than or less than $b(1 - \sin\alpha)$. If slipping should take place, show that it will be such that the points of contact move up the plane.

29. Two uniform cog wheels have masses m_1, m_2, radii a_1, a_2, and moments of inertia I_1, I_2 about their axes. Show that if they rotate with angular speeds ω_1, ω_2 on the same axle and are suddenly locked together, the final angular speed is $(I_1\omega_1 + I_2\omega_2)/(I_1 + I_2)$. Show that if they rotate on parallel axes, and are suddenly meshed at their circumference, the final angular speeds are v/a_1, v/a_2, where $v = (I_1\omega_1 a_1 a_2^2 - I_2\omega_2 a_2 a_1^2)/(I_1 a_2^2 + I_2 a_1^2)$.

30. One end A of a uniform rod AB of mass $3m$ and length $2a$ is freely pivoted at a fixed point on a smooth horizontal plane. The other end B of the rod rests against a face of a smooth uniform cube of mass m free to move with one face in contact with the plane, the vertical plane through AB being a plane of symmetry. If the system is released from rest, show that in the subsequent motion $2a\dot\theta^2(1 + \sin^2\theta) = 3g(\sin\alpha - \sin\theta)$, where θ is the inclination of the rod to the horizontal and α is the initial value of θ. Show also that the initial reaction at B is $(3mg\sin\alpha\cos\alpha/[2(1 + \sin^2\alpha)]$.

31. A smooth solid cube of mass M and edge a rests on a smooth horizontal table. In the plane through the center and parallel to two vertical faces of the cube, a uniform rod, of mass $M/3$ and length $4a$, is placed with one end on the table and the other end leaning against a vertical face of the cube just at its highest edge. If the system is released from rest in this position, prove that at a subsequent instant when the rod makes an angle θ with the horizontal and is still in contact with the cube, $\dot\theta^2 = 3g(1 - 4\sin\theta)/[a(16 - 3\sin^2\theta)]$. Show also that the rod and cube separate when $\sin\theta$ has the value determined by the equation $3\sin^3\theta - 48\sin\theta + 8 = 0$.

32. A uniform solid rough sphere of radius a and mass m rests on a uniform rough board of mass M that lies on a smooth horizontal table. If the board is projected with a velocity v along the table, show that the sphere will slip on the board for a time $2Mv/[\mu g(7M + 2m)]$, where μ is the coefficient of friction between the sphere and the board. Find the velocity of the board at this instant and the angular velocity with which the sphere starts to roll.

33. A sphere of mass m rolls down the rough face of an inclined plane of mass M and angle α that is free to slide on a smooth horizontal plane in a direction perpendicular to its edge. Show that the normal force between the sphere and the inclined plane is $mg[(2m + 7M)\cos\alpha]/[(2 + 5\sin^2\alpha)m + 7M]$.

34. A wedge of mass M rests with one face on a horizontal table and one face inclined at an angle α to the horizontal. A uniform sphere, of mass

m, is projected without rotation down a line of greatest slope of the inclined face. The coefficients of static and kinetic friction between the wedge and the table and between the wedge and the sphere are $\tan\lambda$. Prove that if $2m\cos\alpha\sin(\alpha - 2\lambda) > M\sin 2\lambda$, the wedge will move and its acceleration, while the sphere is sliding, is $\{[2m\cos\alpha\sin(\alpha - 2\lambda) - M\sin 2\lambda]g\}/\{2[M\cos^2\lambda + m\sin\alpha\sin(\alpha - 2\lambda)]\}$.

38

The Inertia Tensor

INTRODUCTION

In the chapters that follow we take up the problem of the general motion of a rigid body under the action of a set of forces. One of our most important discoveries will be that we do not need to know the exact mass distribution in the body to determine the motion of the body; rather it is sufficient to know the total mass, the location of the center of mass, and the value, for some point in the body, of a quantity called the inertia tensor. In this chapter we consider the nature of the inertia tensor. Its importance will become evident later.

THE MOMENT OF INERTIA

Let us consider a system of particles. The *moment of inertia* of the system with respect to an axis that passes through a point o and is in the direction of the unit vector $\mathbf{n}$ is defined as follows:

$$I(o,\mathbf{n}) \equiv \sum_p m(p)[\rho(p)]^2 \tag{1}$$

where $m(p)$ is the mass of the pth particle and $\rho(p)$ is the distance of the pth particle from the axis. If we let $\mathbf{r}(op)$ be the position of the pth particle with respect to the point o and note for a given particle that

$$\rho^2 = (\mathbf{n} \times \mathbf{r}) \cdot (\mathbf{n} \times \mathbf{r}) \tag{2}$$

then we can write Eq. (1) as follows:

$$I(o,\mathbf{n}) \equiv \sum_p m(p)[\mathbf{n} \times \mathbf{r}(op)] \cdot [\mathbf{n} \times \mathbf{r}(op)] \tag{3}$$

In terms of Cartesian coordinates

$$\mathbf{n} = \mathbf{e}_1 n_1 + \mathbf{e}_2 n_2 + \mathbf{e}_3 n_3 \tag{4}$$

$$\mathbf{r}(op) = \mathbf{e}_1 x_1(op) + \mathbf{e}_2 x_2(op) + \mathbf{e}_3 x_3(op) \tag{5}$$

Substituting Eqs. (4) and (5) in Eq. (2) we obtain

$$I(o,\mathbf{n}) \equiv \sum_p m(p) \sum_i \sum_j \left[r^2(op)\delta_{ij} - x_i(op)x_j(op) \right] n_i n_j \tag{6}$$

Equations (1), (3), and (6) provide us with three equivalent ways of expressing the moment of inertia with respect to a particular axis.

THE INERTIA TENSOR

If we define

$$I_{ij}(o) \equiv \sum_p m(p) \left[r^2(op)\delta_{ij} - x_i(op)x_j(op) \right] \tag{7}$$

we can write Eq. (6) as follows:

$$I(o,\mathbf{n}) \equiv \sum_i \sum_j I_{ij}(o) n_i n_j \tag{8}$$

The set of quantities $I_{ij}(o)$ is called the *inertia tensor* for the point o.

If we use Eq. (8) to calculate the moment of inertia about an axis through o and in the direction of the i coordinate axis, that is, $I(o, \mathbf{e}_i)$, we obtain $I(o, \mathbf{e}_i) = I_{ii}(o)$. Thus the terms $I_{11}(o)$, $I_{22}(o)$, and $I_{33}(o)$ are the respective moments of inertia about axes through o and in the 1, 2, and 3 directions. This could also have been seen directly from Eq. (7).

The negatives of the terms $I_{12}(o)$, $I_{21}(o)$, $I_{13}(o)$, $I_{31}(o)$, $I_{23}(o)$, and $I_{32}(o)$ are usually called *products of inertia* with respect to the point o and are designated by the notation $P_{ij}(o)$, that is, for $i \neq j$ we have $P_{ij}(o) \equiv -I_{ij}(o) \equiv \sum_p m(p) x_i(op)x_j(op)$.

NOTATION: We frequently suppress the notation (o) indicating the point through which a given axis passes, or the point for which the inertia tensor is calculated. However this shortcut should not make one lose sight of the dependence of the moment of inertia about a given axis not only on the direction but also on the location of the axis, and the dependence of the inertia tensor on the point with respect to which it is calculated. With this simplification of notation we have

$$I_{ij} = \sum_p m(p) \left[r^2(p)\delta_{ij} - x_i(p)x_j(p) \right] \tag{9}$$

$$I(\mathbf{n}) = \sum_i \sum_j I_{ij} n_i n_j \tag{10}$$

It is sometimes helpful to write Eqs. (7) and (8) in matrix form, thus:

$$[I_{ij}] = \sum_p m(p) \begin{bmatrix} r^2(p) - x_1^2(p) & -x_1(p)x_2(p) & -x_1(p)x_3(p) \\ -x_1(p)x_2(p) & r^2(p) - x_2^2(p) & -x_2(p)x_3(p) \\ -x_1(p)x_3(p) & -x_2(p)x_3(p) & r^2(p) - x_3^2(p) \end{bmatrix} \quad (11)$$

$$I(\mathbf{n}) = \begin{bmatrix} n_1 n_2 n_3 \end{bmatrix} \begin{bmatrix} I_{11} & I_{12} & I_{13} \\ I_{21} & I_{22} & I_{23} \\ I_{31} & I_{32} & I_{33} \end{bmatrix} \begin{bmatrix} n_1 \\ n_2 \\ n_3 \end{bmatrix} \quad (12)$$

TRANSFORMATION OF THE INERTIA TENSOR FROM ONE FRAME TO ANOTHER

If we know the components I_{ij} of the inertia tensor with respect to one Cartesian frame, the components with respect to a second Cartesian frame can be found by making use of the following theorem.

Theorem 1. Let S be a Cartesian frame with axes 1, 2, and 3 in the directions of the unit vectors $\mathbf{e}_1$, $\mathbf{e}_2$, and $\mathbf{e}_3$; and let S' be a second Cartesian frame with axes $1'$, $2'$, and $3'$ in the directions of the unit vectors $\mathbf{e}_{1'}$, $\mathbf{e}_{2'}$, and $\mathbf{e}_{3'}$. The components I_{ij} in the frame S, and the components $I_{i'j'}$ in the frame S' of the inertia tensor with respect to a given point, are related as follows:

$$I_{i'j'} = \sum_r \sum_s e_{i'r} e_{j's} I_{rs} \quad (13)$$

where

$$e_{i'j} \equiv \mathbf{e}_{i'} \cdot \mathbf{e}_j \quad (14)$$

NOTE 1: The quantity $e_{i'j} \equiv \mathbf{e}_{i'} \cdot \mathbf{e}_j$ is just the jth component in the frame S of the vector $\mathbf{e}_{i'}$, or alternatively the i'th component in the frame S' of the vector $\mathbf{e}_j$. Thus the notation is a very natural notation, since it is customary to designate the ith component of a vector $\mathbf{A}$ as simply A_i.

NOTE 2: Any set of quantities $A_{i_1 i_2 \ldots i_n}$ defined with respect to a Cartesian frame that under a rotation, translation, or inversion of axes obeys the transformation law

$$A_{i'_1 i'_2 \ldots i'_n} = \sum_{j_1} \sum_{j_2} \cdots \sum_{j_n} e_{i'_1 j_1} e_{i'_2 j_2} \ldots e_{i'_n j_n} A_{j_1 j_2 \ldots j_n} \quad (15)$$

is called a Cartesian tensor of rank n (see Appendix 6). Thus the set of quantities I_{ij} is a Cartesian tensor of rank 2. This justifies the terminology "inertia *tensor*."

PROOF 1: We note first from the definition of I_{rs} that

$$\sum_r \sum_s e_{i'r} e_{j's} I_{rs} = \sum_r \sum_s e_{i'r} e_{j's} \sum_p m(p)[r^2(p)\delta_{rs} - x_r(p)x_s(p)]$$

$$= \sum_p m(p)\left[r^2(p)\sum_r \sum_s e_{i'r} e_{j's} \delta_{rs} - \sum_r e_{i'r} x_r(p) \sum_s e_{j's} x_s(p)\right] \quad (16)$$

But

$$\sum_r \sum_s e_{i'r} e_{j's} \delta_{rs} = \sum_r e_{i'r} e_{j'r} = \mathbf{e}_{i'} \cdot \mathbf{e}_{j'} = \delta_{i'j'} \quad (17)$$

and

$$\sum_r e_{i'r} x_r(p) = \mathbf{e}_{i'} \cdot \mathbf{r}(p) = x_{i'}(p) \quad (18)$$

Using Eqs. (17) and (18) in Eq. (16) we obtain

$$\sum_r \sum_s e_{i'r} e_{j's} I_{rs} = \sum_p m(p)[r^2(p)\delta_{i'j'} - x_{i'}(p)x_{j'}(p)] = I_{i'j'} \quad (19)$$

which completes the proof of the theorem.

PROOF 2: The theorem could also have been proved using the properties of Cartesian tensors (see Appendix 6). The proof proceeds as follows: x_i is a tensor of rank 1; $x_i x_j$ is the exterior product of two tensors and is therefore a tensor; $r^2 = \sum_i x_i x_i$ is the contraction of a tensor and is therefore a tensor; δ_{ij} is a tensor of rank 2; $r^2 \delta_{ij}$ is the exterior product of two tensors and is therefore a tensor; $r^2 \delta_{ij} - x_i x_j$ is the sum of two tensors, hence is a tensor; and $I_{ij} \equiv \sum_p m(p)[r^2(p)\delta_{ij} - x_i(p)x_j(p)]$ is a linear combination of a set of tensors, hence is a tensor.

THE RELATIONSHIP BETWEEN INERTIA TENSORS AT DIFFERENT POINTS

If we know the inertia tensor at one point, we can find it at a second point by making use of the following theorem.

Theorem 2. The inertia tensor of a dynamical system with respect to an arbitrary point a is given by

$$I_{ij}(a) = I_{ij}(c) + M[R^2(a)\delta_{ij} - X_i(a)X_j(a)] \quad (20)$$

where $I_{ij}(c)$ is the inertia tensor with respect to the center of mass c, M is the total mass of the system, and $\mathbf{R}(a) \equiv \mathbf{e}_1 X_1(a) + \mathbf{e}_2 X_2(a) + \mathbf{e}_3 X_3(a)$ is the vector position of the center of mass c of the system with respect to the point a.

NOTE 1: The second term on the right hand side of Eq. (20) is the inertia tensor with respect to the point a of a system consisting of a single particle of mass M located at the center of mass c. Thus the result above is analogous to the results we obtained for the relationships between the angular momentum and kinetic energy with respect to a point a and the angular momentum and kinetic energy with respect to the center of mass.

NOTE 2: The parallel axis theorem introduced in the preceding chapter is a special case of Theorem 2.

PROOF: From the definition of $I_{ij}(a)$ we have

$$I_{ij}(a) = \sum_p m(p)\big[r^2(ap)\delta_{ij} - x_i(ap)x_j(ap)\big] \tag{21}$$

where $\mathbf{r}(ap) \equiv \mathbf{e}_1 x_1(ap) + \mathbf{e}_2 x_2(ap) + \mathbf{e}_3 x_3(ap)$ is the position of particle p with respect to point a. But

$$\mathbf{r}(ap) = \mathbf{r}(ac) + \mathbf{r}(cp) \tag{22}$$

where $\mathbf{r}(ac)$ is the position of the center of mass c with respect to the point a and $\mathbf{r}(cp)$ is the position of particle p with respect to the center of mass c. Substituting Eq. (22) in Eq. (21) we obtain

$$\begin{aligned}
I_{ij}(a) = \sum_p m(p)\{&[\mathbf{r}(ac) + \mathbf{r}(cp)] \cdot [\mathbf{r}(ac) + \mathbf{r}(cp)]\delta_{ij} \\
&- [x_i(ac) + x_i(cp)][x_j(ac) + x_j(cp)]\} \\
= \sum_p m(p)&\big[r^2(cp)\delta_{ij} - x_i(cp)x_j(cp)\big] \\
+ \Big[\sum_p m(p)\Big]&\big[r^2(ac)\delta_{ij} - x_i(ac)x_j(ac)\big] \\
+ \Big[\sum_p m(p)\mathbf{r}(cp)\Big]&\cdot \mathbf{r}(ac)\delta_{ij} - \Big[\sum_p m(p)x_i(cp)\Big]x_j(ac) \\
+ \mathbf{r}(ac)\cdot\Big[\sum_p m(p)\mathbf{r}(cp)\Big]&\delta_{ij} - x_i(ac)\Big[\sum_p m(p)x_j(cp)\Big] \quad (23)
\end{aligned}$$

But

$$\sum_p m(p)\big[r^2(cp)\delta_{ij} - x_i(cp)x_j(cp)\big] \equiv I(c) \tag{24}$$

$$\Big[\sum_p m(p)\Big]\big[r^2(ac)\delta_{ij} - x_i(ac)x_j(ac)\big] \equiv M\big[R^2(a)\delta_{ij} - X_i(a)X_j(a)\big] \tag{25}$$

And from the definition of the center of mass

$$\sum_p m(p)\mathbf{r}(cp) \equiv M\mathbf{r}(cc) = 0 \tag{26}$$

Substituting Eqs. (24)–(26) in Eq. (23) we obtain Eq. (20). This completes the proof of the theorem.

It follows from the theorem above that if we know the inertia tensor with respect to any point, together with the total mass and the location of the center of mass, then we can find the inertia tensor with respect to any other point.

THE INERTIA ELLIPSOID

A vector may be represented analytically by its components and geometrically by a directed line segment. In an analogous fashion the inertia tensor may be represented analytically by its components and geometrically by a certain ellipsoid, which we call the inertia ellipsoid.

The inertia ellipsoid for the point o is the ellipsoid defined by the equation

$$\sum_i \sum_j I_{ij}(o) x_i x_j = \alpha^2 \qquad (27)$$

where α^2 is an arbitrary positive constant with the dimensions ML^4. The value of this constant is usually chosen to be unity.

The ellipsoid so defined can be shown to be independent of our choice of coordinate frame. Hence for fixed α there is a unique ellipsoid associated with a given intertia tensor. The inertia ellipsoid provides us with a geometrical representation of the inertia tensor. Given the inertia tensor, we can construct the inertia ellipsoid. Conversely given the inertia ellipsoid, we can determine the inertia tensor.

The inertia tensor with respect to a point o can be used to determine the moment of inertia about any axis through o. In a similar fashion the inertia ellipsoid with respect to a point o can be used geometrically to determine the moment of inertia about any axis through o. The method for doing this is explained in the following theorem.

Theorem 3. Given the inertia ellipsoid

$$\sum_i \sum_j I_{ij}(o) x_i x_j = \alpha^2 \qquad (28)$$

the moment of inertia $I(\mathbf{n})$ for arbitrary $\mathbf{n}$ is given by

$$I(\mathbf{n}) = \frac{\alpha^2}{[r(\mathbf{n})]^2} \qquad (29)$$

where $r(\mathbf{n})$ is the distance along the direction $\mathbf{n}$ from the point o to the surface of the inertia ellipsoid, which we assume to be centered at the point o.

PROOF: The coordinates of a point on the surface of the inertia ellipsoid in the direction $\mathbf{n}$ from o are given by

$$x_i = n_i r(\mathbf{n}) \qquad (30)$$

Substituting Eq. (30) in Eq. (28) we obtain

$$\sum_i \sum_j I_{ij}(o) n_i r(\mathbf{n}) n_j r(\mathbf{n}) = \alpha^2 \qquad (31)$$

which can be rewritten

$$\sum_i \sum_j I_{ij}(o) n_i n_j = \frac{\alpha^2}{[r(\mathbf{n})]^2} \qquad (32)$$

But

$$\sum_i \sum_j I_{ij}(o) n_i n_j = I(\mathbf{n}) \qquad (33)$$

Substituting Eq. (33) in Eq. (32) we obtain Eq. (29), which completes the proof of the theorem.

PRINCIPAL AXES

It is always possible to choose a frame $\bar{S}$, with axes $\bar{1}$, $\bar{2}$, and $\bar{3}$, such that the equation for the inertia ellipsoid takes the form

$$\sum_i I_{\bar{i}} x_{\bar{i}}^2 = \alpha^2 \qquad (34)$$

or equivalently such that the inertia tensor is diagonal, that is,

$$I_{\bar{i}\bar{j}} = I_{\bar{i}} \delta_{\bar{i}\bar{j}} \qquad (35)$$

We use the term *principal axes* for the three axes $\bar{1}$, $\bar{2}$, and $\bar{3}$ for which this is true. The corresponding directions are called *principal directions* and are designated by the unit vectors $\mathbf{e}_{\bar{1}}$, $\mathbf{e}_{\bar{2}}$, and $\mathbf{e}_{\bar{3}}$. The moments of inertia associated with the principal axes are called *principal moments of inertia*. From the equations above it is apparent that the quantitites $I_{\bar{1}}$, $I_{\bar{2}}$, and $I_{\bar{3}}$ are the principal moments of inertia.

The following theorem tells us how to find the principal axes and the principal moments of inertia from knowledge of the inertia tensor with respect to some frame.

Theorem 4. Consider a dynamical system for which the inertia tensor I_{ij} for a given point is known.

   a. The principal moments of inertia $I_{\bar{1}}$, $I_{\bar{2}}$, and $I_{\bar{3}}$ are equal to the three roots of the cubic equation

$$|I_{ij} - I_{\bar{k}} \delta_{ij}| = 0 \qquad (36)$$

   where the notation $|\cdots|$ stands for the determinant of the matrix $[\cdots]$.

   b. Given a principal moment of inertia $I_{\bar{k}}$, the direction $\mathbf{e}_{\bar{k}}$ of the corresponding principal axis can be obtained by solving the set of simultaneous equations

$$\sum_j (I_{ij} - I_{\bar{k}}\delta_{ij})e_{\bar{k}j} = 0 \qquad i = 1, 2, 3 \tag{37}$$

for the components $e_{\bar{k}j}$ of the vector $\mathbf{e}_{\bar{k}}$.

NOTE: We give two proofs of Theorem 4, a geometrical proof and an analytical proof.

GEOMETRICAL PROOF: Consider the inertia ellipsoid. Let $r(\mathbf{n})$ be the distance along the direction $\mathbf{n}$ from the origin of the ellipsoid to the surface. The principal directions are the directions for which $r(\mathbf{n})$ has a stationary value. Since the moment of inertia $I(\mathbf{n})$ associated with the direction $\mathbf{n}$ is proportional to $1/r^2(\mathbf{n})$, it follows that the principal directions also are the directions for which $I(\mathbf{n})$ has a stationary value.

We therefore want to find the values of n_1, n_2, and n_3 consistent with the constraint

$$\sum_i n_i^2 = 1 \tag{38}$$

for which the function

$$I(n_1, n_2, n_3) \equiv \sum_i \sum_j I_{ij} n_i n_j \tag{39}$$

has a stationary value. This is equivalent (see Appendix 15) to finding the values of n_1, n_2, n_3 for which the function

$$F(n_1, n_2, n_3, \lambda) \equiv \sum_i \sum_j I_{ij} n_i n_j - \lambda \left(\sum_i n_i^2 - 1 \right) \tag{40}$$

has a stationary value. The necessary and sufficient conditions for this to be true are

$$\frac{\partial F(\mathbf{n}, \lambda)}{\partial n_i} = 0 \qquad i = 1, 2, 3 \tag{41}$$

$$\frac{\partial F(\mathbf{n}, \lambda)}{\partial \lambda} = 0 \tag{42}$$

Equation (42) is equivalent to the condition represented by Eq. (38). Equation (41) gives us

$$\sum_j (I_{ij} - \lambda \delta_{ij}) n_j = 0 \qquad i = 1, 2, 3 \tag{43}$$

The set of equations (43) is a set of three simultaneous equations in the variables n_1, n_2, and n_3. This set will have a nontrivial solution if and only if the determinant of the coefficients of the n_i, vanish, that is,

$$|I_{ij} - \lambda \delta_{ij}| = 0 \tag{44}$$

Solving Eq. (44) we obtain three values of λ. If the three values are distinct, each value when substituted into the set of simultaneous equations (43) will yield a unique set of values of n_1, n_2, and n_3 that will determine one of the directions in which $I(\mathbf{n})$ has a stationary value. We label the three values of λ as $\lambda_{\bar{1}}$, $\lambda_{\bar{2}}$, and $\lambda_{\bar{3}}$ and the corresponding values of $\mathbf{n}$, which are the principal directions, as $\mathbf{e}_{\bar{1}}$, $\mathbf{e}_{\bar{2}}$, and $\mathbf{e}_{\bar{3}}$, respectively. If two of the λ's have the same value, the resulting set of simultaneous equations corresponding to this value will determine only that the direction $\mathbf{n}$ lies in a particular plane. In this case we simply pick any two orthogonal unit vectors in this plane for the directions corresponding to the two λ's. If all three of the λ's have the same value, any $\mathbf{n}$ will satisfy the set of simultaneous equations corresponding to this value. In this case we pick any three orthogonal unit vectors for the directions corresponding to the three λ's.

To complete the proof we now show that the quantity $\lambda_{\bar{k}}$ is just the principal moment of inertia corresponding to the principal direction $\mathbf{e}_{\bar{k}}$. We note first that the components $e_{\bar{k}}$ of the vector $\mathbf{e}_{\bar{k}}$ satisfy the set of simultaneous equations

$$\sum_j (I_{ij} - \lambda_{\bar{k}} \delta_{ij}) e_{\bar{k}j} = 0 \tag{45}$$

Multiplying Eq. (45) by $e_{\bar{k}i}$ and summing over i we obtain

$$\sum_i \sum_j I_{ij} e_{\bar{k}i} e_{\bar{k}j} = \lambda_{\bar{k}} \sum_i \sum_j e_{\bar{k}i} e_{\bar{k}j} \delta_{ij} \tag{46}$$

But

$$\sum_i \sum_j I_{ij} e_{\bar{k}i} e_{\bar{k}j} = I(\mathbf{e}_{\bar{k}}) \equiv I_{\bar{k}} \tag{47}$$

and

$$\sum_i \sum_j e_{\bar{k}i} e_{\bar{k}j} \delta_{ij} = \sum_i e_{\bar{k}i} e_{\bar{k}i} = \mathbf{e}_{\bar{k}} \cdot \mathbf{e}_{\bar{k}} = 1 \tag{48}$$

Substituting Eqs. (47) and (48) in Eq. (46) we obtain

$$\lambda_{\bar{k}} = I_{\bar{k}} \tag{49}$$

This completes the proof of the theorem.

ANALYTICAL PROOF. Given the inertia tensor I_{ij} in the frame S, we wish to find a frame $\bar{S}$ such that

$$I_{\bar{i}\bar{j}} \equiv \sum_k \sum_l e_{\bar{i}k} e_{\bar{j}l} I_{kl} = I_{\bar{i}} \delta_{\bar{i}\bar{j}} \tag{50}$$

Multiplying Eq. (50) by $e_{r\bar{i}}$ and summing over $\bar{i}$ we obtain

$$\sum_{\bar{i}} \sum_k \sum_l e_{r\bar{i}} e_{\bar{i}k} e_{\bar{j}l} I_{kl} = \sum_{\bar{i}} e_{r\bar{i}} I_{\bar{i}} \delta_{\bar{i}\bar{j}} \tag{51}$$

Noting that

$$\sum_i e_{ri}e_{ik} = \sum_i e_{ri}e_{ki} = \mathbf{e}_r \cdot \mathbf{e}_k = \delta_{rk} \quad (52)$$

we can write the left hand side of Eq. (51) as follows:

$$\sum_i \sum_k \sum_l e_{ri}e_{ik}e_{jl}I_{kl} = \sum_k \sum_l \delta_{rk}e_{jl}I_{kl} = \sum_l e_{jl}I_{rl} \quad (53)$$

The right hand side of Eq. (51) can be written

$$\sum_i e_{ri}I_j\delta_{ij} = e_{rj}I_j = \sum_l I_j\delta_{rl}e_{jl} \quad (54)$$

Substituting Eqs. (53) and (54) in Eq. (51) and rearranging we obtain

$$\sum_l (I_{rl} - I_j\delta_{rl})e_{jl} = 0 \quad (55)$$

For each value of j we obtain from Eq. (55) a set of simultaneous equations for the variables e_{j1}, e_{j2}, and e_{j3}. A necessary and sufficient condition that this set have a solution is

$$|I_{rl} - I_j\delta_{rl}| = 0 \quad (56)$$

The solution of Eq. (56) gives us the three values for I_j. Substituting a particular I_j back in Eq. (55) gives us a set of simultaneous equations for the components e_{jk} of the vector $\mathbf{e}_j$, hence enables us to find $\mathbf{e}_j$. This completes the proof of the theorem.

SOME AUXILIARY THEOREMS FOR THE DETERMINATION OF PRINCIPAL AXES

It is frequently possible to determine one or more principal axes without the tedious machinery introduced in the preceding section. Here we gather a number of theorems that are helpful in this respect.

Theorem 5. If the distribution of masses in a dynamical system is unaltered by a rotation through an angle θ about an axis and if θ is not an integer multiple of 2π, the axis is a principal axis with respect to all points on the axis. If furthermore the angle θ is not an integer multiple of π, any pair of orthogonal axes that are perpendicular to the axis of rotation will be principal axes with respect to all points on the axis of rotation.

PROOF: When a body is rotated about a line, the ellipsoid of inertia at any point on the line turns with the body. If the mass distribution is unaltered by the rotation, the inertia ellipsoid at the point must also be unaltered by the rotation. But an ellipsoid can be rotated about an axis through its center through an angle θ that is not an integer multiple of 2π and remain unchanged only if the axis of rotation is one of the principal axes of the ellipsoid, hence the axis of rotation will be a principal axis. If furthermore the angle θ is not an integer multiple of π the ellipsoid must be an ellipsoid of revolution about this axis; hence any pair of orthogonal axes that are perpendicular to the axis of rotation will be principal axes.

Theorem 6. If the distribution of masses in a dynamical system is unaltered by a reflection in a plane, then any axis perpendicular to the plane is a principal axis for the point at which the axis passes through the plane.

PROOF: When a body is reflected in a plane, the ellipsoid of inertia at any point on the plane is also reflected in the plane. If the mass distribution is unaltered by the reflection, the inertia ellipsoid must also be unaltered by the reflection. But an ellipsoid can be reflected in a plane passing through its center and remain unchanged only if the plane of reflection contains two of the principal axes of the ellipsoid, hence any axis perpendicular to the plane of reflection will be a principal axis.

Theorem 7. If we know the direction of one of the principal axes, which for convenience we let be the $\bar{3}$ axis, and if we orient the 3 axis of our coordinate frame in the $\bar{3}$ direction, the directions of the other two principal axes are perpendicular to the 3 axis and make angles $\theta_{\bar{1}}$ and $\theta_{\bar{2}}$, respectively, with the 1 axis, where $\theta_{\bar{1}}$ and $\theta_{\bar{2}}$ are the two values of θ that satisfy the equation

$$\tan 2\theta = \frac{2I_{12}}{I_{11} - I_{22}} \tag{57}$$

PROOF: The moment of inertia about an axis in an arbitrary direction **n** is given by

$$I(\mathbf{n}) = I_{11}n_1^2 + I_{22}n_2^2 + I_{33}n_3^2 + 2I_{12}n_1n_2 + 2I_{13}n_1n_3 + 2I_{23}n_2n_3 \tag{58}$$

Since one of the principal axes is in the 3 direction, we know that the other two are perpendicular to the 3 direction; hence for these axes

$$n_3 = 0 \tag{59}$$

If we let θ be the angle between one of these principal axes and the 1 axis, then for this axis

$$n_1 = \cos\theta \qquad n_2 = \sin\theta \tag{60}$$

Substituting Eqs. (59) and (60) in (58) we obtain

$$I = I_{11}\cos^2\theta + I_{22}\sin^2\theta + 2I_{12}\sin\theta\cos\theta \tag{61}$$

The value of θ for which Eq. (61) has a stationary value, hence the value of θ

for the principal axis, is determined by the condition

$$\frac{dI}{d\theta} = -2I_{11}\cos\theta\sin\theta + 2I_{22}\sin\theta\cos\theta + 2I_{12}(\cos^2\theta - \sin^2\theta)$$
$$= -(I_{11} - I_{22})\sin 2\theta + 2I_{12}\cos 2\theta = 0 \tag{62}$$

Eq. (57) follows immediately from Eq. (62). This completes the proof of the theorem.

Theorem 8. A coordinate axis is a principal axis if and only if all the off diagonal terms in the inertia tensor that are associated with that coordinate axis vanish. For example the 3 axis is a principal axis if and only if $I_{13} = I_{23} = 0$.

PROOF: The moment of inertia in the direction **n** is given by

$$I(n_1, n_2, n_3) = I_{11}n_1^2 + I_{22}n_2^2 + I_{33}n_3^2 + 2I_{12}n_1n_2$$
$$+ 2I_{13}n_1n_3 + 2I_{23}n_2n_3 \tag{63}$$

If the 3 axis is a principal axis, then from the symmetry of the inertia ellipsoid with respect to this axis it is apparent that the moment of inertia about an axis in the direction n_1, n_2, n_3 will be the same as for an axis in the direction $-n_1, -n_2, n_3$. Thus

$$I(n_1, n_2, n_3) = I(-n_1, -n_2, n_3) \tag{64}$$

Substituting Eq. (63) in Eq. (64) we obtain

$$I_{11}n_1^2 + I_{22}n_2^2 + I_{33}n_3^2 + 2I_{12}n_1n_2 + 2I_{13}n_1n_3 + 2I_{23}n_2n_3$$
$$= I_{11}n_1^2 + I_{22}n_2^2 + I_{33}n_3^2 + 2I_{12}n_1n_2 - 2I_{13}n_1n_3 - 2I_{23}n_2n_3 \tag{65}$$

and thus

$$I_{13}n_1n_3 + I_{23}n_2n_3 = 0 \tag{66}$$

Equation (66) can be true for arbitrary n_1 and n_2 if and only if

$$I_{13} = I_{23} = 0 \tag{67}$$

This completes the proof of the theorem.

Theorem 9. If an axis that passes through the center of mass c is a principal axis with respect to the center of mass, it is a principal axis with respect to any point on the axis.

PROOF: Let c be the center of mass of the system and a any other point. Then from Theorem 2

$$I_{13}(a) = I_{13}(c) - Mx_1(ac)x_3(ac) \tag{68}$$
$$I_{23}(a) = I_{23}(c) - Mx_2(ac)x_3(ac) \tag{69}$$

If the 3 axis is a principal axis with respect to the center of mass, it passes through the center of mass and

$$I_{13}(c) = I_{23}(c) = 0 \tag{70}$$

If the 3 axis also passes through the point a, then

$$x_1(ac) = x_2(ac) = 0 \tag{71}$$

Substituting Eqs. (70) and (71) in Eqs. (68) and (69) we obtain

$$I_{13}(a) = I_{23}(a) = 0 \tag{72}$$

It follows that the 3 axis is a principal axis with respect to the point a. This completes the proof of the theorem.

Theorem 10. If an axis is a principal axis with respect to two different points on the axis, it is a principal axis with respect to all points on the axis, and it passes through the center of mass.

PROOF: Suppose that a and b are two different points. Then

$$I_{13}(a) = I_{13}(c) - Mx_1(ac)x_3(ac) \tag{73}$$

$$I_{13}(b) = I_{13}(c) - Mx_1(bc)x_3(bc) \tag{74}$$

If the 3 axis is a principal axis with respect to both these points, then

$$I_{13}(a) = 0 \qquad I_{13}(b) = 0 \tag{75}$$

Since a and b are both on the 3 axis

$$x_1(ac) = x_1(bc) \tag{76}$$

Combining Eqs. (73)–(76) we obtain

$$x_1(ac)[x_3(ac) - x_3(bc)] = 0 \tag{77}$$

Since a and b are different points, $x_3(ac) \neq x_3(bc)$, hence Eq. (77) can be true only if

$$x_1(ac) = 0 \tag{78}$$

Substituting Eqs. (75) and (78) in Eq. (73) we obtain

$$I_{13}(c) = 0 \tag{79}$$

In a similar fashion we can show

$$I_{23}(c) = 0 \tag{80}$$

From Eqs. (79) and (80) it follows that the 3 axis is a principal axis with respect to the center of mass, hence a principal axis for all other points on the axis.

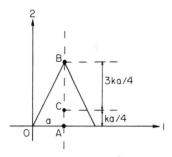

Fig. 1

EXAMPLES

Example 1. Determine the inertia tensor with respect to the origin of a uniform solid cone of mass M whose base is the circle defined by the two equations $(x_1 - a)^2 + x_3^2 = a^2$ and $x_2 = 0$, and whose vertex is at the point $(a, ka, 0)$.

Solution. The intersection of the cone and the 1-2 plane is shown in Fig. 1. The point C is the center of mass and is located at the point $(a, ka/4, 0)$ (refer to Table 4, in the section following the appendices). From symmetry the principal axes for the center of mass C are in the 1, 2 and 3 directions, respectively, and the principal moments of inertia $I_{11}(C)$ and $I_{33}(C)$ are equal. From Table 5

$$I_{11}(C) = I_{33}(C) = \frac{3Ma^2}{20}\left(1 + \frac{k^2}{4}\right) \tag{81}$$

$$I_{22}(C) = \frac{3Ma^2}{10} \tag{82}$$

It follows that the inertia tensor for the center of mass is

$$[I_{ij}(C)] = \frac{3Ma^2}{80}\begin{bmatrix} 4+k^2 & 0 & 0 \\ 0 & 8 & 0 \\ 0 & 0 & 4+k^2 \end{bmatrix} \tag{83}$$

We can now use Theorem 2 to obtain the inertia tensor with respect to the origin. The vector position of the center of mass with respect to the origin is given by $\mathbf{R} = \mathbf{e}_1 a + \mathbf{e}_2(ka/4)$, hence the inertia tensor with respect to the origin of a particle of mass M located at the center of mass is

$$M[R^2\delta_{ij} - X_i X_j] = \frac{Ma^2}{16}\begin{bmatrix} k^2 & -4k & 0 \\ -4k & 16 & 0 \\ 0 & 0 & 16+k^2 \end{bmatrix} \tag{84}$$

Adding Eqs. (83) and (84) we obtain

$$[I_{ij}(O)] = \frac{Ma^2}{20}\begin{bmatrix} 3+2k^2 & -5k & 0 \\ -5k & 26 & 0 \\ 0 & 0 & 23+2k^2 \end{bmatrix} \quad (85)$$

Example 2. The inertia tensor at a given point for a certain system is given by

$$[I_{ij}] = \begin{bmatrix} \alpha & -\beta & -\beta \\ -\beta & \alpha & -\beta \\ -\beta & -\beta & \alpha \end{bmatrix} \quad (86)$$

Determine the principal moments of inertia and the principal axes for the given point.

Solution. From Theorem 4 the principal moments of inertia are the roots of the equation

$$\begin{vmatrix} \alpha - I & -\beta & -\beta \\ -\beta & \alpha - I & -\beta \\ -\beta & -\beta & \alpha - I \end{vmatrix} = 0 \quad (87)$$

Solving Eq. (87) for I and designating the three values as $I_{\bar{1}}$, $I_{\bar{2}}$, and $I_{\bar{3}}$, respectively, we obtain

$$I_{\bar{1}} = \alpha - 2\beta \qquad I_{\bar{2}} = \alpha + \beta \qquad I_{\bar{3}} = \alpha + \beta \quad (88)$$

The components x, y, z of the unit vector corresponding to a principal moment of inertia I satisfy the equation

$$\begin{bmatrix} \alpha - I & -\beta & -\beta \\ -\beta & \alpha - I & -\beta \\ -\beta & -\beta & \alpha - I \end{bmatrix} \begin{bmatrix} x \\ y \\ z \end{bmatrix} = \begin{bmatrix} 0 \\ 0 \\ 0 \end{bmatrix} \quad (89)$$

Setting $I = I_{\bar{1}} = \alpha - 2\beta$ and solving for x, y, and z, we find that any set of values of x, y, z that satisfy the condition

$$x : y : z = 1 : 1 : 1 \quad (90)$$

will satisfy Eq. (90). It follows that the unit vector $\mathbf{e}_{\bar{1}}$ is given by

$$\mathbf{e}_{\bar{1}} = \frac{1}{\sqrt{3}}(\mathbf{e}_1 + \mathbf{e}_2 + \mathbf{e}_3) \quad (91)$$

If we set $I = I_{\bar{2}} = \alpha + \beta$ in Eq. (89) we find that because of the double root, only one of the resulting equations is independent, and any set of values of x, y, and z that satisfies the condition

$$x + y + z = 0 \quad (92)$$

will satisfy Eq. (89) with $I = \alpha + \beta$. This condition simply assures us that the unit vector in question is perpendicular to $e_{\bar{1}}$. If we choose

$$e_{\bar{2}} = \frac{1}{\sqrt{2}}(e_1 - e_2) \qquad (93)$$

then from the orthogonality conditions of the unit vectors $e_{\bar{i}}$ the vector $e_{\bar{3}}$ must be

$$e_{\bar{3}} = \frac{1}{\sqrt{6}}(-e_1 - e_2 + 2e_3) \qquad (94)$$

PROBLEMS

1. Show that none of the principal moments of inertia can exceed the sum of the other two.

2. Show that the sum of the diagonal terms in the inertia tensor at a point is equal to the sum of the principal moments of inertia at the point irrespective of whether the coordinate axes are principal axes.

3. Show that if the particles in a system S all lie in a plane, the moment of inertia of S about a line L perpendicular to the plane is equal to the sum of the moments of inertia of S about any two lines that lie in the plane, are perpendicular to each other, and intersect L.

4. A system consists of a particle of mass $4m$ located at the point $(a, -a, a)$, a particle of mass $3m$ located at the point $(-a, a, a)$, and a particle of mass $2m$ located at the point (a, a, a).
   a. Determine the inertia tensor for the origin of coordinates.
   b. Use the result in **a** to determine the moment of inertia with respect to the axis that passes through the origin in the direction of the unit vector $\mathbf{n} = (1/\sqrt{2})e_1 + (1/\sqrt{2})e_2$.

5. A system consists of a particle of mass $4m$ located at the point $(a, -a, 0)$, a particle of mass $3m$ located at the point $(-a, a, 0)$, and a particle of mass $2m$ located at the point $(a, a, 0)$.
   a. Determine the inertia tensor for the origin of coordinates.
   b. Determine the principal moments of inertia and the principal axes for the origin.

6. A plane rectangular lamina of mass m is located with its corners at the point $(0,0,0,), (a,0,0,), (a,b,0),$ and $(0,b,0)$, respectively.
   a. Determine the inertia tensor for the origin.
   b. Use the result in **a** to determine the moment of inertia with respect to an axis along the diagonal through the origin.
   c. Determine the principal axes and the principal moments of inertia for the case in which $b = 2a$.

The Inertia Tensor

7. A uniform rectangular plate of sides $4a$ and $6a$, respectively, has a circular hole of radius a cut out of its center. Determine the smallest angle between the side of length $6a$ and any principal axis for a corner of the plate.

8. Show that the location of the center of mass and the value of the inertia tensor with respect to any point for a uniform triangular lamina of mass m are the same as those for a system consisting of three particles, each of mass $m/3$, located at the midpoints of the sides of the triangle.

9. Show that the inertial ellipsoid of a cube at the center of the cube is a sphere, and determine the moment of inertia of the cube about any axis passing through the center of the cube.

10. A cube of mass M and side $2a$ is located with one corner at the origin O of a Cartesian reference frame, and the three edges that meet at this corner along the positive x, y, and z axes, respectively.
    a. Show that the equation of the inertial ellipsoid at O is $4(x^2 + y^2 + z^2) - 3(yz + zx + xy) = $ constant.
    b. Show that the moment of inertia about a diagonal is $2Ma^2/3$.
    c. Find the principal axes and the principal moments of inertia for the point O.

11. Calculate the principal moments of inertia of a cone of vertical height h and base radius a, about its vertex. For what value of the ratio h/a is every axis through the vertex a principal axis? For this case find the position of the center of mass and the principal moments of inertia about it.

12. A uniform solid right circular cylinder is of radius a and height $a\sqrt{3}$. Show that one of its principal axes at any point A on the circumference of one end passes through the center of mass of the cylinder and find the principal moments of inertia of the cylinder for the point A.

13. Find the principal moments of inertia of a uniform solid hemisphere of mass M and radius a at its center of mass.

14. Prove that the principal axes of inertia of a thin uniform hemispherical shell of mass M and radius a, at a point of its rim make angles $3\pi/8$, $\pi/8$, and $\pi/2$, respectively, with the axis of the hemisphere. Find the associated principal moments of inertia.

15. Show that the inertial ellipsoid at the center of a uniform solid ellipsoid of mass M and semiaxes a, b, and c is $(b^2 + c^2)x^2 + (c^2 + a^2)y^2 + (a^2 + b^2)z^2 = \epsilon^4$, where ϵ is any constant. Note that the longest and shortest axes of the inertial ellipsoid coincide in direction with the longest and shortest axes, respectively, of the material ellipsoid.

39

Rigid Body Kinematics

INTRODUCTION

Before we discuss the effect of a set of forces on the motion of a rigid body it will be helpful to consider in some detail the preliminary problem of how best to describe the configuration and the motion of a rigid body, and the related problem of how to determine kinematical quantities such as the angular momentum and the kinetic energy of a rigid body. A thorough familiarity with these topics will provide us with the language we need to be able to discuss rigid body dynamics.

THE CONFIGURATION OF A RIGID BODY

Six coordinates are required to specify the configuration of a rigid body with no point fixed: three to specify the position of one point in the body, and three to specify the orientation of the body. The techniques for specifying the position of a point are already quite familiar to us. We therefore concentrate now on the specification of the orientation of a rigid body. To make the discussion as simple as possible, we assume throughout this chapter that we are dealing with a rigid body in which one point is fixed.

EULER ANGLES

There are many sets of coordinates that can be used to specify the orientation of a rigid body One of the simplest sets, and the set most often used, is the set of Euler angles described in this section.

Let us consider a rigid body that is free to rotate about a fixed point O. Such a situation can be achieved by mounting the body on a suspension such

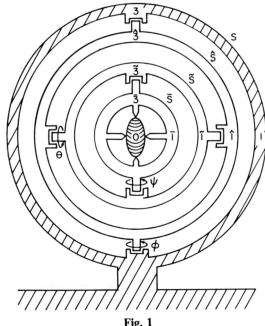

Fig. 1

as the one shown in Fig. 1. The four circular rings S, $\hat{S}$, $\tilde{S}$, and $\bar{S}$ have a common fixed center O, and each can be used to define a Cartesian coordinate frame with origin at O and one and three axes in the directions shown. In the configuration shown the axes 1, $\hat{1}$, $\tilde{1}$ and $\bar{1}$ are all in the same direction, and the axes 3, $\hat{3}$, $\tilde{3}$, and $\bar{3}$ are all in the same direction. It follows that in the configuration shown the axes 2, $\hat{2}$, $\tilde{2}$, and $\bar{2}$ are also in the same direction, pointing into the paper. We assume that the ring S is fixed in space and the other rings are mounted so that $\hat{S}$ is free to rotate through an angle of 360° about the 3 axis of S, $\tilde{S}$ is free to rotate through an angle of 180° about the $\hat{1}$ axis of $\hat{S}$, and $\bar{S}$ is free to rotate through an angle of 360° about the $\tilde{3}$ axis of $\tilde{S}$. We assume further that the rigid body is fixed to the ring $\bar{S}$ and the axes $\bar{1}$, $\bar{2}$, and $\bar{3}$ are principal axes for the point O. The coordinate frames S, $\hat{S}$, $\tilde{S}$, and $\bar{S}$ attached to the rings can be defined analytically as follows:

S A frame fixed in space with origin at O

$\bar{S}$ A frame fixed in the body with origin at O and axes in the principal directions for the point O

$\hat{S}$ A frame with origin at O, axis $\hat{3}$ coincident with axis 3, and axis $\hat{1}$ in the direction of the vector $\mathbf{e}_3 \times \mathbf{e}_{\bar{3}}$

$\tilde{S}$ A frame with origin at O, axis $\tilde{3}$ coincident with axis $\bar{3}$, and axis $\tilde{1}$ in the direction of the vector $\mathbf{e}_3 \times \mathbf{e}_{\bar{3}}$

By rotating the rings or frames $\hat{S}$, $\tilde{S}$, and $\bar{S}$ the rigid body can be brought into any arbitrary orientation. Our objective is to choose a set of three coordinates that uniquely determine the orientation of the rigid body, or equivalently the orientation of frame $\bar{S}$ with respect to frame S. Examination of the suspension system will reveal that for each orientation of the rigid body, except for the case in which the $\bar{3}$ and 3 axes coincide, there is a unique configuration for the set of frames, and conversely with each configuration of the set of frames, except for the case noted above, there is a unique orientation of the rigid body. It follows that any three coordinates that uniquely fix the configuration of the set of frames can be used to specify the orientation of the rigid body. The most obvious coordinates are the angle ϕ that the ring $\hat{S}$ makes with the ring S, the angle θ that the ring $\tilde{S}$ makes with the ring $\hat{S}$, and the angle ψ that the ring $\bar{S}$ makes with the ring $\tilde{S}$. These three angles are called the Euler angles. More precisely the Euler angles ϕ, θ, and ψ are defined as follows:

ϕ The angle that the $\hat{1}$ (or $\tilde{1}$) axis makes with the 1 axis, where the positive sense of ψ corresponds to the positive sense of rotation about the 3 (or $\hat{3}$) axis, and ϕ ranges between 0 and 2π.

θ The angle that the $\tilde{3}$ (or $\bar{3}$) axis makes with the 3 axis, where the positive sense of θ corresponds to the positive sense of rotation about the $\hat{1}$ (or $\tilde{1}$) axis, and θ ranges between 0 and π.

ψ The angle that the $\bar{1}$ axis makes with the $\tilde{1}$ (or $\hat{1}$) axis, where the positive sense of ψ corresponds to the positive sense of rotation about the $\tilde{3}$ (or $\bar{3}$) axis, and ψ ranges between 0 and 2π.

Another suspension that accomplishes the same result as the suspension in Fig. 1 is illustrated in Fig. 2. A rigid body $\bar{S}$ is free to rotate about a diameter of a ring $\tilde{S}$, which is free to rotate inside a second coplanar and concentric ring $\hat{S}$, which is free to rotate about a diameter fixed in space S. The rigid body $\bar{S}$, the ring $\tilde{S}$, the ring $\hat{S}$, and the fixed suspension S can be used to define four coordinate frames with common origin O and the two and three axes in the directions shown. We assume that the axes $\bar{1}$, $\bar{2}$, and $\bar{3}$ are principal axes for the rigid body. The frames S, $\hat{S}$, $\tilde{S}$, and $\bar{S}$ are essentially the same frames we defined previously. In Fig. 2 the four frames are in their reference configuration. By manipulating the suspension system, the rigid body can be brought into any arbitrary orientation. By comparing this suspension with the previous suspension, it can be shown that the Euler angle ϕ is the angle through which the ring $\hat{S}$ has been rotated, the Euler angle θ is the angle through which the ring $\tilde{S}$ has been rotated relative to the ring $\hat{S}$, and the Euler angle ψ is the angle through which the rigid body $\bar{S}$ has been rotated relative to the ring $\tilde{S}$.

From Fig. 2 it is apparent that the pair of angles θ and $\phi - \frac{1}{2}\pi$ is just the pair of spherical angles that fixes the orientation of the principal axis $\bar{3}$ of the

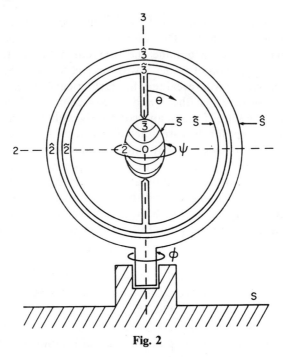

Fig. 2

rigid body with respect to the $\bar{S}$ frame, that is, θ is the angle that the $\bar{3}$ axis makes with the 3 axis, and $\phi - \tfrac{1}{2}\pi$ is the angle that the projection of the $\bar{3}$ axis on the 1-2 plane makes with the 1 axis as shown in Fig. 3.

There is still another way of viewing the Euler angles that is frequently of help. If we define the line of *nodes* as the line of intersection of the 1-2 plane and the $\bar{1}\text{-}\bar{2}$ plane, or equivalently the line lying along the direction of the vector $\mathbf{e}_3 \times \mathbf{e}_{\bar{3}}$ then ϕ is the angle that the line of nodes makes with the 1 axis, θ

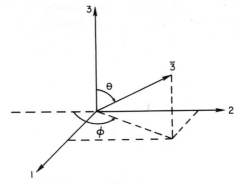

Fig. 3

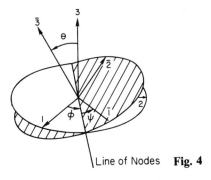

Line of Nodes **Fig. 4**

is the angle that the $\bar{3}$ axis makes with the 3 axis, and ψ is the angle that the $\bar{1}$ axis makes with the line of nodes. This point of view is illustrated in Fig. 4.

THE TRANSFORMATION MATRICES

As we see later in this chapter and in the next chapter, many results can be simplified appreciably by viewing them from one or another of the frames S, $\hat{S}$, $\tilde{S}$, or $\bar{S}$, or possibly from some other frame. It is therefore helpful to consider the techniques one must employ to transform results from one arbitrary frame S' to another arbitrary frame S''. We assume throughout this chapter that we are dealing only with rectangular Cartesian frames, any one of which can be obtained from any other one by a rotation, translation, inversion, or some combination of these operations. From the results of Appendix 6 we can write down the following theorems immediately.

Theorem 1. Let S' and S'' be two different rectangular Cartesian frames. If $A_{i'}$ are the components of a vector in the frame S', then the components $A_{i''}$ of the vector in the frame S'' are given by

$$A_{i''} = \sum_{j'} e_{i''j'} A_{j'} \tag{1}$$

or in matrix form by

$$\begin{bmatrix} A_{1''} \\ A_{2''} \\ A_{3''} \end{bmatrix} = \begin{bmatrix} e_{1''1'} & e_{1''2'} & e_{1''3'} \\ e_{2''1'} & e_{2''2'} & e_{2''3'} \\ e_{3''1'} & e_{3''2'} & e_{3''3'} \end{bmatrix} \begin{bmatrix} A_{1'} \\ A_{2'} \\ A_{3'} \end{bmatrix} \tag{2}$$

where

$$e_{i''j'} \equiv \mathbf{e}_{i''} \cdot \mathbf{e}_{j'} \tag{3}$$

Theorem 2. Let S' and S'' be two different rectangular Cartesian frames. If $A_{i'j'}$ are the components in the frame S' of a Cartesian tensor of rank 2, then the components $A_{i''j''}$ of the tensor in the frame S'' are given by

$$A_{i''j''} = \sum_{k'} \sum_{l'} e_{i''k'} e_{j''l'} A_{k'l'} \tag{4}$$

or in matrix form by

$$\begin{bmatrix} A_{1''1''} & A_{1''2''} & A_{1''3''} \\ A_{2''1''} & A_{2''2''} & A_{2''3''} \\ A_{3''1''} & A_{3''2''} & A_{3''3''} \end{bmatrix}$$

$$= \begin{bmatrix} e_{1''1'} & e_{1''2'} & e_{1''3'} \\ e_{2''1'} & e_{2''2'} & e_{2''3'} \\ e_{3''1'} & e_{3''2'} & e_{3''3'} \end{bmatrix} \begin{bmatrix} A_{1'1'} & A_{1'2'} & A_{1'3'} \\ A_{2'1'} & A_{2'2'} & A_{2'3'} \\ A_{3'1'} & A_{3'2'} & A_{3'3'} \end{bmatrix} \begin{bmatrix} e_{1'1''} & e_{1'2''} & e_{1'3''} \\ e_{2'1''} & e_{2'2''} & e_{2'3''} \\ e_{3'1''} & e_{3'2''} & e_{3'3''} \end{bmatrix} \tag{5}$$

where

$$e_{i''j'} = e_{j'i''} = \mathbf{e}_{i''} \cdot \mathbf{e}_{j'} \tag{6}$$

It follows from the theorems above that if we want to transform the components of a vector or tensor in the frame S' to the components in the frame S'', we need to know the set of quantities $e_{i''j'}$, or equivalently the matrix $[e_{i''j'}]$. The quantities $e_{i''j'}$ are called the *transformation coefficients* for the transformation from the frame S' to the frame S'', and the matrix $[e_{i''j'}]$ is called the *transformation matrix* for the transformation from the frame S' to the frame S''. The following theorem, which is proved in Appendix 6, is useful in dealing with the transformation coefficients or equivalently the transformation matrix.

Theorem 3. Let S', S'', and S''' be three different rectangular Cartesian frames. If we know the transformation coefficients $e_{j''k'}$ and the transformation coefficients $e_{i'''j''}$ then the transformation coefficients $e_{i'''k'}$ are given by

$$e_{i'''k'} = \sum_{j''} e_{i'''j''} e_{j''k'} \tag{7}$$

or in matrix form

$$\begin{bmatrix} e_{1'''1'} & e_{1'''2'} & e_{1'''3'} \\ e_{2'''1'} & e_{2'''2'} & e_{2'''3'} \\ e_{3'''1'} & e_{3'''2'} & e_{3'''3'} \end{bmatrix} = \begin{bmatrix} e_{1'''1''} & e_{1'''2''} & e_{1'''3''} \\ e_{2'''1''} & e_{2'''2''} & e_{2'''3''} \\ e_{3'''1''} & e_{3'''2''} & e_{3'''3''} \end{bmatrix} \begin{bmatrix} e_{1''1'} & e_{1''2'} & e_{1''3'} \\ e_{2''1'} & e_{2''2'} & e_{2''3'} \\ e_{3''1'} & e_{3''2'} & e_{3''3'} \end{bmatrix} \tag{8}$$

THE TRANSFORMATION MATRICES AS FUNCTIONS OF THE EULER ANGLES

In dealing with rigid bodies it is very helpful to be able to express the transformation matrices for the transformations between the frames S, $\hat{S}$, $\tilde{S}$,

The Transformation Matrices as Functions of the Euler Angles

and $\bar{S}$ in terms of the Euler angles ϕ, θ, and ψ. From the definitions of these frames and the definition of the Euler angles given earlier, we can immediately write down the following transformation matrices:

$$[e_{ij}] = \begin{bmatrix} \cos\phi & \sin\phi & 0 \\ -\sin\phi & \cos\phi & 0 \\ 0 & 0 & 1 \end{bmatrix} \quad (9a) \quad [e_{i\hat{j}}] = \begin{bmatrix} \cos\phi & -\sin\phi & 0 \\ \sin\phi & \cos\phi & 0 \\ 0 & 0 & 1 \end{bmatrix} \quad (9b)$$

$$[e_{\hat{i}\hat{j}}] = \begin{bmatrix} 1 & 0 & 0 \\ 0 & \cos\theta & \sin\theta \\ 0 & -\sin\theta & \cos\theta \end{bmatrix} \quad (10a) \quad [e_{\hat{i}\tilde{j}}] = \begin{bmatrix} 1 & 0 & 0 \\ 0 & \cos\theta & -\sin\theta \\ 0 & \sin\theta & \cos\theta \end{bmatrix} \quad (10b)$$

$$[e_{\tilde{i}\tilde{j}}] = \begin{bmatrix} \cos\psi & \sin\psi & 0 \\ -\sin\psi & \cos\psi & 0 \\ 0 & 0 & 1 \end{bmatrix} \quad (11a) \quad [e_{\tilde{i}\bar{j}}] = \begin{bmatrix} \cos\psi & -\sin\psi & 0 \\ \sin\psi & \cos\psi & 0 \\ 0 & 0 & 1 \end{bmatrix} \quad (11b)$$

By taking products of the matrices above we can obtain the transformation matrices for the transformations between frames $\hat{S}$ and $\bar{S}$, between frames S and $\tilde{S}$, and between frames S and $\bar{S}$. Doing this we obtain:

$$[e_{\hat{i}j}] = \begin{bmatrix} \cos\psi & \cos\theta\sin\psi & \sin\theta\sin\psi \\ -\sin\psi & \cos\theta\cos\psi & \sin\theta\cos\psi \\ 0 & -\sin\theta & \cos\theta \end{bmatrix} \quad (12a)$$

$$[e_{\hat{i}\bar{j}}] = \begin{bmatrix} \cos\psi & -\sin\psi & 0 \\ \cos\theta\sin\psi & \cos\theta\cos\psi & -\sin\theta \\ \sin\theta\sin\psi & \sin\theta\cos\psi & \cos\theta \end{bmatrix} \quad (12b)$$

$$[e_{\tilde{i}j}] = \begin{bmatrix} \cos\phi & \sin\phi & 0 \\ -\sin\phi\cos\theta & \cos\phi\cos\theta & \sin\theta \\ \sin\phi\sin\theta & -\cos\phi\sin\theta & \cos\theta \end{bmatrix} \quad (13a)$$

$$[e_{i\tilde{j}}] = \begin{bmatrix} \cos\phi & -\sin\phi\cos\theta & \sin\phi\sin\theta \\ \sin\phi & \cos\phi\cos\theta & -\cos\phi\sin\theta \\ 0 & \sin\theta & \cos\theta \end{bmatrix} \quad (13b)$$

$$[e_{\tilde{i}j}] = \begin{bmatrix} \cos\phi\cos\psi - \sin\phi\cos\theta\sin\psi & \sin\phi\cos\psi + \cos\phi\cos\theta\sin\psi & \sin\theta\sin\psi \\ -\cos\phi\sin\psi - \sin\phi\cos\theta\cos\psi & -\sin\phi\sin\psi + \cos\phi\cos\theta\cos\psi & \sin\theta\cos\psi \\ \sin\phi\sin\theta & -\cos\phi\sin\theta & \cos\theta \end{bmatrix}$$

$$(14a)$$

$$[e_{ij}] = \begin{bmatrix} \cos\phi\cos\psi - \sin\phi\cos\theta\sin\psi & -\cos\phi\sin\psi - \sin\phi\cos\theta\cos\psi & \sin\phi\sin\theta \\ \sin\phi\cos\psi + \cos\phi\cos\theta\sin\psi & -\sin\phi\sin\psi + \cos\phi\cos\theta\cos\psi & -\cos\phi\sin\theta \\ \sin\theta\sin\psi & \sin\theta\cos\psi & \cos\theta \end{bmatrix}$$

(14b)

THE ANGULAR VELOCITIES AS FUNCTIONS OF THE EULER ANGLES

The Euler angles ϕ, θ, and ψ provide us with a set of coordinates that can be used to describe the orientation of a rigid body. It follows that the rotational motion of a rigid body can be described by the variables $\dot\phi$, $\dot\theta$, and $\dot\psi$. However as we shall see it is often much more convenient to describe the rotational motion of a rigid body by using the angular velocities of the frames S, $\hat{S}$, $\tilde{S}$, and $\bar{S}$ with respect to one another. We therefore derive the relationships between the various angular velocities and the Euler angles.

From the definition of the Euler angles in terms of the frames S, $\hat{S}$, $\tilde{S}$, and $\bar{S}$ it follows that

$$\omega(S\hat{S}) = \mathbf{e}_3\dot\phi \tag{15}$$

$$\omega(\hat{S}\tilde{S}) = \hat{\mathbf{e}}_1\dot\theta \tag{16}$$

$$\omega(\tilde{S}\bar{S}) = \tilde{\mathbf{e}}_3\dot\psi \tag{17}$$

Given the components of a particular angular velocity in one frame, we can find the components in any other frame by using the results of Theorem 1. Table 1a lists the components in the frames S, $\hat{S}$, $\tilde{S}$, and $\bar{S}$ of the angular velocities above. The components of the remaining angular velocities can be obtained by simply noting that

$$\omega(S\tilde{S}) = \omega(S\hat{S}) + \omega(\hat{S}\tilde{S}) \tag{18}$$

$$\omega(\hat{S}\bar{S}) = \omega(\hat{S}\tilde{S}) + \omega(\tilde{S}\bar{S}) \tag{19}$$

$$\omega(S\bar{S}) = \omega(S\hat{S}) + \omega(\hat{S}\tilde{S}) + \omega(\tilde{S}\bar{S}) \tag{20}$$

Table 1b lists the components in the frames S, $\hat{S}$, $\tilde{S}$, and $\bar{S}$ of the angular velocities above.

NOTE: To be able to use the quantities $\dot\phi$, $\dot\theta$, and $\dot\psi$ to describe arbitrary rotational motion of a rigid body, it is necessary that the components of the vector $\omega(S\bar{S})$ when considered as functions of $\dot\phi$, $\dot\theta$, and $\dot\psi$ not be functionally

Table 1a

Component	$\omega(S\hat{S})$	$\omega(\hat{S}\tilde{S})$	$\omega(\tilde{S}\bar{S})$
1	0	$\dot\theta\cos\phi$	$\dot\psi\sin\phi\sin\theta$
2	0	$\dot\theta\sin\phi$	$-\dot\psi\cos\phi\sin\theta$
3	$\dot\phi$	0	$\dot\psi\cos\theta$
$\hat{1}$	0	$\dot\theta$	0
$\hat{2}$	0	0	$-\dot\psi\sin\theta$
$\hat{3}$	$\dot\phi$	0	$\dot\psi\cos\theta$
$\tilde{1}$	0	$\dot\theta$	0
$\tilde{2}$	$\dot\phi\sin\theta$	0	0
$\tilde{3}$	$\dot\phi\cos\theta$	0	$\dot\psi$
$\bar{1}$	$\dot\phi\sin\theta\sin\psi$	$\dot\theta\cos\psi$	0
$\bar{2}$	$\dot\phi\sin\theta\cos\psi$	$-\dot\theta\sin\psi$	0
$\bar{3}$	$\dot\phi\cos\theta$	0	$\dot\psi$

Table 1b

Component	$\omega(S\tilde{S})$	$\omega(\hat{S}\bar{S})$	$\omega(S\bar{S})$
1	$\dot\theta\cos\phi$	$\dot\theta\cos\phi + \dot\psi\sin\phi\sin\theta$	$\dot\theta\cos\phi + \dot\psi\sin\phi\sin\theta$
2	$\dot\theta\sin\phi$	$\dot\theta\sin\phi - \dot\psi\cos\phi\sin\theta$	$\dot\theta\sin\phi - \dot\psi\cos\phi\sin\theta$
3	$\dot\phi$	$\dot\psi\cos\theta$	$\dot\phi + \dot\psi\cos\theta$
$\hat{1}$	$\dot\theta$	$\dot\theta$	$\dot\theta$
$\hat{2}$	0	$-\dot\psi\sin\theta$	$-\dot\psi\sin\theta$
$\hat{3}$	$\dot\phi$	$\dot\psi\cos\theta$	$\dot\phi + \dot\psi\cos\theta$
$\tilde{1}$	$\dot\theta$	$\dot\theta$	$\dot\theta$
$\tilde{2}$	$\dot\phi\sin\theta$	0	$\dot\phi\sin\theta$
$\tilde{3}$	$\dot\phi\cos\theta$	$\dot\psi$	$\dot\phi\cos\theta + \dot\psi$
$\bar{1}$	$\dot\phi\sin\theta\sin\psi + \dot\theta\cos\psi$	$\dot\theta\cos\psi$	$\dot\phi\sin\theta\sin\psi + \dot\theta\cos\psi$
$\bar{2}$	$\dot\phi\sin\theta\cos\psi - \dot\theta\sin\psi$	$-\dot\theta\sin\psi$	$\dot\phi\sin\theta\cos\psi - \dot\theta\sin\psi$
$\bar{3}$	$\dot\phi\cos\theta$	$\dot\psi$	$\dot\phi\cos\theta + \dot\psi$

dependent (see Appendix 14). The components will be functionally dependent if the Jacobian $J[\omega_1(S\bar{S}), \omega_2(S\bar{S}), \omega_3(S\bar{S})/\dot\phi, \dot\theta, \dot\psi]$ is zero. Evaluating this Jacobian we find that it is $-\sin\theta$. We find essentially the same result using components of $\omega(S\bar{S})$ in the frames $\hat{S}$, $\tilde{S}$, or $\bar{S}$. It follows that the Euler angles are not well suited for describing the motion when θ is zero or π. Specifically if

θ is zero, the results of Table 1b would require that $\omega_1(S\bar{S})/\omega_2(S\bar{S}) = \cot\phi$, which would put restrictions on the possible values of $\omega(S\bar{S})$.

ANGULAR MOMENTUM AND KINETIC ENERGY

The inertia tensor is important because it enables us to express the angular momentum and the kinetic energy of a rigid body with respect to a point fixed in the rigid body in a particularly simple form, as is shown in the following theorems.

Theorem 4. Consider a rigid body. Let $\bar{o}$ be a point fixed in the rigid body, $\bar{S}$ a frame fixed in the rigid body with origin at $\bar{o}$ and axes in the principal directions for the point $\bar{o}$, S an inertial frame, and S' any arbitrary frame.

    a. The components with reference to the frame S', of the angular momentum of the rigid body with respect to the point $\bar{o}$ are given by

$$H_{i'}(\bar{o}) = \sum_{j'} I_{i'j'}(\bar{o})\omega_{j'}(S\bar{S}) \qquad (21)$$

or in matrix form by

$$\begin{bmatrix} H_{1'}(\bar{o}) \\ H_{2'}(\bar{o}) \\ H_{3'}(\bar{o}) \end{bmatrix} = \begin{bmatrix} I_{1'1'}(\bar{o}) & I_{1'2'}(\bar{o}) & I_{1'3'}(\bar{o}) \\ I_{2'1'}(\bar{o}) & I_{2'2'}(\bar{o}) & I_{2'3'}(\bar{o}) \\ I_{3'1'}(\bar{o}) & I_{3'2'}(\bar{o}) & I_{3'3'}(\bar{o}) \end{bmatrix} \begin{bmatrix} \omega_{1'}(S\bar{S}) \\ \omega_{2'}(S\bar{S}) \\ \omega_{3'}(S\bar{S}) \end{bmatrix} \qquad (22)$$

which can be written in more abbreviated form as simply

$$[H] = [I][\omega] \qquad (22a)$$

where the $I_{i'j'}(\bar{o})$ are the components with reference to the frame S' of the inertia tensor for the point $\bar{o}$, and the $\omega_{i'}(S\bar{S})$ are the components with reference to the frame S' of the angular velocity of frame $\bar{S}$ with respect to frame S.

    b. The components with reference to the frame $\bar{S}$ of the angular momentum of the rigid body with respect to the point $\bar{o}$ are given by

$$H_{\bar{i}}(\bar{o}) = I_{\bar{i}}(\bar{o})\omega_{\bar{i}}(S\bar{S}) \qquad (23)$$

where $I_{\bar{i}}(\bar{o})$ is the moment of inertia associated with the principal axis $\bar{i}$, and the $\omega_{\bar{i}}(S\bar{S})$ are the components with reference to the frame $\bar{S}$ of the angular velocity of frame $\bar{S}$ with respect to frame S.

PROOF: By definition

$$\mathbf{H}(\bar{o}) = \sum_p \mathbf{r}(\bar{o}p) \times [m(p)\dot{\mathbf{r}}(\bar{o}p)] \qquad (24)$$

where $\dot{\mathbf{r}}(\bar{o}p)$ is the time rate of change of the vector $\mathbf{r}(\bar{o}p)$ as noted by an observer fixed in frame S. If we let $\bar{\dot{\mathbf{r}}}(\bar{o}p)$ be the corresponding time rate of change as noted by an observer fixed in $\bar{S}$, then from the results of Chapter 6

$$\dot{\mathbf{r}}(\bar{o}p) = \bar{\dot{\mathbf{r}}}(\bar{o}p) + \omega(S\bar{S}) \times \mathbf{r}(\bar{o}p) \tag{25}$$

Since the points $\bar{o}$ and p are fixed in the rigid body, it follows that

$$\bar{\dot{\mathbf{r}}}(\bar{o}p) = 0 \tag{26}$$

Combining Eqs. (24)–(26) we obtain

$$\mathbf{H}(\bar{o}) = \sum_p \mathbf{r}(\bar{o}p) \times \left[m(p)\omega(S\bar{S}) \times \mathbf{r}(\bar{o}p) \right] \tag{27}$$

Expressing $\mathbf{r}(\bar{o}p)$ and $\omega(S\bar{S})$ in terms of their components in the frame S' we obtain

$$\mathbf{r}(\bar{o}p) = \mathbf{e}_{1'}x_{1'}(\bar{o}p) + \mathbf{e}_{2'}x_{2'}(\bar{o}p) + \mathbf{e}_{3'}x_{3'}(\bar{o}p) \tag{28}$$

$$\omega(S\bar{S}) = \mathbf{e}_{1'}\omega_{1'}(S\bar{S}) + \mathbf{e}_{2'}\omega_{2'}(S\bar{S}) + \mathbf{e}_{3'}\omega_{3'}(S\bar{S}) \tag{29}$$

Substituting Eqs. (28) and (29) in Eq. (27) and carrying out the indicated multiplications, we obtain Eq. (21), which completes the proof of part a of the theorem. The proof of part b follows immediately from the fact that for the frame $\bar{S}$

$$I_{\bar{i}\bar{j}} = I_{\bar{i}}(\bar{o})\delta_{\bar{i}\bar{j}} \tag{30}$$

NOTE: The reduction by substitution of the quantity $\mathbf{A} \equiv \mathbf{r} \times (\omega \times \mathbf{r})$, which appears in the preceding theorem, to the form $A_i = \sum_j (r^2 \delta_{ij} - x_i x_j)\omega_j$ is a little inelegant and tedious. The result could have been obtained neatly and quickly if we had used the techniques developed in Appendix 6. Employing the techniques of Appendix 6, and using the summation convention, we have

$$A_i = \epsilon_{ijk} x_j \epsilon_{klm} \omega_l x_m = \epsilon_{kij} \epsilon_{klm} x_j \omega_l x_m$$
$$= (\delta_{il}\delta_{jm} - \delta_{im}\delta_{jl}) x_j x_m \omega_l = x_j x_j \omega_i - x_j x_i \omega_j$$
$$= r^2 \omega_i - x_i x_j \omega_j = r^2 \delta_{ij} \omega_j - x_i x_j \omega_j = (r^2 \delta_{ij} - x_i x_j)\omega_j \tag{31}$$

which is the desired result.

Theorem 5. Consider a rigid body. Let $\bar{o}$ be a point fixed in the rigid body, $\bar{S}$ a frame fixed in the rigid body with origin at $\bar{o}$ and axes in the principal directions for the point $\bar{o}$, S an inertial frame, and S' any arbitrary frame.

a. The kinetic energy with respect to the point $\bar{o}$ is given by

$$T(\bar{o}) = \tfrac{1}{2} \sum_{i'} \sum_{j'} I_{i'j'}(\bar{o}) \omega_{i'}(S\bar{S}) \omega_{j'}(S\bar{S}) \tag{32}$$

314   Rigid Body Kinematics

or in matrix form by

$$T(\bar{o}) = \frac{1}{2}\begin{bmatrix} \omega_{1'}(S\bar{S}) & \omega_{2'}(S\bar{S}) & \omega_{3'}(S\bar{S}) \end{bmatrix}$$
$$\times \begin{bmatrix} I_{1'1'}(\bar{o}) & I_{1'2'}(\bar{o}) & I_{1'3'}(\bar{o}) \\ I_{2'1'}(\bar{o}) & I_{2'2'}(\bar{o}) & I_{2'3'}(\bar{o}) \\ I_{3'1'}(\bar{o}) & I_{3'2'}(\bar{o}) & I_{3'3'}(\bar{o}) \end{bmatrix} \begin{bmatrix} \omega_{1'}(S\bar{S}) \\ \omega_{2'}(S\bar{S}) \\ \omega_{3'}(S\bar{S}) \end{bmatrix} \quad (33)$$

which can be written in a more abbreviated form as simply

$$T = \tfrac{1}{2}(\omega)[I][\omega] \quad (33a)$$

where the $I_{i'j'}(\bar{o})$ are the components with reference to the frame S' of the inertia tensor with respect to the point $\bar{o}$, and the $\omega_{i'}(S\bar{S})$ are the components with reference to the frame S' of the angular velocity of frame $\bar{S}$ with respect to frame S.

b. The kinetic energy with respect to the point $\bar{o}$, in terms of quantities with reference to the frame $\bar{S}$, is given by

$$T(\bar{o}) = \tfrac{1}{2}\sum_{\bar{i}} I_{\bar{i}}(\bar{o})\left[\omega_{\bar{i}}(S\bar{S})\right]^2 \quad (34)$$

where $I_{\bar{i}}(\bar{o})$ is the moment of inertia associated with the principal axis $\bar{i}$, and the $\omega_{\bar{i}}(S\bar{S})$ are the components with reference to the frame $\bar{S}$ of the angular velocity of frame $\bar{S}$ with respect to frame S.

PROOF: By definition

$$T(\bar{o}) = \tfrac{1}{2}\sum_{p} m(p)\dot{\mathbf{r}}(\bar{o}p) \cdot \dot{\mathbf{r}}(\bar{o}p) \quad (35)$$

Substituting Eqs. (25) and (26) in Eq. (35) we obtain

$$T(\bar{o}) = \tfrac{1}{2}\sum_{p} m(p)\left[\omega(S\bar{S}) \times \mathbf{r}(\bar{o}p)\right] \cdot \left[\omega(S\bar{S}) \times \mathbf{r}(\bar{o}p)\right] \quad (36)$$

Substituting Eqs. (28) and (29) in Eq. (36) and carrying out the indicated multiplications we obtain Eq. (32). This completes the proof of part a. The proof of part b follows immediately from Eq. (30).

NOTE: The reduction of the quantity $B \equiv (\omega \times \mathbf{r}) \cdot (\omega \times \mathbf{r})$, which appears in the preceding theorem, to the form $B = \sum_i \sum_j (r^2 \delta_{ij} - x_i x_j)\omega_i \omega_j$ could have been obtained neatly and quickly if we had used the techniques developed in Appendix 6. Employing the techniques of Appendix 6 and using the summation convention, we have

$$B = \epsilon_{ijk}\omega_j x_k \epsilon_{ilm}\omega_l x_m = [\delta_{jl}\delta_{km} - \delta_{jm}\delta_{kl}]x_k x_m \omega_j \omega_l$$

$$= x_m x_m \omega_j \omega_j - x_l x_j \omega_j \omega_l = r^2 \omega_j \omega_j - x_j x_l \omega_j \omega_l$$

$$= r^2 \delta_{jl}\omega_j \omega_l - x_j x_l \omega_j \omega_l = (r^2 \delta_{jl} - x_j x_l)\omega_j \omega_l \qquad (37)$$

which is the desired result.

THE KINEMATICS OF A RIGID BODY WITH NO POINT FIXED

If we are interested in the motion of a rigid body in which no point is fixed, we simply break the motion down into the motion of the center of mass and the motion with respect to the center of mass. The motion of the center of mass is simply the motion of a point, and can be handled by the techniques we have developed in our study of the motion of a point particle. The motion with respect to the center of mass can be readily handled by a simple reinterpretation of the preceding results. Wherever the point o occurs, we simply replace it by the center of mass c; and instead of the frame S being an inertial frame, it becomes a frame with origin at c and axes parallel to the axes of an inertial frame.

EXAMPLES

Example 1. A cone of height h and angle 2α rolls without slipping inside a fixed cone of angle 2β ($\beta > \alpha$). The axis of the inner cone rotates about the axis of the outer cone with constant angular speed Ω.
  a. Find the angular velocity of the cone.
  b. Find the angular momentum of the cone.
  c. Find the kinetic energy of the cone.

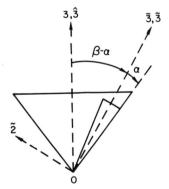

Fig. 5

Solution.

a. Let S be a frame fixed to the outer cone with origin at the tip and 3 axis along the symmetry axis. Let $\bar{S}$ be a frame fixed to the inner cone with origin at the tip and $\bar{3}$ axis along the symmetry axis. The frames $\hat{S}$ and $\tilde{S}$ are then chosen as discussed in this chapter (see Fig. 5). From the conditions of the problem we then have the following information about the Euler angles θ and ϕ:

$$\theta = \beta - \alpha = \text{constant} \tag{38}$$

$$\dot{\phi} = \Omega \tag{39}$$

Using Eqs. (38) and (39) and the results of Table 1b we obtain

$$\omega(S\bar{S}) = \mathbf{e}_{\bar{2}}\Omega \sin(\beta - \alpha) + \mathbf{e}_{\bar{3}}\left[\Omega \cos(\beta - \alpha) + \dot{\psi}\right] \tag{40}$$

The only unknown in Eq. (49) is $\dot{\psi}$. To determine $\dot{\psi}$ we note that the instantaneous axis of rotation of the inner cone is along the line of contact of the inner and outer cones and therefore

$$\omega(S\bar{S}) = -\mathbf{e}_{\bar{2}}A \sin\alpha + \mathbf{e}_{\bar{3}}A \cos\alpha \tag{41}$$

where A is the unknown amplitude of $\omega(S\bar{S})$. Equating Eqs. (40) and (41) we obtain two equations in the two unknowns A and $\dot{\psi}$. Solving for A and substituting the result in Eq. (41) we obtain

$$\omega(S\bar{S}) = \left[\frac{\Omega \sin(\beta - \alpha)}{\sin\alpha}\right]\left[\mathbf{e}_{\bar{2}} \sin\alpha - \mathbf{e}_{\bar{3}} \cos\alpha\right] \tag{42}$$

b. The inertia tensor of the inner cone with respect to the point o when expressed with reference to the frame $\tilde{S}$ is given by

$$[I_{\tilde{i}\tilde{j}}(o)] = \frac{3mh^2}{20} \begin{bmatrix} \tan^2\alpha + 4 & 0 & 0 \\ 0 & \tan^2\alpha + 4 & 0 \\ 0 & 0 & 2\tan^2\alpha \end{bmatrix} \tag{43}$$

Substituting Eqs. (42) and (43) in Eq. (22), letting the S' frame be the $\tilde{S}$ frame, we obtain

$$\begin{bmatrix} H_{\bar{1}}(o) \\ H_{\bar{2}}(o) \\ H_{\bar{3}}(o) \end{bmatrix} = \left(\frac{3mh^2}{20}\right)\left(\frac{\Omega \sin(\beta - \alpha)}{\sin\alpha}\right) \begin{bmatrix} \tan^2\alpha + 4 & 0 & 0 \\ 0 & \tan^2\alpha + 4 & 0 \\ 0 & 0 & 2\tan^2\alpha \end{bmatrix} \begin{bmatrix} 0 \\ \sin\alpha \\ -\cos\alpha \end{bmatrix}$$

$$= \frac{3mh^2\Omega \sin(\beta - \alpha)}{20} \begin{bmatrix} 0 \\ \tan^2\alpha + 4 \\ -2\tan\alpha \end{bmatrix} \tag{44}$$

c. In a similar fashion, if we substitute Eqs. (42) and (43) in Eq. (33) we obtain

$$T(o) = \frac{3mh^2\Omega^2 \sin^2(\beta - \alpha)}{40\cos^2\alpha}(1 + 5\cos^2\alpha)$$

PROBLEMS

1. Show that any finite displacement of a rigid body is equivalent to a translation together with a rotation about an axis parallel to the translation.

2. Show that the angular momentum of a rigid body with respect to its center of mass c is parallel to the angular velocity ω of the body if and only if ω is parallel to a principal axis of inertia for c.

3. A solid right circular cylinder whose radius is one-quarter of its altitude rotates with an angular velocity Ω about an axis that passes through the center of mass and the periphery of the base circles. Find the angle between the polar axis of symmetry and the angular momentum vector.

4. A uniform cube of mass M and edge $2a$ is spinning with angular speed ω about a fixed diagonal of one face. Show that the magnitude of the angular momentum about one of the fixed corners is $(43)^{1/2}Ma^2\omega/3$.

5. A homogeneous ellipsoid whose major axes are of lengths $2a$, $2b$, and $2c$, respectively, rotates about the axis of length $2a$ with an angular speed $\dot{\alpha}$. The axis of length $2a$ itself rotates about a fixed line perpendicular to it and passing through the center of the ellipsoid with an angular speed $\dot{\beta}$. Determine the kinetic energy of the ellipsoid.

6. A homogeneous ellipsoid of revolution whose semimajor axes are of lengths a, a, and c, respectively, rotates about the axis of revolution with an angular velocity $\dot{\alpha}$, while that axis rotates with angular velocity $\dot{\beta}$ about a line that makes an angle γ with the axis and passes through the center of the ellipsoid. Find the kinetic energy of the ellipsoid.

7. OA is a light rod of length b that turns with angular velocity Ω about an axis OB perpendicular to it. A is the middle point of a rod CD of mass m and length $2a$, hinged to OA at A in such a way that CD is always coplanar with OB. If α denotes the angle OAC, prove that the components of the angular momentum of CD about O in the directions OA, OB and a direction perpendicular to them are, respectively, $ma^2\Omega\sin\alpha\cos\alpha/3$, $m\Omega[b^2 + (a^2\cos^2\alpha/3)]$, and $ma^2\dot{\alpha}/3$.

8. A cone of mass m, height h, and semivertical angle α rolls without slipping on a horizontal plane, and a point on its axis describes one

complete circle in time τ. Show that the kinetic energy of the cone is $(3\pi^2 mh^2/10\tau^2)(1 + 5\cos^2\alpha)$.

9. The base of a cone of mass m, height h, and semivertical angle α rolls without slipping on a plane, and the vertex is fixed at a height above the plane equal to the radius of the base, so that the axis of the cone is parallel to the plane. The axis of the cone makes one revolution about the fixed point in a time τ.
   a. Determine the angular velocity of the cone.
   b. Determine the angular momentum of the cone.
   c. Determine the kinetic energy of the cone.

10. A thin uniform disk of mass M, radius a, and center C rolls without slipping on a horizontal plane surface with the plane of the disk making a constant angle with the vertical and C moving with constant speed V in a horizontal circle of radius $2a$. If T is the kinetic energy of the disk, prove that $\frac{3}{4} < T/MV^2 < \frac{25}{32}$.

11. A uniform disk of radius a rolls without slipping on a horizontal plane and makes a constant angle β with the plane, the center of the disk describing a circle of radius λa with velocity V.
   a. Determine the angular velocity of the disk.
   b. Determine the angular momentum of the disk with respect to its center of mass.
   c. Determine the kinetic energy of the disk with respect to a point fixed on the horizontal plane.

12. Two conical shells of semivertical angles β and $\beta + 2\alpha$, respectively, have a common axis and vertex. Between them is pressed a solid cone of semivertical angle α. The shells are made to rotate in opposite senses with angular speed Ω about their common axes.
   a. Determine the angular velocity of the cone.
   b. Determine the angular momentum of the cone with respect to its vertex.
   c. Determine the kinetic energy of the cone.

13. A uniform cube of mass M and side $2a$ is free to rotate about its center O, which is fixed. The cube is set in motion from rest by an impulse applied at one corner, the impulse being such that the components of the velocity, parallel to the sides of the cube, are u_1, u_2, and u_3, respectively. Show that the kinetic energy of the cube is

$$\frac{M}{22}\left[5(u_1^2 + u_2^2 + u_3^2) + 6(u_2 u_3 + u_3 u_1 + u_1 u_2)\right]$$

14. A homogeneous rigid body of mass m is in the shape of a rectangular cube of sides $2a$, $2a$, and $4a$. The body is free to rotate about its center

of mass which is fixed. Show that the kinetic energy of the body in terms of the Euler angles is

$$\frac{ma^2}{6}\left[5(\dot\phi^2\sin^2\theta + \dot\theta^2) + 2(\dot\psi^2 + \dot\phi^2\cos^2\theta + 2\dot\psi\dot\phi\cos\theta)\right]$$

15. A homogeneous lamina in the form of an equilateral triangle ABC of side $2a$ moves on a fixed table with only the side BC in contact with it. If BC makes an angle α with a fixed line in the table, and the plane of the lamina is inclined at an angle β to the table, show that the kinetic energy of the lamina is given by

$$\frac{MV^2}{2} + \frac{Ma^2}{12}\left[(1 + 2\cos^2\beta)\dot\beta^2 + (1 + \cos^2\beta)\dot\alpha^2\right]$$

where M is its mass and V is the horizontal component of the velocity of the center of mass.

40

Rigid Body Dynamics

INTRODUCTION

If a rigid body is free to rotate about a fixed point o, three coordinates are required to specify the orientation of the rigid body. It follows that three independent equations of motion are required to determine the rotational motion of a rigid body acted on by a set of forces. A sufficient set of equations is provided by the vector equation

$$\mathbf{G}(o) = \dot{\mathbf{H}}(o) \tag{1}$$

where $\mathbf{G}(o)$ is the net external torque with respect to the point o, $\mathbf{H}(o)$ is the angular momentum with respect to the point o, and the dot represents the time rate of change as noted by an observer fixed in an inertial frame.

If no point in a rigid body is fixed, six coordinates are required to specify the configuration of the system, hence six independent equations of motion are required to determine the behavior of the system. In this case the motion can be broken down into the motion of the center of mass, and the motion with respect to the center of mass. The motion of the center of mass is governed by the equation

$$\mathbf{F} = M\ddot{\mathbf{R}} \tag{2}$$

where $\mathbf{F}$ is the net external force, M is the total mass, and $\mathbf{R}$ is the vector position of the center of mass with respect to a point fixed in an inertial frame. The motion with respect to the center of mass is governed by the equation

$$\mathbf{G}(c) = \dot{\mathbf{H}}(c) \tag{3}$$

where $\mathbf{G}(c)$ is the torque with respect to the center of mass, and $\mathbf{H}(c)$ the angular momentum with respect to the center of mass.

THE EQUATIONS OF MOTION OF A RIGID BODY WITH ONE POINT FIXED

Equation (1) together with the results of the preceding chapter provide us in principle with the equations necessary to determine the motion of a rigid body that is rotating about a fixed point o. However, the use of Eq. (1) requires a delicate touch if we want to avoid impossibly complex results. For example, a straightforward approach might suggest that we first write Eq. (1) as $G_i(o) = \dot{H}_i(o)$, where the components are with reference to an inertial frame, and then express $H_i(o)$ in terms of the Euler angles ϕ, θ, and ψ. If this is done, we find that the resulting equations are exceedingly messy. The trick to avoiding such a situation is to express the equations of motion with reference to some frame other than an inertial frame, and the frame most often used is a frame fixed in the body with origin at o and principal axes in the principal directions for the point o. In expressing Eq. (1) with reference to a noninertial frame S', the important fact to remember is that although the components in an inertial frame S of a vector $\dot{\mathbf{A}}$ are $\dot{A}_i$, the components in the frame S' of $\dot{\mathbf{A}}$ are *not* $\dot{A}_{i'}$, but instead are given by $\dot{A}_{i'} + \sum_{j'}\sum_{k'}\epsilon_{i'j'k'}\omega_{j'}(SS')A_{k'}$, where $\epsilon_{i'j'k'}$ is the Levi-Civita symbol (see Appendix 6) and $\omega(SS')$ is the angular velocity of S' with respect to S. This latter result follows from the fact that $\dot{\mathbf{A}} = \dot{\mathbf{A}}' + \omega(SS') \times \mathbf{A}$ as shown in Chapter 6. The technique for handling the motion of a rigid body with one point fixed is spelled out in the following theorem.

Theorem 1. Consider a rigid body that is constrained to rotate about a fixed point o and is acted on by an external torque $\mathbf{G}(o)$. Let S be an inertial frame with origin at o; $\bar{S}$ a frame fixed in the rigid body with origin at o and axes in the principal directions for the point o; and S' any arbitrary frame. Let $\omega_{i'}(SS')$ be the i' component of the angular velocity of frame S' with respect to frame S, $\omega_{i'}(S\bar{S})$ the i' component of the angular velocity of frame $\bar{S}$ with respect to frame S, and $I_{i'j'}(o)$ the $i'j'$ component of the inertia tensor with respect to the point o. If $\omega_{i'}(SS')$, $\omega_{i'}(S\bar{S})$, and $I_{i'j'}(o)$ are known functions of the Euler angles and their time derivatives, the motion of the body can be determined by starting with the equation

$$\mathbf{G}(o) = \dot{\mathbf{H}}(o) + \omega(SS') \times \mathbf{H}(o) \tag{4}$$

and then expressing it in terms of components in the S' frame, to obtain

$$G_{i'}(o) = \dot{H}_{i'}(o) + \sum_{j'}\sum_{k'}\epsilon_{i'j'k'}\omega_{j'}(SS')H_{k'}(o) \tag{5}$$

where $\epsilon_{i'j'k'}$ is the Levi-Civita symbol, and $H_{i'}(o)$ is the i' component of the angular momentum with respect to the point o, and finally by making use of the relation

$$H_{i'}(o) = \sum_{j'} I_{i'j'}(o)\omega_{j'}(S\bar{S}) \tag{6}$$

In particular, if the frame S' is chosen to be the frame $\bar{S}$, the equations above give us

$$G_{\bar{i}}(o) = I_{\bar{i}}\dot{\omega}_{\bar{i}}(S\bar{S}) + \sum_{j}\sum_{k}\epsilon_{\bar{i}\bar{j}\bar{k}}\omega_{\bar{j}}(S\bar{S})\omega_{\bar{k}}(S\bar{S})I_{\bar{k}}(o) \qquad (7)$$

or equivalently

$$G_{\bar{1}}(o) = I_{\bar{1}}(o)\dot{\omega}_{\bar{1}}(S\bar{S}) - \omega_{\bar{2}}(S\bar{S})\omega_{\bar{3}}(S\bar{S})\left[I_{\bar{2}}(o) - I_{\bar{3}}(o)\right] \qquad (8a)$$

$$G_{\bar{2}}(o) = I_{\bar{2}}(o)\dot{\omega}_{\bar{2}}(S\bar{S}) - \omega_{\bar{3}}(S\bar{S})\omega_{\bar{1}}(S\bar{S})\left[I_{\bar{3}}(o) - I_{\bar{1}}(o)\right] \qquad (8b)$$

$$G_{\bar{3}}(o) = I_{\bar{3}}(o)\dot{\omega}_{\bar{3}}(S\bar{S}) - \omega_{\bar{1}}(S\bar{S})\omega_{\bar{2}}(S\bar{S})\left[I_{\bar{1}}(o) - I_{\bar{2}}(o)\right] \qquad (8c)$$

where $I_{\bar{i}}(o)$ is the $\bar{i}$ principal moment of inertia for the point o. The latter set of equations is called the Euler equations.

PROOF: From the results of Chapter 6, the equation

$$\mathbf{G}(o) = \dot{\mathbf{H}}(o) \qquad (9)$$

can be expressed in the form given by Eq. (4). Expressing Eq. (4) in terms of its components with reference to the frame S' we obtain Eq. (5). Equation (6) follows immediately from the results of Chapter 39. With reference to the frame $\bar{S}$

$$I_{\bar{i}\bar{j}}(o) = I_{\bar{i}}(o)\delta_{\bar{i}\bar{j}} \qquad (10)$$

Letting S' be $\bar{S}$ in Eqs. (5) and (6), and combining the resulting equations with Eq. (10), we obtain Eq. (7), which completes the proof of the theorem.

NOTE: In applying the theorem above to a specific problem, one tries to choose the frame S' in such a way as to make the resulting equations as simple as possible. The best choice usually is one in which $I_{i'j'}(o)$ is of the form $A_{i'}\delta_{i'j'}$, where the $A_{i'}$ are constants. If the rigid body is completely asymmetrical, this will be true only if S' is identical with $\bar{S}$. However, if two of the principal moments of inertia are equal, say $I_{\bar{1}}(o)$ and $I_{\bar{2}}(o)$, then any frame that always has one axis pointing in the $\bar{3}$ direction will satisfy the condition; and if all three of the principal moments of inertia are equal, any frame will satisfy the condition. If there is a choice of frames for which the condition is satisfied, as there will be when two or more of the principal moments of inertia are equal, the best choice is the one that results in the simplest expressions in terms of the Euler angles for the quantities $\omega_{i'}(S\bar{S})$ and $\omega_{i'}(SS')$.

THE EQUATIONS OF MOTION OF A RIGID BODY WITH NO POINT FIXED

The extension of the preceding results to a rigid body with no point fixed is straightforward. In this case we simply break the motion down into the motion of the center of mass and the motion with respect to the center of mass. The motion of the center of mass is the same as that of a point particle acted on by a force equal to the net force acting on the rigid body. The motion with respect to the center of mass is the same as it would be if the center of mass were fixed and it was acted on by the same torques with respect to the center of mass. In converting the results of the preceding theorem to results applicable to rotation about the center of mass we simply replace o wherever it occurs by c, and replace the frame S by a frame with origin at c and axes parallel to the axes of an inertial frame.

EQUIMOMENTAL SYSTEMS

Two distributions of matter that have the same total mass and the same principal moments of inertia with respect to the center of mass are said to be *equimomental systems*.

From the results we have obtained in this chapter, it follows that two equimomental systems that are acted on by the same forces have the same dynamical behavior.

EXAMPLES

Example 1. Freely Rotating Symmetric Top. A rigid body, having an axis of symmetry, is constrained to rotate about a fixed point o on this axis. There are no external torques with respect to the point o. Describe the motion.

NOTE: This problem is *not* meant to illustrate Theorem 1, but to show how the proper choice of frames can simplify the results.

Solution. Given

$$G_i(o) = 0 \tag{11}$$

it follows that

$$H_i(o) = \text{constant} \tag{12}$$

Note that this is true only for the components of the angular momentum in an inertial frame. The components of the angular momentum with reference to some other frame are not necessarily constants, as can be seen by setting the torque equal to zero in Eq. (5). If we choose our 3 axis in the direction of $\mathbf{H}(o)$ then

$$\begin{bmatrix} H_1(o) \\ H_2(o) \\ H_3(o) \end{bmatrix} = \begin{bmatrix} 0 \\ 0 \\ H \end{bmatrix} \tag{13}$$

where H is a constant equal to the magnitude of the angular momentum. We now choose the axes of the body frame $\bar{S}$ such that the $\bar{3}$ axis is along the symmetry axis of the top, and we define frames $\hat{S}$ and $\tilde{S}$ as in Chapter 39. It then follows from Chapter 38 that

$$[I_{\tilde{i}\,\tilde{j}}(o)] = \begin{bmatrix} A & 0 & 0 \\ 0 & A & 0 \\ 0 & 0 & C \end{bmatrix} \tag{14}$$

where A, A, and C are the principal moments of inertia of the top with respect to the point o. We next note that

$$H_{\tilde{i}}(o) = \sum_j e_{\tilde{i}j} H_j(o) \tag{15}$$

where the $e_{\tilde{i}j}$ are the transformation coefficients for the transformation from the frame S to the frame $\tilde{S}$. Combining Eq. (15) with the relation

$$H_{\tilde{i}}(o) = \sum_j I_{\tilde{i}\,\tilde{j}}(o)\omega_{\tilde{j}}(S\overline{S}) \tag{16}$$

we obtain

$$\sum_j I_{\tilde{i}\,\tilde{j}}(o)\omega_{\tilde{j}}(S\overline{S}) = \sum_j e_{\tilde{i}j} H_j(o) \tag{17}$$

Substituting Eqs. (13) and (14) in Eq. (17), and expressing $\omega_{\tilde{j}}(S\overline{S})$ and $e_{\tilde{i}j}$ as functions of the Euler angles ϕ, θ, and ψ, we obtain

$$\begin{bmatrix} A & 0 & 0 \\ 0 & A & 0 \\ 0 & 0 & C \end{bmatrix} \begin{bmatrix} \dot{\theta} \\ \dot{\phi}\sin\theta \\ \dot{\phi}\cos\theta + \dot{\psi} \end{bmatrix} = \begin{bmatrix} \cos\phi & \sin\phi & 0 \\ -\sin\phi\cos\theta & \cos\phi\cos\theta & \sin\theta \\ \sin\phi\sin\theta & -\cos\phi\sin\theta & \cos\theta \end{bmatrix} \begin{bmatrix} 0 \\ 0 \\ H \end{bmatrix} \tag{18}$$

which yields the three equations

$$A\dot{\theta} = 0 \tag{19}$$

$$A\dot{\phi}\sin\theta = H\sin\theta \tag{20}$$

$$C(\dot{\phi}\cos\theta + \dot{\psi}) = H\cos\theta \tag{21}$$

Solving these equations for $\dot{\theta}$, $\dot{\phi}$, and $\dot{\psi}$ we obtain

$$\dot{\theta} = 0 \tag{22}$$

$$\dot{\phi} = \frac{H}{A} \tag{23}$$

$$\dot{\psi} = H\cos\theta \frac{A-C}{AC} \tag{24}$$

It follows that the angle θ between the axis of symmetry and the direction of the angular momentum vector is a constant, the axis of symmetry precesses at a constant rate H/A around the direction of the angular momentum, and the top spins around its symmetry axis at a constant rate.

NOTE: The *spin* of a symmetric top is defined as the component of the angular velocity of the top in the direction of the axis of symmetry. If the $\bar{3}$ axis is chosen to be the symmetry axis, then the spin is equal to $\dot{\phi}\cos\theta + \dot{\psi}$.

Example 2. Symmetric Top in a Gravitational Field. A rigid body, having an axis of symmetry, is constrained to rotate about a fixed point o on this axis. The body is in a uniform gravitational field. Describe the motion.

Solution 1. Choose the 3 axis in the vertical direction and the $\bar{3}$ direction along the axis of symmetry. The most convenient frame for our analysis is the $\tilde{S}$ frame defined in Chapter 39. With reference to this frame

$$[G_{\tilde{i}}(o)] = \begin{bmatrix} mgh\sin\theta \\ 0 \\ 0 \end{bmatrix} \quad (25)$$

$$[I_{\tilde{i}\tilde{j}}(o)] = \begin{bmatrix} A & 0 & 0 \\ 0 & A & 0 \\ 0 & 0 & C \end{bmatrix} \quad (26)$$

$$[\omega_{\tilde{i}}(S\tilde{S})] = \begin{bmatrix} \dot{\theta} \\ \dot{\phi}\sin\theta \\ \dot{\phi}\cos\theta \end{bmatrix} \quad (27)$$

$$[\omega_{\tilde{i}}(S\bar{S})] = \begin{bmatrix} \dot{\theta} \\ \dot{\phi}\sin\theta \\ \dot{\phi}\cos\theta + \dot{\psi} \end{bmatrix} \quad (28)$$

where A, A, and C are the principal moments of inertia for the point o, and h is the distance from the point o to the center of mass. Letting S' be $\tilde{S}$ in Eqs. (5) and (6), and substituting Eqs. (25)–(28) in the resulting equations, we obtain

$$mgh\sin\theta = A\frac{d}{dt}(\dot{\theta}) + (\dot{\phi}\sin\theta)(\dot{\phi}\cos\theta + \dot{\psi})C - (\dot{\phi}\cos\theta)(\dot{\phi}\sin\theta)A \quad (29)$$

$$0 = A\frac{d}{dt}(\dot{\phi}\sin\theta) + (\dot{\phi}\cos\theta)(\dot{\theta})A - (\dot{\theta})(\dot{\phi}\cos\theta + \dot{\psi})C \quad (30)$$

$$0 = C\frac{d}{dt}(\dot{\phi}\cos\theta + \dot{\psi}) \quad (31)$$

From Eq. (31) we obtain

$$\dot{\phi}\cos\theta + \dot{\psi} = s \quad (32)$$

where s is a constant.

Substituting Eq. (32) in Eqs. (29) and (30) and defining

$$\beta \equiv \frac{Cs}{A} \quad (33)$$

we obtain

$$\ddot{\theta} - \dot{\phi}^2 \sin\theta\cos\theta + \beta\dot{\phi}\sin\theta = \frac{mgh\sin\theta}{A} \tag{34}$$

$$\ddot{\phi}\sin\theta + 2\dot{\phi}\dot{\theta}\cos\theta - \beta\dot{\theta} = 0 \tag{35}$$

Equation (35) is equivalent to the equation

$$\frac{d}{dt}\left[\dot{\phi}\sin^2\theta + \beta\cos\theta\right] = 0 \tag{36}$$

from which it follows that

$$\dot{\phi}\sin^2\theta + \beta\cos\theta = \alpha \tag{37}$$

where α is a constant. Solving Eq. (37) for $\dot{\phi}$ we obtain

$$\dot{\phi} = \frac{\alpha - \beta\cos\theta}{\sin^2\theta} \tag{38}$$

Substituting Eq. (38) in Eq. (34) we obtain

$$\ddot{\theta} = \frac{mgh\sin\theta}{A} + \frac{(\alpha - \beta\cos\theta)^2\cos\theta}{\sin^3\theta} - \frac{\beta(\alpha - \beta\cos\theta)}{\sin\theta} \tag{39}$$

which can be rewritten

$$\ddot{\theta} = -\frac{dU}{d\theta} \tag{40}$$

where

$$U(\theta) = \frac{mgh\cos\theta}{A} + \frac{(\alpha - \beta\cos\theta)^2}{2\sin^2\theta} \tag{41}$$

Equation (40) can be solved to obtain $\theta(t)$. Knowing $\theta(t)$ we can use Eq. (38) to obtain $\phi(t)$. Knowing $\theta(t)$ and $\phi(t)$ we can use Eq. (32) to obtain $\psi(t)$. The solutions we thus obtain do not provide a good geometrical feel for the motion. It is possible however without solving the problem completely to get some insight into the nature of the motion by noting that Eq. (40) is perfectly analogous to the equation $\ddot{x} = -dV/dx$, which is the equation of motion of a particle of unit mass that is moving in one dimension in a potential $V(x)$. Hence the θ motion may be thought of as the motion of a point mass in the effective potential $U(\theta)$.

A closer look at the effective potential $U(\theta)$ reveals that if $\alpha \neq \beta$ then the potential is ∞ at $\theta = 0$ and $\theta = \pi$ and has a single minimum between 0 and π. Hence the θ motion will consist of an oscillatory motion between two values θ_1 and θ_2. This motion is called *nutation*. As the top nutates between θ_1 and θ_2 the axis of the top is precessing around the vertical direction with an angular speed $\dot{\phi}$. If the angle $\cos^{-1}(\alpha/\beta) \equiv \theta_0$ falls between θ_1 and θ_2, then from Eq. (38) it follows that the direction of the precessional motion will reverse every time θ passes through the value θ_0, hence in this case the top will precess with a looping motion. If $\alpha = \beta$, as would be the case if at some instant $\theta = 0$, and if $(2mghA/C^2s^2) > \frac{1}{2}$, the potential has a finite value at $\theta = 0$, drops to a minimum between $\theta = 0$ and $\theta = \pi$, and goes to ∞ at $\theta = \pi$. If $\alpha = \beta$ and $(2mghA/C^2s^2) < \frac{1}{2}$, the potential has a finite value at $\theta = 0$ and rises from this value to ∞ as θ goes to π.

Solution 2. The solution could also have been obtained by hunting for constants of the motion. Thus:

a. $G_3 = 0$ and so

$$H_3(o) = \sum_i e_{3i} H_i(o) = \sum_i \sum_j e_{3i} I_{ij}(o) \omega_j(S\bar{S})$$

$$= A\dot{\phi} \sin^2\theta + C\cos\theta (\dot{\phi}\cos\theta + \dot{\psi}) = \text{constant} \qquad (42)$$

b. The forces are conservative, hence

$$T + V = \frac{1}{2} \sum_i \sum_j I_{ij} \omega_i(S\bar{S}) \omega_j(S\bar{S}) + mgh\cos\theta$$

$$= \frac{1}{2} A\dot{\theta}^2 + \frac{1}{2} A(\dot{\phi}\sin\theta)^2 + \frac{1}{2} C(\dot{\phi}\cos\theta + \dot{\psi})^2 + mgh\cos\theta$$

$$= \text{constant} \qquad (43)$$

c. $G_{\bar{3}} = 0$, hence from the third of Euler's equations

$$\omega_{\bar{3}}(S\bar{S}) = \dot{\phi}\cos\theta + \dot{\psi} = \text{constant} \qquad (44)$$

These three equations can be shown to be equivalent to the equations of motion obtained in Solution 1.

Example 3. Freely Rotating Asymmetrical Top. A rigid body is constrained to rotate about a fixed point o. There are no external torques with respect to the point o. Describe the motion.

Solution. To simplify the notation we do the following:

a. Designate the three principal moments of inertia with respect to the point o as A, B, and C, respectively, that is,

$$I_{\bar{1}}(o) = A \qquad I_{\bar{2}}(o) = B \qquad I_{\bar{3}}(o) = C \qquad (45)$$

b. Label our principal axes in such a way that

$$A > B > C \qquad (46)$$

c. Let

$$\omega \equiv \omega(S\bar{S}) \qquad (47)$$

and designate the components of ω in the frame $\bar{S}$ as p, q, and r, respectively, that is,

$$p \equiv \omega_{\bar{1}} \qquad q \equiv \omega_{\bar{2}} \qquad r \equiv \omega_{\bar{3}} \qquad (48)$$

d. Suppress all reference to the point o where this leads to no ambiguity. For example, we write **H** instead of **H**(o), and we speak of the "angular momentum" instead of the "angular momentum with respect to the point o."

We carry out the solution in two stages. In the first stage we use Euler's equations to obtain p, q, and r in terms of the Euler angles θ, ϕ, and ψ. This provides us with three first order differential equations in θ, ϕ, and ψ, which can be solved to obtain θ, ϕ, and ψ as functions of the time.

Stage 1. Euler's equations for the problem considered are:

$$A\dot{p} - qr(B - C) = 0 \tag{49}$$

$$B\dot{q} - rp(C - A) = 0 \tag{50}$$

$$C\dot{r} - pq(A - B) = 0 \tag{51}$$

Since there are no external torques, the angular momentum **H** is a constant of the motion; hence H, the magnitude of the angular momentum, is a constant of the motion. We thus have

$$A^2 p^2 + B^2 q^2 + C^2 r^2 = H^2 = \text{constant} \tag{52}$$

Since the only forces acting on the body act at the point o and these forces do no work, the kinetic energy T is a constant of the motion. We thus have

$$A p^2 + B q^2 + C r^2 = 2T = \text{constant} \tag{53}$$

We now define

$$D \equiv \frac{H^2}{2T} \tag{54}$$

$$\Omega \equiv \frac{2T}{H} \tag{55}$$

The quantity D can be shown to be greater than C and less than A, that is,

$$A > D > C \tag{56}$$

To prove this we note that

$$\frac{D}{A} = \frac{H^2}{2AT} = \frac{A^2 p^2 + B^2 q^2 + C^2 r^2}{A^2 p^2 + ABq^2 + ACr^2} < 1 \tag{57}$$

$$\frac{D}{C} = \frac{H^2}{2CT} = \frac{A^2 p^2 + B^2 q^2 + C^2 r^2}{CAp^2 + CBq^2 + C^2 r^2} > 1 \tag{58}$$

Substituting Eqs. (54) and (55) in Eqs. (52) and (53) we obtain

$$A p^2 + B q^2 + C r^2 = D \Omega^2 \tag{59}$$

$$A^2 p^2 + B^2 q^2 + C^2 r^2 = D^2 \Omega^2 \tag{60}$$

Solving Eqs. (59) and (60) for p^2 and r^2 in terms of q^2 we obtain

$$p^2 = \frac{B(B-C)}{A(A-C)}(\alpha^2 - q^2) \tag{61}$$

$$r^2 = \frac{B(A-B)}{C(A-C)}(\beta^2 - q^2) \tag{62}$$

where

$$\alpha^2 \equiv \frac{D(D-C)}{B(B-C)}\Omega^2 \tag{63}$$

$$\beta^2 \equiv \frac{D(A-D)}{B(A-B)}\Omega^2 \tag{64}$$

Substituting Eqs. (61) and (62) in Eq. (50) we obtain

$$\dot{q}^2 = \frac{(A-B)(B-C)}{AC}(\alpha^2 - q^2)(\beta^2 - p^2) \tag{65}$$

Multiplying Eq. (65) by $AC/[(A-B)(B-C)\alpha^2\beta^2]$ we obtain

$$\frac{AC\dot{q}^2}{(A-B)(B-C)\alpha^2\beta^2} = \left(1 - \frac{q^2}{\alpha^2}\right)\left(1 - \frac{q^2}{\beta^2}\right) \tag{66}$$

Let us for convenience assume

$$\alpha^2 > \beta^2 \tag{67}$$

and define

$$Q \equiv \frac{q}{\beta} \tag{68}$$

If we substitute Eq. (68) in Eq. (66) we obtain

$$\frac{AC\dot{Q}^2}{(A-B)(B-C)\alpha^2} = (1 - Q^2)(1 - k^2 Q^2) \tag{69}$$

where

$$k^2 \equiv \frac{\beta^2}{\alpha^2} < 1 \tag{70}$$

We now define

$$\tau \equiv \left[\frac{(A-B)(B-C)\alpha^2}{AC}\right]^{1/2} t \tag{71}$$

Substituting Eq. (71) in Eq. (69) we obtain

$$\left(\frac{dQ}{d\tau}\right)^2 = (1 - Q^2)(1 - k^2 Q^2) \tag{72}$$

The solution to Eq. (72) is

$$Q = \text{sn}(\tau - \tau_0, k) \tag{73}$$

where sn is the Jacobian elliptic function (see Appendix 16) and τ_0 is an integration constant.

Equation (73) provides us with $q(t)$. If we substitute Eq. (73) in Eqs. (61) and (62) we obtain $p(t)$ and $q(t)$. We have thus formally completed Stage 1 of the solution. The solutions we have obtained however do not provide one with a particularly good feel for the results. Therefore we consider the same problem from a geometrical point of view.

We note first that the instantaneous value of the angular velocity can be represented by a point or vector $\boldsymbol{\omega}$ in angular velocity space. If we now draw a line through the origin of coordinates in the direction of $\mathbf{H}$, the line will remain at rest with respect to an inertial frame of reference, since $\mathbf{H}$ is a constant of the motion. We call this line the *invariable line*. The projection of the angular velocity $\boldsymbol{\omega}$ onto the invariable line is given by

$$\frac{\boldsymbol{\omega} \cdot \mathbf{H}}{H} = \frac{2T}{H} \tag{74}$$

The projection of $\boldsymbol{\omega}$ onto the invariable line is thus a constant. Hence the point $\boldsymbol{\omega}$ lies on a plane that is perpendicular to the invariable line at a fixed point, a plane that is thus fixed in the inertial frame. This plane is called the *invariable plane*.

The angular velocity must also satisfy the condition

$$\boldsymbol{\omega} \cdot \mathbf{H} = 2T \tag{75}$$

In the body frame this condition is equivalent to the equation

$$Ap^2 + Bq^2 + Cr^2 = 2T \tag{76}$$

which is the equation of an ellipsoid fixed in the body frame with center at the origin. We call this ellipsoid the *Poinsot ellipsoid*. Thus the point $\boldsymbol{\omega}$ lies on the surface of the Poinsot ellipsoid. The Poinsot ellipsoid is fixed in the body frame but in the inertial frame is rotating about the origin with the same angular velocity as the body. Finally we note, using Theorem 7 in Appendix 5, that the normal to the Poinsot ellipsoid at a surface point (p, q, r) in the body frame is in the direction of the vector whose components in the body frame are (Ap, Bq, Cr). These are just the components in the body frame of the angular momentum vector $\mathbf{H}$. Hence the normal to the Poinsot ellipsoid at the point (p, q, r) is in the direction of $\mathbf{H}$.

The point $\boldsymbol{\omega}$ can simultaneously lie on the invariable plane and the

Poinsot ellipsoid, while the normals to both surfaces at the point ω are in the same direction (i.e., the direction of **H**), only if the Poinsot ellipsoid is tangent to the invariable plane and ω lies at the point of contact.

Finally since the direction of ω is in the direction of the instantaneous axis of rotation of the body, it follows that the point of contact of the Poinsot ellipsoid and the invariable plane lies on the instantaneous axis of rotation, hence is at rest in both frames. The Poinsot ellipsoid thus moves on the invariable plane as if it were rolling on a perfectly rough plane.

Gathering results we see that if we construct the invariable plane defined by Eq. (74) and the Poinsot ellipsoid defined by Eq. (75) and allow the ellipsoid to roll without slipping on the plane while its center remains fixed, the path traced out by the point of contact will be the path of the point ω. Since the moments of inertia A, B, and C are all different, the Poinsot ellipsoid will wobble as it rolls and thus the angular velocity vector will precess and nutate about the direction of the angular momentum vector.

Stage 2. We assume that we know $p(t)$, $q(t)$, and $r(t)$ and wish to find $\theta(t)$, $\phi(t)$, and $\psi(t)$. The relations between p, q, and r and θ, ϕ, and ψ (given in Table 1*b* in Chapter 39) are

$$p = \dot{\phi} \sin\theta \sin\psi + \dot{\theta} \cos\psi \tag{77}$$

$$q = \dot{\phi} \sin\theta \cos\psi - \dot{\theta} \sin\psi \tag{78}$$

$$r = \dot{\phi} \cos\theta + \dot{\psi} \tag{79}$$

In principle we can substitute $p(t)$, $q(t)$, and $r(t)$ into Eqs. (77)–(79) and solve for $\theta(t)$, $\phi(t)$, and $\psi(t)$. However our task is simplified because we know that **H** is a constant of the motion. If we choose our inertial 3 axis in the direction of the angular momentum, the components of the angular momentum in the inertial frame are given by

$$\begin{bmatrix} H_1 \\ H_2 \\ H_3 \end{bmatrix} = \begin{bmatrix} 0 \\ 0 \\ H \end{bmatrix} \tag{80}$$

It follows from the results of the preceding chapter that the components of **H** in the body frame are

$$\begin{bmatrix} H_{\bar{1}} \\ H_{\bar{2}} \\ H_{\bar{3}} \end{bmatrix} = \begin{bmatrix} e_{\bar{1}1} & e_{\bar{1}2} & e_{\bar{1}3} \\ e_{\bar{2}1} & e_{\bar{2}2} & e_{\bar{2}3} \\ e_{\bar{3}1} & e_{\bar{3}2} & e_{\bar{3}3} \end{bmatrix} \begin{bmatrix} 0 \\ 0 \\ H \end{bmatrix} = \begin{bmatrix} H \sin\theta \sin\psi \\ H \sin\theta \cos\psi \\ H \cos\theta \end{bmatrix} \tag{81}$$

But

$$\begin{bmatrix} H_{\bar{1}} \\ H_{\bar{2}} \\ H_{\bar{3}} \end{bmatrix} = \begin{bmatrix} Ap \\ Bq \\ Cr \end{bmatrix} \tag{82}$$

Equating Eqs. (81) and (82) we obtain
$$Ap = H \sin\theta \sin\psi \qquad (83)$$
$$Bq = H \sin\theta \cos\psi \qquad (84)$$
$$Cr = H \cos\theta \qquad (85)$$

From Eq. (85) we obtain
$$\cos\theta = \frac{Cr}{H} \qquad (86)$$

Dividing Eq. (83) by Eq. (84) we obtain
$$\tan\psi = \frac{Ap}{Bq} \qquad (87)$$

Since we know $p(t)$, $q(t)$, and $r(t)$, Eqs. (86) and (87) enable us to obtain $\theta(t)$ and $\psi(t)$. If we now multiply Eq. (77) by $\sin\psi$ and Eq. (78) by $\cos\psi$ and add the resulting equations, we obtain
$$\dot\phi \sin\theta = p \sin\psi + q \cos\psi \qquad (88)$$

Since we know $p(t)$, $q(t)$, $\theta(t)$, and $\psi(t)$, we can use Eq. (88) to obtain $\dot\phi(t)$. We can then integrate to obtain $\phi(t)$. We have thus in principle completed the solution of the problem.

Example 4. A homogeneous sphere of mass M and radius a moves on a perfectly rough horizontal plane under the action of a force **f** whose line of action passes through the center of mass. Show that the motion of the center of the sphere is the same as if the plane were smooth and the force **f** reduced to five-sevenths of its former value.

Solution. The motion of the sphere is governed by the equations
$$\mathbf{F} = M\ddot{\mathbf{R}} \qquad (89)$$
$$\mathbf{G}(c) = \dot{\mathbf{H}}(c) \qquad (90)$$

If we choose our inertial frame with origin on the horizontal plane and z axis vertically up, and let **S** be the force the surface exerts on the sphere, then the net force acting on the sphere is
$$\mathbf{F} = \mathbf{f} + \mathbf{S} + M\mathbf{g} = \mathbf{i}(f_x + S_x) + \mathbf{j}(f_y + S_y) + \mathbf{k}(f_z + S_z - Mg) \qquad (91)$$

The only force that exerts a torque with respect to the center of mass is **S**. Hence the net torque with respect to the center of mass is given by
$$\mathbf{G}(c) = (-\mathbf{k}a) \times \mathbf{S} = \mathbf{i}aS_y - \mathbf{j}aS_x \qquad (92)$$

The angular momentum with respect to the center of mass is
$$\mathbf{H}(c) = \frac{2Ma^2}{5}\left[\mathbf{i}\omega_x + \mathbf{j}\omega_y + \mathbf{k}\omega_z\right] \qquad (93)$$

Substituting Eqs. (91)–(93) into Eqs. (89) and (90) we obtain

$$f_x + S_x = M\ddot{X} \tag{94}$$

$$f_y + S_y = M\ddot{Y} \tag{95}$$

$$f_z + S_z - Mg = M\ddot{Z} \tag{96}$$

$$aS_y = \frac{2Ma^2}{5}\dot{\omega}_x \tag{97}$$

$$-aS_x = \frac{2Ma^2}{5}\dot{\omega}_y \tag{98}$$

$$0 = \frac{2Ma^2}{5}\dot{\omega}_z \tag{99}$$

In addition to the equations of motion above, we have a number of constraint conditions. The center of the sphere must remain a distance a above the surface, hence

$$Z = a \tag{100}$$

Since the surface is rough, the point on the sphere that is in contact with the surface must be moving with zero velocity. Thus

$$\dot{\mathbf{R}} + \boldsymbol{\omega} \times (-\mathbf{k}a) = \mathbf{i}\dot{X} + \mathbf{j}\dot{Y} + \mathbf{k}\dot{Z} + (\mathbf{i}\omega_x + \mathbf{j}\omega_y + \mathbf{k}\omega_z) \times (-\mathbf{k}a)$$

$$= \mathbf{i}(\dot{X} - \omega_y a) + \mathbf{j}(\dot{Y} + \omega_x a) + \mathbf{k}(\dot{Z}) = 0 \tag{101}$$

from which it follows that

$$\dot{X} = \omega_y a \tag{102}$$

$$\dot{Y} = -\omega_x a \tag{103}$$

Combining Eqs. (94)–(99), (102), and (103) we obtain

$$\tfrac{5}{7} f_x = M\ddot{X} \tag{104}$$

$$\tfrac{5}{7} f_y = M\ddot{Y} \tag{105}$$

These are the same equations that would be obtained if the sphere were smooth and the force $\mathbf{f}$ replaced by $\tfrac{5}{7}\mathbf{f}$.

PROBLEMS

1. Show by the use of Euler's equations that for any free motion of a lamina, the component of the angular velocity in the plane of the lamina is constant in magnitude.

2. A thin uniform disk of radius a and mass m is rotating with a uniform angular velocity ω about a fixed axis passing through its center but inclined at an angle α to the axis of symmetry. Find the direction and magnitude of the torque that is exerted on the bearings.

334 Rigid Body Dynamics

3. An S-shaped crankshaft, constructed of two semicircles, each of radius a and mass $m/2$, is free to rotate about a smooth rod that passes through the center and both ends of the crankshaft. The crankshaft is accelerated at a constant rate from rest to the angular velocity Ω in a time τ. Determine the couple in the direction of the rod that must be applied to the crankshaft to produce this rotation, and the reaction couple exerted by the rod on the crankshaft.

4. A square sheet is constrained to rotate with an angular velocity Ω about an axis passing through its center and making an angle α with the axis through the center of mass and normal to the sheet. At the instant the axis of rotation lies in the plane determined by this axis of symmetry and a diagonal, the body is released. Find the rate at which the axis of symmetry precesses about the constant direction of the angular momentum.

5. A uniform double right circular cone of semivertical angle α, symmetrical about the common vertex, about which it is free to rotate, is set rotating about a generator. Prove that the axis of the body describes in space a cone of semivertical angle $\tan^{-1}(\frac{1}{2}\tan\alpha + 2\cot\alpha)$.

6. A solid homogeneous cuboid of edges $2a$, $2a$, and $4a$ can turn freely under no forces about its center, which is fixed. It is set spinning with angular velocity Ω about a diagonal. Find the semivertical angle of the cone described in space by the line through the center parallel to the longer edges, and show that the time taken by this line to move once around the cone is $10\pi/[\Omega(11)^{1/2}]$.

7. A homogeneous circular disk of mass m and radius a is set rotating with an angular velocity Ω about an axis passing through the center of the disk and making an angle α with the normal. The disk is suddenly released.
   a. Show that the axis of rotation describes a cone in space, and also describes a cone relative to the disk.
   b. Find the time taken by the axis of rotation to describe a cone in space.
   c. Find the time taken by the axis of rotation to describe a cone relative to the disk.

8. A uniform solid right circular cone of height b, mass m, and semivertical angle $\pi/4$ moves freely about its vertex O, which is fixed, under no forces other than the reaction at O. If the cone is set spinning with angular velocity Ω about a generator L, find the components of the angular velocity at any subsequent time t with respect to a frame $\bar{S}$ fixed in the body with origin at O, the $\bar{3}$ axis along the axis of the cone, and the $\bar{1}$ axis in the plane determined by the generator L and the axis of the

cone. Find the reaction at O and show that its magnitude is $(3/40)(29)^{1/2}mb\Omega^2$.

9. A plane lamina moves under no forces with a point O fixed. The moments of inertia about the principal axes at O lying in the plane are A and B, where $A > B$. The component of the angular velocity in the plane of the lamina is Ω. Show that Ω is a constant of the motion, and that if the motion is started in any manner that makes the total kinetic energy $A\Omega^2/2$ the lamina will ultimately rotate about the principal axis associated with A.

10. A uniform cone of mass m, height h, and semivertical angle α is rotating with angular velocity Ω about a generator when suddenly the generator is loosed and the diameter of the base that intersects the generator is fixed. Prove that the new angular velocity is $[1 - (h^2/4k^2)]\Omega \sin \alpha$, where mk^2 is the moment of inertia of the cone about a diameter of the base.

11. A uniform rod of mass m and length a is smoothly pivoted at one end to a fixed support. The rod rests in equilibrium with the other end in contact with a smooth vertical wall. The distance of the fixed support from the wall is b, where $b < a$. Show that if the rod is gently displaced from this position of equilibrium, contact with the wall will cease when the rod makes an acute angle $\cos^{-1}[2(a^2 - b^2)^{1/2}/(3a)]$ with the upward vertical.

12. A top consists of a uniform circular disk of mass $4m$ and radius $3a$, at the center of which is rigidly attached the end O of a uniform rod, OX, of length $4a$ and mass m, the rod being perpendicular to the disk. Show that if the top can precess in steady motion with the end fixed and the lowest point of the disk at the same level as X, the spin about the axis must be not less than $[94g/(9a)]^{1/2}$. The spin of a symmetric top is the component of the angular velocity of the top in the direction of the axis of symmetry.

13. A circular disk of radius a and mass m spins on a smooth table about a vertical diameter. Find the condition that this motion is stable.

14. Consider a top of mass m, height of center of mass h, axial moment of inertia C, and transverse moment of inertia A, which is spinning on a rough plane with its axis vertical, n being the spin. If $C^2n^2 < 4Amgh$, show that a possible motion in the neighborhood of the vertical is one in which the projection of the center of mass of the top onto the horizontal plane describes approximately the equiangular spiral $r = r_0 \exp\{[(4Amgh - C^2n^2)^{1/2}/Cn]\theta\}$.

15. A symmetrical top that is precessing steadily can be described by the condition $\dot{\theta} = \ddot{\theta} = 0$. Develop a relationship among the parameters that must be satisfied to ensure this type of motion.

16. An axially symmetrical top of mass m is freely pivoted about a point O on its axis. Its center of mass G is at a distance h from O and its principal moments of inertia at O are A, A, and C. Initially the top is set in motion with its axis horizontal and with a small angular velocity n about its axis. Show that when G is at its lowest point in the subsequent path, the axis of the top makes an angle approximately $Cn/(2mghA)^{1/2}$ with the downward vertical.

17. A top in the form of a homogeneous right circular cone of height $3a/2$ and radius of base a spins steadily about a vertical axis. The top is suddenly subjected to a transverse impulse such that it acquires a nutational angular velocity of one-fifth the spin velocity. Find the maximum nutational excursion of the ensuing motion if the constant spin velocity is equal to $(15/4)(2g/a)^{1/2}$, where g is the acceleration due to gravity.

18. A uniform circular hoop of radius a rolls uniformly on a rough horizontal plane so that the point of contact describes a circle of radius r in time $2\pi/\Omega$, the plane of the hoop being inclined inward at an angle α to the horizontal. Show that $g\cos\alpha = \Omega^2 \sin\alpha [2r - \frac{3}{2}a\cos\alpha]$. Find also the least value of the coefficient of friction if the hoop is not to sideslip.

19. A solid cone of mass M, height b, and semivertical angle α rolls in steady motion on a rough horizontal table, the line of contact rotating with angular velocity Ω. Show that the reaction of the table on the cone is equivalent to a single force that cuts the line of contact at a distance $(3b\cos\alpha/4) + (k^2\Omega^2\cot\alpha/g)$ from the vertex, where Mk^2 is the moment of inertia of the cone about the line of contact. Show that the greatest possible value that Ω can have without the cone overturning is $[gb\sin\alpha(1+3\sin^2\alpha)]^{1/2}/(2k\cos\alpha)$.

20. A uniform circular disk of mass M and radius a is so mounted that it can turn freely about its center, which is fixed. The disk is spinning in a horizontal plane with an angular velocity ω, when a particle of mass m falling vertically with a speed v hits the disk near the edge and sticks to it. Prove that immediately after the impact the particle is moving in a direction inclined to the horizontal at an angle α, where $\tan\alpha = 4\{[m(M+2m)]/[M(M+4m)]\}(v/a\omega)$.

21. A spinning symmetrical top is sliding on a smooth horizontal floor with its symmetry axis vertical. The center of mass of the top is a distance h above the floor and is moving with a speed $(gh)^{1/2}$. The angular velocity of the top is $20(g/h)^{1/2}$. The principal moments of inertia of the top at the bottom tip are $mh^2/4$, $5mh^2/4$, and $5mh^2/4$, respectively. Suddenly the bottom point of the top is fixed, in such a way that the top is free to move about this fixed point. Determine the maximum angle that the symmetry axis makes with the vertical in the ensuing motion.

22. A uniform solid sphere rolls without slipping inside a stationary inverted cone of semiangle $\pi/4$ with its axis vertical. If the center of the sphere describes a horizontal circle of radius a with angular velocity $(5g/9a)^{1/2}$, prove that the sphere will have no component of angular velocity about the vertical. Show that the coefficient of friction between sphere and cone is not less than $\frac{2}{7}$.

23. A uniform sphere rolls without slipping on a horizontal plane that is kept rotating with constant angular velocity Ω about a fixed vertical axis. Show that the center of the sphere describes a circle fixed in space and that it makes a complete revolution in time $7\pi/\Omega$.

24. A sphere of radius a rolls without slipping on a perfectly rough vertical plane. Gravity is acting vertically downward. The plane rotates with a uniform angular velocity s about a vertical axis that lies in the plane itself. If the sphere is initially at rest with respect to the plane, find the vertical distance of the sphere from its starting point as a function of time. Show that the maximum distance the sphere is able to descend is $5g/s^2$. Assume that the sphere is constrained to remain in contact with the plane.

25. A rigid body has principal moments of inertia A, A, and C at a point about which it moves under no forces except a resisting couple equal to k times the resultant angular velocity. If at time $t = 0$ the instantaneous axis of rotation makes an angle α with the axis of symmetry, prove that at time t it makes an angle ϕ with this axis where $\tan\phi = \tan\alpha \exp\{kt[(A - C)/(AC)]\}$. Also show that the plane containing the instantaneous axis and the axis of symmetry has rotated about the latter through an angle $[(C - A)C\omega \cos\alpha/(Ak)][1 - \exp(-kt/C)]$, where ω is the magnitude of the initial angular velocity.

41

Impulsive Motion of a Rigid Body

INTRODUCTION

In Chapter 18 we discussed the effect of an instantaneous impulse on the motion of a particle. In this chapter we extend these results, and consider the effect of an instantaneous impulse on the motion of a rigid body.

THE EFFECT OF AN INSTANTANEOUS IMPULSE ON THE MOTION OF A RIGID BODY

The following theorem summarizes the effect of an instantaneous impulse on the motion of a rigid body. This straightforward extension of the results of Chapter 18 is stated without proof.

Theorem. If a rigid body is acted upon by an instantaneous impulse $\hat{\mathbf{f}}$, the configuration of the rigid body will be unchanged but the momentum **P** with respect to any point a whose velocity does not change discontinuously at the instant of the impulse will undergo a finite discontinuous change $\Delta \mathbf{P}$ equal to the instantaneous impulse, that is,

$$\Delta \mathbf{P} = \hat{\mathbf{f}} \tag{1}$$

and the angular momentum with respect to any point a whose velocity does not change discontinuously at the instant of the impulse, or with respect to the center of mass c, will undergo a finite discontinuous change $\Delta \mathbf{H}$ equal to the instantaneous angular impulse $\hat{\mathbf{g}} \equiv \mathbf{r} \times \hat{\mathbf{f}}$, that is,

$$\Delta \mathbf{H} = \hat{\mathbf{g}} \equiv \mathbf{r} \times \hat{\mathbf{f}} \tag{2}$$

where **r** is the distance from the point a or from the center of mass c to the point of application of the impulse.

RIGID BODY COLLISIONS

If two rigid bodies collide the effect of the collision frequently can be determined to a good degree of approximation by making use of the following hypothesis, which is a straightforward extension of the results of Chapter 18.

Hypothesis 1 (Poisson's Hypothesis). Consider a collision between a rigid body A and another rigid body $\bar{A}$. Let a be the point of contact on body A, $\bar{a}$ the point of contact on body $\bar{A}$, $\mathbf{n}(a\bar{a})$ a unit vector normal to the surface of contact between the two bodies, the positive sense being from a to $\bar{a}$, and $\mathbf{v}(a\bar{a})$ the velocity of point $\bar{a}$ with respect to point a. The collision can be broken down into two stages. In the first stage, called the period of compression, the colliding surfaces are compressed until $\mathbf{n} \cdot \mathbf{v}(a\bar{a}) = 0$. In the second stage, called the period of restitution, the surfaces expand and the bodies fly apart. The impulse imparted by body A to body $\bar{A}$ during the period of compression is equal to an instantaneous impulse $\hat{\mathbf{f}}'$ and the impulse imparted by body A to body $\bar{A}$ during the period of restitution is equivalent to an instantaneous impulse $\hat{\mathbf{f}}''$. The normal component of $\hat{\mathbf{f}}''$ is equal to e times the normal component of $\hat{\mathbf{f}}'$, where e is an experimental constant called the *coefficient of restitution*, which depends only on the natures of the two surfaces and whose value lies between 0 and 1. If the surface is rough there is little experimental evidence that would enable one to predict beforehand the value of the tangential components of $\hat{\mathbf{f}}'$ and $\hat{\mathbf{f}}''$. However, in the absence of any other information it is customary to assume that the tangential components of $\hat{\mathbf{f}}'$ and $\hat{\mathbf{f}}''$ bear the same relation to the normal components as they would if they were ordinary forces and the ordinary laws of friction were in effect. If the surface is smooth and $e = 1$, we say that the collision is an *elastic collision*. If the surface is rough or $e \neq 1$, we say that the collision is an *inelastic collision*. If the surface is perfectly rough, or sufficiently rough to reduce the tangential velocity of the particle to zero, and $e = 0$, we say that the collision is a *perfectly inelastic collision*.

The preceding hypothesis is theoretically the most satisfying starting point for handling the collision of rigid bodies. However one can sometimes shorten the calculations by making use of the following theorem, which applies to a free collision between two smooth rigid bodies.

Theorem 1. (Newton's Law). Consider a collision between a smooth rigid body A and another smooth rigid body $\bar{A}$. Let a be the point of contact on A, $\bar{a}$ the point of contact on $\bar{A}$, $\mathbf{n}(a\bar{a})$ a unit vector normal to the surface of contact between the two bodies, the positive sense being from a to $\bar{a}$; $\mathbf{v}'(a\bar{a})$ the velocity of point $\bar{a}$ with respect to point a just before the impact, and $\mathbf{v}''(a\bar{a})$ the velocity of point $\bar{a}$ with respect to point a just after the impact. If during the collision the two bodies are not acted upon by any external constraints or impulsive forces other than that at the point of contact, then

$$\mathbf{v}''(a\bar{a}) \cdot \mathbf{n}(a\bar{a}) = -e\mathbf{v}'(a\bar{a}) \cdot \mathbf{n}(a\bar{a}) \tag{3}$$

That is, the normal component of the relative velocity of point $\bar{a}$ with respect to point a after the collision is equal to $-e$ times its value before the collision.

PROOF: Let M be the mass of body A, $\overline{M}$ the mass of body $\overline{A}$, $\mathbf{R}$ the position of the center of mass c of body A with respect to the origin O of an inertial frame, $\overline{\mathbf{R}}$ the position of the center of mass $\bar{c}$ of $\overline{A}$ with respect to O, $\mathbf{r}$ the position of a with respect to c, $\bar{\mathbf{r}}$ the position of $\bar{a}$ with respect to $\bar{c}$, $\mathbf{n}$ the unit vector $\mathbf{n}(a\bar{a})$, $\mathbf{V}$ the velocity of c with respect to O, $\overline{\mathbf{V}}$ the velocity of $\bar{c}$ with respect to O, ω the angular velocity of A, $\bar{\omega}$ the angular velocity of $\overline{A}$, $\hat{\mathbf{f}}'$ the impulse applied to body $\overline{A}$ during the compressive stage, and $\hat{\mathbf{f}}''$ the impulse applied to body $\overline{A}$ during the restitution stage. We designate the precollision velocities with a prime, the postcollision velocities with a double prime, and the velocities at the instant when the compressive stage is ending and the restitution stage is beginning with an asterisk. For a given orientation the angular momentum $\mathbf{H}(c)$ of body A with respect to the center of mass c is a linear vector function of ω, which for the orientation at the moment of the collision we designate $\phi(\omega)$, that is, at the moment of the collision $\mathbf{H}(c) = \phi(\omega)$, where $\phi(\alpha\omega + \beta\Omega) = \alpha\phi(\omega) + \beta\phi(\Omega)$. Similarly the angular momentum $\overline{\mathbf{H}}(\bar{c})$ of body $\overline{A}$ with respect to the center of mass $\bar{c}$ is at the moment of collision a linear vector function of $\bar{\omega}$, which we designate $\bar{\phi}(\bar{\omega})$. Applying the impulsive equations of motion to each stage in the collision and to each body we obtain

$$M\mathbf{V}^* - M\mathbf{V}' = -\hat{\mathbf{f}}' \tag{4}$$

$$\overline{M}\overline{\mathbf{V}}^* - \overline{M}\overline{\mathbf{V}}' = \hat{\mathbf{f}}' \tag{5}$$

$$\phi(\omega^*) - \phi(\omega') = -\mathbf{r} \times \hat{\mathbf{f}}' \tag{6}$$

$$\bar{\phi}(\bar{\omega}^*) - \bar{\phi}(\bar{\omega}') = \bar{\mathbf{r}} \times \hat{\mathbf{f}}' \tag{7}$$

$$M\mathbf{V}'' - M\mathbf{V}^* = -\hat{\mathbf{f}}'' \tag{8}$$

$$\overline{M}\overline{\mathbf{V}}'' - \overline{M}\overline{\mathbf{V}}^* = \hat{\mathbf{f}}'' \tag{9}$$

$$\phi(\omega'') - \phi(\omega^*) = -\mathbf{r} \times \hat{\mathbf{f}}'' \tag{10}$$

$$\bar{\phi}(\bar{\omega}'') - \bar{\phi}(\bar{\omega}^*) = \bar{\mathbf{r}} \times \hat{\mathbf{f}}'' \tag{11}$$

Solving for the pre- and postcollision velocities we obtain

$$V' = V^* + \frac{\hat{f}'}{M} \tag{12}$$

$$\bar{V}' = \bar{V}^* - \frac{\hat{f}'}{M} \tag{13}$$

$$\omega' = \omega^* + \phi^{-1}(r \times \hat{f}') \tag{14}$$

$$\bar{\omega}' = \bar{\omega}^* - \bar{\phi}^{-1}(\bar{r} \times \hat{f}') \tag{15}$$

$$V'' = V^* - \frac{\hat{f}''}{M} \tag{16}$$

$$\bar{V}'' = \bar{V}^* + \frac{\hat{f}''}{M} \tag{17}$$

$$\omega'' = \omega^* - \phi^{-1}(r \times \hat{f}'') \tag{18}$$

$$\bar{\omega}'' = \bar{\omega}^* - \bar{\phi}^{-1}(\bar{r} \times \hat{f}'') \tag{19}$$

The velocity of point a is $V + \omega \times r$, the velocity of point $\bar{a}$ is $\bar{V} + \bar{\omega} \times \bar{r}$, and thus $v(a\bar{a}) = (\bar{V} + \bar{\omega} \times \bar{r}) - (V + \omega \times r)$. It follows that

$$v'(a\bar{a}) = \bar{V}' + \bar{\omega}' \times \bar{r} - V' - \omega' \times r \tag{20}$$

$$v^*(a\bar{a}) = \bar{V}^* + \bar{\omega}^* \times \bar{r} - V^* - \omega^* \times r \tag{21}$$

$$v''(a\bar{a}) = \bar{V}'' + \bar{\omega}'' \times \bar{r} - V'' - \omega'' \times r \tag{22}$$

Substituting Eqs. (12)–(15) in Eq. (20), and Eqs. (16)–(19) in Eq. (22), and making use of Eq. (21), we obtain

$$v'(a\bar{a}) = v^*(a\bar{a}) - \frac{\hat{f}'}{M} - \frac{\hat{f}'}{M} - \bar{\phi}^{-1}(\bar{r} \times \hat{f}') \times \bar{r} - \phi^{-1}(r \times \hat{f}') \times r \tag{23}$$

$$v''(a\bar{a}) = v^*(a\bar{a}) + \frac{\hat{f}''}{M} + \frac{\hat{f}''}{M} + \bar{\phi}^{-1}(\bar{r} \times \hat{f}'') \times \bar{r} + \phi^{-1}(r \times \hat{f}'') \times r \tag{24}$$

If the bodies are smooth and the collision satisfies Hypothesis 1, then

$$\hat{f}' = \hat{f}'n \tag{25}$$

$$\hat{f}'' = e\hat{f}'n \tag{26}$$

$$n \cdot v^*(a\bar{a}) = 0 \tag{27}$$

Taking the dot product of Eq. (23) with en, and the dot product of Eq. (24) with n, adding the results, and making use of Eqs. (25)–(27), we obtain

$$en \cdot v'(a\bar{a}) + n \cdot v''(a\bar{a}) = 0 \tag{28}$$

which is the same as Eq. (3). This completes the proof of the theorem.

PROBLEMS

1. A uniform square lamina $ABCD$ of mass m and edge $2a$ is free to turn about the corner A, which is fixed. The corner C is struck by an impulse of magnitude $\hat{f}$ normal to the lamina. Show that the lamina begins to turn about the line through A parallel to BD with angular velocity $6\sqrt{2}\,\hat{f}/7ma$. If the support breaks immediately before the blow, show that the lamina begins to turn with angular velocity $3\sqrt{2}\,\hat{f}/ma$ about a line parallel to BD dividing AC in the ratio $5:7$.

2. A lamina of mass m and moment of inertia mk^2 with respect to its center of mass is falling vertically without rotation in a vertical plane and strikes a fixed horizontal plane with speed V. At the moment of impact the distance of its center of mass from the point of contact is a and from the plane is b. The coefficient of restitution is e and the plane is sufficiently rough to prevent sliding. Show from Hypothesis 1 that the normal component of the impulse on the lamina is $mV(1 + e)$. Show that if one had assumed Theorem 1 to be true for collisions of rough bodies, the normal component would have been $m(1 + e)V[(b^2 + k^2)/(a^2 + k^2)]$.

3. A uniform sphere of mass m and radius a, which is rotating with an angular velocity ω about a horizontal axis through its center, is gently placed on a perfectly rough horizontal plane. Assume that there is no normal impulse and show that the sphere receives a tangential impulse of magnitude $2ma\omega/7$.

4. A circular lamina resting on a smooth horizontal table receives an impulse $\hat{\mathbf{f}}$ in the plane of the table and begins to turn around a point on its circumference. Determine the distance of the line of action of the impulse from the center of the lamina.

5. Two equal uniform square plates $OABC$, $ODEF$, each of mass m, are smoothly hinged at O and placed at rest on a smooth horizontal plane with the diagonals EO, and OB in line. An impulse $\hat{\mathbf{f}}$ is applied at B in the plane of the plates. If ω and ω' are the angular velocities imparted to $OABC$ and $ODEF$, respectively, and are measured in opposite senses, show that $\omega = 3\omega'$ and that $\hat{f}^2/4m \leq T \leq 7\hat{f}^2/4m$, where T is the kinetic energy of the system after the impulse.

6. A uniform circular loop of mass M and radius a can turn freely about a point O of its circumference. When it is at rest, it is struck at right angles to its plane by a blow of impulse $\hat{f}$ at another point P of its circumference. Show that the angular velocity generated by the blow cannot exceed $3\hat{f}/(\sqrt{2}\,Ma)$, and find the distance OP for which this greatest angular velocity occurs.

7. A uniform sphere moving without spin on a smooth horizontal table hits a rough vertical wall, e being the coefficient of restitution, μ the coefficient of friction, and α the angle between the incident velocity and the normal.
   a. Prove that rolling during compression implies rolling during restitution and requires $7\mu \geqslant 2\tan\alpha$.
   b. If sliding in compression is followed by rolling in restitution, prove that $[(2\tan\alpha)/(1+e)] \leqslant 7\mu < 2\tan\alpha$.
   c. If β is the angle of reflection, prove that $\tan\beta = (5\tan\alpha)/(7e)$ if $7\mu(1+e) \geqslant 2\tan\alpha$ and $\tan\beta = [\tan\alpha - \mu(1+e)]/e$ if $7\mu(1+e) \leqslant 2\tan\alpha$.

PART 3

AN INTRODUCTION TO LAGRANGIAN AND HAMILTONIAN MECHANICS

SECTION 1

Lagrangian Mechanics

42

Holonomic Constraint Forces

INTRODUCTION

The configuration of a dynamical system consisting of N particles can be described by the set of coordinates $\mathbf{x} \equiv x_1, x_2, \ldots, x_{3N}$, where the set $\mathbf{x}$ has been defined in Chapter 31. Geometrically the configuration $\mathbf{x}$ of the system can be represented by a point in the Cartesian configuration space $\mathbf{x}$, which is the space whose coordinates are the members of the set of coordinates $\mathbf{x}$. The motion of this point is governed by Newton's law, which in terms of the set of coordinates $\mathbf{x}$ can be written $f_i = m_i \ddot{x}_i$.

If the force $\mathbf{f} \equiv f_1, f_2, \ldots, f_{3N}$ is a known function of the set of coordinates $\mathbf{x}$, the set of velocities $\dot{\mathbf{x}}$, and the time t, we can use Newton's law to determine $\mathbf{x}$ and $\dot{\mathbf{x}}$ as functions of the time. In many situations, however, a force is described not in terms of what it does to a system but in terms of what it prevents the system from doing. For example, if a particle is confined to remain on a particular surface, we are not ordinarily given the value of the force that the surface exerts on the particle, but rather we are simply told that the surface exerts whatever force is necessary to constrain the particle to remain on it. A force exerted by an agent, called a constraint, which responds to the action of a dynamical system in such a way as to prevent it from assuming certain configurations or making certain displacements, is called a *constraint force*. This chapter considers one important class of constraint forces.

VIRTUAL DISPLACEMENTS AND VIRTUAL WORK

In discussing constraints, the concept of a virtual displacement will be helpful.

A *virtual displacement* of a dynamical system is a hypothetical change in the configuration of the system at some fixed time t. The notation $\delta \mathbf{x} \equiv \delta x_1, \delta x_2, \ldots, \delta x_{3N}$ is used to designate an infinitesimal virtual displacement.

Two points should be particularly noticed in the foregoing definition. First a virtual displacement occurs at a particular time t, and does not involve a change in the time. We can imagine the system to be initially frozen in its configuration at time t, at which point we step in and alter its configuration. The second point to notice is that a virtual displacement does not in general correspond to the actual displacement the system undergoes as a result of the forces acting on it. It is for this reason that we use the notation $\delta \mathbf{x}$ rather than $d\mathbf{x}$, since the notation $d\mathbf{x}$ usually designates the actual displacement of the system in a time dt.

The *virtual work* δW of a force in a virtual displacement is the work that would be done by the force if the system underwent the virtual displacement.

HOLONOMIC CONSTRAINTS

A constraint that restricts the possible configurations of a system to those that satisfy an equation of the form

$$\phi(\mathbf{x}, t) = 0 \tag{1}$$

and is such that there is no work done by the constraint force in any virtual displacement consistent with the constraint is called a *holonomic constraint force*.

If, for example, a particle is constrained to remain in contact with a smooth surface, the force exerted by the surface on the object is a holonomic constraint force. Note that if the surface is moving, the actual work done by the constraint force is not in general zero, and yet the work done by the constraint force in any virtual displacement consistent with the constraint is zero. Hence the constraint is holonomic even if the surface is moving.

Other examples of holonomic constraint forces are the internal forces that maintain the interparticle distance in a rigid body fixed, and the forces acting at a smooth hinge joining two bodies.

In the definition above we have assumed that the constraint force limits the configuration to the configurations that satisfy a single equation $\phi(\mathbf{x}, t) = 0$. It is possible of course that a constraint might limit the configurations to those that satisfy a set of equations $\phi_1(\mathbf{x}, t) = 0$, $\phi_2(\mathbf{x}, t) = 0, \ldots$. To handle this case we simply break the force the constraint is exerting into component forces, and associate one force with each constraint condition. Thus if there are M constraint equations there will be M constraint forces. For example if a bead is constrained to remain on a smooth wire, the configuration of the wire can in general be represented by the intersection of two surfaces $\phi_1(\mathbf{x}, t) = 0$ and $\phi_2(\mathbf{x}, t) = 0$; hence the particle can be assumed to be constrained to remain on the smooth surface $\phi_1(\mathbf{x}, t) = 0$ and also on the smooth surface

$\phi_2(\mathbf{x}, t) = 0$. Thus a smooth wire generates two constraint forces, rather than one.

HOLONOMIC SYSTEMS

In Chapters 43–46 we restrict our attention to systems of particles in which the only constraints acting are holonomic constraints. Such systems are called *holonomic systems*.

43

Generalized Coordinates for Holonomic Systems

INTRODUCTION

The configuration of a dynamical system consisting of N particles can as we have seen be described by the set of coordinates $\mathbf{x} \equiv x_1, x_2, \ldots, x_{3N}$. There are however many other sets of coordinates besides the set $\mathbf{x}$ that could be used to describe the system. Furthermore if the system is acted on by a set of M holonomic constraint forces and we know the M equations of constraint, it is not necessary to use a full $3N$ coordinates to specify the configuration of the system. By appropriately choosing our coordinates it is possible to get by with only $3N - M$ coordinates. In this chapter we consider some of the modifications involved if we take advantage of these possibilities.

DEGREES OF FREEDOM

A holonomic system consisting of N particles that is subjected to M holonomic constraint forces is said to have $3N - M$ *degrees of freedom*.

GENERALIZED COORDINATES AND VELOCITIES

Given a holonomic system of f degrees of freedom, it is always possible to find a set of f coordinates $\mathbf{q} \equiv q_1, q_2, \ldots, q_f$ that together with the constraint conditions, will uniquely specify the configuration of the system. The coordinates $q_1, q_2, \ldots, q_f$ are called generalized coordinates, and their time rates of change, $\dot{q}_1, \dot{q}_2, \ldots, \dot{q}_f$ are called generalized velocities.

The position of a particle constrained to remain on the spherical surface defined by the equation $x^2 + y^2 + z^2 = a^2$ could, for example, be specified by the pair of coordinates (θ, ϕ) from the set of spherical coordinates (r, θ, ϕ) or by the pair of coordinates (ϕ, z) from the set of cylindrical coordinates (ρ, ϕ, z). Hence either of these pairs would constitute a set of generalized coordinates. The position of the particle would however not be uniquely specified by the pair of coordinates (x, y), since there are in general two possible values of z consistent with the constraint equation for each acceptable value of (x, y). Hence this pair would ordinarily not constitute a suitable set of generalized coordinates.

TRANSFORMATION BETWEEN CARTESIAN AND GENERALIZED COORDINATES

By making use of the constraint conditions it is possible to express the $3N$ Cartesian coordinates $x_1, x_2, \ldots, x_{3N}$ as functions of the f generalized coordinates $q_1, q_2, \ldots, q_f$ and the time t, that is,

$$x_i = x_i(\mathbf{q}, t) \tag{1}$$

When this is done, the transformation equations (1) implicitly contain the constraint conditions, since any value for the set $\mathbf{q} \equiv q_1, q_2, \ldots, q_f$ that is substituted into Eq. (1) will yield a value for the set $\mathbf{x} \equiv x_1, x_2, \ldots, x_{3N}$ that is consistent with the constraint conditions.

The following theorems relating to the transformation of Eq. (1) are useful in later chapters.

Theorem 1. Given the transformation

$$x_i = x_i(\mathbf{q}, t) \tag{2}$$

between the set of generalized coordinates $\mathbf{q}$ and the set of Cartesian coordinates $\mathbf{x}$, then

$$\dot{x}_i = \dot{x}_i(\mathbf{q}, \dot{\mathbf{q}}, t) \tag{3}$$

$$\frac{\partial \dot{x}_i(\mathbf{q}, \dot{\mathbf{q}}, t)}{\partial q_j} = \frac{d}{dt}\left[\frac{\partial x_i(\mathbf{q}, t)}{\partial q_j}\right] \tag{4}$$

$$\frac{\partial \dot{x}_i(\mathbf{q}, \dot{\mathbf{q}}, t)}{\partial \dot{q}_j} = \frac{\partial x_i(\mathbf{q}, t)}{\partial q_j} \tag{5}$$

PROOF: The total time derivative of Eq. (2) is given by

$$\dot{x}_i = \sum_k \frac{\partial x_i(\mathbf{q}, t)}{\partial q_k} \dot{q}_k + \frac{\partial x_i(\mathbf{q}, t)}{\partial t} \tag{6}$$

Equation (3) follows from Eq. (6). Taking the derivative of Eq. (6) with respect to q_j we obtain

$$\frac{\partial \dot{x}_i(\mathbf{q},\dot{\mathbf{q}},t)}{\partial q_j} = \sum_k \frac{\partial^2 x_i(\mathbf{q},t)}{\partial q_j \partial q_k} \dot{q}_k + \frac{\partial^2 x_i(\mathbf{q},t)}{\partial q_j \partial t}$$

$$= \sum_k \frac{\partial}{\partial q_k}\left[\frac{\partial x_i(\mathbf{q},t)}{\partial q_j}\right] \dot{q}_k + \frac{\partial}{\partial t}\left[\frac{\partial x_i(\mathbf{q},t)}{\partial q_j}\right]$$

$$= \frac{d}{dt}\left[\frac{\partial x_i(\mathbf{q},t)}{\partial q_j}\right] \tag{7}$$

which is Eq. (4). Finally, taking the derivative of Eq. (6) with respect to $\dot{q}_j$ we obtain Eq. (5).

GENERALIZED COMPONENTS OF A FORCE

In the chapters that follow we reformulate the equations of motion for a dynamical system in terms of generalized coordinates. In this reformulation we find that the concept of generalized components of a force, introduced in Theorem 2, is very useful.

Theorem 2. If a holonomic system of f degrees of freedom, whose configuration can be defined in terms of a set of generalized coordinates $\mathbf{q} \equiv q_1, q_2, \ldots, q_f$, is acted on by a force, the virtual work δW done by the force in an arbitrary virtual displacement $\delta \mathbf{q}$, or equivalently any virtual displacement consistent with the constraints, can be written in the form

$$\delta W = \sum_i Q_i \, \delta q_i \tag{8}$$

The quantities $Q_1, Q_2, \ldots, Q_f$ are called the generalized components of the force. A particular generalized component Q_i is called the q_i generalized component of the force or simply the q_i component of the force. The set of generalized components is designated $\mathbf{Q}$. The force whose generalized components are the set $\mathbf{Q}$ is called the force $\mathbf{Q}$.

PROOF: At a given time the virtual work δW done by a force in an infinitesimal virtual displacement consistent with the constraint depends only on the force and the virtual displacement $\delta \mathbf{q}$. If we expand δW in powers of $\delta q_1, \delta q_2, \ldots, \delta q_f$ and note that the higher order terms can be dropped, since the δq_i are infinitesimals, we obtain Eq. (8).

NOTE 1: The q_i component of a force does not in general have the dimensions of a force. For example if q_i is an angle, the q_i component of the force will have the dimensions of a torque, not a force.

NOTE 2: The q_i component of a force is not in general the component of the force in the q_i direction.

NOTE 3: The q_i component of a force depends not only on the coordinate q_i but also on the other coordinates in the set **q**, in the same sense that a partial derivative depends not only on the quantity that is varied but also on the quantities that are held constant.

DETERMINATION OF THE GENERALIZED COMPONENTS OF A FORCE

Method 1. Since Eq. (8) must be true for arbitrary infinitesimal virtual displacements $\delta\mathbf{q}$ consistent with the constraints, it follows that to determine a particular q_i component it is necessary only to calculate the work done in a virtual displacement in which all the variables $q_1, q_2, \ldots, q_f$ are held constant except the particular q_i whose generalized component we wish to determine. Then the work done is simply $\delta W = Q_i \delta q_i$ and Q_i can be determined by dividing δW by δq_i.

Method 2. If we know the Cartesian components f_i of a force and the transformation equation $\mathbf{x} = \mathbf{x}(\mathbf{q}, t)$, then the generalized components of the force can be determined by using the following theorem.

Theorem 3. Given a holonomic system that is acted on by a force **f**, let $\mathbf{q} \equiv q_1, q_2, \ldots, q_f$ be a set of generalized coordinates for the system, and $\mathbf{x} \equiv x_1, x_2, \ldots, x_{3N}$ the set of Cartesian coordinates that determine the configuration of the N particles making up the system. If we know the set of equations $\mathbf{x} = \mathbf{x}(\mathbf{q}, t)$ relating the set of coordinates $\mathbf{x}$ to the set of coordinates $\mathbf{q}$ and the time t, and the $3N$ components f_i of the force **f**, the q_i generalized component of the force **f** is given by the equation

$$Q_i = \sum_j f_j \frac{\partial x_j(\mathbf{q}, t)}{\partial q_i} \tag{9}$$

PROOF: The work done by the force **f** in an arbitrary virtual displacement $\delta\mathbf{x}$ is given by

$$\delta W = \sum_j f_j \delta x_j \tag{10}$$

But the values of x_j that are consistent with the constraint satisfy the equation

$$x_j = x_j(\mathbf{q}, t) \tag{11}$$

From Eq. (11) it follows that

$$\delta x_j = \sum_i \frac{\partial x_j(\mathbf{q}, t)}{\partial q_i} \delta q_i \tag{12}$$

Combining Eqs. (10) and (12) we obtain

$$\delta W = \sum_i \left[\sum_j f_j \frac{\partial x_j(\mathbf{q}, t)}{\partial q_i} \right] \delta q_i \tag{13}$$

But from Theorem 2

$$\delta W = \sum_i Q_i \, \delta q_i \tag{14}$$

Comparing Eqs. (13) and (14) we obtain Eq. (9).

Method 3. If the Cartesian components f_i of a force are derivable from a potential $V(\mathbf{x}, t)$ [i.e., $f_i = -\partial V(\mathbf{x}, t)/\partial x_i$], the generalized components of the force can be determined by using the following theorem.

Theorem 4. Given a holonomic system that is acted on by a force $\mathbf{f}$, let $\mathbf{q} \equiv q_1, q_2, \ldots, q_f$ be a set of generalized coordinates for the system and $\mathbf{x} \equiv x_1, x_2, \ldots, x_{3N}$ the set of Cartesian coordinates that determine the configuration of the N particles making up the system. If the force $\mathbf{f}$ is derivable from a potential, that is,

$$f_i = -\frac{\partial V(\mathbf{x}, t)}{\partial x_i} \tag{15}$$

the generalized components Q_i of the force are given by

$$Q_i = -\frac{\partial V(\mathbf{q}, t)}{\partial q_i} \tag{16}$$

PROOF: From Theorem 3

$$Q_i = \sum_j f_j \frac{\partial x_j(\mathbf{q}, t)}{\partial q_i} \tag{17}$$

Substituting Eq. (15) in Eq. (17) we obtain

$$Q_i = -\sum_j \frac{\partial V(\mathbf{x}, t)}{\partial x_j} \frac{\partial x_j(\mathbf{q}, t)}{\partial q_i} = -\frac{\partial V(\mathbf{q}, t)}{\partial q_i} \tag{18}$$

THE GENERALIZED COMPONENTS OF A HOLONOMIC CONSTRAINT FORCE

For a constraint to be holonomic, there must be no work done by the constraint force in any virtual displacement consistent with the constraint. From this condition and from the definition of the generalized components of a force it follows that the generalized components of a holonomic constraint force are zero.

EXAMPLES

Example 1. A rod BC of mass m and length $2a$ is suspended from a fixed point A by an elastic filament AB, of unstretched length b and spring constant k, which is attached to the rod at end B as shown in Fig. 1a. The motion of the system is restricted to a vertical plane containing the rod and the filament. Considering the rod to be the system of interest, choose a suitable set of generalized coordinates, and find the generalized components of the forces acting on the system.

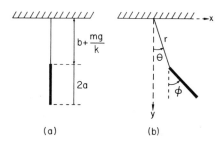

Fig. 1

Solution. The rod has three degrees of freedom. A suitable set of generalized coordinates is: r, the length of the filament; θ, the angle that the filament makes with the vertical; and ϕ, the angle that the rod makes with the vertical. These coordinates are shown in Fig. 1b. The forces acting on the rod are: (1) a force of magnitude $T = k(r - b)$ exerted by the filament along the line shown in Fig. 2, and (2) the force of gravity, which is of magnitude mg acting vertically down through the center of mass. Let Q'_r, Q'_θ, and Q'_ϕ be the generalized components of the force due to the filament, Q''_r, Q''_θ, and Q''_ϕ the generalized components of the force due to gravity and Q_r, Q_θ, and Q_ϕ the generalized components of the net force acting on the system. We obtain the generalized components above by each of the three methods discussed in the body of the chapter.

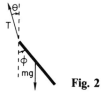

Fig. 2

Method 1. The virtual work done by the force exerted by the filament in an arbitrary virtual displacement consistent with the constraints is

$$\delta W = -k(r - b)\delta r \qquad (19)$$

It follows that

$$Q'_r = -k(r-b) \tag{20}$$
$$Q'_\theta = Q'_\phi = 0 \tag{21}$$

The virtual work done by the gravitational force in an arbitrary virtual displacement consistent with the constraints is

$$\delta W = mg\,\delta(r\cos\theta + a\cos\phi)$$
$$= mg\cos\theta\,\delta r - mgr\sin\theta\,\delta\theta - mga\sin\phi\,\delta\phi \tag{22}$$

It follows that

$$Q''_r = mg\cos\theta \tag{23}$$
$$Q''_\theta = -mgr\sin\theta \tag{24}$$
$$Q''_\phi = -mga\sin\phi \tag{25}$$

The generalized components of the net force are

$$Q_r = Q'_r + Q''_r = -k(r-b) + mg\cos\theta \tag{26}$$
$$Q_\theta = Q'_\theta + Q''_\theta = -mgr\sin\theta \tag{27}$$
$$Q_\phi = Q'_\phi + Q''_\phi = -mga\sin\phi \tag{28}$$

Method 2. If we choose an inertial system with x and y axes as shown in Fig. 1b, the x and y coordinates of the point of application of the force due to the filament are

$$x' = r\sin\theta \tag{29}$$
$$y' = r\cos\theta \tag{30}$$

and the x and y components of the force are

$$F'_x = -k(r-b)\sin\theta \tag{31}$$
$$F'_y = -k(r-b)\cos\theta \tag{32}$$

It follows that the generalized components of this force are

$$Q'_r = F'_x \frac{\partial x'}{\partial r} + F'_y \frac{\partial y'}{\partial r} = -k(r-b) \tag{33}$$

$$Q'_\theta = F'_x \frac{\partial x'}{\partial \theta} + F'_y \frac{\partial y'}{\partial \theta} = 0 \tag{34}$$

$$Q'_\phi = F'_x \frac{\partial x'}{\partial \phi} + F'_y \frac{\partial y'}{\partial \phi} = 0 \tag{35}$$

Similarly the x and y components of the coordinates of the point of application of the gravitational force are

$$x'' = r\sin\theta + a\sin\phi \tag{36}$$
$$y'' = r\cos\theta + a\cos\phi \tag{37}$$

and the x and y components of the force are
$$F_x'' = 0 \tag{38}$$
$$F_y'' = mg \tag{39}$$
hence
$$Q_r'' = F_x'' \frac{\partial x''}{\partial r} + F_y'' \frac{\partial y''}{\partial r} = mg \cos\theta \tag{40}$$
$$Q_\theta'' = F_x'' \frac{\partial x''}{\partial \theta} + F_y'' \frac{\partial y''}{\partial \theta} = -mgr \sin\theta \tag{41}$$
$$Q_\phi'' = F_x'' \frac{\partial x''}{\partial \phi} + F_y'' \frac{\partial y''}{\partial \phi} = -mga \sin\phi \tag{42}$$

The generalized components of the net force are obtained, as in Method 1, by summing the contributions from both forces.

Method 3. The forces acting on the system are conservative. The potential energy of the system is
$$V = \tfrac{1}{2}k(r-b)^2 - mg(r\cos\theta + a\cos\phi) + \text{constant} \tag{43}$$
where the value of the constant depends on the choice of reference configuration. It follows that
$$Q_r = -\frac{\partial V}{\partial r} = -k(r-b) + mg\cos\theta \tag{44}$$
$$Q_\theta = -\frac{\partial V}{\partial \theta} = -mgr\sin\theta \tag{45}$$
$$Q_\phi = -\frac{\partial V}{\partial \phi} = -mga\sin\phi \tag{46}$$

The generalized components Q_r', Q_θ', Q_ϕ', Q_r'', Q_θ'', and Q_ϕ'' are readily obtained by considering individually the contributions to the potential due to the force of the filament and due to the force of gravity.

PROBLEMS

1. A particle is moving in a uniform gravitational field. The motion of the particle is described by a set of spherical coordinates r, θ, and ϕ, with the polar axis directed vertically up. Find the generalized components Q_r, Q_θ, and Q_ϕ of the gravitational force.

2. A straight smooth wire is forced to rotate in a vertical plane with a constant angular velocity ω about a point O of its length. A bead of mass m can slide on the wire. Considering the bead to be the system of interest, choose a suitable set of generalized coordinates, and find the generalized components of the forces acting on the system.

3. A uniform rod of mass m and length $2a$ has a small ring of mass m pivoted at one end. The ring slides on a smooth horizontal wire. The system consisting of the rod and the ring is constrained to remain in the vertical plane containing the wire. Choosing as generalized coordinates s, the distance of the ring from a fixed point on the wire, and θ, the angle that the rod makes with the downward vertical, find the generalized components Q_s and Q_θ for all the forces acting on the system.

4. A cylinder of mass m and radius a rolls without slipping on a plane that is inclined at an angle α with respect to the horizontal. The motion is uniplanar. Choose a suitable set of generalized coordinates, and find the generalized components of the forces acting on the system.

5. A uniform thin hollow circular cylinder of radius a and mass m rolls without slipping inside another such cylinder of radius $2a$ and mass M. The latter rolls without slipping on a horizontal table. Choosing as generalized coordinates θ, the angular displacement of the larger cylinder, and ϕ, the angle that the plane of the axes makes with the vertical, find the generalized components for all the forces acting on the system.

6. A circular cylinder of radius a and mass m rolls without slipping inside a horizontal semicircular groove of radius b cut in a block of mass M. The block is constrained to move without friction in a vertical guide and is supported by a spring of spring constant k. Taking as coordinates the vertical displacement x of the block and the angular displacement θ of the center of the cylinder, both measured from the position of static equilibrium, determine the generalized components of the net force acting on the system.

7. A hoop of mass M and radius a rolls without slipping down a plane inclined at an angle α with the horizontal. A particle of mass m is suspended from the center of the hoop by an elastic filament of spring constant k and unstretched length b. The motion is uniplanar. Considering the system made up of the hoop, the filament, and the particle to be the system of interest, choose a suitable set of generalized coordinates, and find the generalized components of the forces acting on the system.

44

Lagrange's Equations of Motion for a Holonomic System

INTRODUCTION

The motion of a holonomic system could in principle be determined by making use of the $3N$ equations $f_i = m_i \ddot{x}_i$ together with the f constraint conditions $\phi_1(\mathbf{x}, t) = 0$, $\phi_2(\mathbf{x}, t) = 0, \ldots, \phi_f(\mathbf{x}, t) = 0$. In many cases this is practically impossible, or at least highly impractical. If for example we are dealing with a rigid body containing a large number N of particles, the number of equations generated by the approach above would be prohibitively large. Even in the case of a system consisting of a few particles subjected to a reasonable number of constraints, the number of equations (i.e., $3N + f$) that is required in the approach can be awkward. Furthermore the Cartesian coordinates x_i may not be very well suited to the particular problem at hand. The analysis of central force motion, for example, would be quite tedious if we stuck to Cartesian coordinates.

In this chapter we show how the equations of motion for a holonomic system can be reformulated in terms of generalized coordinates. Not only does this enable us to shift to a set of coordinates more convenient than Cartesian coordinates, but it reduces the number of equations from $3N + f$ to simply f and eliminates the necessity of explicitly including the constraint forces in the analysis.

LAGRANGE'S EQUATIONS OF MOTION

If we are able to write down the kinetic energy T of a holonomic system in terms of a set of generalized coordinates $\mathbf{q}$, the associated set of generalized velocities $\dot{\mathbf{q}}$, and the time t, and if we know the generalized components Q_i of the nonconstraint forces, then the motion of the system can be determined by making use of the following theorem.

Theorem 1. Consider a holonomic system having f degrees of freedom, whose configuration is defined by the set of generalized coordinates $\mathbf{q} \equiv q_1, q_2, \ldots, q_f$. If in addition to the constraint forces acting on the system, the system is acted on by a force $\mathbf{Q}$, the equations of motion of the system are

$$\frac{d}{dt}\left[\frac{\partial T(\mathbf{q}, \dot{\mathbf{q}}, t)}{\partial \dot{q}_i}\right] - \frac{\partial T(\mathbf{q}, \dot{\mathbf{q}}, t)}{\partial q_i} = Q_i \tag{1}$$

where $T(\mathbf{q}, \dot{\mathbf{q}}, t)$ is the kinetic energy of the constrained system. We refer to these equations as Lagrange's equations for holonomic systems.

PROOF: Let $\mathbf{f}$ be the set of Cartesian components of the force $\mathbf{Q}$ and $\mathbf{f}'$ be the set of Cartesian components of the total constraint force. By virtue of Newton's law

$$m_j \ddot{x}_j = f_j + f_j' \tag{2}$$

Multiplying Eq. (2) by $\partial x_j(\mathbf{q}, t)/\partial q_i$ and summing over j we obtain

$$\sum_j m_j \ddot{x}_j \frac{\partial x_j}{\partial q_i} = \sum_i f_j \frac{\partial x_j}{\partial q_i} + \sum_i f_j' \frac{\partial x_j}{\partial q_i} \tag{3}$$

From Theorem 3 in Chapter 43, it follows that

$$\sum_j f_j \frac{\partial x_j}{\partial q_i} = Q_i \tag{4}$$

The second term on the right hand side is the q_i generalized component of the constraint force, which as shown in Chapter 43, is zero. Hence

$$\sum_j f_j' \frac{\partial x_j}{\partial q_i} = 0 \tag{5}$$

To reformulate the left hand side of Eq. (3) in terms of generalized coordinates and velocities, we note first that

$$m_j \ddot{x}_j = \frac{d}{dt}(m_j \dot{x}_j) = \frac{d}{dt}\left[\frac{\partial}{\partial \dot{x}_j}\left(\frac{1}{2}m_j \dot{x}_j^2\right)\right]$$

$$= \frac{d}{dt}\left[\frac{\partial}{\partial \dot{x}_j}\left(\sum_k \frac{1}{2}m_k \dot{x}_k^2\right)\right] = \frac{d}{dt}\left[\frac{\partial T(\dot{\mathbf{x}})}{\partial \dot{x}_j}\right] \tag{6}$$

Substituting Eq. (6) in Eq. (3) and making use of the fact that $(dA/dt)B =$

$d(AB)/dt - A(dB/dt)$, and Theorem 1 in Chapter 43, we obtain

$$\sum_j m_j \ddot{x}_j \frac{\partial x_j}{\partial q_i} = \sum_j \frac{d}{dt}\left[\frac{\partial T(\dot{\mathbf{x}})}{\partial \dot{x}_j}\right]\frac{\partial x_j}{\partial q_i}$$

$$= \sum_j \frac{d}{dt}\left[\frac{\partial T(\dot{\mathbf{x}})}{\partial \dot{x}_j}\frac{\partial x_j}{\partial q_i}\right] - \sum_j \frac{\partial T(\dot{\mathbf{x}})}{\partial \dot{x}_j}\frac{d}{dt}\left(\frac{\partial x_j}{\partial q_i}\right)$$

$$= \sum_j \frac{d}{dt}\left[\frac{\partial T(\dot{\mathbf{x}})}{\partial \dot{x}_j}\frac{\partial \dot{x}_j}{\partial \dot{q}_i}\right] - \sum_j \frac{\partial T(\dot{\mathbf{x}})}{\partial \dot{x}_j}\frac{\partial \dot{x}_j}{\partial q_i}$$

$$= \frac{d}{dt}\left[\sum_j \frac{\partial T(\dot{\mathbf{x}})}{\partial \dot{x}_j}\frac{\partial \dot{x}_j}{\partial \dot{q}_i}\right] - \sum_j \frac{\partial T(\dot{\mathbf{x}})}{\partial \dot{x}_j}\frac{\partial \dot{x}_j}{\partial q_i}$$

$$= \frac{d}{dt}\left[\frac{\partial T(\mathbf{q},\dot{\mathbf{q}},t)}{\partial \dot{q}_i}\right] - \frac{\partial T(\mathbf{q},\dot{\mathbf{q}},t)}{\partial q_i} \tag{7}$$

Substituting Eqs. (4), (5), and (7) in Eq. (3) we obtain Eq. (1).

Lagrange's equations (1) provide us with f second order differential equations in the f unknowns $q_1, q_2, \ldots, q_f$. Hence if we know the configuration $\mathbf{q}$ and velocity $\dot{\mathbf{q}}$ of the system at some instant, we can determine it at any other arbitrary time t.

THE LAGRANGIAN

If the nonconstraint forces are derivable from a potential $V(\mathbf{q}, t)$, Lagrange's equations become

$$\frac{d}{dt}\left[\frac{\partial T(\mathbf{q},\dot{\mathbf{q}},t)}{\partial \dot{q}_i}\right] - \frac{\partial T(\mathbf{q},\dot{\mathbf{q}},t)}{\partial q_i} = -\frac{\partial V}{\partial q_i} \tag{8}$$

If we define a quantity

$$L(\mathbf{q},\dot{\mathbf{q}},t) \equiv T(\mathbf{q},\dot{\mathbf{q}},t) - V(\mathbf{q},t) \tag{9}$$

called the *Lagrangian*, Eq. (8) can be written in the more compact form

$$\frac{d}{dt}\left[\frac{\partial L(\mathbf{q},\dot{\mathbf{q}},t)}{\partial \dot{q}_i}\right] - \frac{\partial L(\mathbf{q},\dot{\mathbf{q}},t)}{\partial q_i} = 0 \tag{10}$$

In working problems it is usually a little simpler to use Eq. (8) rather than Eq. (10). However the introduction of the Lagrangian is the starting point for some very important theoretical developments, which we consider in later chapters.

CONSTANTS OF THE MOTION

In Chapter 33 we saw that solving the equations of motion for a dynamical system is equivalent to finding the independent constants of the motion. Each new formulation of the laws of dynamics tends to provide a different perspective that frequently is useful in finding constants of the motion. The following theorems illustrate this fact.

Theorem 2. If the behavior of a system is describable by a Lagrangian function

$$L(\mathbf{q},\dot{\mathbf{q}},t) = T(\mathbf{q},\dot{\mathbf{q}},t) - V(\mathbf{q},t) \tag{11}$$

and if L is not an explicit function of the time then the quantity

$$H(\mathbf{q},\dot{\mathbf{q}}) \equiv \sum_i \frac{\partial L(\mathbf{q},\dot{\mathbf{q}})}{\partial \dot{q}_i} \dot{q}_i - L(\mathbf{q},\dot{\mathbf{q}}) \tag{12}$$

is a constant of the motion. Furthermore if the equations of transformation between the set of Cartesian coordinates $\mathbf{x}$ and the set of generalized components $\mathbf{q}$ are not functions of the time, that is, the constraints are time independent, or if the kinetic energy $T(\mathbf{q},\dot{\mathbf{q}},t)$ is a homogeneous quadratic function of the set of generalized velocities $\dot{\mathbf{q}}$, then

$$H = T + V \tag{13}$$

PROOF: If L is not an explicit function of the time, the time rate of change of the quantity H is given by

$$\frac{dH}{dt} = \sum_i \frac{d}{dt}\left[\frac{\partial L(\mathbf{q},\dot{\mathbf{q}})}{\partial \dot{q}_i}\right]\dot{q}_i + \sum_i \frac{\partial L(\mathbf{q},\dot{\mathbf{q}})}{\partial \dot{q}_i}\ddot{q}_i$$

$$- \sum_i \frac{\partial L(\mathbf{q},\dot{\mathbf{q}})}{\partial q_i}\dot{q}_i - \sum_i \frac{\partial L(\mathbf{q},\dot{\mathbf{q}})}{\partial \dot{q}_i}\ddot{q}_i$$

$$= \sum_i \left\{\frac{d}{dt}\left[\frac{\partial L(\mathbf{q},\dot{\mathbf{q}})}{\partial \dot{q}_i}\right] - \frac{\partial L(\mathbf{q},\dot{\mathbf{q}})}{\partial q_i}\right\}\dot{q}_i \tag{14}$$

But from Lagrange's equations of motion the term in the braces is zero and therefore

$$\frac{dH}{dt} = 0 \tag{15}$$

It follows that H is a constant of the motion. We have therefore proved the first part of the theorem. To prove the second part, we note first, by combining Eqs. (11) and (12), that

$$H = \sum_i \frac{\partial L}{\partial \dot{q}_i}\dot{q}_i - L = \sum_i \frac{\partial T}{\partial \dot{q}_i}\dot{q}_i - T + V \tag{16}$$

If $\mathbf{x} = \mathbf{x}(\mathbf{q})$, then

$$T = \frac{1}{2}\sum_i m_i \dot{x}_i^2 = \frac{1}{2}\sum_i m_i \left[\sum_j \frac{\partial x_i}{\partial q_j}\dot{q}_j\right]\left[\sum_k \frac{\partial x_i}{\partial q_k}\dot{q}_k\right]$$

$$= \frac{1}{2}\sum_j \sum_k \left[\sum_i m_i \frac{\partial x_i}{\partial q_j}\frac{\partial x_i}{\partial q_k}\right]\dot{q}_j \dot{q}_k$$

$$\equiv \frac{1}{2}\sum_j \sum_k a_{jk}(\mathbf{q})\dot{q}_j \dot{q}_k \tag{17}$$

Hence T is a homogeneous quadratic function of the set of generalized velocities $\dot{\mathbf{q}}$. Taking the derivative of T with respect to $\dot{q}_i$ we obtain

$$\frac{\partial T}{\partial \dot{q}_i} = \sum_j a_{ij}\dot{q}_j \tag{18}$$

Substituting Eq. (18) in Eq. (16) and then making use of Eq. (17) we obtain

$$H = \sum_i \sum_j a_{ij}\dot{q}_i\dot{q}_j - T + V = 2T - T + V = T + V \tag{19}$$

This result follows if either $\mathbf{x} = \mathbf{x}(\mathbf{q})$ or $T(\mathbf{q}, \dot{\mathbf{q}}, t)$ is a homogeneous quadratic function of the set of generalized velocities $\dot{\mathbf{q}}$. This completes the proof of the theorem.

Theorem 3. If L is not an explicit function of one of the generalized coordinates, q_i, then the quantity $\partial L(\mathbf{q}, \dot{\mathbf{q}}, t)/\partial \dot{q}_i$ is a constant of the motion.

PROOF: If L is not a function of q_i then

$$\frac{\partial L(\mathbf{q}, \dot{\mathbf{q}}, t)}{\partial q_i} = 0 \tag{20}$$

Substituting this result in Lagrange's equation of motion we obtain

$$\frac{d}{dt}\left[\frac{\partial L(\mathbf{q}, \dot{\mathbf{q}}, t)}{\partial \dot{q}_i}\right] = 0 \tag{21}$$

Hence $\partial L(\mathbf{q}, \dot{\mathbf{q}}, t)/\partial \dot{q}_i$ is a constant of the motion.

EXAMPLES

Example 1. A small bead of mass M is initially at rest on a smooth horizontal wire and is attached to a point on the wire by a massless spring of spring constant k and unstretched length a. A rod of mass m and length $2b$ is freely suspended from the bead. The system is set in motion in the vertical plane containing the wire. Determine the equations of motion.

Solution. We choose as generalized coordinates the displacement x of the bead from its equilibrium position (Fig. 1) and the angle θ that the rod makes

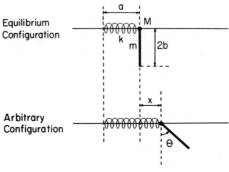

Fig. 1

with the vertical. The kinetic energy of the system is

$$T = \frac{1}{2} M\dot{x}^2 + \frac{1}{2} m \left\{ \left[\frac{d}{dt}(x + b\sin\theta) \right]^2 + \left[\frac{d}{dt}(b\cos\theta) \right]^2 \right\} + \frac{1}{2} \frac{mb^2}{3} \dot{\theta}^2$$

$$= \frac{M}{2} \dot{x}^2 + \frac{m}{2}(\dot{x}^2 + 2b\cos\theta\,\dot{x}\dot{\theta} + b^2\dot{\theta}^2) + \frac{m}{6} b^2\dot{\theta}^2$$

$$= \frac{M + m}{2} \dot{x}^2 + \frac{2m}{3} b^2\dot{\theta}^2 + mb\cos\theta\,\dot{x}\dot{\theta} \qquad (22)$$

The potential energy of the system is

$$V = \frac{k}{2} x^2 - mgb\cos\theta \qquad (23)$$

The Lagrangian is thus

$$L = \frac{M + m}{2} \dot{x}^2 + \frac{2m}{3} b^2\dot{\theta}^2 + mb\cos\theta\,\dot{x}\dot{\theta} - \frac{k}{2} x^2 + mgb\cos\theta \qquad (24)$$

Lagrange's equations are

$$\frac{d}{dt}\left(\frac{\partial L}{\partial \dot{x}}\right) - \frac{\partial L}{\partial x} = 0 \qquad (25)$$

$$\frac{d}{dt}\left(\frac{\partial L}{\partial \dot{\theta}}\right) - \frac{\partial L}{\partial \theta} = 0 \qquad (26)$$

Substituting Eq. (24) in Eqs. (25) and (26) we obtain

$$(M + m)\ddot{x} + mb\cos\theta\,\ddot{\theta} - mb\sin\theta\,\dot{\theta}^2 + kx = 0 \qquad (27)$$

$$\frac{4mb^2}{3}\ddot{\theta} + mb\cos\theta\,\ddot{x} - mb\sin\theta\,\dot{x}\dot{\theta} + mgb\sin\theta = 0 \qquad (28)$$

Since the Lagrangian is not an explicit function of the time, and the kinetic energy is a homogeneous quadratic function of the generalized velocities $\dot{x}$ and $\dot{\theta}$, the total energy $T + V$ is a constant of the motion. This result provides us with one step in the solution of Eqs. (27) and (28).

PROBLEMS

1. A particle of mass m is constrained to remain in a plane and moves under the action of a conservative force directed toward a point O in the plane. The potential energy of the particle is $V(r)$, where r is the distance of the particle from the point O. Write down Lagrange's equations of motion of the particle in polar coordinates.

2. A spherical pendulum consists of a particle of mass m suspended from a fixed point O by a massless rod of length a. Write the Lagrangian equations of motion of the particle in terms of the spherical coordinates θ and ϕ of the particle. Choose the origin of coordinates at the point O, and the polar axis vertically down.

3. A particle of mass M is free to slide on a smooth horizontal wire. A second particle of mass m is suspended from the first particle by a rigid massless rod of length a. The system is constrained to move in the vertical plane containing the wire. Choosing as generalized coordinates the distance x of the mass M from a fixed point on the wire, and the angle θ that the rod makes with the vertical, find the equations of motion of the system.

4. A particle of mass m is suspended from a fixed point by a string of length a. A second particle of mass M is suspended from the first particle by a string of length b. The system is constrained to move in a vertical plane. Choosing as generalized coordinates the angles θ and ϕ, which the two strings make respectively with the vertical, obtain the equations of motion. Show that the same equations can be obtained by a direct application of Newton's law.

5. A bead is constrained to move along a smooth conical spiral defined in cylindrical coordinates ρ, ϕ, and z by the equations $\rho = az$ and $\phi = -bz$, where a and b are constants. Gravity is acting in the negative z direction. Show that the equation of motion is $\ddot{z}(a^2 + 1 + a^2 b^2 z^2) + a^2 b^2 z \dot{z}^2 = -g$.

6. The configuration of a certain conservative holonomic system is uniquely determined by the generalized coordinates q_1 and q_2. The kinetic energy is given by $T = \dot{q}_1^2/[2(a + bq_2)] + (q_2 \dot{q}_2)^2/2$ and the potential energy is given by $V = c + dq_2$, where a, b, c, and d are constants. Show that the dependence of q_2 on the time t can be obtained from an equation of the form $(q_2 - k)(q_2 - 2k)^2 = h(t - t_0)^2$.

7. A bead moves on a smooth wire that is bent in the form of a vertical circle of radius a. The wire rotates about a fixed vertical diameter with a uniform angular velocity ω. Let θ be the angular distance from the lowest point. Initially the particle is at rest relative to the wire with $\theta = \alpha$, where $\omega \cos(\alpha/2) = (g/a)^{1/2}$. Show that $\ddot{\theta} - \omega^2 \sin\theta \cos\theta + (g/a)\sin\theta = 0$ and $\dot{\theta}^2 = \omega^2(1 - \cos\theta)(\cos\theta - \cos\alpha)$.

8. A particle is free to slide on a smooth circular loop that is constrained to rotate in its plane with a constant angular velocity ω about a point on its circumference. Find the differential equation of motion of the particle.

9. An Atwood's machine is built as follows: A string of length a passes over a light fixed pulley, supporting a mass m_1 on one end and a pulley of mass m_2 and negligible moment of inertia on the other. Over the second pulley passes a string of length b supporting a mass m_3 on one end and m_4 on the other, where $m_3 \neq m_4$. Set up Lagrange's equations of motion for the system, using two appropriate generalized coordinates. From these show that the mass m_1 remains in equilibrium if $m_1 = m_2 + m_3 + m_4 - [(m_4 - m_3)^2/(m_3 + m_4)]$.

10. A light string passes over a pulley of mass $4m$ and radius a that can rotate freely about a fixed horizontal axis. A particle A of mass $2m$ is attached to one end and a pulley of mass m and radius b to the other. The second pulley is also free to rotate about a horizontal axis and carries a light string with a particle B of mass $2m$ at one end and a particle C of mass $11m$ at the other end. The system moves in a vertical plane with all strings vertical. Assuming the pulleys to be uniform disks, calculate the accelerations of particles A, B, and C.

11. A uniform rod of mass m and length $2a$ has a small light ring attached to one end. The ring is threaded onto a smooth vertical rigid wire and the rod is free to move in a vertical plane containing the wire. Let x be the distance of the ring below a fixed point on the wire, and θ the angle made in a downward direction by the rod with the horizontal. Write down Lagrange's equations of motion using x and θ as generalized coordinates, and show that $\dot{x} = gt - a\dot{\theta}\cos\theta + \text{constant}$ and $(4 - 3\cos^2\theta)\dot{\theta}^2 = \text{constant}$.

12. A ring of mass m slides on a smooth straight wire inclined at an angle α to the horizontal. A particle of mass M is attached to the ring by a string of length a. Show that a motion is possible in which the string is inclined at a constant angle α to the vertical and the ring slides down the wire with constant acceleration $g\sin\alpha$. Show that if the motion is slightly perturbed, the resulting perturbed motion will be simple harmonic with frequency $[(m + M)g\cos\alpha/ma]^{1/2}$.

13. A massless flat plate rests on a frictionless plane that is inclined to the horizontal at an angle α. A massless rod of length b is fixed at right angles to the plate. From the top of the rod hangs a simple pendulum of length a, where $a < b$. Introduce suitable generalized coordinates, and find the Lagrangian of the system. Derive the equations of motion.

14. The point of suspension of a simple pendulum of length a and mass m is free to move along a horizontal line. A spring of spring constant k has one end attached to the point of suspension of the pendulum and the

other attached to a point on the horizontal line. Write the Lagrangian equations of motion, using as generalized coordinates x, the displacement of the point of suspension of the pendulum from its equilibrium position, and θ, the angle that the pendulum makes with the vertical. Assuming small oscillations, show that this system is equivalent to a simple pendulum of length $a + (mg/k)$.

15. A solid uniform disk of mass M and radius R has attached to its face a small mass m at a distance r from its center. The disk is free to roll without slipping along a horizontal straight line. Show that if the disk is slightly displaced from its equilibrium position and released it will oscillate with a period $2\pi\{[(3MR^2/2) + m(R-r)^2]/mgr\}^{1/2}$.

16. A smooth uniform rod BC of mass M is supported in a horizontal position by two parallel massless inextensible strings, each of length a, which are attached respectively to the ends B and C of the rod and to two fixed supports A and D. A bead of mass m is free to slide on the rod, and the system is free to move in the vertical plane containing the supports A and D. The system is released from rest with the bead at the center of the rod, and the strings AB and DC each making an angle α with the vertical. Using x, the displacement of the bead from the center of the rod, and θ, the angle the strings make with the vertical, as generalized coordinates, write down Lagrange's equations of motion and show that $a(M + m\sin^2\theta)\dot\theta^2 = 2g(M + m)(\cos\theta - \cos\alpha)$.

17. A uniform ladder of mass m and length a has one end on a smooth horizontal floor and the other end against a smooth vertical wall. Choose the angle θ that the ladder makes with the wall as a generalized coordinate and find the equation of motion.

18. A uniform bar of mass M and length $2b$ is suspended from one end by a spring of force constant k and unstretched length a. The bar can swing freely in one vertical plane, and the spring is constrained to move in the vertical direction only. Obtain Lagrange's equations of motion for some set of generalized coordinates.

19. A hoop of mass M and radius b is free to rotate about a vertical diameter. A rod of mass m and length $2a$, where $a < b$, slides freely inside the hoop with its ends on the hoop. Choose an appropriate set of generalized coordinates, and obtain Lagrange's equations of motion.

20. A rigid straight rail is fixed in the plane of a lamina, and a small engine of mass m moves itself by friction drive along the rail. The lamina is horizontal and is free to rotate about a vertical axis that intersects the lamina at a point O, which is a perpendicular distance a from the rail. The moment of inertia of the lamina and rail about the axis of rotation is mb^2. The engine and the lamina are initially at rest, and the engine is located at the point on the rail that is closest to the point O. The engine

then moves and its motion relative to the rail is known. Let x be the displacement along the rail of the engine from its initial position, and ϕ the angle through which the lamina has turned. Determine ϕ as a function of x.

21. Four uniform rods, each of mass m and length $2a$, are smoothly joined together at their ends to form a rhombus $ABCD$. The rod AB is smoothly pivoted at its center and the system is free to move in a vertical plane with rod CD below rod AB. Let θ be the angle made by rods AB and DC with the horizontal, and ϕ the angle made by rods AD and BC with the vertical. Show that the Lagrangian for the system is given by $L = (2ma^2/3)(2\dot\theta^2 + 5\dot\phi^2) + 4mga\cos\phi$. Show that AB and DC rotate with constant angular velocity, and that AD and BC oscillate in synchrony with a simple pendulum of length $5a/3$.

22. Two identical rods AC and BC, each of mass m and length $2a$, are smoothly hinged at C. A light ring attached to end A of rod AC is threaded on a smooth horizontal wire and a similar ring attached to end B of rod BC is threaded on a smooth vertical wire that intersects the horizontal wire at a point O. The system moves in the plane of the wires with C higher than O. Obtain Lagrange's equations of motion for the system using as generalized coordinates θ, the angle the rod AC makes with the horizontal, and ϕ, the angle rod BC makes with the horizontal, where θ and ϕ are both measured in the same sense. If initially the system is released from rest with $\theta = \pi/3$ and $\phi = \pi/6$, show that at this instant, the angular accelerations of the bars are $9g/11a$ and $3(3)^{1/2}g/22a$.

SECTION 2

Applications of Lagrangian Mechanics

45

Vibrating Systems

INTRODUCTION

In dynamics we frequently deal with systems that are vibrating about a configuration of stable equilibrium, that is, a configuration in which the system can remain permanently and stably at rest. In this chapter we consider certain types of such systems. We are particularly interested in systems for which the energy is small enough that the system does not depart appreciably from the equilibrium configuration.

NATURAL MODES OF MOTION

A Lagrangian system is *harmonic with respect to the set of coordinates* q if the Lagrangian in terms of the set of coordinates q is of the form $L = \frac{1}{2}\sum_i\sum_j m_{ij}\dot{q}_i\dot{q}_j - \frac{1}{2}\sum_i\sum_j k_{ij}q_iq_j$, where the matrices $[m_{ij}]$ and $[k_{ij}]$ are real, constant, symmetric, and positive definite (see Appendix 12). In this section we investigate the motion of harmonic systems. We summarize the results in the following theorem.

Theorem 1. Consider a Lagrangian system of f degrees of freedom that is harmonic with respect to the set of variables q. Let the Lagrangian be given by

$$L = \tfrac{1}{2}\sum_i\sum_j m_{ij}\dot{q}_i\dot{q}_j - \tfrac{1}{2}\sum_i\sum_j k_{ij}q_iq_j \tag{1}$$

Such a system has the following properties.
  a. There is a set of possible modes of motion in which the point in the configuration space q representing the instantaneous configuration of the system oscillates in simple harmonic motion about the origin along a straight line passing through the origin. These modes of motion are called *natural modes of motion*. The directions in q-space of the straight

lines along which such motion is possible are called *modal directions*. A vector **c** whose direction is a modal direction is called a *modal vector*. In a natural mode of motion all the particles in the system are oscillating with the same frequency and are either in phase or 180° out of phase with one another.

b. Associated with each modal direction there is a single angular frequency of oscillation ω, called a natural or *modal frequency*. The value of a particular modal frequency must be one of the f positive roots of the equation

$$|k_{ij} - \omega^2 m_{ij}| = 0 \qquad (2)$$

where $|k_{ij} - \omega^2 m_{ij}|$ is the determinant of the matrix $[k_{ij} - \omega^2 m_{ij}]$. We label the f positive roots of Eq. (2) as $\omega_{\bar{1}}, \omega_{\bar{2}}, \ldots, \omega_{\bar{f}}$, respectively.

c. If we substitute a particular one of the roots of Eq. (2) into the equation

$$\sum_j (k_{ij} - \omega^2 m_{ij}) c_j = 0 \qquad (3)$$

then any vector **c** whose components satisfy the resulting set of equations will be a modal vector with which the given frequency is associated. We label a modal vector associated with the frequency $\omega_{\bar{j}}$ as $\mathbf{c}_{\bar{j}}$. If the frequency $\omega_{\bar{j}}$ is a unique frequency Eqs. (3) will determine a unique one dimensional subspace within which $\mathbf{c}_{\bar{j}}$ must lie; that is, the vectors will all lie along a single direction. If the frequency $\omega_{\bar{j}}$ is one of a set of s identical frequencies Eqs. (3) will determine a unique s dimensional subspace within which $\mathbf{c}_{\bar{j}}$ must lie.

d. A set of modal directions is linearly independent if the corresponding modal vectors are linearly independent. From the set of all modal directions it is always possible to choose a set of f linearly independent modal directions.

PROOF: The motion of the system is determined by Lagrange's equations

$$\frac{d}{dt}\left[\frac{\partial L(\mathbf{q},\dot{\mathbf{q}},t)}{\partial \dot{q}_i}\right] - \frac{\partial L(\mathbf{q},\dot{\mathbf{q}},t)}{\partial q_i} = 0 \qquad (4)$$

Substituting Eq. (1) in Eq. (4) we obtain

$$\sum_j m_{ij} \ddot{q}_j + \sum_j k_{ij} q_j = 0 \qquad (5)$$

We assume a solution of the form

$$q_j = c_j \cos(\omega t + \phi) \qquad (6)$$

These are the equations of a point in **q**-space that moves on a line that passes through the origin and is executing simple harmonic motion about the origin. The angular frequency is ω, the amplitude is $(\sum_j c_j^2)^{1/2}$, the direction of the line along which the oscillations take place is the direction of the vector **c** whose components are c_j, and the phase of the oscillation is ϕ. If we substitute Eq.

(6) in Eq. (5) we obtain

$$-\omega^2 \sum_j m_{ij} c_j \cos(\omega t + \phi) + \sum_j k_{ij} c_j \cos(\omega t + \phi) = 0 \tag{7}$$

Dividing Eq. (7) by $\cos(\omega t + \phi)$ we obtain

$$\sum_j (k_{ij} - \omega^2 m_{ij}) c_j = 0 \tag{8}$$

The set of Eqs. (8) constitute a set of simultaneous equations for the constants c_j. This set of equations will have a solution if and only if

$$|k_{ij} - \omega^2 m_{ij}| = 0 \tag{9}$$

It can be shown that if the matrices $[k_{ij}]$ and $[m_{ij}]$ are real, symmetric, and positive definite, there will be f real positive roots of Eq. (9). Given a particular root ω_r the components of the vector $\mathbf{c}_r$ can be obtained by letting $\omega = \omega_r$ in Eq. (8) and solving for the c_j. The proof of the rest of the theorem follows immediately from the theory of simultaneous equations.

There are not a great number of real systems for which the Lagrangian can be written in the form demanded by Theorem 1. However, if a system is moving under the action of a force that is derivable from a potential having a minimum at some point, and if the total energy of the system is such that the system is confined to a small region in the neighborhood of the minimum, it is possible to choose generalized coordinates for which the Lagrangian is approximately of the desired form. This result is stated in the following theorem.

Theorem 2. Consider a Lagrangian system with f degrees of freedom, for which the potential is a function of the configuration only and has a minimum value for a particular configuration.

If we choose a set of generalized coordinates $\mathbf{q} \equiv q_1, q_2, \ldots, q_f$ such that:
a. The kinetic energy function is of the form

$$T(\mathbf{q}, \dot{\mathbf{q}}) = \tfrac{1}{2} \sum_i \sum_j a_{ij}(\mathbf{q}) \dot{q}_i \dot{q}_j \tag{10}$$

where $a_{ij} = a_{ji}$. This will be true if the equations for the transformation between the set of generalized coordinates $\mathbf{q}$ and the set of Cartesian coordinates $\mathbf{x}$ do not contain the time.
b. The minimum value of the potential energy to which we assign the value zero occurs at $\mathbf{q} = 0$, that is,

$$[V(\mathbf{q})]_{\text{minimum}} = V(0) = 0 \tag{11}$$

Then for small displacements from the configuration $\mathbf{q} = 0$, the Lagrangian function is given approximately by

$$L(\mathbf{q}, \dot{\mathbf{q}}) = \tfrac{1}{2} \sum_i \sum_j m_{ij} \dot{q}_i \dot{q}_j - \tfrac{1}{2} \sum_i \sum_j k_{ij} q_i q_j \tag{12}$$

where

$$m_{ij} = a_{ij}(0) \tag{13}$$

$$k_{ij} = \left[\frac{\partial^2 V(\mathbf{q})}{\partial q_i \, \partial q_j}\right]_{\mathbf{q}=0} \tag{14}$$

and the matrices $[m_{ij}]$ and $[k_{ij}]$ are real, symmetric, and positive definite. Hence, the system is harmonic with respect to the coordinates $\mathbf{q}$.

PROOF: If we expand $a_{ij}(\mathbf{q})$ and $V(\mathbf{q})$ in a Taylor series about the configuration $\mathbf{q} = 0$ we obtain

$$a_{ij}(\mathbf{q}) = a_{ij}(0) + \sum_k \left[\frac{\partial a_{ij}(\mathbf{q})}{\partial q_k}\right]_{\mathbf{q}=0} q_k + \cdots \tag{15}$$

$$V(\mathbf{q}) = V(0) + \sum_i \left[\frac{\partial V(\mathbf{q})}{\partial q_i}\right]_{\mathbf{q}=0} q_i + \tfrac{1}{2} \sum_i \sum_j \left[\frac{\partial^2 V(\mathbf{q})}{\partial q_i \, \partial q_j}\right]_{\mathbf{q}=0} q_i q_j + \cdots \tag{16}$$

But

$$V(0) = 0 \tag{17}$$

and since $V(\mathbf{q})$ is a minimum at $\mathbf{q} = 0$, it follows that

$$\left[\frac{\partial V(\mathbf{q})}{\partial q_i}\right]_{\mathbf{q}=0} = 0 \tag{18}$$

Thus if we approximate $a_{ij}(\mathbf{q})$ and $V(\mathbf{q})$ by the first nonvanishing terms in the expansions we obtain

$$a_{ij}(\mathbf{q}) \approx a_{ij}(0)$$
$$V(\mathbf{q}) \approx \tfrac{1}{2} \sum_i \sum_j \left[\frac{\partial^2 V(\mathbf{q})}{\partial q_i \, \partial q_j}\right]_{\mathbf{q}=0} q_i q_j \tag{19}$$

It follows immediately that the Lagrangian is given approximately by Eq. (12). By virtue of their definitions the m_{ij} and the k_{ij} are constants, and $m_{ij} = m_{ji}$ and $k_{ij} = k_{ji}$. Since the kinetic energy is always positive, it follows that $[m_{ij}]$ is positive definite; and since the potential energy is zero at $\mathbf{q} = 0$ and greater than zero for other points in the neighborhood of $\mathbf{q} = 0$, it follows that $[k_{ij}]$ is positive definite. This completes the proof of the theorem.

NATURAL COORDINATES

If a system is harmonic with respect to a set of coordinates $\mathbf{q}$, it is possible to choose a new set of coordinates that reduces the motion of the system to the motion of a set of independent harmonic oscillators. This result is contained in the following theorem.

Theorem 3. Consider a Lagrangian system of f degrees of freedom that is harmonic with respect to the set of coordinates **q**. Let the Lagrangian of the system be given by

$$L = \tfrac{1}{2} \sum_i \sum_j m_{ij} \dot{q}_i \dot{q}_j - \tfrac{1}{2} \sum_i \sum_j k_{ij} q_i q_j \tag{20}$$

Let $\mathbf{c}_{\bar{1}}, \mathbf{c}_{\bar{2}}, \ldots, \mathbf{c}_{\bar{f}}$ be a set of f linearly independent modal vectors in the configuration space **q** and $\omega_{\bar{1}}, \omega_{\bar{2}}, \ldots, \omega_{\bar{f}}$ the corresponding modal frequencies. If we define a quantity $q_{\bar{r}}$, which we shall refer to as the natural coordinate associated with the modal vector $\mathbf{c}_{\bar{r}}$, as follows:

$$q_{\bar{r}} = \sum_i \sum_j q_i m_{ij} c_{\bar{r}j} \equiv \sum_i \sum_j c_{\bar{r}i} m_{ij} q_j \tag{21}$$

or equivalently in matrix form as follows:

$$q_{\bar{r}} = [q_1 \cdots q_f] \begin{bmatrix} m_{11} & \cdots & m_{1f} \\ \vdots & & \vdots \\ m_{f1} & \cdots & m_{ff} \end{bmatrix} \begin{bmatrix} c_{\bar{r}1} \\ \vdots \\ c_{\bar{r}f} \end{bmatrix}$$

$$\equiv [c_{\bar{r}1} \cdots c_{\bar{r}f}] \begin{bmatrix} m_{11} & \cdots & m_{1f} \\ \vdots & & \vdots \\ m_{f1} & \cdots & m_{ff} \end{bmatrix} \begin{bmatrix} q_1 \\ \vdots \\ q_f \end{bmatrix} \tag{22}$$

then the equation of motion for $q_{\bar{r}}$ is

$$\ddot{q}_{\bar{r}} + \omega_{\bar{r}}^2 q_{\bar{r}} = 0 \tag{23}$$

which is the equation of motion of a simple harmonic oscillator. It follows that if we choose as generalized coordinates the set of natural coordinates $\bar{\mathbf{q}} \equiv q_{\bar{1}}, q_{\bar{2}}, \ldots, q_{\bar{f}}$ then: (1) the motion of each of the natural coordinates is independent of the motion of the other coordinates, (2) a motion in which the natural coordinate $q_{\bar{r}}$ is varying, but the remaining natural coordinates are not, corresponds to the natural mode of motion associated with the vector $\mathbf{c}_{\bar{r}}$, and (3) the modal directions in the $\bar{\mathbf{q}}$ configuration space are along the coordinate axes. Finally, since the complete motion of the system can be described in terms of the set of coordinates $q_{\bar{1}}, q_{\bar{2}}, \ldots, q_{\bar{f}}$ and each of these coordinates corresponds to a particular natural mode of motion, it follows that the complete motion of the system can be considered to be a superposition of natural modes of motion.

PROOF: The modal frequency $\omega_{\bar{r}}$ and the components of the modal vector $\mathbf{c}_{\bar{r}}$ satisfy the equation

$$\sum_j \left[k_{ij} - \omega_{\bar{r}}^2 m_{ij} \right] c_{\bar{r}j} = 0 \tag{24}$$

If we multiply Eq. (24) by q_i and sum over i we obtain

$$\sum_i \sum_j k_{ij} c_{\bar{r}j} q_i - \omega_{\bar{r}}^2 \sum_i \sum_j m_{ij} c_{\bar{r}j} q_i = 0 \tag{25}$$

From Eq. (5) we have

$$\sum_i k_{ji} q_i = -\sum_i m_{ji} \ddot{q}_i \qquad (26)$$

Since k_{ij} and m_{ij} are symmetric, we can rewrite Eq. (26)

$$\sum_i k_{ij} q_i = -\sum_i m_{ij} \ddot{q}_i \qquad (27)$$

Substituting Eq. (27) in Eq. (25) we obtain

$$\sum_i \sum_j m_{ij} c_{\bar{r}j} \ddot{q}_i + \omega_{\bar{r}}^2 \sum_i \sum_j m_{ij} c_{\bar{r}j} q_i = 0 \qquad (28)$$

If we define

$$q_{\bar{r}} \equiv \sum_i \sum_j m_{ij} c_{\bar{r}j} q_i \qquad (29)$$

then Eq. (28) can be rewritten

$$\ddot{q}_{\bar{r}} + \omega_{\bar{r}}^2 q_{\bar{r}} = 0 \qquad (30)$$

The remainder of the theorem follows immediately from this result.

NORMAL COORDINATES

From the set of all possible modal vectors it is possible to choose a set of vectors that are orthogonalized and normalized in such a way that when one writes the kinetic and potential energies in terms of the corresponding natural coordinates, the associated matrices $[m_{ij}]$ and $[k_{ij}]$ are both diagonal matrices. In this case the natural coordinates are called normal coordinates. We consider normal coordinates in a later chapter.

EXAMPLES

Example 1. A particle A of mass $2m$ is suspended from a fixed point O by a string of length $3a$. A second particle B of mass m is suspended from particle A by a string of length $2a$. The motion of the system is confined to a vertical plane. The configuration of the system (Fig. 1) at a particular instant of time is described by the angle θ which the string joining O and A makes with the vertical, and the angle ϕ that the string joining A and B makes with the vertical.

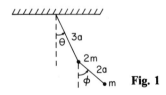

Fig. 1

a. Determine the modal frequencies for small oscillations about the equilibrium position.
b. For each of the modal frequencies determine a set of initial conditions that would result in periodic motion with that frequency.
c. Find a set of natural coordinates.
d. If the system is initially released from rest with $\theta = 0$ and $\phi = \epsilon$, determine $\theta(t)$ and $\phi(t)$.

Solution.

a. The kinetic and potential energies of the system are:

$$T = \tfrac{1}{2} ma^2 \left[27\dot{\theta}^2 + 12\dot{\theta}\dot{\phi} \cos(\theta - \phi) + 4\dot{\phi}^2 \right] \tag{31}$$

$$V = mga \left[11 - 9\cos\theta - 2\cos\phi \right] \tag{32}$$

Note that the kinetic energy is a homogeneous quadratic function of $\dot{\theta}$ and $\dot{\phi}$, and the potential energy has a minimum at $\theta = \phi = 0$, and is zero when $\theta = \phi = 0$. Thus the conditions of Theorem 2 are satisfied. For small oscillations

$$T \approx \tfrac{1}{2} ma^2 (27\dot{\theta}^2 + 12\dot{\theta}\dot{\phi} + 4\dot{\phi}^2) \tag{33}$$

$$V \approx \tfrac{1}{2} mga (9\theta^2 + 2\phi^2) \tag{34}$$

The matrices $[m_{ij}]$ and $[k_{ij}]$ are thus

$$[m_{ij}] = ma^2 \begin{bmatrix} 27 & 6 \\ 6 & 4 \end{bmatrix} \tag{35}$$

$$[k_{ij}] = mga \begin{bmatrix} 9 & 0 \\ 0 & 2 \end{bmatrix} \tag{36}$$

Setting the determinant of the matrix $[k_{ij} - \omega^2 m_{ij}]$ equal to zero we obtain

$$\begin{vmatrix} 9mga - 27ma^2\omega^2 & -6ma^2\omega^2 \\ -6ma^2\omega^2 & 2mga - 4ma^2\omega^2 \end{vmatrix} = 0 \tag{37}$$

which yields two positive roots for ω, namely,

$$\omega_{\bar{1}} = \left(\frac{g}{a} \right)^{1/2} \tag{38}$$

$$\omega_{\bar{2}} = \left(\frac{g}{4a} \right)^{1/2} \tag{39}$$

b. The components $c_{\bar{1}1}$ and $c_{\bar{1}2}$ of the modal vector associated with the modal frequency $\omega_{\bar{1}}$ are determined by solving the equation

$$\begin{bmatrix} 9mga - 27ma^2(g/a) & -6ma^2(g/a) \\ -6ma^2(g/a) & 2mga - 4ma^2(g/a) \end{bmatrix} \begin{bmatrix} c_{\bar{1}1} \\ c_{\bar{1}2} \end{bmatrix} = \begin{bmatrix} 0 \\ 0 \end{bmatrix} \tag{40}$$

Any vector of the form

$$\begin{bmatrix} c_{\bar{1}1} \\ c_{\bar{1}2} \end{bmatrix} = \begin{bmatrix} \alpha \\ -3\alpha \end{bmatrix} \quad (41)$$

where α is arbitrary, will satisfy Eq. (40). Similarly, determining the modal vectors associated with $\omega_{\bar{2}}$ we find

$$\begin{bmatrix} c_{\bar{2}1} \\ c_{\bar{2}2} \end{bmatrix} = \begin{bmatrix} 2\beta \\ 3\beta \end{bmatrix} \quad (42)$$

where β is arbitrary. If follows that if at time $t = 0$, $\theta = \alpha$ and $\phi = -3\alpha$, the system will oscillate with angular frequency $(g/a)^{1/2}$ and the ratio θ/ϕ will remain $-\frac{1}{3}$. And if at time $t = 0$, $\theta = 2\beta$ and $\phi = 3\beta$, the system will oscillate with angular frequency $(g/4a)^{1/2}$ and the ratio θ/ϕ will remain $2/3$.

c. The natural coordinate associated with the modal vector $[c_{\bar{1}1} \ c_{\bar{1}2}] = [\alpha \ -3\alpha]$ is

$$q_{\bar{1}} = [\alpha \ -3\alpha] \begin{bmatrix} 27ma^2 & 6ma^2 \\ 6ma^2 & 4ma^2 \end{bmatrix} \begin{bmatrix} \theta \\ \phi \end{bmatrix} \quad (43)$$

If we choose $\alpha = (3ma^2)^{-1}$ we obtain

$$q_{\bar{1}} = 3\theta - 2\phi \quad (44)$$

Similarly if we choose $\beta = (24ma^2)^{-1}$ we obtain

$$q_{\bar{2}} = 3\theta + \phi \quad (45)$$

The pair of generalized coordinates $q_{\bar{1}}$ and $q_{\bar{2}}$ constitute a set of natural coordinates.

d. The equations of motion for $q_{\bar{1}}$ and $q_{\bar{2}}$ are, respectively, $\ddot{q}_{\bar{1}} + \omega_{\bar{1}}^2 q_{\bar{1}} = 0$ and $\ddot{q}_{\bar{2}} + \omega_{\bar{2}}^2 q_{\bar{2}} = 0$. It follows that

$$q_{\bar{1}} = A \cos \omega_{\bar{1}} t + B \sin \omega_{\bar{1}} t \quad (46)$$

$$q_{\bar{2}} = C \cos \omega_{\bar{2}} t + D \sin \omega_{\bar{2}} t \quad (47)$$

where A, B, C, and D are determined by the initial conditions. At $t = 0$, $\theta = 0$, $\phi = \epsilon$, $\dot{\theta} = 0$, $\dot{\phi} = 0$; hence from Eqs. (44) and (45), $q_{\bar{1}} = -2\epsilon$, $q_{\bar{2}} = \epsilon$, $\dot{q}_{\bar{1}} = 0$, $\dot{q}_{\bar{2}} = 0$. Substituting these initial conditions in Eqs. (46) and (47) and solving for A, B, C, and D we obtain

$$q_{\bar{1}} = -2\epsilon \cos \omega_{\bar{1}} t \quad (48)$$

$$q_{\bar{2}} = \epsilon \cos \omega_{\bar{2}} t \quad (49)$$

Solving Eqs. (44) and (45) for θ and ϕ we obtain

$$\theta = \frac{q_{\bar{1}} + 2q_{\bar{2}}}{9} \quad (50)$$

$$\phi = \frac{-q_{\bar{1}} + q_{\bar{2}}}{3} \quad (51)$$

Substituting Eqs. (48), (49), (38), and (39) in Eqs. (50) and (51) we obtain

$$\theta = \frac{-2\epsilon \cos\left[(g/a)^{1/2}t\right] + 2\epsilon \cos\left[(g/4a)^{1/2}t\right]}{9} \quad (52)$$

$$\phi = \frac{2\epsilon \cos\left[(g/a)^{1/2}t\right] + \epsilon \cos\left[(g/4a)^{1/2}t\right]}{3} \quad (53)$$

NOTE: The computations involved in solving the problem above could have been considerably condensed if after obtaining Eqs. (33) and (34) we had defined a new set of generalized coordinates

$$q_1 \equiv (mga)^{1/2}\theta \quad (54)$$

$$q_2 \equiv (mga)^{1/2}\phi \quad (55)$$

With this change the matrices $[m_{ij}]$ and $[k_{ij}]$ become

$$[m_{ij}] = \frac{a}{g}\begin{bmatrix} 27 & 6 \\ 6 & 4 \end{bmatrix} \quad (56)$$

$$[k_{ij}] = \begin{bmatrix} 9 & 0 \\ 0 & 2 \end{bmatrix} \quad (57)$$

and the basic equation from which the modal frequencies and modal vectors are determined becomes simply

$$\begin{bmatrix} 9 - 27\lambda & -6\lambda \\ -6\lambda & 2 - 4\lambda \end{bmatrix} \begin{bmatrix} c_1 \\ c_2 \end{bmatrix} = \begin{bmatrix} 0 \\ 0 \end{bmatrix} \quad (58)$$

where

$$\lambda \equiv \frac{a}{g}\omega^2 \quad (59)$$

We leave it to the reader to demonstrate the utility of these shortcuts.

PROBLEMS

1. A particle moves on the inside of a smooth surface the equation of which is $2z - (x^2/a) - (y^2/b) = 0$, where a and b are positive constants and the positive z axis is vertically up. Determine the modal frequencies for small oscillations about the equilibrium position.

2. A uniform rod AB of mass $5m$ and length $2a$ turns freely about a fixed horizontal axis through A, perpendicular to the rod. A light rod BC, of length $2a$, is freely jointed to the rod AB at B and carries a mass m at the other end C. If in a displaced position the rods AB and BC make angles θ and ϕ, respectively, with the downward vertical, show that the modal

frequencies for small oscillations in the plane ABC are given by $\omega_{\bar{1}} = (3g/2a)^{1/2}$ and $\omega_{\bar{2}} = (7g/20a)^{1/2}$. The system is initially held at rest with $\theta = \theta_0$ and $\phi = \phi_0$, and then released. If the resulting motion is a natural mode of motion with frequency $\omega_{\bar{1}}$, what is the ratio of θ_0 to ϕ_0?

3. A thin hoop of radius a and mass m is allowed to oscillate in its own plane with one point of the hoop fixed. Attached to the hoop is a small mass m, which is constrained to move in a frictionless manner along the hoop.
   a. Determine the modal frequencies for small oscillations about the equilibrium position.
   b. For each of the modal frequencies determine a set of initial conditions that would result in periodic motion with that frequency.
   c. Determine a set of natural coordinates.

4. A light string OAB is tied to a fixed point at O, and carries a particle of mass $2m$ at point A and a particle of mass m at point B. The lengths OA, AB are $a/2$, $3a/4$, respectively. The string is free to move in a vertical plane, and the system oscillates about the position of equilibrium. The inclinations of OA and AB to the vertical are denoted by θ and ϕ, respectively. The system is held with the string straight and inclined at a small angle α to the vertical, and is let go from rest in this position at the instant $t = 0$. Show that at any subsequent time $\theta = (\alpha/3)(2\cos nt + \cos 2nt)$ and $\phi = (\alpha/3)(4\cos nt - \cos 2nt)$, where $n = (g/a)^{1/2}$.

5. A smooth circular wire, of mass $8m$ and radius a, swings in a vertical plane, being suspended by an inextensible string of length a attached to one point of it; a particle P, of mass m, can slide on the wire. If the system is released from rest with the ring in its equilibrium position and P is displaced through a small angle β from its equilibrium position, show that at time t the angle that the string makes with the vertical is $(\beta/21)[\cos(nt/2\sqrt{2}) - \cos(nt)]$, where $n^2 = 3g/a$.

6. A uniform bar CD of mass $10m$ and length b is supported in a horizontal position by two vertical strings AC and BD, each of length $2a$. The ends A and B of the strings are attached to fixed supports. From the ends C and D of the bar are hung simple pendula CE and DF, each consisting of a string of length a and a bob of mass $3m$. When the system is displaced from its equilibrium position, the strings AC and BD make an angle θ with the vertical, the string CE makes an angle ϕ with the vertical, and the string DF makes an angle ψ with the vertical. Show that the modal frequencies for small oscillations of the system about its equilibrium are $\omega_{\bar{1}} = (g/a)^{1/2}$, $\omega_{\bar{2}} = (2g/a)^{1/2}$, and $\omega_{\bar{3}} = (2g/5a)^{1/2}$. If the system is released from rest in the configuration $(\theta_0, \phi_0, \psi_0)$ show that the resulting motion will be a natural mode of motion if $\theta_0 = 0$ and

$\phi_0/\psi_0 = -1$, or if $\phi_0/\theta_0 = \psi_0/\theta_0 = -4$, or if $\phi_0/\theta_0 = \psi_0/\theta_0 = \frac{4}{3}$. Show that the coordinates $q_{\bar{1}} = \phi - \psi$, $q_{\bar{2}} = 8\theta - 3\phi - 3\psi$, and $q_{\bar{3}} = 8\theta + \phi + \psi$ are natural coordinates.

7. A light elastic string is stretched between two fixed points A and B. The points A and B are a distance $4a$ apart and the tension in the string is T. A particle of mass $3m$ is attached to the midpoint C of the string and particles of mass $4m$ are attached to the midpoints of AC and CB. When the system is at rest small transverse velocities u are suddenly imparted in the same direction to both the particles of mass $4m$. Prove that the displacement of the particle of mass $3m$ at time t is given by $(4u/5\omega)[\sqrt{6}\sin(\omega t/\sqrt{6}) - \sin \omega t]$, where $\omega^2 = T/ma$.

8. Three particles of masses m, M, and m, respectively, are free to move along a straight line. The middle particle M is joined to each of the outer particles m by a spring of spring constant k and unstretched length a. The particles are initially at rest. A fourth particle, of mass m, moving with speed v along the line formed by the three particles, strikes one of the end particles and rebounds elastically. Determine the subsequent motion of the particle of mass M.

9. A uniform triangular lamina OAB can turn freely in a vertical plane about the point O, which is fixed. The lamina has mass $3m$ and is isosceles with $OA = OB = 2a$, and the angle AOB is $120°$. A particle of mass m is suspended from A by a string of length a and an equal particle is suspended from B by a string also of length a. Small vibrations take place in the vertical plane OAB. Show that the modal frequencies are in the ratio $\sqrt{3} : 1 : 2$. If θ is the angle made by AB with the horizontal and ϕ and ψ are the angles made by the strings with the downward vertical, all angles being in the same sense, show that the coordinates $\phi - \psi$, $8\theta + \phi + \psi$, and $\theta - \phi - \psi$ are natural coordinates.

10. Two equal uniform rods AB and BC, each of mass m and length $2a$, are smoothly hinged together at B. They are supported at A, B and C by light vertical springs such that in equilibrium A, B, and C are at the same level. The spring constants for the springs at A and C are each k, and the spring constant for the spring at B is $2k$. Let x, y, and z be the vertical displacements of the points A, B, and C, respectively, from their equilibrium positions. Determine the modal frequencies, the modal vectors, and a set of natural coordinates.

SECTION 3

Hamiltonian Mechanics

46

Hamilton's Equations of Motion

INTRODUCTION

In this chapter we consider the Hamiltonian formulation of the equations of motion. We restrict our attention to systems for which the constraints are holonomic and the nonconstraint forces are derivable from a potential.

THE HAMILTONIAN

The behavior of a system for which there exists a Lagrangian L is defined if we know the Lagrangian as a function of the generalized coordinates $\mathbf{q} \equiv q_1, q_2, \ldots, q_f$, the generalized velocities $\dot{\mathbf{q}} \equiv \dot{q}_1, \dot{q}_2, \ldots, \dot{q}_f$, and the time t.

There are in addition to the Lagrangian other quantities that when known as functions of particular sets of variables, determine the behavior of a system. One very important such quantity is the Hamiltonian, which we consider in the following theorem.

Theorem 1. Given a Lagrangian function $L(\mathbf{q}, \dot{\mathbf{q}}, t)$. If we define a set of quantities p_i, called generalized momenta, as follows:

$$p_i \equiv \frac{\partial L(\mathbf{q}, \dot{\mathbf{q}}, t)}{\partial \dot{q}_i} \qquad (1)$$

and a quantity H, called the Hamiltonian, as follows:

$$H = \sum_i p_i \dot{q}_i - L \qquad (2)$$

then the function $H(\mathbf{q}, \mathbf{p}, t)$ contains the same information as the function $L(\mathbf{q}, \dot{\mathbf{q}}, t)$.

PROOF: From Eq. (2) it follows that

$$\frac{\partial H(\mathbf{q},\mathbf{p},t)}{\partial p_i} = \dot{q}_i + \sum_j p_j \frac{\partial \dot{q}_j(\mathbf{q},\mathbf{p},t)}{\partial p_i} - \sum_j \frac{\partial L(\mathbf{q},\dot{\mathbf{q}},t)}{\partial \dot{q}_j} \frac{\partial \dot{q}_j(\mathbf{q},\mathbf{p},t)}{\partial p_i} \quad (3)$$

Substituting Eq. (1) in Eq. (3) we obtain

$$\dot{q}_i = \frac{\partial H(\mathbf{q},\mathbf{p},t)}{\partial p_i} \quad (4)$$

Rearranging Eq. (2) we obtain

$$L = \sum_i \dot{q}_i p_i - H \quad (5)$$

Comparing Eqs. (1) and (2) with Eqs. (4) and (5) we see that there is a symmetrical relationship between them. We now note that if we are given $L(\mathbf{q},\dot{\mathbf{q}},t)$ we can use Eq. (1) to find $\mathbf{p}(\mathbf{q},\dot{\mathbf{q}},t)$ which can then be inverted to obtain $\dot{\mathbf{q}}(\mathbf{q},\mathbf{p},t)$. If we then substitute $\dot{\mathbf{q}}(\mathbf{q},\mathbf{p},t)$ into Eq. (2) we obtain $H(\mathbf{q},\mathbf{p},t)$. Conversely if we are given $H(\mathbf{q},\mathbf{p},t)$ we can use Eq. (4) to find $\dot{\mathbf{q}}(\mathbf{q},\mathbf{p},t)$, which can then be inverted to obtain $\mathbf{p}(\mathbf{q},\dot{\mathbf{q}},t)$. If we then substitute $\mathbf{p}(\mathbf{q},\dot{\mathbf{q}},t)$ into Eq. (5) we obtain $L(\mathbf{q},\dot{\mathbf{q}},t)$. It follows that if we know $L(\mathbf{q},\dot{\mathbf{q}},t)$ we can find $H(\mathbf{q},\mathbf{p},t)$, and if we know $H(\mathbf{q},\mathbf{p},t)$ we can find $L(\mathbf{q},\dot{\mathbf{q}},t)$. Hence both functions contain the same information. This completes the proof of the theorem.

Although Eq. (2) is the fundamental definition of the Hamiltonian, there are many situations in which the Hamiltonian is equal to the sum of the kinetic energy T and the potential energy V. The conditions for which this is true are given in the following theorem.

Theorem 2. Consider a Lagrangian system whose configuration is defined by the set of generalized coordinates $\mathbf{q} \equiv q_1, q_2, \ldots, q_f$ and whose behavior is defined by the Lagrangian function $L(\mathbf{q},\dot{\mathbf{q}},t) = T(\mathbf{q},\dot{\mathbf{q}},t) - V(\mathbf{q},t)$. If the equations of transformation between the set of Cartesian coordinates $\mathbf{x}$ and the set of generalized coordinates $\mathbf{q}$ are not functions of the time, that is, $\mathbf{x} = \mathbf{x}(\mathbf{q})$, or the kinetic energy $T(\mathbf{q},\dot{\mathbf{q}},t)$ is a homogeneous quadratic function of the set of generalized velocities $\dot{\mathbf{q}}$, then the Hamiltonian H is the sum of the kinetic energy T and the potential energy V, that is,

$$H = T + V \quad (6)$$

PROOF: The proof of this theorem is contained in the proof of Theorem 2, Chapter 44.

HAMILTON'S EQUATIONS OF MOTION

If we are given the Hamiltonian function $H(\mathbf{q},\mathbf{p},t)$ and wish to find $\mathbf{q}(t)$, we could proceed by finding the Lagrangian function $L(\mathbf{q},\dot{\mathbf{q}},t)$ and then making use of Lagrange's equations of motion, that is,

$$\frac{d}{dt}\left[\frac{\partial L(\mathbf{q},\dot{\mathbf{q}},t)}{\partial \dot{q}_i}\right] - \frac{\partial L(\mathbf{q},\dot{\mathbf{q}},t)}{\partial q_i} = 0 \qquad (7)$$

There is however a more direct way.

Theorem 3. Consider a Lagrangian system whose configuration is defined by the set of generalized coordinates $\mathbf{q} \equiv q_1, q_2, \ldots, q_f$ and whose behavior is defined by the Hamiltonian function $H(\mathbf{q},\mathbf{p},t)$. The behavior of the system can be determined from the equations

$$\dot{q}_i = \frac{\partial H(\mathbf{q},\mathbf{p},t)}{\partial p_i} \qquad (8a)$$

$$\dot{p}_i = -\frac{\partial H(\mathbf{q},\mathbf{p},t)}{\partial q_i} \qquad (8b)$$

These equations are called *Hamilton's equations of motion*.

PROOF: Equation (8a) was derived in the proof of Theorem 1. To obtain Eq. (8b) we note from the definitions of the Hamiltonian, Eq. (2), that

$$\frac{\partial H(\mathbf{q},\mathbf{p},t)}{\partial q_j} = \sum_i p_i \frac{\partial \dot{q}_i(\mathbf{q},\mathbf{p},t)}{\partial q_j} - \frac{\partial L(\mathbf{q},\dot{\mathbf{q}},t)}{\partial q_j} - \sum_i \frac{\partial L(\mathbf{q},\dot{\mathbf{q}},t)}{\partial \dot{q}_i} \frac{\partial \dot{q}_i(\mathbf{q},\mathbf{p},t)}{\partial q_j} \qquad (9)$$

Substituting Eq. (1) in Eq. (9) we obtain

$$\frac{\partial H(\mathbf{q},\mathbf{p},t)}{\partial q_j} = -\frac{\partial L(\mathbf{q},\dot{\mathbf{q}},t)}{\partial q_j} \qquad (10)$$

Making use of Lagrange's equation of motion, Eq. (7), in Eq. (10), we obtain

$$\frac{\partial H(\mathbf{q},\mathbf{p},t)}{\partial q_j} = -\frac{d}{dt}\left[\frac{\partial L(\mathbf{q},\dot{\mathbf{q}},t)}{\partial \dot{q}_i}\right] \qquad (11)$$

Substituting Eq. (1) in Eq. (11) we obtain Eq. (8b). Equations (8a) and (8b) provide us with $2f$ equations in the $2f$ unknowns $q_1, q_2, \ldots, q_f, p_1, p_2, \ldots, p_f$. Hence they, together with the initial conditions, enable us to solve for $\mathbf{p}(t)$ and $\mathbf{q}(t)$, and are thus sufficient to determine the behavior of the system. This completes the proof of the theorem.

NOTE: The equation obtained when $H(\mathbf{q},\mathbf{p},t)$ is substituted in Eq. (8a) and the partial derivative carried out is essentially the same equation one obtains when $L(\mathbf{q},\dot{\mathbf{q}},t)$ is substituted in Eq. (1) and the partial derivative carried out. Hence Eq. (8a) is nothing more than the definition of p_i. If we are not initially given the Lagrangian function $L(\mathbf{q},\dot{\mathbf{q}},t)$, then Eq. (8a) is needed, along with Eq. (8b), to obtain $\mathbf{q}(t)$. But if we know the Lagrangian function $L(\mathbf{q},\dot{\mathbf{q}},t)$ we can obtain the same equation from the definition of p_i.

THE IMPORTANCE OF THE HAMILTONIAN APPROACH

There is in general no advantage to be obtained in the solution of a specific well-defined problem by using the Hamiltonian approach rather than the Lagrangian approach. To see this, we note that to obtain the Hamiltonian function $H(\mathbf{q},\mathbf{p},t)$ we obtain the Lagrangian function $L(\mathbf{q},\dot{\mathbf{q}},t)$, and then make use of the definitions $p_i \equiv \partial L(\mathbf{q},\dot{\mathbf{q}},t)/\partial \dot{q}_i$ and $H \equiv \sum_i p_i \dot{q}_i - L$. Thus before we can obtain the Hamiltonian function $H(\mathbf{q},\mathbf{p},t)$ we need to find the Lagrangian function $L(\mathbf{q},\dot{\mathbf{q}},t)$. Furthermore, if we substitute $H(\mathbf{q},\mathbf{p},t)$ in Hamilton's equations of motion, Eqs. (8a) and (8b), and carry out the indicated derivatives, we will find that the first equation is nothing more than the equation we obtained from the definition $p_i \equiv \partial L(\mathbf{q},\dot{\mathbf{q}},t)/\partial \dot{q}_i$, and when this definition is substituted in the second equation the resulting equation is simply Lagrange's equation of motion. Thus after considerable effort we obtain the same equation we could have obtained directly by using the Lagrangian function $L(\mathbf{q},\dot{\mathbf{q}},t)$ in Lagrange's equation of motion.

Despite the foregoing disclaimer, the Hamiltonian approach is of tremendous importance in physics, not because it provides us directly and immediately with a new technique for solving problems, but because by offering a new formulation of the laws of dynamics it opens doors and reveals new directions in which one can proceed in the quest to understand the consequences of Newton's laws of motion. These further developments are considered in detail in Volume 2.

CONSTANTS OF THE MOTION IN THE HAMILTONIAN FORMULATION

In Chapter 33 we saw that solving the equations of motions for a dynamical system is equivalent to finding the independent constants of the motion. Each new formulation of the laws of dynamics tends to provide a different perspective that frequently is useful in finding constants of the motion. The following two theorems, emanating from Hamilton's equations of motion, point out some possible constants of the motion.

Theorem 4. If the Hamiltonian function, $H(\mathbf{q},\mathbf{p},t)$, is not an explicit function of the time, the Hamiltonian H is a constant of the motion.

PROOF: The time rate of change of H is given by

$$\frac{dH}{dt} = \sum_i \frac{\partial H(\mathbf{q},\mathbf{p},t)}{\partial q_i} \dot{q}_i + \sum_i \frac{\partial H(\mathbf{q},\mathbf{p},t)}{\partial p_i} \dot{p}_i + \frac{\partial H(\mathbf{q},\mathbf{p},t)}{\partial t} \tag{12}$$

Substituting Eqs. (8a) and (8b) in Eq. (9) we obtain

$$\frac{dH}{dt} = \frac{\partial H(\mathbf{q},\mathbf{p},t)}{\partial t} \tag{13}$$

If $H(\mathbf{q},\mathbf{p},t)$ is not an explicit function of the time then $\partial H(\mathbf{q},\mathbf{p},t)/\partial t = 0$, hence $dH/dt = 0$. The theorem follows immediately.

Theorem 5. If the Hamiltonian function $H(\mathbf{q},\mathbf{p},t)$ is not an explicit function of a particular generalized coordinate q_i, the conjugate generalized momentum p_i is a constant of the motion. If the Hamiltonian function $H(\mathbf{q},\mathbf{p},t)$ is not an explicit function of a particular generalized momentum p_i, the conjugate generalized coordinate q_i is a constant of the motion.

PROOF: The proof of this theorem follows immediately from Hamilton's equations of motion, Eqs. (8a) and (8b).

PROBLEMS

1. Determine, by means of Hamilton's equations, the path of a projectile of mass m that is fired at speed v_0 at an angle α with the horizontal.

2. A uniform bar of mass m and length $2a$ is suspended from one end by a spring of spring constant k. The spring is constrained to remain vertical and the bar is constrained to remain in a vertical plane. Choose a set of generalized coordinates and obtain Lagrange's equations of motion and Hamilton's equations of motion for the system. Show that they both reduce to the same set of equations of motion.

3. A particle of mass m is constrained to move on a smooth sphere of radius a and is acted on by gravity. Choose as generalized coordinates the spherical coordinates θ and ϕ, with the polar axis vertically up. Show that the Hamiltonian function for the particle is given by $H = [(p_\theta^2 + p_\phi^2 \csc^2\theta)/(2ma^2)] + mga\cos\theta$. Obtain Hamilton's equations of motion and show that $a\ddot{\theta} = [(h^2\cot\theta\csc^2\theta)/(m^2a^3)] + g\sin\theta$, where h is the constant value of p_ϕ.

4. A uniform circular wire of radius a and mass $2m$ is free to rotate about a fixed vertical diameter. A smooth bead of mass m is threaded on the wire. Show that the Hamiltonian for the system is $H = (2ma^2)^{-1}\{p_\theta^2 + [p_\phi^2/(1 + \sin^2\theta)]\} - mga\cos\theta$, where ϕ is the angle through which the wire has rotated and θ is the angle made by the radius to the bead with the downward vertical. Obtain Hamilton's equations and derive from these the equations $\dot{\phi}(1 + \sin^2\theta) = $ constant and $\ddot{\theta} - \dot{\phi}^2\sin\theta\cos\theta = -g\sin\theta/a$. Show that steady motion is possible with $\dot{\phi} = \omega$, $\theta = \alpha$, and $a\omega^2 = g\sec\alpha$.

5. A particle of mass m is constrained to remain on a smooth wire. The configuration of the wire is changing with time and given in cylindrical coordinates by the equations $\rho = a$, $z = bt + c\phi$. Gravity is acting in the

negative z direction. Choose a suitable set of generalized coordinates. Obtain the Hamiltonian function for the chosen generalized coordinates, and determine the corresponding equations of motion.

6. A particle in a uniform gravitational field is constrained to the surface of a sphere. The radius r of the sphere varies in time, that is, $r = r(t)$. Choosing the spherical coordinates θ and ϕ as generalized coordinates with the origin at the center of the sphere and the polar axis vertically up, obtain the Hamiltonian function and the equations of motion. Is the Hamiltonian the total energy? Discuss energy conservation.

7. A mass m is suspended by means of a string that passes through a small hole in a table. The particle is set in motion in a vertical plane and the string is drawn up through the hole at a constant rate α.
    a. Choosing the angle θ that the string makes with the vertical as the generalized coordinate, find the Hamiltonian function for the system.
    b. Is the Hamiltonian equal to the total energy?
    c. Is the Hamiltonian function a constant of the motion?

8. A particle of mass m can slide freely along a rod AB. A rod OC of length a is attached at right angle to the rod AB at a point C. The two rods are constrained to remain in a vertical plane and are made to rotate in this plane at a constant angular speed ω about the point O, which is fixed. Let s be the distance of the particle from the point C, and assume that when $t = 0$ then $s = \dot{s} = 0$ and OC is horizontal. Choosing s as the generalized coordinate, obtain and solve Hamilton's equations of motion.

9. A particle of mass m moves in three dimensions under the action of a conservative force with potential energy $V(r)$. Using the spherical coordinates r, θ, and ϕ as generalized coordinates, obtain the Hamiltonian function for the system. Show that the quantities p_ϕ, $(p_r^2/2m) + (p_\theta^2/2mr^2) + (p_\phi^2/2mr^2\sin^2\theta) + V(r)$, and $p_\theta^2 + (p_\phi^2/\sin^2\theta)$ are constants of the motion.

10. Two particles of masses m and M interact. The potential energy of interaction is $V(r)$, where r is the distance between the two particles. Choosing the Cartesian coordinates X, Y, Z of the center of mass of the system, and the spherical coordinates r, θ, ϕ, which locate the particle m with respect to the particle M, find the Hamiltonian function and obtain six constants of the motion.

SUPPLEMENTARY MATERIAL

Appendices

APPENDIX 1

Analytical Representation of a Sine Function

Given an arbitrary sine function, we can represent it in any one of the following equivalent ways:

$$y = a\cos(bx - c)$$
$$y = a'\cos(bx) + a''\sin(bx)$$
$$y = Re[\bar{a}\exp(-ibx)]$$
$$y = Re\{a\exp[-i(bx - c)]\}$$

where a, b, c, a', and a'' are real constants and $\bar{a}$ is a complex constant. Equating the expressions above it is easy to show:

$$\bar{a} = a' + ia''$$
$$\bar{a} = a\exp(ic)$$
$$a' = a\cos c$$
$$a'' = a\sin c$$
$$a = (\bar{a}\bar{a}^*)^{1/2}$$
$$a = [(a')^2 + (a'')^2]^{1/2}$$
$$c = \tan^{-1}\left(\frac{a''}{a'}\right)$$

APPENDIX 2

Partial Differentiation

If f is a function of the set of independent variables $\mathbf{x}, y \equiv x_1, x_2, \ldots, x_n, y$, the quantity that is obtained by taking the derivative of the function $f(\mathbf{x}, y)$ with respect to the variable y while holding the set of variables $\mathbf{x}$ fixed is called the partial derivative of the function $f(\mathbf{x}, y)$ with respect to y and is designated $\partial f(\mathbf{x}, y)/\partial y$ or $(\partial f/\partial y)_\mathbf{x}$. The value of a partial derivative depends not only on the variable with respect to which the derivative is taken, but also on the variables that are held constant. Suppose for example that the quantity f can be expressed as a function of the set of variables $\mathbf{x}, z \equiv x_1, x_2, \ldots, x_n, z$ or as a function of the set of variables $\mathbf{y}, z \equiv y_1, y_2, \ldots, y_n, z$; then the partial derivatives $\partial f(\mathbf{x}, z)/\partial z$ and $\partial f(\mathbf{y}, z)/\partial z$ are different quantities. It is therefore a dangerous practice to abbreviate the partial derivative $\partial f(\mathbf{x}, z)/\partial z$ as $\partial f/\partial z$, unless there is no chance of ambiguity. Failure to observe this caution can be a source of numerous and annoying errors.

We now consider four very useful theorems.

Theorem 1. If $f_1, f_2, \ldots, f_{n+1}$ and $g_1, g_2, \ldots, g_{n+1}$ are functions of the set of independent variables $\mathbf{x} \equiv x_1, x_2, \ldots, x_n$, and if these functions satisfy the relation

$$\sum_i f_i \, dg_i = 0 \tag{1}$$

it follows that

$$\sum_i f_i \frac{\partial g_i(\mathbf{x})}{\partial x_j} = 0 \tag{2}$$

PROOF: We note first that

$$dg_i = \sum_j \frac{\partial g_i(\mathbf{x})}{\partial x_j} \, dx_j \tag{3}$$

Substituting Eq. (3) in Eq. (1) we obtain

$$\sum_i \sum_j f_i \frac{\partial g_i(\mathbf{x})}{\partial x_j} dx_j = 0 \tag{4}$$

Since the x_j are independent variables, Eq. (4) must be true for arbitrary values of dx_j. Setting the coefficients of dx_j in Eq. (4) equal to zero, we obtain Eq. (2).

Theorem 2. If x is a function of the set of independent variables $y, \mathbf{z} \equiv y, z_1, z_2, \ldots, z_n$, then

$$\frac{\partial x(y, \mathbf{z})}{\partial y} = \left[\frac{\partial y(x, \mathbf{z})}{\partial x} \right]^{-1} \tag{5}$$

PROOF: We note first that

$$dx = \frac{\partial x(y, \mathbf{z})}{\partial y} dy + \sum_i \frac{\partial x(y, \mathbf{z})}{\partial z_i} dz_i \tag{6}$$

From Theorem 1 it follows that

$$\frac{\partial x(x, \mathbf{z})}{\partial x} = \frac{\partial x(y, \mathbf{z})}{\partial y} \frac{\partial y(x, \mathbf{z})}{\partial x} + \sum_i \frac{\partial x(y, \mathbf{z})}{\partial z_i} \frac{\partial z_i(x, \mathbf{z})}{\partial x} \tag{7}$$

The term on the left-hand side of Eq. (7) is equal to unity, and the partial derivatives $\partial z_i(x, \mathbf{z})/\partial x$ are all zero. Thus we can rewrite Eq. (7)

$$1 = \frac{\partial x(y, \mathbf{z})}{\partial y} \frac{\partial y(x, \mathbf{z})}{\partial x} \tag{8}$$

Equation (5) follows immediately from Eq. (8).

Theorem 3. If x is a function of the set of independent variables $y, z, \mathbf{w} \equiv y, z, w_1, w_2, \ldots, w_n$ then

$$\frac{\partial x(y, z, \mathbf{w})}{\partial y} = - \frac{\partial x(y, z, \mathbf{w})}{\partial z} \frac{\partial z(x, y, \mathbf{w})}{\partial y} \tag{9}$$

PROOF: We note first that

$$dx = \frac{\partial x(y, z, \mathbf{w})}{\partial y} dy + \frac{\partial x(y, z, \mathbf{w})}{\partial z} dz + \sum_i \frac{\partial x(y, z, \mathbf{w})}{\partial w_i} dw_i \tag{10}$$

From Theorem 1 it follows that

$$\frac{\partial x(x, y, \mathbf{w})}{\partial y} = \frac{\partial x(y, z, \mathbf{w})}{\partial y} \frac{\partial y(x, y, \mathbf{w})}{\partial y} + \frac{\partial x(y, z, \mathbf{w})}{\partial z} \frac{\partial z(x, y, \mathbf{w})}{\partial y}$$

$$+ \sum_i \frac{\partial x(y, z, \mathbf{w})}{\partial w_i} \frac{\partial w_i(x, y, \mathbf{w})}{\partial y} \tag{11}$$

Noting that $\partial x(x, y, \mathbf{w})/\partial y = 0$, $\partial y(x, y, \mathbf{w})/\partial y = 1$, and $\partial w_i(x, y, \mathbf{w})/\partial y = 0$, we obtain

$$0 = \frac{\partial x(y, z, \mathbf{w})}{\partial y} + \frac{\partial x(y, z, \mathbf{w})}{\partial z} \frac{\partial z(x, y, \mathbf{w})}{\partial y} \tag{12}$$

Equation (9) follows immediately from Eq. (12).

Theorem 4. If x and y are functions of the set of independent variables z, $\mathbf{w} \equiv z, w_1, w_2, \ldots, w_n$, then

$$\frac{\partial x(y, \mathbf{w})}{\partial y} = \frac{\partial x(z, \mathbf{w})/\partial z}{\partial y(z, \mathbf{w})/\partial z} \tag{13}$$

PROOF: We note first that

$$dx = \frac{\partial x(z, \mathbf{w})}{\partial z} dz + \sum_i \frac{\partial x(z, \mathbf{w})}{\partial w_i} dw_i \tag{14}$$

From Theorem 1 it follows that

$$\frac{\partial x(y, \mathbf{w})}{\partial y} = \frac{\partial x(z, \mathbf{w})}{\partial z} \frac{\partial z(y, \mathbf{w})}{\partial y} + \sum_i \frac{\partial x(z, \mathbf{w})}{\partial w_i} \frac{\partial w_i(y, \mathbf{w})}{\partial y}$$

Noting that $\partial w_i(y, \mathbf{w})/\partial y = 0$ and from Theorem 2

$$\frac{\partial z(y, \mathbf{w})}{\partial y} = \left[\frac{\partial y(z, \mathbf{w})}{\partial z}\right]^{-1}$$

we obtain Eq. (13).

APPENDIX 3

Jacobians

In situations involving two or more independent variables, the manipulation of the various partial derivatives that arise often can be handled best by employing Jacobians.

Suppose the set of quantities $\mathbf{y} \equiv y_1, y_2, \ldots, y_n$ are functions of the set of independent variables $\mathbf{x} \equiv x_1, x_2, \ldots, x_n$. The Jacobian $J(y_1, y_2, \ldots, y_n/x_1, x_2, \ldots, x_n) \equiv J(\mathbf{y}/\mathbf{x})$ is then defined as follows

$$J\left(\frac{\mathbf{y}}{\mathbf{x}}\right) \equiv \begin{vmatrix} \dfrac{\partial y_1(\mathbf{x})}{\partial x_1} & \dfrac{\partial y_1(\mathbf{x})}{\partial x_2} & \cdots & \dfrac{\partial y_1(\mathbf{x})}{\partial x_n} \\ \vdots & \vdots & \vdots & \vdots \\ \dfrac{\partial y_n(\mathbf{x})}{\partial x_1} & \dfrac{\partial y_n(\mathbf{x})}{\partial x_2} & \cdots & \dfrac{\partial y_n(\mathbf{x})}{\partial x_n} \end{vmatrix}$$

A number of theorems involving Jacobians can be derived by using the properties of determinants and the results of Appendix 2. We simply state them here without proof.

Theorem 1.

$$\frac{\partial x(y, \mathbf{z})}{\partial y} = J\left(\frac{x, \mathbf{z}}{y, \mathbf{z}}\right)$$

Theorem 2.

$$J\left(\frac{\mathbf{y}}{\mathbf{x}}\right) = \left[J\left(\frac{\mathbf{x}}{\mathbf{y}}\right)\right]^{-1}$$

Theorem 3.

$$J\left(\frac{\mathbf{y}}{\mathbf{x}}\right) = J\left(\frac{\mathbf{y}}{\mathbf{z}}\right) J\left(\frac{\mathbf{z}}{\mathbf{x}}\right)$$

APPENDIX 4

Vector Algebra

VECTORS

A scalar property of an object is a property that can be represented by a single number. A person's mass is an example of a scalar property.

A vector property of an object is a property that can be represented geometrically by a directed line segment (Fig. 1). The speed and direction of a moving point can for example be represented by the length and direction, respectively, of a line segment; therefore these two properties taken together constitute a vector property, which we call the velocity of the point.

We refer to the directed line segment that is used to represent a particular vector property of an object as simply a vector, and we designate it by a boldface letter (e.g., **a**). We designate the magnitude of the vector **a** as a. Stating this result formally we have

Fig. 1

Definition 1. A vector is a directed line segment.

To exploit the concept of a vector, it is helpful to introduce a number of additional definitions.

Definition 2. Two vectors **a** and **b** are equal if and only if their directions and magnitudes are equal.

Definition 3. The negative of a vector **a**, which we designate as −**a**, is the vector whose magnitude is the same as the magnitude of **a** but whose direction is exactly opposite.

Definition 4. The zero vector, which we designate as **0**, is a vector of zero magnitude.

Definition 5. A unit vector, which we designate as **e**, is a vector of unit magnitude.

Definition 6. If two vectors **a** and **b** are joined by placing the tail of vector **b** on the head of vector **a**, then the vector obtained by drawing a line from the tail of **a** to the head of **b** is called the sum of the vectors **a** and **b** and is designated as **a** + **b** (Fig. 2).

Fig. 2

Definition 7. The product of a scalar h and a vector **a**, which we designate h**a**, is a vector whose direction is the same as that of **a** but whose magnitude is equal to h times the magnitude of **a**.

Definition 8. The scalar or dot product of two vectors **a** and **b**, which we designate as **a** · **b**, is the scalar obtained by taking the product of the magnitude of **a**, the magnitude of **b**, and the cosine of the angle between **a** and **b**. That is,

$$\mathbf{a} \cdot \mathbf{b} = ab \cos \theta \tag{1}$$

where θ is the angle between **a** and **b**.

Definition 9. The vector or cross product of two vectors **a** and **b**, which we designate as **a** × **b**, is the vector whose magnitude is equal to the product of the magnitude of **a**, the magnitude of **b**, and the sine of the smallest angle θ between **a** and **b**, and whose direction is oriented perpendicular to the directions of both **a** and **b** and is such that when the tails of **a**, **b**, and **a** × **b** are joined, **a**, **b**, and **a** × **b**, in that order, form a right handed system; that is, if the fingers of the right hand are pointed in the direction of **a** and then rotated through the smallest angle θ to the direction **b**, the thumb when extended at right angles to the fingers will give the sense of **a** × **b**. We can express this mathematically as follows:

$$\mathbf{a} \times \mathbf{b} = ab \sin \theta \, \mathbf{e} \tag{2}$$

where θ is the angle between **a** and **b**, and **e** is a unit vector normal to **a** and **b** and in sense is such that **a**, **b**, and **e**, in that order, form a right handed system.

From the definitions above it follows that:

$$\mathbf{a} + \mathbf{b} = \mathbf{b} + \mathbf{a} \tag{3}$$

$$\mathbf{a} + (\mathbf{b} + \mathbf{c}) = (\mathbf{a} + \mathbf{b}) + \mathbf{c} \tag{4}$$

$$\mathbf{a} + (-\mathbf{a}) \equiv \mathbf{a} - \mathbf{a} = \mathbf{0} \tag{5}$$

$$\mathbf{a} + \mathbf{0} = \mathbf{a} \tag{6}$$

$$(h_1 h_2)\mathbf{a} = h_1(h_2 \mathbf{a}) \tag{7}$$

$$(h_1 + h_2)\mathbf{a} = h_1 \mathbf{a} + h_2 \mathbf{a} \tag{8}$$

$$h(\mathbf{a} + \mathbf{b}) = h\mathbf{a} + h\mathbf{b} \tag{9}$$

$$\mathbf{a} \cdot \mathbf{b} = \mathbf{b} \cdot \mathbf{a} \tag{10}$$

$$\mathbf{a} \cdot (\mathbf{b} + \mathbf{c}) = \mathbf{a} \cdot \mathbf{b} + \mathbf{a} \cdot \mathbf{c} \tag{11}$$

$$\mathbf{a} \cdot (h\mathbf{b}) = (h\mathbf{a}) \cdot \mathbf{b} = h(\mathbf{a} \cdot \mathbf{b}) \tag{12}$$

$$\mathbf{a} \cdot \mathbf{a} = a^2 \tag{13}$$

$$\mathbf{a} \times \mathbf{b} = -\mathbf{b} \times \mathbf{a} \tag{14}$$

$$\mathbf{a} \times (\mathbf{b} + \mathbf{c}) = \mathbf{a} \times \mathbf{b} + \mathbf{a} \times \mathbf{c} \tag{15}$$

$$\mathbf{a} \times (h\mathbf{b}) = (h\mathbf{a}) \times \mathbf{b} = h(\mathbf{a} \times \mathbf{b}) \tag{16}$$

$$\mathbf{a} \times \mathbf{a} = \mathbf{0} \tag{17}$$

NOTE 1: Vector multiplication is not commutative, that is, $\mathbf{a} \times \mathbf{b} \neq \mathbf{b} \times \mathbf{a}$.

NOTE 2: Vector multiplication is not associative, that is, $\mathbf{a} \times (\mathbf{b} \times \mathbf{c}) \neq (\mathbf{a} \times \mathbf{b}) \times \mathbf{c}$.

THE COMPONENTS OF A VECTOR

Let us consider a set of rectangular Cartesian axes 1, 2, and 3. Any vector **a** can be decomposed into the sum of three vectors, one of which is in the 1 direction, one of which is in the 2 direction, and one of which is in the 3 direction. We designate these vectors as $\mathbf{a}_1$, $\mathbf{a}_2$, and $\mathbf{a}_3$, respectively. It follows that

$$\mathbf{a} = \mathbf{a}_1 + \mathbf{a}_2 + \mathbf{a}_3 \tag{18}$$

If we let $\mathbf{e}_1$, $\mathbf{e}_2$, and $\mathbf{e}_3$ be unit vectors in the 1, 2, and 3 directions, respectively, then

$$\mathbf{a}_1 = a_1 \mathbf{e}_1 \qquad \mathbf{a}_2 = a_2 \mathbf{e}_2 \qquad \mathbf{a}_3 = a_3 \mathbf{e}_3 \tag{19}$$

and therefore

$$\mathbf{a} = a_1 \mathbf{e}_1 + a_2 \mathbf{e}_2 + a_3 \mathbf{e}_3 \tag{20}$$

The scalars a_1, a_2, and a_3 are called the components of **a** in the 1, 2, and 3 directions, respectively.

If we take the scalar product of Eq. (20) with $\mathbf{e}_i$ and note that
$$\mathbf{e}_i \cdot \mathbf{e}_j = \delta_{ij} \tag{21}$$
where δ_{ij} is the Kronecker delta, defined as follows
$$\delta_{ij} = 1 \quad \text{if } i = j$$
$$\delta_{ij} = 0 \quad \text{if } i \neq j \tag{22}$$
we obtain
$$\mathbf{a} \cdot \mathbf{e}_i = a_i \tag{23}$$
But from the definition of the dot product
$$\mathbf{a} \cdot \mathbf{e}_i = a \cos \theta_i \tag{24}$$
where θ_i is the angle between the vector $\mathbf{a}$ and the i axis. It follows that the a_i component of a vector is just the projection of the vector onto the i axis. This result could also have been obtained directly from a geometric consideration of the decomposition of the vector $\mathbf{a}$.

Gathering results, we see that a vector $\mathbf{a}$ can be written in component form in any of the following equivalent ways:
$$\mathbf{a} = a_1 \mathbf{e}_1 + a_2 \mathbf{e}_2 + a_3 \mathbf{e}_3 \tag{25}$$
$$\mathbf{a} = (\mathbf{a} \cdot \mathbf{e}_1)\mathbf{e}_1 + (\mathbf{a} \cdot \mathbf{e}_2)\mathbf{e}_2 + (\mathbf{a} \cdot \mathbf{e}_3)\mathbf{e}_3 \tag{26}$$
$$\mathbf{a} = a \cos \theta_1 \mathbf{e}_1 + a \cos \theta_2 \mathbf{e}_2 + a \cos \theta_3 \mathbf{e}_3 \tag{27}$$
We shall use whichever form is most convenient for the problem at hand.

Decomposition of a vector into its components frequently makes the calculation of sums and products considerably simpler. Thus
$$\mathbf{a} + \mathbf{b} = (a_1 \mathbf{e}_1 + a_2 \mathbf{e}_2 + a_3 \mathbf{e}_3) + (b_1 \mathbf{e}_1 + b_2 \mathbf{e}_2 + b_3 \mathbf{e}_3)$$
$$= (a_1 + b_1)\mathbf{e}_1 + (a_2 + b_2)\mathbf{e}_2 + (a_3 + b_3)\mathbf{e}_3 \tag{28}$$
$$\mathbf{a} \cdot \mathbf{b} = (a_1 \mathbf{e}_1 + a_2 \mathbf{e}_2 + a_3 \mathbf{e}_3) \cdot (b_1 \mathbf{e}_1 + b_2 \mathbf{e}_2 + b_3 \mathbf{e}_3)$$
$$= + a_1 b_1 \mathbf{e}_1 \cdot \mathbf{e}_1 + a_1 b_2 \mathbf{e}_1 \cdot \mathbf{e}_2 + a_1 b_3 \mathbf{e}_1 \cdot \mathbf{e}_3$$
$$+ a_2 b_1 \mathbf{e}_2 \cdot \mathbf{e}_1 + a_2 b_2 \mathbf{e}_2 \cdot \mathbf{e}_2 + a_2 b_3 \mathbf{e}_2 \cdot \mathbf{e}_3$$
$$+ a_3 b_1 \mathbf{e}_3 \cdot \mathbf{e}_1 + a_3 b_2 \mathbf{e}_3 \cdot \mathbf{e}_2 + a_3 b_3 \mathbf{e}_3 \cdot \mathbf{e}_3$$
$$= a_1 b_1 + a_2 b_2 + a_3 b_3 \tag{29}$$
$$\mathbf{a} \times \mathbf{b} = (a_1 \mathbf{e}_1 + a_2 \mathbf{e}_2 + a_3 \mathbf{e}_3) \times (b_1 \mathbf{e}_1 + b_2 \mathbf{e}_2 + b_3 \mathbf{e}_3)$$
$$= + a_1 b_1 \mathbf{e}_1 \times \mathbf{e}_1 + a_1 b_2 \mathbf{e}_1 \times \mathbf{e}_2 + a_1 b_3 \mathbf{e}_1 \times \mathbf{e}_3$$
$$+ a_2 b_1 \mathbf{e}_2 \times \mathbf{e}_1 + a_2 b_2 \mathbf{e}_2 \times \mathbf{e}_2 + a_2 b_3 \mathbf{e}_2 \times \mathbf{e}_3$$
$$+ a_3 b_1 \mathbf{e}_3 \times \mathbf{e}_1 + a_3 b_2 \mathbf{e}_3 \times \mathbf{e}_2 + a_3 b_3 \mathbf{e}_3 \times \mathbf{e}_3$$
$$= +(a_2 b_3 - a_3 b_2)\mathbf{e}_2 \times \mathbf{e}_3$$
$$+ (a_3 b_1 - a_1 b_3)\mathbf{e}_3 \times \mathbf{e}_1$$
$$+ (a_1 b_2 - a_2 b_1)\mathbf{e}_1 \times \mathbf{e}_2 \tag{30}$$

If e_1, e_2, and e_3 form a right handed system (RHS) then

$$e_2 \times e_3 = e_1 \quad \text{(RHS)} \tag{31}$$

$$e_3 \times e_1 = e_2 \quad \text{(RHS)} \tag{32}$$

$$e_1 \times e_2 = e_3 \quad \text{(RHS)} \tag{33}$$

If e_1, e_2, and e_3 form a left handed system (LHS) then

$$e_2 \times e_3 = -e_1 \quad \text{(LHS)} \tag{34}$$

$$e_3 \times e_1 = -e_2 \quad \text{(LHS)} \tag{35}$$

$$e_1 \times e_2 = -e_3 \quad \text{(LHS)} \tag{36}$$

It follows that

$$\mathbf{a} \times \mathbf{b} = (a_2 b_3 - a_3 b_2) e_1 + (a_3 b_1 - a_1 b_3) e_2 + (a_1 b_2 - a_2 b_1) e_3 \quad \text{(RHS)} \tag{37}$$

$$\mathbf{a} \times \mathbf{b} = -(a_2 b_3 - a_3 b_2) e_1 - (a_3 b_1 - a_1 b_3) e_2 - (a_1 b_2 - a_2 b_1) e_3 \quad \text{(LHS)} \tag{38}$$

It thus follows that the cross product as we have defined it cannot be expressed unambiguously in terms of its components unless we restrict ourselves to right or left handed systems. To avoid this difficulty it is frequently customary to redefine the cross product of the vectors $\mathbf{a}$ and $\mathbf{b}$ as simply the quantity

$$\mathbf{a} \times \mathbf{b} = (a_2 b_3 - a_3 b_2) e_1 + (a_3 b_1 - a_1 b_3) e_2 + (a_1 b_2 - a_2 b_1) e_3 \tag{39}$$

When this is done the cross product is no longer a unique vector but one of two possible vectors. Which vector we end up with depends on whether our coordinate system is a right or left handed system. A cross product that is so defined is referred to as an axial vector rather than simply as a vector.

As long as one always uses right handed coordinate systems, the two definitions can be freely interchanged. Since we always use right handed coordinate systems, the distinction is now dropped.

A convenient way of expressing the cross product is as follows:

$$\mathbf{a} \times \mathbf{b} = \begin{vmatrix} e_1 & e_2 & e_3 \\ a_1 & a_2 & a_3 \\ b_1 & b_2 & b_3 \end{vmatrix} \tag{40}$$

The following notation is frequently used to express a vector in component form:

$$a_1 e_1 + a_2 e_2 + a_3 e_3 \equiv (a_1, a_2, a_3) \tag{41}$$

In this notation the sum, scalar product, and vector product become

$$(a_1, a_2, a_3) + (b_1, b_2, b_3) = (a_1 + b_1, a_2 + b_2, a_3 + b_3) \tag{42}$$

$$(a_1, a_2, a_3) \cdot (b_1, b_2, b_3) = a_1 b_1 + a_2 b_2 + a_3 b_3 \tag{43}$$

$$(a_1, a_2, a_3) \times (b_1, b_2, b_3) = (a_2 b_3 - a_3 b_2, a_3 b_1 - a_1 b_3, a_1 b_2 - a_2 b_1) \tag{44}$$

When speaking of the vector whose components are a_1, a_2, a_3 we usually abbreviate the statement by simply saying the vector a_i. With this abbreviation the sum of a vector a_i and a vector b_i is simply a vector c_i, where

$$c_i = a_i + b_i \tag{45}$$

The dot product of a vector a_i and a vector b_i is the scalar c, where

$$c = \sum_i a_i b_i \tag{46}$$

To obtain the corresponding expression for the cross product we define a permutation symbol ϵ_{ijk}, called the Levi-Civita symbol, as follows

$$\begin{aligned} \epsilon_{ijk} &= +1 \quad \text{when} \quad ijk \text{ is an even permutation of 123} \\ &= -1 \quad \text{when} \quad ijk \text{ is an odd permutation of 123} \\ &= 0 \quad \text{otherwise} \end{aligned} \tag{47}$$

In terms of the Levi-Civita symbol the cross product of a vector a_i and a vector b_i is the vector c_i, where

$$c_i = \sum_j \sum_k \epsilon_{ijk} a_j b_k \tag{48}$$

APPENDIX 5

Vector Calculus

DERIVATIVE OF A VECTOR

If with every value of a variable x there is associated a vector $\mathbf{u}$, then $\mathbf{u}$ is said to be a *vector function* of x, and we write

$$\mathbf{u} = \mathbf{u}(x) \tag{1}$$

Definition 1. The derivative of a vector function $\mathbf{u}(x)$ with respect to the variable x is defined as follows:

$$\frac{d\mathbf{u}(x)}{dx} = \lim_{\Delta x \to 0} \frac{\mathbf{u}(x + \Delta x) - \mathbf{u}(x)}{\Delta x} \tag{2}$$

Theorem 1. If $\phi(x)$ is a scalar function of x, and $\mathbf{u}(x)$ and $\mathbf{v}(x)$ are vector functions of x, then

$$\frac{d}{dx}[\mathbf{u} + \mathbf{v}] = \frac{d\mathbf{u}}{dx} + \frac{d\mathbf{v}}{dx} \tag{3}$$

$$\frac{d}{dx}[\phi \mathbf{u}] = \frac{d\phi}{dx}\mathbf{u} + \phi \frac{d\mathbf{u}}{dx} \tag{4}$$

$$\frac{d}{dx}[\mathbf{u} \cdot \mathbf{v}] = \frac{d\mathbf{u}}{dx} \cdot \mathbf{v} + \mathbf{u} \cdot \frac{d\mathbf{v}}{dx} \tag{5}$$

$$\frac{d}{dx}[\mathbf{u} \times \mathbf{v}] = \frac{d\mathbf{u}}{dx} \times \mathbf{v} + \mathbf{u} \times \frac{d\mathbf{v}}{dx} \tag{6}$$

PROOF: The proof of the relations above follows directly from Definition 1 and is left to the reader.

THE GRADIENT, DIVERGENCE, AND CURL

If with every point (x_1, x_2, x_3) in a region of space R there is associated a scalar ϕ, we say that the function $\phi(x_1, x_2, x_3)$ defines a *scalar field*. If with every point (x_1, x_2, x_3) in a region R there is associated a vector $\mathbf{u}$, we say that the function $\mathbf{u}(x_1, x_2, x_3)$ defines a *vector field*.

Definition 2. The gradient of the scalar field $\phi(x_1, x_2, x_3)$ is the vector field defined as follows:

$$\text{grad}\,\phi = \mathbf{e}_1 \frac{\partial \phi}{\partial x_1} + \mathbf{e}_2 \frac{\partial \phi}{\partial x_2} + \mathbf{e}_3 \frac{\partial \phi}{\partial x_3} \tag{7}$$

Definition 3. The divergence of the vector field $\mathbf{u}(x_1, x_2, x_3)$ is the scalar field defined as follows:

$$\text{div}\,\mathbf{u} \equiv \frac{\partial \phi_1}{\partial x_1} + \frac{\partial \phi_2}{\partial x_2} + \frac{\partial \phi_3}{\partial x_3} \tag{8}$$

Definition 4. The curl of the vector field $\mathbf{u}(x_1 x_2, x_3)$ is the vector field defined as follows:

$$\text{curl}\,\mathbf{u} \equiv \mathbf{e}_1 \left(\frac{\partial u_3}{\partial x_2} - \frac{\partial u_2}{\partial x_3} \right) + \mathbf{e}_2 \left(\frac{\partial u_1}{\partial x_3} - \frac{\partial u_3}{\partial x_1} \right) + \mathbf{e}_3 \left(\frac{\partial u_2}{\partial x_1} - \frac{\partial u_1}{\partial x_2} \right) \tag{9}$$

THE DEL OPERATOR

If we define an operator ∇, called del, as follows:

$$\nabla \equiv \mathbf{e}_1 \frac{\partial}{\partial x_1} + \mathbf{e}_2 \frac{\partial}{\partial x_2} + \mathbf{e}_3 \frac{\partial}{\partial x_3} \tag{10}$$

then

$$\text{grad}\,\phi \equiv \nabla \phi \tag{11}$$

$$\text{div}\,\mathbf{u} \equiv \nabla \cdot \mathbf{u} \tag{12}$$

$$\text{curl}\,\mathbf{u} \equiv \nabla \times \mathbf{u} \tag{13}$$

VECTOR INTEGRATION

There are three general types of integral encountered in vector calculus: line integrals, surface integrals, and volume integrals. To define each of these, let us let $\phi(x_1, x_2, x_3)$ be an arbitrary scalar field and $\mathbf{u}(x_1, x_2, x_3)$ an arbitrary vector field in a given region R of space. We can then make the following definitions:

Definition 5. Line Integrals. Let L be a particular curved line joining two points in the region R. If we divide the line into N equal segments, each of length ΔL, which we label 1 to N, and if we let $P_i \equiv [x_1(i), x_2(i), x_3(i)]$ be a point on the ith segment and $\mathbf{t}(P_i)$ be the unit tangent vector to the line at this point, we can define the following integrals along the line L.

$$\int_L \phi \, d\mathbf{L} \equiv \lim_{N \to \infty} \sum_{i=1}^N \phi(P_i) \mathbf{t}(P_i) \Delta L \tag{14}$$

$$\int_L \mathbf{u} \cdot d\mathbf{L} \equiv \lim_{N \to \infty} \sum_{i=1}^\infty \mathbf{u}(P_i) \cdot \mathbf{t}(P_i) \Delta L \tag{15}$$

$$\int_L \mathbf{u} \times d\mathbf{L} \equiv \lim_{N \to \infty} \sum_{i=1}^\infty \mathbf{u}(P_i) \times \mathbf{t}(P_i) \Delta L \tag{16}$$

Definition 6. Surface Integrals. Let S be a particular surface within the region R. If we divide the surface into N equal segments, each of area ΔS, which we label 1 to N, and if we let $P_i \equiv [x_1(i), x_2(i), x_3(i)]$ be a point on the ith segment and $\mathbf{n}(P_i)$ be the unit normal to the surface at this point, we can define the following integrals over the surface:

$$\int_S \phi \, d\mathbf{S} \equiv \lim_{N \to \infty} \sum_{i=1}^N \phi(P_i) \mathbf{n}(P_i) \Delta S \tag{17}$$

$$\int_S \mathbf{u} \cdot d\mathbf{S} \equiv \lim_{N \to \infty} \sum_{i=1}^N \mathbf{u}(P_i) \cdot \mathbf{n}(P_i) \Delta S \tag{18}$$

$$\int_S \mathbf{u} \times d\mathbf{S} \equiv \lim_{N \to \infty} \sum_{i=1}^N \mathbf{u}(P_i) \times \mathbf{n}(P_i) \Delta S \tag{19}$$

Definition 7. Volume Integrals. Let V be a particular volume within the region R. If we divide the volume into N equal segments, each of volume ΔV, which we label 1 to N, and if we let $P_i \equiv [x_1(i), x_2(i), x_3(i)]$ be a point in the ith segment, we can define the following integrals over the volume:

$$\int_V \phi \, dV = \lim_{N \to \infty} \sum_{i=1}^N \phi(P_i) \Delta V \tag{20}$$

$$\int_V \mathbf{u} \, dV = \lim_{N \to \infty} \sum_{i=1}^N \mathbf{u}(P_i) \Delta V \tag{21}$$

GREEN'S THEOREMS

Suppose that $\phi(x)$ is a single valued continuous and differentiable function in the region $a \leqslant x \leqslant b$, it then follows that

$$\int_a^b \frac{d\phi}{dx} \, dx = \phi(b) - \phi(a) \tag{22}$$

Thus the integral of the function $d\phi/dx$ over the range a to b depends only on the value of $\phi(x)$ at the endpoints. As we shall see, an analogous result holds in two and three dimensions.

Theorem 1. Let $\phi(x_1, x_2)$ be a function defined in a two dimensional region R, and let S be an area in the region R bounded by the closed line L. If $\phi(x_1, x_2)$ is single valued, continuous, and differentiable over the area S, then

$$\int_S \frac{\partial \phi}{\partial x_i} \, dS = \int_L \phi n_i \, dL \tag{23}$$

where $\mathbf{n} \equiv (n_1, n_2)$ is a unit vector normal to the line L and pointing in the outward sense.

PROOF: Consider the area S bounded by the line L (Fig. 1). Let A be the point on L for which x_1 has its minimum value and B the point on L for which x_1 has its maximum value. Let $x_2 = f(x_1)$ be the lower curve and $x_2 = g(x_1)$ the upper curve joining the points A and B. It then follows that

$$\int_S \frac{\partial \phi}{\partial x_2} \, dS = \int \int_S \frac{\partial \phi}{\partial x_2} \, dx_1 \, dx_2$$

$$= \int_a^b \int_{f(x_1)}^{g(x_1)} \frac{\partial \phi}{\partial x_2} \, dx_2 \, dx_1$$

$$= \int_a^b \left[\phi(x_1, g) - \phi(x_1, f) \right] dx_1$$

$$= \int_a^b \phi(x_1, g) \, dx_1 - \int_a^b \phi(x_1, f) \, dx_1 \tag{24}$$

Along the upper curve $dx_1 = n_2 \, dL$, and along the lower curve $dx_1 = -n_2 \, dL$. It follows that

$$\int_S \frac{\partial \phi}{\partial x_2} \, dS = \int_g \phi n_2 \, dL + \int_f \phi n_2 \, dL$$

$$= \int_L \phi n_2 \, dL \tag{25}$$

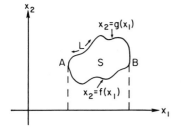

Fig. 1

In a similar fashion we can show

$$\int_S \frac{\partial \phi}{\partial x_1} dS = \int_L \phi n_1 \, dL \qquad (26)$$

This completes the proof of the theorem.

Theorem 2. Let $\phi(x_1, x_2, x_3)$ be a function defined in a three dimensional region R, and let V be a volume in the region R bounded by the closed surface S. If $\phi(x_1, x_2, x_3)$ is single valued, continuous and differentiable in the volume V, then

$$\int_V \frac{\partial \phi}{\partial x_i} dV = \int_S \phi n_i \, dS \qquad (27)$$

where $\mathbf{n} \equiv (n_1, n_2, n_3)$ is a unit vector normal to the surface S and pointing in the outward sense.

PROOF: The proof of this theorem is perfectly analogous to the proof of the preceding theorem and is left to the reader.

We refer to Theorem 1 as the two dimensional Green's theorem and to Theorem 2 as the three dimensional Green's theorem. The form in which these theorems are presented in this appendix is not the usual form found in most textbooks. However for our purposes it is the most useful form.

SOME USEFUL INTEGRAL THEOREMS

Green's theorems enable us to derive a number of useful integral theorems.

Theorem 3. Let $\phi(x_1, x_2, x_3)$ be an arbitrary scalar field and $\mathbf{u}(x_1, x_2, x_3)$ an arbitrary vector field defined in a region R of space. Let V be a volume in R bounded by the closed surface S. If $\phi(x_1, x_2, x_3)$ and $\mathbf{u}(x_1, x_2, x_3)$ are single valued, continuous, and differentiable in the volume V, then

$$\int_V \nabla \cdot \mathbf{u} \, dV = \int_S \mathbf{n} \cdot \mathbf{u} \, dS \qquad (28)$$

$$\int_V \nabla \phi \, dV = \int_S \mathbf{n} \phi \, dS \qquad (29)$$

$$\int_V \nabla \times \mathbf{u} \, dV = \int_S \mathbf{n} \times \mathbf{u} \, dS \qquad (30)$$

where $\mathbf{n}$ is a unit vector normal to the surface element and pointing in the outward sense.

Equation (28), which can be written in the equivalent form

$$\int_V \nabla \cdot \mathbf{u} \, dV = \int_S \mathbf{u} \cdot d\mathbf{S} \qquad (31)$$

is called *Gauss's theorem*.

PROOF: The proof of Eqs. (28)–(30) can be obtained directly from Green's theorem. Thus

$$\int_V \nabla \cdot \mathbf{u}\, dV = \sum_i \int_V \frac{\partial u_i}{\partial x_i}\, dV = \sum_i \int_S u_i n_i\, dS = \int_S \mathbf{n} \cdot \mathbf{u}\, dS \tag{32}$$

$$\int_V \nabla \phi\, dV = \sum_i \mathbf{e}_i \int_V \frac{\partial \phi}{\partial x_i}\, dV = \sum_i \mathbf{e}_i \int_S \phi n_i\, dS = \int_S \mathbf{n}\phi\, dS \tag{33}$$

$$\int_V \nabla \times \mathbf{u}\, dV = \sum_i \sum_j \sum_k \mathbf{e}_i \epsilon_{ijk} \int_V \left[\frac{\partial u_k}{\partial x_j} - \frac{\partial u_j}{\partial x_k} \right] dV$$

$$= \sum_i \sum_j \sum_k \mathbf{e}_i \epsilon_{ijk} \int_S \left[n_j u_k - n_k u_j \right] dS = \int_S \mathbf{n} \times \mathbf{u}\, dS \tag{34}$$

Theorem 4. Let $\phi(x_1, x_2, x_3)$ be an arbitrary scalar field and $\mathbf{u}(x_1, x_2, x_3)$ an arbitrary vector field defined in a region R of space. Let S be a surface in R bounded by a closed line. If $\phi(x_1, x_2, x_3)$ and $\mathbf{u}(x_1, x_2, x_3)$ are single valued, continuous, and differentiable in the volume V, then

$$\int_S (\mathbf{n} \times \nabla) \cdot \mathbf{u}\, dS = \int_L \mathbf{t} \cdot \mathbf{u}\, dL \tag{35}$$

$$\int_S (\mathbf{n} \times \nabla)\phi\, dS = \int_L \mathbf{t}\phi\, dL \tag{36}$$

$$\int_S (\mathbf{n} \times \nabla) \times \mathbf{u}\, dS = \int_L \mathbf{t} \times \mathbf{u}\, dL \tag{37}$$

where $\mathbf{n}$ is a unit vector normal to the surface element dS, and $\mathbf{t}$ is a unit vector tangent to the line element dL. If the positive sense of $\mathbf{t}$ is chosen arbitrarily, the positive sense of $\mathbf{n}$ is the direction in which a right handed screw would move if it were turned in the positive $\mathbf{t}$ sense. Equation (35) can be written in the equivalent form

$$\int_S \nabla \times \mathbf{u} \cdot d\mathbf{S} = \int_L \mathbf{u} \cdot d\mathbf{L} \tag{38}$$

and in this form is called *Stokes' theorem*.

PROOF: If our surface lies in a plane, and if we let $\mathbf{n}'$ be the unit vector normal to the element dL, then from the two dimensional Green's theorem we have

$$\int_S (\mathbf{n} \times \nabla) \cdot \mathbf{u}\, dS = \sum_i \sum_j \sum_k \epsilon_{ijk} \int_S n_j \frac{\partial u_i}{\partial x_k}\, dS$$

$$= \sum_i \sum_j \sum_k \epsilon_{ijk} \int_L n_j u_i n'_k\, dL$$

$$= \int_L \mathbf{u} \cdot (\mathbf{n} \times \mathbf{n}')\, dL$$

$$= \int_L \mathbf{u} \cdot \mathbf{t}\, dL \tag{39}$$

It follows that Eq. (35) is true if the surface lies in a plane. If the surface does not lie in a plane, the surface S can be partitioned into a large number N of small pieces $S_1, S_2, \ldots, S_N$, bounded by the closed lines $L_1, L_2, \ldots, L_N$, respectively. We then note first that

$$\int_S (\mathbf{n} \times \nabla) \cdot \mathbf{u}\, dS = \sum_{i=1}^{N} \int_{S_i} (\mathbf{n} \times \nabla) \cdot \mathbf{u}\, dS_i \tag{40}$$

In the limit as N approaches infinity, the pieces become flat surfaces, hence in this limit

$$\sum_{i=1}^{N} \int_{S_i} (\mathbf{n} \times \nabla) \cdot \mathbf{u}\, dS_i = \sum_{i=1}^{N} \int_{L_i} \mathbf{t} \cdot \mathbf{u}\, dL_i \tag{41}$$

Since the portions of the contours $L_1, L_2, \ldots, L_N$ that lie within the surface S are traversed twice, once in each direction, it follows that the only contribution to the last integral comes from the portions of the contours $L_1, L_2, \ldots, L_N$ that lie on the boundary of the surface S, that is, on the contour L. Hence

$$\sum_{i=1}^{N} \int_{L_i} \mathbf{t} \cdot \mathbf{u}\, dL_i = \int_L \mathbf{t} \cdot \mathbf{u}\, dL \tag{42}$$

If we combine Eqs. (40)–(42) we obtain Eq. (35), which is the desired result. This completes the proof of Eq. (35).

The proof of Eqs. (36) and (37) is perfectly similar to the proof of Eq. (35) and is left to the reader.

INTEGRAL DEFINITIONS OF THE GRADIENT, DIVERGENCE, AND CURL

If we take the limit of Eq. (28) as V approaches zero we obtain

$$\lim_{V \to 0} \int_V \nabla \cdot \mathbf{u}\, dV = \lim_{V \to 0} \int_S \mathbf{n} \cdot \mathbf{u}\, dS \tag{43}$$

The left hand side approaches the value $(\nabla \cdot \mathbf{u})V$; hence

$$\nabla \cdot \mathbf{u} = \lim_{V \to 0} \frac{1}{V} \int_S \mathbf{n} \cdot \mathbf{u}\, dS \tag{44}$$

In a similar fashion we obtain from Eqs. (29) and (30)

$$\nabla \phi = \lim_{V \to 0} \frac{1}{V} \int_S \mathbf{n} \phi\, dS \tag{45}$$

$$\nabla \times \mathbf{u} = \lim_{V \to 0} \frac{1}{V} \int_S \mathbf{n} \times \mathbf{u}\, dS \tag{46}$$

Equations (44), (45), and (46) provide us with alternate definitions of the divergence, gradient, and curl, respectively.

THE GRADIENT

A definition of the gradient was given earlier. The following theorems concerning the gradient should provide the reader not only with some useful properties of the gradient but also with a deeper understanding of the nature and significance of the gradient.

Theorem 5. Let $\phi(x_1, x_2, x_3)$ be a scalar field defined over a region R. The change $d\phi$ in the function $\phi(x_1, x_2, x_3)$ in an infinitesimal displacement $d\mathbf{x} \equiv (dx_1, dx_2, dx_3)$ is given by

$$d\phi = \nabla\phi \cdot d\mathbf{x}$$

PROOF: If we take the dot product of the vector

$$\nabla\phi \equiv \mathbf{e}_1 \frac{\partial \phi}{\partial x_1} + \mathbf{e}_2 \frac{\partial \phi}{\partial x_2} + \mathbf{e}_3 \frac{\partial \phi}{\partial x_3} \tag{47}$$

and the vector

$$d\mathbf{x} \equiv \mathbf{e}_1 dx_1 + \mathbf{e}_2 dx_2 + \mathbf{e}_3 dx_3 \tag{48}$$

We obtain

$$\nabla\phi \cdot d\mathbf{x} = \frac{\partial \phi}{\partial x_1} dx_1 + \frac{\partial \phi}{\partial x_2} dx_2 + \frac{\partial \phi}{\partial x_3} dx_3 = d\phi \tag{49}$$

Theorem 6. Let $\phi(x_1, x_2, x_3)$ be a scalar field defined over a region R. The direction of $\nabla\phi$ at an arbitrary point (x_1, x_2, x_3) is in the direction of the maximum rate of increase of the function $\phi(x_1, x_2, x_3)$ at the point, and the magnitude of $\nabla\phi$ is equal to the rate of increase of the function in this direction.

PROOF: Let $d\mathbf{x}$ be an arbitrary displacement in space. Let $\mathbf{n}$ be a unit vector in the direction of $d\mathbf{x}$ and let dx be the magnitude of $d\mathbf{x}$, that is,

$$d\mathbf{x} \equiv \mathbf{n}\, dx \tag{50}$$

It then follows that

$$d\phi = \nabla\phi \cdot d\mathbf{x} = \nabla\phi \cdot \mathbf{n}\, dx \tag{51}$$

hence

$$\frac{d\phi}{dx} = \nabla\phi \cdot \mathbf{n} \tag{52}$$

The maximum value of $d\phi/dx$ occurs when $\mathbf{n}$ is in the same direction as $\nabla\phi$. Hence $\nabla\phi$ is in the direction of the maximum rate of increase of ϕ. Furthermore when $\mathbf{n}$ is in the same direction as $\nabla\phi$ then $\nabla\phi \cdot \mathbf{n}$ is equal to the magnitude of $\nabla\phi$ and $d\phi/dx$ has its maximum value. Hence the magnitude of $\nabla\phi$ is equal to the maximum value of $d\phi/dx$.

Theorem 7. If $\phi(x_1, x_2, x_3) = c$ defines a surface in space, the normal to the surface at an arbitrary point on the surface is in the same direction as the gradient of ϕ at the point.

PROOF: For a displacement $d\mathbf{x}$ that lies on the surface $\phi(x_1, x_2, x_3) = c$, the change in ϕ will be zero. Hence for such a displacement

$$d\phi = \nabla\phi \cdot d\mathbf{x} = 0 \qquad (53)$$

It follows that $\nabla\phi$ must be perpendicular to $d\mathbf{x}$. Since the only restriction on $d\mathbf{x}$ is that it must lie on the surface, it follows that $\nabla\phi$ must be perpendicular to the surface.

THE DIVERGENCE

A grasp of the geometrical significance of the divergence can be obtained by a consideration of the velocity field of an incompressible fluid. If ρ is the density and $\mathbf{v}$ the velocity of the fluid, the net flow of fluid out of a volume V, which is enclosed by a surface S, is given by

$$\text{net flow out of } V = \int_S \rho \mathbf{v} \cdot d\mathbf{S} \qquad (54)$$

where $d\mathbf{S} \equiv \mathbf{n}\, dS$ and $\mathbf{n}$ is a unit vector normal to the surface and directed out of the volume. Since the fluid is supposed to be incompressible, there can be a net flow out of V only if there is a source within V. The net flow out of V is therefore a measure of the source strength within V. Thus,

$$\text{source strength in } V = \int_S \rho \mathbf{v} \cdot d\mathbf{S} \qquad (55)$$

If we now combine Eq. (55) with the integral definition of the divergence of $\rho\mathbf{v}$ [see Eq. (44)] we have

$$\nabla \cdot \rho\mathbf{v} = \lim_{V \to 0} \frac{1}{V} \int_S \rho \mathbf{v} \cdot d\mathbf{S} = \text{source strength per unit volume} \qquad (56)$$

The divergence of $\rho\mathbf{v}$ at a particular point is thus a measure of the source strength per unit volume at the point.

THE CURL

A grasp of the geometrical significance of the curl can be obtained by considering the motion of a rigid body that is rotating about a fixed point with a constant angular velocity ω. The velocity $\mathbf{v}$ of an arbitrary point $\mathbf{r}$ in the rigid body is given by

$$\mathbf{v} = \boldsymbol{\omega} \times \mathbf{r} \qquad (57)$$

It follows that

$$\nabla \times \mathbf{v} = \nabla \times [\boldsymbol{\omega} \times \mathbf{r}] = 2\boldsymbol{\omega} \qquad (58)$$

Hence the curl of $\mathbf{v}$ is a measure of the angular velocity of the rigid body.

APPENDIX 6

Cartesian Tensors

INTRODUCTION

Physical quantities often can be described or represented by a set of numbers referred to a particular coordinate frame. The velocity of a particle for example is usually described by giving its components in a Cartesian frame. When a quantity is described in this manner, we refer to the set of numbers as a representation of the quantity in the given coordinate frame. Frequently we are given the representation of a quantity in a particular frame and we need or would like to know what its representation is in some other frame. To do this we must know how the representation of the quantity transforms when we go from the one frame to the other. In considering this problem we shall find that there are certain quantities whose representations have particularly convenient transformation properties.

EUCLIDEAN SET OF COORDINATE FRAMES

One of the first things we need to know when talking about representations and transformations is the set of coordinate frames that is being considered.

In this appendix we assume that we are dealing with a set of coordinate frames defined by the following conditions: (1) each member in the set is a rectangular, Cartesian coordinate frame, and (2) any member in the set can be obtained from any other member in the set by a rotation, translation, inversion, or some combination of these operations. For convenience we refer to such a set as a *Euclidean set* of coordinate frames (see Appendix 17). A Euclidean set of coordinate frames can be generated by starting with a particular rectangular Cartesian frame, then subjecting it to every possible rotation, translation, inversion, and combination of these operations.

NOTE: There are many other sets of frames, but we do not consider them in this appendix. One could for example consider a set of rectangular

Cartesian coordinate frames in which any member in the set could be obtained from any other member in the set by a rotation only. Such a set would be a subset of the Euclidean set. Or one could consider a set of coordinate systems not necessarily Cartesian in which any member in the set could be obtained from any other member in the set by any transformation, as long as it would allow one to pass freely from the one system to the other and back again. In this case the Euclidean set would be a subset of such a set.

We designate a particular rectangular Cartesian frame by the letter S; the axes of the frame S by the numbers 1, 2, and 3; and the unit vectors in the 1, 2, and 3 directions by the symbols $\mathbf{e}_1$, $\mathbf{e}_2$, and $\mathbf{e}_3$, respectively. If we are dealing with more than one frame we shall designate the frames by the letters S, S', S'', ...; the sets of axes by the sets of numbers $(1,2,3)$, $(1',2',3')$, $(1'',2'',3'')$, ...; and the sets of unit vectors by the sets of symbols $(\mathbf{e}_1, \mathbf{e}_2, \mathbf{e}_3)$, $(\mathbf{e}_{1'}, \mathbf{e}_{2'}, \mathbf{e}_{3'})$, $(\mathbf{e}_{1''}, \mathbf{e}_{2''}, \mathbf{e}_{3''})$, ...

TRANSFORMATION COEFFICIENTS

The concepts introduced in this section are very useful in handling the transformation from one member to another member of a Euclidean set of coordinate frames.

If S and S' are two different frames in the same Euclidean set, then the quantities

$$e_{i'j} \equiv \mathbf{e}_{i'} \cdot \mathbf{e}_j \tag{1}$$

are called the *transformation coefficients* for the transformation from the frame S to the frame S'. Since $\mathbf{e}_{i'} \cdot \mathbf{e}_j$ is just the projection of the vector $\mathbf{e}_{i'}$ in the $\mathbf{e}_j$ direction, it follows that $e_{i'j}$ is simply the jth component in the frame S of the vector $\mathbf{e}_{i'}$. The notation is thus a very natural one, since it is customary to designate the jth component of a vector $\mathbf{A}$ as simply A_j.

Theorem 1. The transformation coefficient $e_{i'j}$ is equal to the cosine of the angle between $\mathbf{e}_{i'}$ and $\mathbf{e}_j$, that is,

$$e_{i'j} = \cos(i', j) \tag{2}$$

PROOF: From the definition of transformation coefficients $e_{i'j} \equiv \mathbf{e}_{i'} \cdot \mathbf{e}_j$. From the definition of the dot product $\mathbf{e}_{i'} \cdot \mathbf{e}_j = 1 \times 1 \times \cos(i', j) = \cos(i', j)$. Combining these results we obtain Eq. (2).

Theorem 2. The transformation coefficients $e_{i'j}$ for the transformation from the frame S to the frame S', and the transformation coefficients $e_{ij'}$ for the transformation from the frame S' to the frame S are related as follows:

$$e_{i'j} = e_{ji'} \tag{3}$$

PROOF: From the definition of a transformation coefficient $e_{i'j} \equiv \mathbf{e}_{i'} \cdot \mathbf{e}_j$ and $e_{ji'} \equiv \mathbf{e}_j \cdot \mathbf{e}_{i'}$. Since the dot product is commutative, $e_{i'j} = e_{ji'}$.

Theorem 3. If $e_{i'j}$ are the transformation coefficients for the transformation from the frame S to the frame S' and $e_{i''j'}$ are the transformation coefficients for the transformation from the frame S' to the frame S'', the transformation coefficients $e_{i''j}$ for the transformation from the frame S to the frame S'' are given by

$$e_{i''j} = \sum_k e_{i''k'} e_{k'j} \qquad (4)$$

In matrix notation

$$\begin{bmatrix} e_{1''1} & e_{1''2} & e_{1''3} \\ e_{2''1} & e_{2''2} & e_{2''3} \\ e_{3''1} & e_{3''2} & e_{3''3} \end{bmatrix} = \begin{bmatrix} e_{1''1'} & e_{1''2'} & e_{1''3'} \\ e_{2''1'} & e_{2''2'} & e_{2''3'} \\ e_{3''1'} & e_{3''2'} & e_{3''3'} \end{bmatrix} \begin{bmatrix} e_{1'1} & e_{1'2} & e_{1'3} \\ e_{2'1} & e_{2'2} & e_{2'3} \\ e_{3'1} & e_{3'2} & e_{3'3} \end{bmatrix} \qquad (5)$$

PROOF: In component form the dot product $\mathbf{a} \cdot \mathbf{b}$ of two vectors is given by $\sum_j a_j b_j$. It follows that the dot product of the vectors $\mathbf{e}_{i''}$ and the vector $\mathbf{e}_j$ can be written

$$\mathbf{e}_{i''} \cdot \mathbf{e}_j = \sum_{k'} e_{i''k'} e_{jk'} \qquad (6)$$

But

$$\mathbf{e}_{i''} \cdot \mathbf{e}_j = e_{i''j} \qquad (7)$$

and

$$\sum_{k'} e_{i''k'} e_{jk'} = \sum_{k'} e_{i''k'} e_{k'j} \qquad (8)$$

Combining Eqs. (6)–(8) we obtain the desired result.

Theorem 4. The transformation coefficients $e_{i'j}$ for the transformation from the frame S to the frame S' satisfy the following relation:

$$\sum_j e_{i'j} e_{k'j} = \delta_{i'k'} \qquad (9)$$

where $\delta_{i'k'}$ is the Kronecker delta.

PROOF: In component form the dot product $\mathbf{a} \cdot \mathbf{b}$ of two vectors $\mathbf{a}$ and $\mathbf{b}$ is given by $\sum_j a_j b_j$. It follows that

$$\sum_j e_{i'j} e_{k'j} = \mathbf{e}_{i'} \cdot \mathbf{e}_{k'} = \delta_{i'k'} \qquad (10)$$

SCALARS

A physical quantity that can be represented by a single number a that has the same value in all coordinate frames is said to be a *scalar quantity*, and the number is said to be a *scalar*.

The distance between two points is an example of a quantity that can be represented by a scalar.

VECTORS

A vector quantity is a quantity that can be represented geometrically by a directed line segment. The directed line segment itself is called a vector.

A vector can be represented analytically by giving its components with reference to a particular rectangular Cartesian frame. If we know the components of a vector with respect to one member of a Euclidean set of frames, we can find its components with respect to some other member of the set by making use of the following theorem.

Theorem 5. If S and S' are two members of a Euclidean set of coordinate frames, and if a_1, a_2, and a_3 are the components of a vector $\mathbf{a}$ in the frame S, the components $a_{1'}$, $a_{2'}$, and $a_{3'}$ of $\mathbf{a}$ in the frame S' are given by

$$a_{i'} = \sum_j e_{i'j} a_j \tag{11}$$

or in matrix form by

$$\begin{bmatrix} a_{1'} \\ a_{2'} \\ a_{3'} \end{bmatrix} = \begin{bmatrix} e_{1'1} & e_{1'2} & e_{1'3} \\ e_{2'1} & e_{2'2} & e_{2'3} \\ e_{3'1} & e_{3'2} & e_{3'3} \end{bmatrix} \begin{bmatrix} a_1 \\ a_2 \\ a_3 \end{bmatrix} \tag{12}$$

where the $e_{i'j}$ are the transformation coefficients for the transformation from the frame S to the frame S'.

PROOF: The component $a_{i'}$ is just the projection of the vector $\mathbf{a}$ on the $\mathbf{e}_{i'}$ direction, that is,

$$a_{i'} = \mathbf{a} \cdot \mathbf{e}_{i'} \tag{13}$$

Writing $\mathbf{a}$ in terms of its components in the S frame we have

$$\mathbf{a} = \sum_j a_j \mathbf{e}_j \tag{14}$$

Combining Eqs. (13) and (14) we obtain

$$a_{i'} = \left(\sum_j a_j \mathbf{e}_j \right) \cdot \mathbf{e}_{i'} = \sum_j \mathbf{e}_{i'} \cdot \mathbf{e}_j a_j = \sum_j e_{i'j} a_j \tag{15}$$

CARTESIAN VECTORS

The representations of a vector in different rectangular Cartesian frames are related by the transformation law given in the preceding theorem. Conversely if we are given a quantity that can be represented by a set of three numbers in a rectangular Cartesian frame, and if the different representations are related in the same manner as the components of a vector are related, then the quantity can be represented by a directed line segment; hence it is a vector quantity. This fact suggests the following generalization of the definition of a vector and a vector quantity.

Definition. A *Cartesian vector* is a mathematical entity that has the following properties:

a. It can be represented in an arbitrary rectangular Cartesian frame S by a set of three numbers a_1, a_2, and a_3, called the components of the vector in the frame S.
b. If S' is any other frame in the Euclidean set containing the frame S, then the components a_i in the frame S and the components $a_{i'}$ in the frame S' are related as follows:

$$a_{i'} = \sum_j e_{i'j} a_j$$

or in matrix form

$$\begin{bmatrix} a_{1'} \\ a_{2'} \\ a_{3'} \end{bmatrix} = \begin{bmatrix} e_{1'1} & e_{1'2} & e_{1'3} \\ e_{2'1} & e_{2'2} & e_{2'3} \\ e_{3'1} & e_{3'2} & e_{3'3} \end{bmatrix} \begin{bmatrix} a_1 \\ a_2 \\ a_3 \end{bmatrix} \quad (16)$$

where the $e_{i'j}$ are the transformation coefficients for the transformation from the frame S to the frame S'.

Definition. A *Cartesian vector quantity* is a quantity that can be represented mathematically by a Cartesian vector.

A Cartesian vector is usually described by giving its components in a particular frame. Thus the Cartesian vector whose components in the frame S are a_i is spoken of as simply the vector a_i.

CARTESIAN TENSORS

The preceding definition of a Cartesian vector suggests the following generalization of the concept.

Definition. A *Cartesian tensor of rank 2* is a mathematical entity that has the following properties:

a. It can be represented in an arbitrary rectangular Cartesian frame S by a set of nine numbers a_{ij}, where i and j run from 1 to 3, called the components of the tensor in the frame S.
b. If S' is any other frame in the Euclidean set containing the frame S, then the components a_{ij} in the frame S and the components $a_{i'j'}$ in the frame S' are related as follows:

$$a_{i'j'} = \sum_k \sum_l e_{i'k} e_{j'l} a_{kl} \quad (17)$$

or in matrix form

$$\begin{bmatrix} a_{1'1'} & a_{1'2'} & a_{1'3'} \\ a_{2'1'} & a_{2'2'} & a_{2'3'} \\ a_{3'1'} & a_{3'2'} & a_{3'3'} \end{bmatrix} = \begin{bmatrix} e_{1'1} & e_{1'2} & e_{1'3} \\ e_{2'1} & e_{2'2} & e_{2'3} \\ e_{3'1} & e_{3'2} & e_{3'3} \end{bmatrix} \begin{bmatrix} a_{11} & a_{12} & a_{13} \\ a_{21} & a_{22} & a_{23} \\ a_{31} & a_{32} & a_{33} \end{bmatrix} \begin{bmatrix} e_{11'} & e_{12'} & e_{13'} \\ e_{21'} & e_{22'} & e_{23'} \\ e_{31'} & e_{32'} & e_{33'} \end{bmatrix}$$

where the $e_{i'j}$ and the $e_{ij'}$ are the transformation coefficients for the transformations from frame S to frame S' and from frame S' to frame

S, respectively. Note that the matrix $[e_{i'j}]$ is just the transpose of the matrix $[e_{ij'}]$.

Continuing the generalization further we are led to the following definition.

Definition. A *Cartesian tensor of rank n* is a mathematical entity that has the following properties:

a. It can be represented in an arbitrary rectangular Cartesian frame S by a set of 3^n numbers $a_{i_1 i_2 \ldots i_n}$, where $i_1, i_2, \ldots,$ and i_n run from 1 to 3, called the components of the tensor in the frame S.

b. If S' is any other frame in the Euclidean set containing the frame S, the components $a_{i_1 i_2 \ldots i_n}$ in the frame S and the components $a_{i'_1 i'_2 \ldots i'_n}$ are related as follows:

$$a_{i'_1, i'_2 \ldots i'_n} = \sum_{j_1} \sum_{j_2} \cdots \sum_{j_n} e_{i'_1 j_1} e_{i'_2 j_2} \cdots e_{i'_n j_n} a_{j_1 j_2 \ldots j_n} \tag{18}$$

where the $e_{i'j}$ are the transformation coefficients for the transformation from the frame S to the frame S'.

From the defintion above it follows that a scalar is a Cartesian tensor of rank 0 and a vector is a Cartesian tensor of rank 1.

ADDITION AND MULTIPLICATION OF CARTESIAN TENSORS

To exploit the concept of a Cartesian tensor we find it helpful to introduce a number of additional definitions and theorems, which we state without proof.

Theorem 6. Given two nth rank Cartesian tensors $a_{i_1 i_2 \ldots i_n}$ and $b_{i_1 i_2 \ldots i_n}$, the quantity

$$c_{i_1 i_2 \ldots i_n} \equiv a_{i_1 i_2 \ldots i_n} + b_{i_1 i_2 \ldots i_n} \tag{19}$$

is called the sum of the two tensors and is itself a tensor of rank n.

Theorem 7. Given an mth rank Cartesian tensor $a_{i_1 i_2 \ldots i_m}$ and an nth rank Cartesian tensor $b_{j_1 j_2 \ldots j_n}$, the quantity

$$c_{i_1 i_2 \ldots i_m j_1 j_2 \ldots j_n} \equiv a_{i_1 i_2 \ldots i_m} b_{j_1 j_2 \ldots j_n} \tag{20}$$

is called the exterior or outer product of the two tensors and is itself a tensor of rank $m + n$.

Example. Given a tensor a_i of rank 1 and a tensor b_{jk} of rank 2, the quantity $c_{ijk} \equiv a_i b_{jk}$ is a tensor of rank 3.

Theorem 8. Given an nth rank Cartesian tensor $a_{i_1 i_2 \ldots i_n}$, the quantity

$$c_{i_1 i_2 \ldots i_r i_{r+3} \ldots i_n} = \sum_j a_{i_1 i_2 \ldots i_r j j i_{r+3} \ldots i_n} \tag{21}$$

is called a contraction of the tensor and is itself a tensor of rank $n - 2$.

Example. Given a tensor a_{ijk} of rank 3, the quantity $c_k = \sum_i a_{iik}$ is a tensor of rank 1.

ORIENTED CARTESIAN TENSORS

A set of coordinate frames defined by the conditions that each member in the set is a rectangular Cartesian coordinate frame and any member in the set can be obtained from any other member in the set by a rotation, translation, or some combination of these two operations is called an *oriented Euclidean set*. Since the foregoing statement rules out inversions, the members of an oriented Euclidean set must be either all right handed coordinate frames or all left handed coordinate frames.

In analogy with the definition of a Cartesian tensor, we can now define an oriented Cartesian tensor as follows.

Definition. An *oriented Cartesian tensor* of rank n is a mathematical entity that has the following properties:

a. It can be represented in an arbitrary rectangular Cartesian frame S by a set of 3^n numbers $a_{i_1 i_2 \ldots i_n}$, where $i_1, i_2, \ldots,$ and i_n run from 1 to 3, called the components in the frame S.

b. If S' is any other frame in the *oriented Euclidean set* containing the frame S, the components $a_{i_1 i_2 \ldots i_n}$ in the frame S and the components $a_{i'_1 i'_2 \ldots i'_n}$ in the frame S' are related as follows:

$$a_{i'_1 i'_2 \ldots i'_n} = \sum_{j_1} \sum_{j_2} \cdots \sum_{j_n} e_{i'_1 j_1} e_{i'_2 j_2} \cdots e_{i'_n j_n} a_{j_1 j_2 \ldots j_n} \qquad (22)$$

where the $e_{i'j}$ are the transformation coefficients for the transformation from the frame S to the frame S'.

From the definition above it follows that a Cartesian tensor is also an oriented Cartesian tensor, but an oriented Cartesian tensor is not necessarily a Cartesian tensor.

The rules for addition and multiplication of oriented Cartesian tensors are the same as for Cartesian tensors.

THE LEVI-CIVITA TENSOR

The set of quantities ϵ_{ijk} defined by the conditions:

$\epsilon_{ijk} = +1$ if ijk is an even permutation of the sequence 123

$\epsilon_{ijk} = -1$ if ijk is an odd permutation of the sequence 123

$\epsilon_{ijk} = 0$ otherwise

can be shown to constitute an oriented Cartesian tensor of rank 3 and is called the *Levi-Civita tensor* or Levi-Civita symbol.

Theorem 9. The Levi-Civita tensor satisfies the following relation

$$\sum_m \epsilon_{mrs}\epsilon_{mpq} = \delta_{rp}\delta_{sq} - \delta_{rq}\delta_{sp} \tag{23}$$

where δ_{ij} is the Kronecker delta.

DOT AND CROSS PRODUCTS

Given two Cartesian vectors a_i and b_i the quantity

$$c \equiv \sum_i a_i b_i \tag{24}$$

is a Cartesian tensor of rank 0, which we call the *dot or scalar product* of the vectors a_i and b_i. The scalar product c is a contraction of the outer product $a_i b_j$.

Given two Cartesian vectors a_i and b_i, the quantity

$$c_i \equiv \sum_j \sum_k \epsilon_{ijk} a_j b_k \tag{25}$$

is an oriented Cartesian tensor of rank 2, which we call the *cross or vector product* of the vectors a_i and b_i. The cross product c_i is a contraction of the outer product $\epsilon_{ijk} a_l b_m$.

DIFFERENTIATION OF A TENSOR

A *tensor field* is a region of space with every point of which a tensor is associated.

Theorem 10. Given an nth rank Cartesian tensor field $\phi_{i_1 i_2 \ldots i_n}(\mathbf{x})$, the partial derivative with respect to the variable x_j which may be written in any one of the following forms:

$$\frac{\partial \phi_{i_1 i_2 \ldots i_n}(\mathbf{x})}{\partial x_j} \equiv \partial_j \phi_{i_1 i_2 \ldots i_n} \equiv \phi_{i_1 i_2 \ldots i_n, j} \tag{26}$$

is a Cartesian tensor field of rank $n + 1$.

Definition. Given a Cartesian scalar field $\phi(\mathbf{x})$, the Cartesian vector field

$$\psi_i \equiv \partial_i \phi \tag{27}$$

is called the *gradient* of ϕ.

Definition. Given a Cartesian vector field $\phi_i(\mathbf{x})$, the Cartesian scalar field

$$\psi \equiv \sum_i \partial_i \phi_i \tag{28}$$

is called the *divergence* of ϕ.

Definition. Given a Cartesian vector field $\phi_i(\mathbf{x})$, the oriented Cartesian vector field

$$\psi_i \equiv \sum_j \sum_k \epsilon_{ijk} \partial_j \phi_k \tag{29}$$

is called the *curl* of ϕ.

Definition. Given the Cartesian scalar field $\phi(\mathbf{x})$, the Cartesian scalar field

$$\psi = \sum_i \partial_i \partial_i \phi \equiv \sum_i \partial_i^2 \phi \tag{30}$$

is called the Laplacian of ϕ.

THE SUMMATION CONVENTION

The manipulation of tensors can frequently be formally simplified if we adopt the summation convention. According to this convention, repeated indices in the same term are always summed over unless otherwise indicated. For example, the dot product of the vectors a_i and b_j is written as $a_i b_i$ rather than $\sum_i a_i b_i$. Similarly, the cross product of a_i and b_j is written $\epsilon_{ijk} a_j b_k$ rather than $\sum_j \sum_k \epsilon_{ijk} a_j b_k$.

EXAMPLES

Example 1. Use tensor notation to prove the vector identity

$$\mathbf{A} \times (\mathbf{B} \times \mathbf{C}) = \mathbf{B}(\mathbf{A} \cdot \mathbf{C}) - \mathbf{C}(\mathbf{A} \cdot \mathbf{B}) \tag{31}$$

Solution. Starting with the left hand side of Eq. (31) in tensor notation, and using the summation convention, we have

$$\begin{aligned}
\epsilon_{ijk} A_j \epsilon_{klm} B_l C_m &= \epsilon_{ijk} \epsilon_{klm} A_j B_l C_m \\
&= \epsilon_{kij} \epsilon_{klm} A_j B_l C_m \\
&= (\delta_{il} \delta_{jm} - \delta_{im} \delta_{jl}) A_j B_l C_m \\
&= \delta_{il} \delta_{jm} A_j B_l C_m - \delta_{im} \delta_{jl} A_j B_l C_m \\
&= A_j B_i C_j - A_j B_j C_i \\
&= B_i (A_j C_j) - C_i (A_j B_j)
\end{aligned} \tag{32}$$

which is just the right hand side of Eq. (31) in tensor notation.

Example 2. Use tensor notation to prove the vector identity

$$\nabla \cdot (\mathbf{A} \times \mathbf{B}) = \mathbf{B} \cdot (\nabla \times \mathbf{A}) - \mathbf{A} \cdot (\nabla \times \mathbf{B}) \tag{33}$$

Solution. Starting with the left hand side of Eq. (33) in tensor notation, and using the summation convention, we have

$$\partial_i(\epsilon_{ijk}A_jB_k) = \epsilon_{ijk}\partial_i(A_jB_k)$$
$$= \epsilon_{ijk}\left[(\partial_i A_j)B_k + A_j(\partial_i B_k)\right]$$
$$= B_k\epsilon_{kij}\partial_i A_j - A_j\epsilon_{jik}\partial_i B_k \qquad (34)$$

which is just the right hand side of Eq. (33) in tensor notation.

APPENDIX 7

Orthogonal Curvilinear Coordinates

ORTHOGONAL CURVILINEAR COORDINATE SYSTEMS

Let q_1, q_2, q_3 be a set of coordinates that are related to the Cartesian coordinates x, y, z by a regular transformation (see Appendix 17):

$$q_1 = q_1(x, y, z) \tag{1a}$$

$$q_2 = q_2(x, y, z) \tag{1b}$$

$$q_3 = q_3(x, y, z) \tag{1c}$$

With each of the coordinates q_i there is a family of surfaces defined by the equation

$$q_i(x, y, z) = \text{constant} \tag{2}$$

If one member from each of the three families of surfaces passes through each point in space, and if each of the surfaces is orthogonal to every member of the other two families, the three families of surfaces are said to constitute an *orthogonal system of surfaces*. If we have an orthogonal system of surfaces and if the three surfaces passing through the point P are given by

$$q_1(x, y, z) = q_1(P) \tag{3a}$$

$$q_2(x, y, z) = q_2(P) \tag{3b}$$

$$q_3(x, y, z) = q_3(P) \tag{3c}$$

where the $q_i(P)$ are constants, the numbers $q_1(P)$, $q_2(P)$, and $q_3(P)$ are called the coordinates of the point P in the q_1, q_2, q_3 system, and the system is called an *orthogonal curvilinear coordinate system*. Our interest in this appendix lies in orthogonal curvilinear coordinate systems only.

UNIT VECTORS IN ORTHOGONAL CURVILINEAR COORDINATES

Consider an orthgonal curvilinear coordinate system q_1, q_2, q_3. Let P be a point, $S_i(P)$ the surface defined by the equation $q_i(x, y, z) = q_i(P)$, and $L_i(P)$

the line defined by the intersection of the surfaces $S_j(P)$ and $S_k(P)$, where $i \neq j \neq k \neq i$. The three surfaces $S_1(P)$, $S_2(P)$, and $S_3(P)$ intersect at the point P, and the three lines $L_1(P)$, $L_2(P)$, and $L_3(P)$ intersect at the point P. We designate the unit vector that is normal to the surface $S_i(P)$ at P, or equivalently is tangent to the line $L_i(P)$ and is pointing in the direction of increasing q_i, as $\mathbf{e}_i(P)$, or simply $\mathbf{e}_i$.

From Theorems 6 and 7 in Appendix 5, the gradient of a function $\phi(x, y, z)$ at a point P is normal to the surface $\phi(x, y, z) = \phi(P)$ at the point P and points in the direction of increasing ϕ. It follows that the unit vectors $\mathbf{e}_i(P)$ are given by

$$\mathbf{e}_i(P) = \left[\frac{\nabla q_i(x, y, z)}{|\nabla q_i(x, y, z)|} \right]_{x = x(P), y = y(P), z = z(P)} \tag{4}$$

or in abbreviated notation

$$\mathbf{e}_i = \frac{\nabla q_i}{|\nabla q_i|} \tag{5}$$

Alternatively we can determine $\mathbf{e}_i(P)$ by noting that the line $L_i(P)$ is defined by the vector equation

$$\mathbf{r} = \left[\mathbf{e}_x x(q_1, q_2, q_3) + \mathbf{e}_y y(q_1, q_2, q_3) + \mathbf{e}_z z(q_1, q_2, q_3) \right]_{q_j = q_j(P), \, q_k = q_k(P)} \tag{6}$$

where $i \neq j \neq k \neq i$, hence the unit vector $\mathbf{e}_i(P)$ is given by

$$\mathbf{e}_i(P) = \left[\frac{\partial \mathbf{r}/\partial q_i}{|\partial \mathbf{r}/\partial q_i|} \right]_{q_1 = q_1(P), \, q_2 = q_2(P), \, q_3 = q_3(P)} \tag{7}$$

or in abbreviated notation

$$\mathbf{e}_i = \frac{\partial \mathbf{r}/\partial q_i}{|\partial \mathbf{r}/\partial q_i|} \tag{8}$$

Since the coordinate system q_1, q_2, q_3 is orthogonal, the three unit vectors $\mathbf{e}_1(P)$, $\mathbf{e}_2(P)$, and $\mathbf{e}_3(P)$ are orthgonal.

TRANSFORMATION FROM A CARTESIAN TO AN ORTHOGONAL CURVILINEAR COORDINATE SYSTEM

Equation (5) or (8) can be used to obtain the unit vectors $\mathbf{e}_1$, $\mathbf{e}_2$, and $\mathbf{e}_3$ at an arbitrary point P in terms of the unit vectors $\mathbf{e}_x$, $\mathbf{e}_y$, and $\mathbf{e}_z$. It follows that if we are given a vector $\mathbf{A}$ in the form

$$\mathbf{A} = A_x \mathbf{e}_x + A_y \mathbf{e}_y + A_z \mathbf{e}_z \tag{9}$$

then we can readily convert it to the form

$$\mathbf{A} = A_1 \mathbf{e}_1 + A_2 \mathbf{e}_2 + A_3 \mathbf{e}_3 \tag{10}$$

and vice versa. The resulting transformation between the components A_x, A_y, and A_z and the components A_1, A_2, and A_3 can be expressed in matrix form as follows:

$$\begin{bmatrix} A_1 \\ A_2 \\ A_3 \end{bmatrix} = \begin{bmatrix} \mathbf{e}_x \cdot \mathbf{e}_1 & \mathbf{e}_y \cdot \mathbf{e}_1 & \mathbf{e}_z \cdot \mathbf{e}_1 \\ \mathbf{e}_x \cdot \mathbf{e}_2 & \mathbf{e}_y \cdot \mathbf{e}_2 & \mathbf{e}_z \cdot \mathbf{e}_2 \\ \mathbf{e}_x \cdot \mathbf{e}_3 & \mathbf{e}_y \cdot \mathbf{e}_3 & \mathbf{e}_z \cdot \mathbf{e}_3 \end{bmatrix} \begin{bmatrix} A_x \\ A_y \\ A_z \end{bmatrix} \qquad (11)$$

SPATIAL DERIVATIVES OF A VECTOR IN ORTHOGONAL CURVILINEAR COORDINATES

If we are given a vector **A** that is a function of the set of variables $\mathbf{q} \equiv q_1, q_2, q_3$ and we wish to determine one of the partial derivatives $\partial \mathbf{A}(\mathbf{q})/\partial q_i$, we can express the vector **A** in the form

$$\mathbf{A} = A_x(\mathbf{q})\mathbf{e}_x + A_y(\mathbf{q})\mathbf{e}_y + A_z(\mathbf{q})\mathbf{e}_z \qquad (12)$$

and evaluate the derivative as follows:

$$\frac{\partial \mathbf{A}(\mathbf{q})}{\partial q_i} = \frac{\partial A_x(\mathbf{q})}{\partial q_i}\mathbf{e}_x + \frac{\partial A_y(\mathbf{q})}{\partial q_i}\mathbf{e}_y + \frac{\partial A_z(\mathbf{q})}{\partial q_i}\mathbf{e}_z \qquad (13)$$

If we wish to obtain the derivative $\partial \mathbf{A}(\mathbf{q})/\partial q_i$ in terms of the unit vectors $\mathbf{e}_1$, $\mathbf{e}_2$, and $\mathbf{e}_3$ rather than in terms of the unit vectors $\mathbf{e}_x$, $\mathbf{e}_y$, and $\mathbf{e}_z$, we can proceed as above and then follow the prescription outlined in the preceding section to convert Eq. (13) to an expression in terms of the unit vectors $\mathbf{e}_1$, $\mathbf{e}_2$, and $\mathbf{e}_3$. Alternatively if we have by this method first obtained and tabulated the derivatives $\partial \mathbf{e}_i(\mathbf{q})/\partial q_j$ in terms of the unit vectors $\mathbf{e}_1$, $\mathbf{e}_2$, and $\mathbf{e}_3$, we can start with the vector **A** in the form

$$\mathbf{A} = A_1(\mathbf{q})\mathbf{e}_1(\mathbf{q}) + A_2(\mathbf{q})\mathbf{e}_2(\mathbf{q}) + A_3(\mathbf{q})\mathbf{e}_3(\mathbf{q}) \qquad (14)$$

and evaluate the derivative as follows:

$$\frac{\partial \mathbf{A}(\mathbf{q})}{\partial q_i} = \frac{\partial A_1(\mathbf{q})}{\partial q_i}\mathbf{e}_1 + \frac{\partial A_2(\mathbf{q})}{\partial q_i}\mathbf{e}_2 + \frac{\partial A_3(\mathbf{q})}{\partial q_i}\mathbf{e}_3$$
$$+ A_1 \frac{\partial \mathbf{e}_1(\mathbf{q})}{\partial q_i} + A_2 \frac{\partial \mathbf{e}_2(\mathbf{q})}{\partial q_i} + A_3 \frac{\partial \mathbf{e}_3(\mathbf{q})}{\partial q_i} \qquad (15)$$

CYLINDRICAL COORDINATES

The cylindrical coordinates, ρ, ϕ, and z are defined by the equations

$$x = \rho \cos \phi \qquad (16a)$$
$$y = \rho \sin \phi \qquad (16b)$$
$$z = z \qquad (16c)$$

or equivalently by the equations

$$\rho = (x^2 + y^2)^{1/2} \tag{17a}$$

$$\phi = \tan^{-1}\left(\frac{y}{x}\right) \tag{17b}$$

$$z = z \tag{17c}$$

If we apply the general results obtained earlier to this particular case we obtain the following results. The unit vectors are given by

$$\mathbf{e}_\rho = \cos\phi\, \mathbf{e}_x + \sin\phi\, \mathbf{e}_y \tag{18}$$

$$\mathbf{e}_\phi = -\sin\phi\, \mathbf{e}_x + \cos\phi\, \mathbf{e}_y \tag{19}$$

$$\mathbf{e}_z = \mathbf{e}_z \tag{20}$$

The transformation equation for converting from x, y, z components to ρ, ϕ, z components is

$$\begin{bmatrix} A_\rho \\ A_\phi \\ A_z \end{bmatrix} = \begin{bmatrix} \cos\phi & \sin\phi & 0 \\ -\sin\phi & \cos\phi & 0 \\ 0 & 0 & 1 \end{bmatrix} \begin{bmatrix} A_x \\ A_y \\ A_z \end{bmatrix} \tag{21}$$

The spatial derivatives of the unit vectors $\mathbf{e}_\rho, \mathbf{e}_\phi, \mathbf{e}_z$ are given by

$$\begin{aligned} \frac{\partial \mathbf{e}_\rho}{\partial \rho} = 0 & & \frac{\partial \mathbf{e}_\rho}{\partial \phi} = \mathbf{e}_\phi & & \frac{\partial \mathbf{e}_\rho}{\partial z} = 0 \\ \frac{\partial \mathbf{e}_\phi}{\partial \rho} = 0 & & \frac{\partial \mathbf{e}_\phi}{\partial \phi} = -\mathbf{e}_\rho & & \frac{\partial \mathbf{e}_\phi}{\partial z} = 0 \\ \frac{\partial \mathbf{e}_z}{\partial \rho} = 0 & & \frac{\partial \mathbf{e}_z}{\partial \phi} = 0 & & \frac{\partial \mathbf{e}_z}{\partial z} = 0 \end{aligned} \tag{22}$$

SPHERICAL COORDINATES

The spherical coordinates r, θ, and ϕ are defined by the equations

$$x = r\sin\theta\cos\phi \tag{23a}$$

$$y = r\sin\theta\sin\phi \tag{23b}$$

$$z = r\cos\theta \tag{23c}$$

or equivalently by the equations

$$r = (x^2 + y^2 + z^2)^{1/2} \tag{24a}$$

$$\theta = \tan^{-1}\left[\frac{(x^2 + y^2)^{1/2}}{z}\right] \tag{24b}$$

$$\phi = \tan^{-1}\left(\frac{y}{x}\right) \tag{24c}$$

The unit vectors $\mathbf{e}_r$, $\mathbf{e}_\theta$, and $\mathbf{e}_\phi$ are given by

$$\mathbf{e}_r = \sin\theta\cos\phi\,\mathbf{e}_x + \sin\theta\sin\phi\,\mathbf{e}_y + \cos\theta\,\mathbf{e}_z \tag{25}$$

$$\mathbf{e}_\theta = \cos\theta\cos\phi\,\mathbf{e}_x + \cos\theta\sin\phi\,\mathbf{e}_y - \sin\theta\,\mathbf{e}_z \tag{26}$$

$$\mathbf{e}_\phi = -\sin\phi\,\mathbf{e}_x + \cos\phi\,\mathbf{e}_y \tag{27}$$

The transformation equations for converting from x, y, z components to r, θ, ϕ components are

$$\begin{bmatrix} A_r \\ A_\theta \\ A_\phi \end{bmatrix} = \begin{bmatrix} \sin\theta\cos\phi & \sin\theta\sin\phi & \cos\theta \\ \cos\theta\cos\phi & \cos\theta\sin\phi & -\sin\theta \\ -\sin\phi & \cos\phi & 0 \end{bmatrix} \begin{bmatrix} A_x \\ A_y \\ A_z \end{bmatrix} \tag{28}$$

The derivatives of the unit vectors $\mathbf{e}_r$, $\mathbf{e}_\theta$, and $\mathbf{e}_\phi$ are

$$\frac{\partial \mathbf{e}_r}{\partial r} = 0 \qquad \frac{\partial \mathbf{e}_r}{\partial \theta} = \mathbf{e}_\theta \qquad \frac{\partial \mathbf{e}_r}{\partial \phi} = \sin\theta\,\mathbf{e}_\phi$$

$$\frac{\partial \mathbf{e}_\theta}{\partial r} = 0 \qquad \frac{\partial \mathbf{e}_\theta}{\partial \theta} = -\mathbf{e}_r \qquad \frac{\partial \mathbf{e}_\theta}{\partial \phi} = \cos\theta\,\mathbf{e}_\phi \tag{29}$$

$$\frac{\partial \mathbf{e}_\phi}{\partial r} = 0 \qquad \frac{\partial \mathbf{e}_\phi}{\partial \theta} = 0 \qquad \frac{\partial \mathbf{e}_\phi}{\partial \phi} = -\sin\theta\,\mathbf{e}_r - \cos\theta\,\mathbf{e}_\theta$$

APPENDIX 8

Ordinary Differential Equations

INTRODUCTION

In this appendix we define a variety of terms associated with ordinary differential equations and state without proof a number of theorems. The purpose of the appendix is not to present methods for solving differential equations, but simply to indicate something of the nature of their solutions.

GENERAL THEORY

Definition 1. An *ordinary differential equation* is an equation involving ordinary derivatives of a dependent variable x with respect to an independent variable t, that is, an equation of the form

$$F(t, x, \dot{x}, \ddot{x}, \ldots,) = 0 \tag{1}$$

where

$$\dot{x} \equiv \frac{dx}{dt}$$

$$\ddot{x} \equiv \frac{d^2x}{dt^2}$$

$$\ldots$$

Throughout this appendix we deal with ordinary differential equations. For simplicity we refer to them simply as differential equations.

Definition 2. The order of the highest ordered derivative involved in a differential equation is called the *order of the differential equation*.

Definition 3. By a *solution* of a differential equation

$$F(t, x, \dot{x}, \ddot{x}, \ldots) = 0 \tag{2}$$

is meant a function $x(t)$ that when substituted into the differential equation reduces the differential equation to an identity in t.

Theorem 1. Given the nth order differential equation

$$F(t, x, \dot{x}, \ldots, \overset{(n)}{x}) = 0 \qquad (3)$$

where $\overset{(n)}{x} \equiv d^n x/dt^n$, if we can write the equation in the form

$$\overset{(n)}{x} = f(t, x, \dot{x}, \ldots, \overset{(n-1)}{x}) \qquad (4)$$

and if f, $\partial f/\partial x$, $\partial f/\partial \dot{x}$, ..., $\partial f/\partial \overset{(n-1)}{x}$ are real, finite, single valued, and continuous at all points $(t, x, \dot{x} \ldots \overset{(n-1)}{x})$ within a region R of the space whose coordinates are $t, x, \dot{x} \ldots \overset{(n-1)}{x}$, there is one and only one solution $x(t)$ in R that passes through any abritrary given point in R.

The conditions listed in Theorem 1 are more restrictive than they have to be, but they generally suffice for problems of physical interest.

We do not give a proof of Theorem 1. A detailed proof of a stronger theorem can be found in most books on differential equations. However some idea of the direction of the proof can be obtained by considering the case of the first order differential equation. In this case we wish to show that there is a unique solution to the differential equation $\dot{x} = f(t, x)$, provided f and $\partial f/\partial x$ are real, finite, single valued, and continuous. If we construct a curve in t-x space by starting at the point (t_0, x_0) moving in the direction of the line whose slope is $f(t_0, x_0)$ to a nearby point (t_1, x_1), proceeding from there in the direction of the line whose slope is $f(t_1, x_1)$ to a new point (t_2, x_2), and continuing the process to points (t_3, x_3), (t_4, x_4), and so forth, the slope of the resulting broken line at any one of its points will differ only slightly from the value of $f(t, x)$ at that point. If $f(t, x)$ and $\partial f(t, x)/\partial x$ are real, finite, single valued, and continuous, then in the limit as the distance between successive points decreases toward zero the broken line will approach a smooth curve that passes through the point (t_0, x_0) and satisfies the differential equation $\dot{x} = f(t, x)$ at all points along its length. Furthermore the limiting curve is unique.

Definition 4. If a function

$$x = x(t, c_1, c_2, \ldots, c_n) \qquad (5)$$

where $c_1, c_2, \ldots, c_n$ is a set of constants, is a solution to a given differential equation regardless of the specific values assigned to the constants, the constants are called *arbitrary constants*. If furthermore the set of constants cannot be replaced by a smaller set that still maintains the same degree of generality for the solution, then the constants are called *essential constants* or *independent constants*.

Definition 5. If

$$x = x(t, c_1, c_2, \ldots) \qquad (6)$$

is a solution to a given differential equation and if any other solution to the differential equation can be obtained by a suitable choice of the arbitrary constants $c_1, c_2, \ldots,$ the solution is said to be a *complete solution*.

If an nth order differential equation $F(t, x, \dot{x}, \ldots, \overset{(n)}{x}) = 0$ has at least one solution passing through each point $(t, x, \dot{x}, \ldots, \overset{(n-1)}{x})$ in a region R of the space whose coordinates are $t, x, \dot{x}, \ldots, \overset{(n-1)}{x}$, a complete solution to the equation will contain at least n arbitrary constants. To see this we note that for a given value of t_0, there is at least one solution for each distinct set of values for the n quantities $x, \dot{x}, \ldots, \overset{(n-1)}{x}$, hence there must be at least n arbitrary constants in a complete solution if we wish to be free to choose the point $(t, x, \dot{x}, \ldots, \overset{(n-1)}{x})$ through which the solution passes.

Definition 6. Let

$$F\left(t, x, \dot{x}, \ldots, \overset{(n)}{x}\right) = 0 \tag{7}$$

be an nth order differential equation

a. A solution of Eq. (7) containing a set of n arbitrary and independent constants is called a *general solution* and the constants are called the integration constants.

b. A solution obtained from a general solution by giving particular values to one or more of the n integration constants is called a *particular solution*.

c. A solution that cannot be obtained from any general solution by any choice of the arbitrary constants is called a *singular solution*.

More often than not the terms "complete solution" and "general solution" mean the same thing in texts on differential equations. We find it convenient however to maintain a distinction. As the following theorem shows, the distinction disappears in most differential equations of physical interest.

Theorem 2. Let

$$F\left(t, x, \dot{x}, \ldots, \overset{(n)}{x}\right) = 0 \tag{8}$$

be an nth order differential equation, and let

$$x = x(t, c_1, c_2, \ldots, c_n) \tag{9}$$

be a general solution to the equation.

If:

a. The differential equation can be written in the form

$$\overset{(n)}{x} = f\left(t, x, \dot{x}, \ldots, \overset{(n-1)}{x}\right) \tag{10}$$

b. The quantities $f, \partial f/\partial x, \partial f/\partial \dot{x}, \ldots,$ are real, finite, single valued, and continuous in a region R of the space whose coordinates are $t, x, \dot{x}, \ldots, \overset{(n-1)}{x}$.

c. By proper choice of the constants $c_1, c_2, \ldots, c_n$ we can obtain a particular solution that passes through an arbitrary given point $(t_0, x_0, \dot{x}_0, \ldots, \overset{(n-1)}{x_0})$ in the region R.

Then: the solution is a complete solution in the region R, and there will be one and only one solution passing through a given point in R.

Definition 7. If we seek a function $x(t)$ that not only satisfies a particular nth order differential equation

$$F(t, x, \dot{x}, \ldots, \overset{(n)}{x}) = 0 \tag{11}$$

but also satisfies some supplementary conditions, the supplementary conditions are called *initial conditions* or *boundary conditions*. Intial conditions usually refer to the specification of the value of $x, \dot{x}, \ddot{x}, \ldots,$ and $\overset{(n-1)}{x}$ at some particular value of t, whereas boundary conditions usually involve conditions at two or more values of t.

SYSTEMS OF DIFFERENTIAL EQUATIONS

Definition 8. A set of n simultaneous differential equations of the form

$$F_i(t, x_1, x_2, \ldots, x_n, \dot{x}_1, \dot{x}_2, \ldots, \dot{x}_n, \ddot{x}_1, \ddot{x}_2, \ldots, \ddot{x}_n, \ldots) = 0 \tag{12}$$

where i ranges from 1 to n is said to be a *system of simultaneous differential equations*.

Definition 9. By a *solution* of the system of n differential equations

$$F_i(t, x_1, x_2, \ldots, x_n, \dot{x}_1, \dot{x}_2, \ldots, \dot{x}_n, \ldots) = 0 \qquad i = 1, 2, \ldots, n \tag{13}$$

is meant a set of functions $x_1(t), x_2(t), \ldots, x_n(t)$ that when substituted into the set of differential equations above reduces it to a set of identities in t.

Analysis of systems of differential equations is greatly simplified by the use of an appropriate notation. If we denote the set of functions $F_1, F_2, \ldots, F_n$ by $\mathbf{F}$, the set of quantities $x_1, x_2, \ldots, x_n$ by $\mathbf{x}$, the set of quantities $\dot{x}_1, \dot{x}_2, \ldots, \dot{x}_n$ by $\dot{\mathbf{x}}$, and so forth, we can represent the system of differential equations by the equation

$$\mathbf{F}(t, \mathbf{x}, \dot{\mathbf{x}}, \ldots) = 0 \tag{14}$$

and a solution as

$$\mathbf{x} = \mathbf{x}(t) \tag{15}$$

The remaining definitions and theorems concerning systems of differential equations are simple generalizations of the results of the preceding section; in most cases they involve simply replacing $F, x, \dot{x}, \ldots$ by $\mathbf{F}, \mathbf{x}, \dot{\mathbf{x}}, \ldots$. We leave this project to the reader. In the generalization the main thing to note is that a solution to a set of m simultaneous differential equations of order n will contain $m \times n$ arbitrary and independent constants, and under suitable conditions a general solution will be a complete solution and the particular solutions unique.

APPENDIX 9

Linear Differential Equations

INTRODUCTION

An ordinary differential equation of the form

$$a_1(t)x + a_2(t)\dot{x} + a_3(t)\ddot{x} + \ldots = g(t) \tag{1}$$

where $a_1(t), a_2(t), \ldots, g(t)$ are functions of t is called a *linear differential equation*. If $g(t) = 0$ the equation is said to be *homogeneous*. If $g(t) \neq 0$ the equation is said to be *inhomogeneous*.

In this appendix we state without proof a number of theorems that apply to linear differential equations, and we obtain solutions to a few important linear differential equations.

SOME USEFUL THEOREMS

Theorem 1. If $x(t)$ is any solution of a linear homogeneous differential equation and C is any constant, then $Cx(t)$ is also a solution.

Theorem 2. If $x_1(t)$ and $x_2(t)$ are solutions of a linear homogeneous differential equation, then $x_1(t) + x_2(t)$ is a solution.

Theorem 3. If $x_i(t)$ is a solution of an inhomogeneous linear differential equation and $x_h(t)$ is a solution of the corresponding homogeneous equation, then $x_i(t) + x_h(t)$ is also a solution of the inhomogeneous equation.

COMPLEX SOLUTIONS

Up till now we have been dealing with differential equations whose solutions are real functions. As the following sections reveal, there are situations in which it is convenient to consider differential equations that have complex solutions and contain complex terms. To clearly distinguish between real and complex quantities, we put a bar over a quantity that is or may be complex.

In dealing with complex quantities it is helpful to remember that a complex number may be written either in the form $r \exp(i\phi)$ or in the form $x + iy$, and the two forms are related as follows:

$$r \exp(i\phi) = r \cos\phi + ir \sin\phi \tag{2}$$

$$x + iy = (x^2 + y^2)^{1/2} \exp\left[i \tan^{-1}\left(\frac{y}{x}\right)\right] \tag{3}$$

Theorem 4. Given a linear differential equation

$$a_1 x + a_2 \dot{x} + a_3 \ddot{x} + \ldots = g \tag{4}$$

if we can find a complex solution $\bar{z}(t) = x(t) + iy(t)$ for the differential equation

$$a_1 \bar{z} + a_2 \dot{\bar{z}} + a_3 \ddot{\bar{z}} + \ldots = \bar{f} \tag{5}$$

where $\bar{f} = g + ih$ is a complex function whose real part is equal to g, then the real part of $\bar{z}(t)$ will be a solution to the differential Eq. (4). That is,

$$x(t) = \text{Re}[\bar{z}(t)] \tag{6}$$

is a solution to the differential Eq. (4).

SECOND ORDER LINEAR HOMOGENEOUS DIFFERENTIAL EQUATION WITH CONSTANT COEFFICIENTS

The solution to the linear differential equation

$$a\ddot{x} + b\dot{x} + cx = 0 \tag{7}$$

where a, b, and c are constants, is most easily determined by hunting for complex solutions to the equation

$$a\ddot{\bar{z}} + b\dot{\bar{z}} + c\bar{z} = 0 \tag{8}$$

To solve Eq. (8) we first guess the solution to be of the form

$$\bar{z} = \bar{C} \exp(\bar{\gamma} t) \tag{9}$$

where $\bar{C}$ and $\bar{\gamma}$ are constants. Substituting Eq. (9) in Eq. (8) we obtain Eq. (10) as a necessary and sufficient condition for Eq. (9) to be a solution of Eq. (8)

$$a\bar{\gamma}^2 + b\bar{\gamma} + c = 0 \tag{10}$$

Solving for $\bar{\gamma}$ we obtain

$$\bar{\gamma} = \frac{-b \pm (b^2 - 4ac)^{1/2}}{2a} \tag{11}$$

There are thus two independent solutions of the desired form, namely,

$$\bar{z}_1 = \bar{C}_1 \exp\left\{\left[\frac{-b + (b^2 - 4ac)^{1/2}}{2a}\right]t\right\} \tag{12}$$

$$\bar{z}_2 = \bar{C}_2 \exp\left\{\left[\frac{-b - (b^2 - 4ac)^{1/2}}{2a}\right]t\right\} \tag{13}$$

If $b^2 \neq 4ac$ the two solutions are independent and we can obtain a complete solution by taking the sum of the two solutions above. The complete solution is thus

$$\bar{z} = \exp\left(-\frac{bt}{2a}\right)\left\{\bar{C}_1 \exp\left[\frac{(b^2 - 4ac)^{1/2}}{2a}t\right] + \bar{C}_2 \exp\left[-\frac{(b^2 - 4ac)^{1/2}}{2a}t\right]\right\} \tag{14}$$

If $b^2 = 4ac$, then the two solutions are not independent. In this case we guess the solution to be of the form $(\bar{C}_1 + \bar{C}_2 t) \exp(\bar{\gamma} t)$, and we find that this will be a solution for arbitrary $\bar{C}_1$ and $\bar{C}_2$, provided $\bar{\gamma}$ is $-b/2a$. The complete solution for this case is thus

$$\bar{z} = (\bar{C}_1 + \bar{C}_2 t) \exp\left(-\frac{bt}{2a}\right) \tag{15}$$

The complete solution for the differential Eq. (7) can now be obtained by taking the real part of the complete solution to the differential Eq. (8). The details of the calculations are straightforward, and we simply state the results. There are three cases, depending on whether $b^2 - 4ac < 0$, $b^2 - 4ac > 0$, or $b^2 - 4ac = 0$. We consider each case separately.

Case 1. If $b^2 - 4ac < 0$, the solution to Eq. (7) can be written in any one of the following equivalent forms (see Appendix 1):

$$\begin{aligned} x &= \exp(-\beta t)[A_1 \cos \omega_1 t + A_2 \sin \omega_1 t] \\ &\equiv \exp(-\beta t)[A \cos(\omega_1 t - \phi)] \\ &\equiv \exp(-\beta t)[A \sin(\omega_1 t - \psi)] \end{aligned} \tag{16}$$

where

$$\beta \equiv \frac{b}{2a} \tag{17}$$

$$\omega_1 \equiv \frac{(4ac - b^2)^{1/2}}{2a} \equiv (\omega_0^2 - \beta^2)^{1/2} \tag{18}$$

and A_1, A_2, A, ϕ, and ψ are constants whose values are determined by the initial conditions. The constants are related as follows:

$$A \equiv (A_1^2 + A_2^2)^{1/2} \tag{19}$$

$$\phi = \psi + \frac{\pi}{2} = \tan^{-1}\left(\frac{A_2}{A_1}\right) \tag{20}$$

Case 2. If $b^2 - 4ac > 0$, the solution to Eq. (7) is

$$x = \exp(-\beta t)\left[A \exp(\omega_2 t) + B \exp(-\omega_2 t)\right] \tag{21}$$

where

$$\beta \equiv \frac{b}{2a} \tag{22}$$

$$\omega_2 \equiv \frac{(b^2 - 4ac)^{1/2}}{2a} \equiv \left[\beta^2 - \omega_0^2\right]^{1/2} \tag{23}$$

and A and B are constants whose values are determined by the initial conditions.

Case 3. If $b^2 - 4ac = 0$, the solution to Eq. (7) is

$$x = (A + Bt)\exp(-\beta t) \tag{24}$$

where

$$\beta \equiv \frac{b}{2a} \tag{25}$$

and A and B are constants whose values are determined by the initial conditions.

SECOND ORDER LINEAR INHOMOGENEOUS DIFFERENTIAL EQUATIONS WITH CONSTANT COEFFICIENTS

The complete solution to the inhomogeneous linear differential equation

$$a\ddot{x} + b\dot{x} + cx = g(t) \tag{26}$$

where a, b, and c are constants and $g(t)$ is some function of t can be obtained by finding a particular solution to the inhomogeneous equation and adding it to the complete solution of the corresponding homogeneous equation, which we obtained in the preceding section.

If the function $g(t)$ can be broken down into a series of terms, that is, $g(t) = \sum_i g_i(t)$, and if we can find a particular solution $x = x_i(t)$ for each of the differential equations $a\ddot{x} + b\dot{x} + cx = g_i(t)$, the function $x = \sum_i x_i(t)$ will be a particular solution of Eq. (26).

If the function $g(t)$ is periodic it can be expressed by means of a Fourier series as a sum of terms of the form $d \cos(\omega t + \delta)$, where d, ω, and δ are constants. It follows that if we can find a particular solution to the equation

$$a\ddot{x} + b\dot{x} + cx = d\cos(\omega t + \delta) \tag{27}$$

then we can find a particular solution for Eq. (26) for arbitrary periodic functions $g(t)$.

A particular solution to Eq. (27) is most easily obtained by hunting for a particular complex solution to the equation

$$a\ddot{z} + b\dot{z} + cz = \bar{d}\exp(i\omega t) \tag{28}$$

where
$$\bar{d} = d\exp(i\delta) \tag{29}$$

To find a particular solution to Eq. (28) we first guess the solution to be of the form
$$\bar{z} = \bar{A}\exp(i\omega t) \tag{30}$$

Substituting Eq. (30) in Eq. (28) we obtain Eq. (31) as a necessary and sufficient condition for Eq. (30) to be a solution to Eq. (28)
$$-a\omega^2\bar{A} + ib\omega\bar{A} + c\bar{A} = \bar{d} \tag{31}$$

Solving for $\bar{A}$ we obtain
$$\bar{A} = \frac{\bar{d}}{(c - a\omega^2) + ib\omega} \tag{32}$$

which can be written in the equivalent form
$$\bar{A} = \frac{(\bar{d}/a)\left[(\omega_0^2 - \omega^2) - i2\beta\omega\right]}{(\omega_0^2 - \omega^2)^2 + 4\beta^2\omega^2} \tag{33}$$

where
$$\omega_0^2 = \frac{c}{a} \tag{34}$$

$$\beta = \frac{b}{2a} \tag{35}$$

It follows that a particular solution to Eq. (28) is given by
$$\bar{z} = \frac{(\bar{d}/a)\left[(\omega_0^2 - \omega^2) - i2\beta\omega\right]}{(\omega_0^2 - \omega^2)^2 + 4\beta^2\omega^2}\exp(i\omega t) \tag{36}$$

A particular solution for Eq. (27) can now be obtained by taking the real part of Eq. (36). If we do this we obtain
$$x = \frac{(d/a)\cos(\omega t + \delta - \phi)}{\left[(\omega_0^2 - \omega^2)^2 + 4\beta^2\omega^2\right]^{1/2}} \tag{37}$$

where
$$\tan\phi = \frac{2\omega\beta}{\omega_0^2 - \omega^2} \tag{38}$$

APPENDIX 10

Differentiation of an Integral

If we are given an integral of the form

$$I(x) = \int_{u(x)}^{v(x)} f(x, y)\, dy \tag{1}$$

where u, v, and f are arbitrary functions of the variables indicated, the derivative with respect to x is given by

$$\frac{dI(x)}{dx} = f[x, v(x)] \frac{dv(x)}{dx} - f[x, u(x)] \frac{du(x)}{dx}$$
$$+ \int_{u(x)}^{v(x)} \frac{\partial f(x, y)}{\partial x}\, dy \tag{2}$$

To prove Eq. (2), let us first define $\phi(x, y)$ as the function whose partial derivative with respect to y is $f(x, y)$, that is,

$$\frac{\partial \phi(x, y)}{\partial y} = f(x, y) \tag{3}$$

If we substitute Eq. (3) in Eq. (1) we obtain

$$I = \int_u^v \frac{\partial \phi(x, y)}{\partial y}\, dy = \phi(x, v) - \phi(x, u) \tag{4}$$

Taking the derivative of Eq. (4) with respect to x we obtain

$$\frac{dI}{dx} = \frac{\partial \phi(x, v)}{\partial v} \frac{dv}{dx} - \frac{\partial \phi(x, u)}{\partial u} \frac{du}{dx} + \frac{\partial \phi(x, v)}{\partial x} - \frac{\partial \phi(x, u)}{\partial x}$$

$$= f(x, v) \frac{dv}{dx} - f(x, u) \frac{du}{dx} + \int_u^v \frac{\partial}{\partial y}\left[\frac{\partial \phi(x, y)}{\partial x}\right] dy$$

$$= f(x, v) \frac{dv}{dx} - f(x, u) \frac{du}{dx} + \int_u^v \frac{\partial f(x, y)}{\partial x}\, dy \tag{5}$$

which is the desired result.

APPENDIX 11

Exact Differentials

In this appendix we consider the definition and properties of exact differentials. We suppose throughout that we are dealing with a set of quantities $f_1, f_2, \ldots, f_n$ that are functions of the variables $x_1, x_2, \ldots, x_n$ and that each of the functions f_i and each of its partial derivatives $\partial f_i/\partial x_j$ is single valued and continuous everywhere in a region R.

Definition. The quantity $\sum_i f_i\, dx_i$ is said to be an exact differential if and only if there exists a function $F(\mathbf{x})$ such that

$$dF = \sum_i f_i\, dx_i \tag{1}$$

everywhere in R or, equivalently, if and only if there exists a function $F(\mathbf{x})$ such that

$$f_i = \frac{\partial F}{\partial x_i} \tag{2}$$

Theorem 1. The quantity $\sum_i f_i\, dx_i$ is an exact differential if and only if for any pair of values of i and j

$$\frac{\partial f_i}{\partial x_j} = \frac{\partial f_j}{\partial x_i} \tag{3}$$

everywhere in the region R.

PROOF: If $\sum_i f_i\, dx_i$ is an exact differential there exists an F such that $dF = \sum_i f_i\, dx_i$. It follows that

$$\frac{\partial f_i}{\partial x_j} = \frac{\partial}{\partial x_j}\left[\frac{\partial F}{\partial x_i}\right] = \frac{\partial}{\partial x_i}\left[\frac{\partial F}{\partial x_j}\right] = \frac{\partial f_j}{\partial x_i} \tag{4}$$

To prove the converse we need to show that if Eq. (3) is true, there exists a function F such that $dF = \sum_i f_i \, dx_i$. Let us consider the function

$$F(\mathbf{x}) \equiv \sum_{i=1}^{n} \int_0^{x_i} f_i(x_1, x_2, \ldots, x_i, 0, 0, \ldots) \, dx_i \tag{5}$$

which is nothing more than the line integral of $\sum_i f_i \, dx_i$ from 0 to $\mathbf{x}$ along the particular path that is obtained by joining the points $(0, 0, \ldots, 0)$, $(x_1, 0, 0, \ldots, 0)$, $(x_1, x_2, 0, 0, \ldots, 0)$, $\ldots$, and $(x_1, x_2, \ldots, x_n)$ with straight lines. We assume of course that the path lies in the region R. If it does not, we can choose another path that does. If we take the partial derivative of $F(\mathbf{x})$ with respect to x_i, and make use of the results of Appendix 10, we obtain

$$\frac{\partial F}{\partial x_i} = f_i(x_1, x_2, \ldots, x_i, 0, 0, \ldots)$$

$$+ \sum_{j=i+1}^{n} \int_0^{x_j} \frac{\partial f_j(x_1, x_2, \ldots, x_j, 0, 0, \ldots)}{\partial x_i} \, dx_j \tag{6}$$

If Eq. (3) is true, we can rewrite Eq. (6)

$$\frac{\partial F}{\partial x_i} = f_i(x_1, x_2, \ldots, x_i, 0, 0, \ldots)$$

$$+ \sum_{j=i+1}^{n} \int_0^{x_j} \frac{\partial f_i(x_1, x_2, \ldots, x_j, 0, 0, \ldots)}{\partial x_j} \, dx_j$$

$$= f_i(x_1, x_2, \ldots, x_i, 0, 0, \ldots)$$

$$+ \sum_{j=i+1}^{n} [f_i(x_1, x_2, \ldots, x_j, 0, 0, \ldots)$$

$$- f_i(x_1, x_2, \ldots, x_{j-1}, 0, 0, \ldots)]$$

$$= f_i(x_1, x_2, \ldots, x_n) \tag{7}$$

From Eq. (7) it follows that

$$dF \equiv \sum_i \frac{\partial F}{\partial x_i} \, dx_i = \sum_i f_i \, dx_i \tag{8}$$

Thus if Eq. (3) is true there is a function F, namely, the function defined by Eq. (5), such that $dF = \sum_i f_i \, dx_i$. This completes the proof.

Theorem 2. The line integral $\int \sum_i f_i \, dx_i$ between an arbitrary pair of points, **a** and **b**, in R depends only on the endpoints **a** and **b**, and not on the path, if and only if the following two conditions are satisfied:

a. The quantity $\sum_i f_i \, dx_i$ is an exact differential; that is, either there exists a function F such that $f_i = \partial F / \partial x_i$ everywhere in R, or equivalently $\partial f_i / \partial x_j = \partial f_j / \partial x_i$ everywhere in R.

b. Either the function $F(\mathbf{x})$ is single valued, or the region R is simply connected, that is, any pair of paths with the same endpoints can be continuously deformed one into the other through intermediate paths all lying in R and all having the same endpoints.

NOTE: If condition a is satisfied and the region R is simply connected, then the function $F(\mathbf{x})$ will be single valued. But if condition a is satisfied and the function $F(\mathbf{x})$ is single valued it does not necessarily follow that the region R is simply connected.

PROOF: If there exists an $F(\mathbf{x})$ such that $dF = \sum_i f_i \, dx_i$, then

$$\int_{\mathbf{a}}^{\mathbf{b}} \left[\sum_i f_i \, dx_i \right] = \int_{\mathbf{a}}^{\mathbf{b}} dF = F(\mathbf{b}) - F(\mathbf{a}) \tag{9}$$

If $F(\mathbf{x})$ is single valued the quantity $F(\mathbf{b}) - F(\mathbf{a})$ depends only on the endpoints. Therefore the line integral $\int [\sum_i f_i \, dx_i]$ depends only on the endpoints. Conversely suppose the line integral $\int \sum_i f_i \, dx_i$ depends only on the endpoints. We can then define a single valued functions $F(\mathbf{x})$ as follows:

$$F(\mathbf{x}) = \int_{\mathbf{a}}^{\mathbf{x}} \sum_i f_i \, dx_i \tag{10}$$

where $\mathbf{a}$ is some fixed point. From Eq. (10) it follows that

$$\frac{\partial F}{\partial x_i} \equiv \lim_{\Delta x_i \to 0} \left\{ \frac{F(x_1, x_2, \ldots, x_i + \Delta x_i, x_{i+1}, \ldots, x_n) - F(x_1, x_2, \ldots, x_n)}{\Delta x_i} \right\}$$

$$= \lim_{\Delta x_i \to 0} \left\{ \frac{\int_{x_1, \ldots, x_i, \ldots, x_n}^{x_1, \ldots, x_i + \Delta x_i, \ldots, x_n} \sum_i f_i \, dx_i}{\Delta x_i} \right\}$$

$$= \lim_{\Delta x_i \to 0} \left\{ \frac{\int_{x_i}^{x_i + \Delta x_i} f_i \, dx_i}{\Delta x_i} \right\} = f_i \tag{11}$$

From Eq. (11) it follows that

$$dF = \sum_i \frac{\partial F}{\partial x_i} \, dx_i = \sum_i f_i \, dx_i \tag{12}$$

and therefore $\sum_i f_i \, dx_i$ is an exact differential.

To complete the proof we show that if condition a is satisfied, and if R is simply connected, the line integral $\int \sum_i f_i \, dx_i$ between $\mathbf{a}$ and $\mathbf{b}$ will be independent of the path. Let I represent the line integral $\int \sum_i f_i \, dx_i$ along some arbitrary path joining $\mathbf{a}$ and $\mathbf{b}$, that is,

$$I = \int_{\mathbf{a}}^{\mathbf{b}} \sum_i f_i \, dx_i \tag{13}$$

If we consider a variation of the path (see Appendix 32) with the endpoints $\mathbf{a}$

and **b** fixed, then

$$\delta I = \delta \int_a^b \sum_i f_i \, dx_i = \int_a^b \left[\sum_i \delta f_i \, dx_i + \sum_i f_i \delta(dx_i) \right]$$

$$= \int_a^b \left[\sum_i \delta f_i \, dx_i + \sum_i f_i d(\delta x_i) \right] \quad (14)$$

Integrating the last term by parts we obtain

$$\delta I = \int_a^b \left[\sum_i \delta f_i \, dx_i - \sum_i df_i \, \delta x_i \right]$$

$$= \int_a^b \left[\sum_i \sum_j \frac{\partial f_i}{\partial x_j} \delta x_j \, dx_i - \sum_i \sum_j \frac{\partial f_i}{\partial x_j} dx_j \, \delta x_i \right]$$

$$= \int_a^b \left\{ \sum_i \sum_j \left[\frac{\partial f_i}{\partial x_j} - \frac{\partial f_j}{\partial x_i} \right] \delta x_j \, dx_i \right\} \quad (15)$$

If condition a is true then $\partial f_i/\partial x_j = \partial f_j/\partial x_i$ everywhere in R, and therefore as long as the variation of the path does not carry the path outside of the region R

$$\delta I = 0 \quad (16)$$

But we have assumed that R is simply connected, and therefore any two paths joining **a** and **b** can be continuously deformed from one into the other without crossing the boundary of R, and since any such deformation can be carried out by a succession of infinitesimal variations in which $\delta I = 0$, it follows that the value of I will be the same for all paths in R joining **a** and **b**. This completes the proof of the theorem.

Theorem 3. The line integral $\int \sum_i f_i \, dx_i$ between any arbitrary pair of points **a** and **b** in R will depend on the endpoints **a** and **b** alone if and only if the line integral $\int \sum_i f_i \, dx_i$ around every closed path in R vanishes.

PROOF: Let us consider an arbitrary closed path that passes successively through the points **a**, **b**, **c**, and **d** and then returns to the point **a**. We represent the integral along this path as $I(\mathbf{a}, \mathbf{b}, \mathbf{c}, \mathbf{d}, \mathbf{a})$. We can break this integral down into components as follows:

$$I(\mathbf{a}, \mathbf{b}, \mathbf{c}, \mathbf{d}, \mathbf{a}) = I(\mathbf{a}, \mathbf{b}, \mathbf{c}) + I(\mathbf{c}, \mathbf{d}, \mathbf{a})$$
$$= I(\mathbf{a}, \mathbf{b}, \mathbf{c}) - I(\mathbf{a}, \mathbf{d}, \mathbf{c}) \quad (17)$$

If the integral I between points **a** and **c** is independent of the path, then $I(\mathbf{a}, \mathbf{b}, \mathbf{c}) - I(\mathbf{a}, \mathbf{d}, \mathbf{c}) = 0$; hence $I(\mathbf{a}, \mathbf{b}, \mathbf{c}, \mathbf{d}, \mathbf{a}) = 0$. Hence if the integral between an arbitrary pair of points is independent of the path, the integral around any closed path is zero. The converse is similarly proved by noting that any arbitrary pair of paths $I(\mathbf{a}, \mathbf{b}, \mathbf{c})$ and $I(\mathbf{a}, \mathbf{d}, \mathbf{c})$ joining two points **a** and **c** can be appropriately combined to form a single closed path $I(\mathbf{a},\mathbf{b},\mathbf{c},\mathbf{d},\mathbf{a})$ by just reversing the procedure represented by Eq. (17). It follows that if the integral around every closed path is zero, $I(\mathbf{a},\mathbf{b},\mathbf{c}) = I(\mathbf{a},\mathbf{d},\mathbf{c})$. Hence if the integral around every path is zero the integral between any arbitrary pair of points is independent of the path.

Example. Show that the quantity

$$\sum_i f_i\, dx_i = \frac{x_2\, dx_1 - x_1\, dx_2}{x_1^2 + x_2^2} \tag{18}$$

is an exact differential in a region R that excludes the origin but that the line integral $\int \sum_i f_i\, dx_i$ between two points $\mathbf{a}$ and $\mathbf{b}$ depends on the path.

Solution. The quantity above is the differential of the function

$$F = \tan^{-1}\left(\frac{x_1}{x_2}\right) \tag{19}$$

which is defined everywhere in R, hence is an exact differential. The function F however is not single valued, nor is the region R simply connected; hence the integral depends on the path. This can also be verified by considering two paths joining $\mathbf{a}$ and $\mathbf{b}$ such that the closed curve formed by the two paths encloses the origin. The values of the integral along these two paths will be different.

APPENDIX 12

Matrices

INTRODUCTION

A *matrix* is a rectangular array of numbers. We use the notation

$$[a] \equiv [a_{ij}] \equiv \begin{bmatrix} a_{11} & a_{12} & \cdots & a_{1n} \\ a_{21} & a_{22} & \cdots & a_{2n} \\ \vdots & \vdots & & \vdots \\ a_{m1} & a_{m2} & \cdots & a_{mn} \end{bmatrix} \qquad (1)$$

to represent a matrix. A matrix having m rows and n columns is said to be of order $m \times n$ (read: m by n) or to be an $m \times n$ matrix. The element in the ith row and the jth column of the matrix $[a]$ is designated a_{ij} and is called the ij element of the matrix. Note carefully that the first subscript refers to the row and the second to the column.

Equality of Matrices

Two matrices are said to be equal if and only if they are of the same order and the corresponding elements of the matrices are equal. Thus if $[a]$ is a matrix of order $m \times n$ and $[b]$ is a matrix of order $r \times s$, then $[a] = [b]$ if and only if $m = r$, $n = s$, and $a_{ij} = b_{ij}$.

MATRIX OPERATIONS

Addition of Matrices

If $[a]$ and $[b]$ are two matrices of the same order, we define the sum $[a] + [b]$ of $[a]$ and $[b]$ as the matrix $[c]$ whose elements are $c_{ij} = a_{ij} + b_{ij}$. Thus

$$[a_{ij}] + [b_{ij}] = [a_{ij} + b_{ij}] \qquad (2)$$

Multiplication of a Matrix by a Number

If λ is a number and $[a]$ is a matrix, we define the product $\lambda[a]$ or $[a]\lambda$ of λ and $[a]$ as the matrix $[c]$ whose elements are $c_{ij} = \lambda a_{ij}$, that is,

$$\lambda[a_{ij}] = [a_{ij}]\lambda = [\lambda a_{ij}] \tag{3}$$

Multiplication of Two Matrices

If $[a]$ is a matrix of order $m \times n$, and $[b]$ is a matrix of order $n \times p$, we define the product $[a][b]$ of $[a]$ and $[b]$ as the matrix $[c]$ of order $m \times p$ whose elements are $c_{ik} = \sum_j a_{ij} b_{jk}$. Thus

$$[a_{ij}][b_{jk}] = \left[\sum_j a_{ij} b_{jk}\right] \tag{4}$$

Note that the ij element in the product $[c]$ is the product of the ith row in $[a]$ and the jth column in $[b]$.

MNEMONIC AID. The following diagram is useful in remembering how an element is formed in a product.

$$\begin{bmatrix} a_{i1} \cdots a_{in} \end{bmatrix} \begin{bmatrix} b_{1j} \\ \vdots \\ b_{nj} \end{bmatrix} = \begin{bmatrix} c_{ij} \end{bmatrix} \tag{5}$$

Transposition of a Matrix

If $[a]$ is a matrix we define the transpose of $[a]$, which we designate $[a]^T$, as the matrix obtained by interchanging the rows and the columns of $[a]$. Thus if

$$[a] = \begin{bmatrix} a_{11} & a_{12} & \cdots & a_{1n} \\ a_{21} & a_{22} & \cdots & a_{2n} \\ \vdots & & & \vdots \\ a_{m1} & a_{m2} & \cdots & a_{mn} \end{bmatrix} \tag{6}$$

then

$$[a]^T = \begin{bmatrix} a_{11} & a_{21} & \cdots & a_{m1} \\ a_{12} & a_{22} & \cdots & a_{m2} \\ \vdots & & & \vdots \\ a_{1n} & a_{2n} & \cdots & a_{mn} \end{bmatrix} \tag{7}$$

Note that if $[a]$ is of order $m \times n$, then $[a]^T$ is of order $n \times m$.

Complex Conjugation of a Matrix

If $[a]$ is a matrix we define the complex conjugate of $[a]$, which we designate as $[a]^*$, as the matrix obtained by replacing each element in $[a]$ by its complex conjugate.

Transposition Plus Complex Conjugation of a Matrix

If $[a]$ is a matrix we designate the matrix that is obtained by taking the transposed complex conjugate of $[a]$ as $[a]^\dagger$. Thus

$$[a]^\dagger \equiv \{[a]^*\}^T \equiv \{[a]^T\}^* \tag{8}$$

TYPES OF MATRIX

A *zero matrix*, which we designate $[0]$, is a matrix all of whose elements are zero. Thus

$$[0] \equiv \begin{bmatrix} 0 & 0 & \cdots & 0 \\ 0 & 0 & \cdots & 0 \\ \vdots & \vdots & & \vdots \\ 0 & 0 & \cdots & 0 \end{bmatrix} \tag{9}$$

A *row matrix* is a matrix consisting of a single row. When we wish to distinguish a row matrix from matrices containing more than one row we use one of the following notations:

$$[a_1 a_2 \cdots a_n] \equiv (a_i] \equiv [a] \tag{10}$$

A row matrix of order $1 \times n$ is spoken of as simply a row matrix of order n.

A *column matrix* is a matrix consisting of a single column. When we wish to distinguish a column matrix from matrices containing more than one column, we use one of the following notations:

$$\begin{bmatrix} a_1 \\ a_2 \\ \vdots \\ a_n \end{bmatrix} \equiv [a_i) \equiv [a) \tag{11}$$

A column matrix of order $n \times 1$ is spoken of as simply a column matrix of order n.

MNEMONIC AID. To remember that the parenthesis is on the *left for a row* matrix and on the *right for a column* matrix, keep in mind that in designating an element a_{ij} in a matrix $[a_{ij}]$ the *left subscript designates the row* and the *right subscript designates the column*.

A *square matrix* is a matrix having the same number of rows as columns. A square matrix of order $n \times n$ is spoken of as simply a square matrix of order n.

The diagonal extending from the upper left corner to the lower right corner of a square matrix is called the *principal diagonal* of the matrix.

The sum of the elements along the principal diagonal of a square matrix $[a]$ is called the *trace* of the matrix and is designated $\text{Tr}[a]$, that is,

$$\text{Tr}[a] \equiv \sum_i a_{ii} \qquad (12)$$

A *diagonal matrix* is a square matrix in which all terms other than those along the principal diagonal are zero. Thus if $[a]$ is a diagonal matrix with diagonal elements $k_1, k_2, \ldots, k_n$, we write

$$[a] = [k_i \delta_{ij}] = \begin{bmatrix} k_1 & 0 & \ldots & 0 \\ 0 & k_2 & \ldots & 0 \\ \vdots & & & \vdots \\ 0 & 0 & \ldots & k_n \end{bmatrix} \qquad (13)$$

A *scalar matrix* is a diagonal matrix in which all the diagonal elements are equal. Thus if $[a]$ is a scalar matrix whose diagonal elements are equal to k, we have

$$[a] = [k \delta_{ij}] = \begin{bmatrix} k & 0 & \ldots & 0 \\ 0 & k & \ldots & 0 \\ \vdots & \vdots & & \vdots \\ 0 & 0 & \ldots & k \end{bmatrix} \qquad (14)$$

A *unit matrix* $[e]$ is a scalar matrix in which the diagonal elements are equal to unity. Thus

$$[e] = [\delta_{ij}] = \begin{bmatrix} 1 & 0 & \ldots & 0 \\ 0 & 1 & \ldots & 0 \\ \vdots & \vdots & & \vdots \\ 0 & 0 & \ldots & 1 \end{bmatrix} \qquad (15)$$

A *symmetric matrix* is a square matrix in which the ij element is equal to the ji element.

A *skew-symmetric* or *antisymmetric matrix* is a square matrix in which the ij element is equal to the negative of the ji element.

A real symmetric matrix $[a]$ is said to be *positive definite* if the quadratic form $(x)[a](x)$ is greater than zero for every nonzero value of $[x]$. A real symmetric matrix is said to be *nonnegative* if the quadratic form $(x)[a](x)$ is greater than or equal to zero for every nonzero value of $[x]$. A real symmetric matrix $[a]$ is said to be *positive semidefinite* if the quadratic form $(x)[a](x)$ is greater than or equal to zero for every nonzero value of $[x]$ and equal to zero for some nonzero value of $[x]$.

PROPERTIES OF MATRICES

In this section we state without proof a number of useful properties of matrices.

Theorem 1. The operations of addition and multiplication obey the following rules:

$$[a] + [b] = [b] + [a] \tag{16}$$

$$[a] + \{[b] + [c]\} = \{[a] + [b]\} + [c] \tag{17}$$

$$k\{[a] + [b]\} = k[a] + k[b] \tag{18}$$

$$\{k_1 + k_2\}[a] = k_1[a] + k_2[a] \tag{19}$$

$$[a]\{[b][c]\} = \{[a][b]\}[c] \tag{20}$$

$$[a]\{[b] + [c]\} = [a][b] + [a][c] \tag{21}$$

NOTE: Matrix multiplication is *not* in general commutative, that is,

$$[a][b] \neq [b][a] \tag{22}$$

Theorem 2. Diagonal matrices commute with one another. Thus if [a] and [b] are two diagonal matrices, [a][b] = [b][a].

Theorem 3. The unit and zero matrices obey the following laws:

$$[a][e] = [a] \quad \text{and} \quad [e][a] = [a] \tag{23}$$

$$[a][0] = [0] \quad \text{and} \quad [0][a] = [0] \tag{24}$$

$$[a] + [0] = [0] + [a] = [a] \tag{25}$$

NOTE 1: We did not write the property represented by Eq. (23) in the form $[a][e] = [e][a] = [a]$ because if [a] is of order $m \times n$, where $m \neq n$, then the premultiplying [e] is of order m and the postmultiplying [e] is of order n. A similar comment applies to the property represented by Eq. (24).

NOTE 2: If $[a][b] = [0]$ it does not follow that either $[a] = [0]$ or $[b] = [0]$.

Theorem 4. The operation of transposition obeys the following rules:

$$\{[a] + [b]\}^T = [a]^T + [b]^T \tag{26}$$

$$\{[a][b]\}^T = [b]^T[a]^T \tag{27}$$

$$\{[a]^T\}^T = [a] \tag{28}$$

DETERMINANTS

If $[a]$ is a square matrix we define the determinant of $[a]$, which we designate

$$|a| \equiv |[a]| \equiv |a_{ij}| \equiv \begin{vmatrix} a_{11} & a_{12} & \cdots & a_{1n} \\ a_{21} & a_{22} & \cdots & a_{2n} \\ \vdots & \vdots & & \vdots \\ a_{m1} & a_{m2} & \cdots & a_{mn} \end{vmatrix} \qquad (29)$$

as the number

$$|a| \equiv \sum_{i_1} \sum_{i_2} \cdots \sum_{i_n} \epsilon_{i_1 i_2 \cdots i_n} a_{i_1 1} a_{i_2 2} \cdots a_{i_n n} \qquad (30)$$

where

$$\begin{aligned}\epsilon_{i_1 i_2 \cdots i_n} &= +1 && \text{if } i_1 i_2 \cdots i_n \text{ is an even permutation of } 12 \cdots n \\ &= -1 && \text{if } i_1 i_2 \cdots i_n \text{ is an odd permutation of } 12 \cdots n \\ &= 0 && \text{otherwise} \end{aligned} \qquad (31)$$

The evaluation of the determinant of a square matrix by means of the definition is a tedious process if the matrix is of order greater than 2. In the following paragraphs we outline without proof a simpler method of evaluating a determinant.

If $[a]$ is a square matrix we define the *complementary minor* $|M_{ij}|$ of the element a_{ij} of $[a]$ as the determinant of the matrix obtained by striking out the ith row and the jth column of the matrix $[a]$, and we define the *cofactor* $|A_{ij}|$ of the element a_{ij} of $[a]$ as its complementary minor prefixed by the sign $(-1)^{i+j}$, that is,

$$|A_{ij}| = (-1)^{i+j} |M_{ij}| \qquad (32)$$

MNEMONIC AID. To determine whether the minor $|M_{ij}|$ of the element a_{ij} is multiplied by $+$ or $-$ to form the cofactor, one can use the following diagram.

$$\begin{bmatrix} + & - & + & - & \cdots \\ - & + & - & + & \cdots \\ + & - & + & - & \cdots \\ - & + & - & + & \cdots \\ \vdots & \vdots & \vdots & \vdots & \end{bmatrix} \qquad (33)$$

Theorem 5. If $[a]$ is a square matrix of order n, the value of the determinant $|a|$ is equal to the sum of the products of the elements $a_{i1}, a_{i2}, \ldots, a_{in}$ in any row i and their corresponding cofactors $|A_{i1}|, |A_{i2}|, \ldots, |A_{in}|$, or the sum of the products of the elements $a_{1j}, a_{2j}, \ldots, a_{nj}$ in any column j and their corresponding cofactors $|A_{1j}|, |A_{2j}|, \ldots, |A_{nj}|$. Thus

$$|a| = \sum_i a_{ij}|A_{ij}| \qquad j = 1, 2, \ldots, n \tag{34}$$

$$|a| = \sum_j a_{ij}|A_{ij}| \qquad i = 1, 2, \ldots, n \tag{35}$$

This theorem enables us to reduce the evaluation of a determinant of order n to the evaluation of a set of determinants of order $n - 1$. These determinants can in turn be expressed in terms of determinants of order $n - 2$. Proceeding in this fashion we can ultimately reduce the problem to the evaluation of determinants of as low an order as we desire.

The following theorems, which we state without proof, are useful in the evaluation and manipulation of determinants.

Theorem 6. The value of a determinant is zero if any of the following conditions are satisfied:
 a. All the elements in one row (or one column) are zero.
 b. Two rows (or two columns) are identical.
 c. Two rows (or two columns) are proportional.

Theorem 7. The value of a determinant is unchanged if we perform any one of the following operations:
 a. We interchange all the rows with the columns.
 b. We add to each element of one row k times the corresponding element in another row, where k is any number.
 c. We add to each element of one column k times the corresponding element in another column, where k is any number.

Theorem 8. The value of a determinant changes its sign but not its magnitude if two rows are interchanged or if two columns are interchanged.

Theorem 9. If each element in one row (or column) of a determinant is multiplied by a number k, the value of the determinant is multiplied by k.

Theorem 10. If the elements $a_{i1}, a_{i2}, \ldots, a_{in}$ of a given row i in a determinant $|a|$ are written as the sum of two terms, that is, $a_{i1}, a_{i2}, \ldots, a_{in} \equiv b_{i1} + c_{i1}, b_{i2} + c_{i2}, \ldots, b_{in} + c_{in}$, the determinant $|a|$ can be written as the sum of the determinant obtained by setting the terms $c_{i1}, c_{i2}, \ldots, c_{in}$ equal to zero in $|a|$ and the determinant obtained by setting the terms $b_{i1}, b_{i2}, \ldots, b_{in}$ equal to zero in $|a|$. Thus

$$\begin{vmatrix} a_{11} & a_{12} & \cdots & a_{1n} \\ a_{21} & a_{22} & \cdots & a_{2n} \\ \vdots & & & \vdots \\ a_{i1} & a_{i2} & & a_{in} \\ \vdots & & & \vdots \\ a_{n1} & a_{n2} & \cdots & a_{nn} \end{vmatrix} = \begin{vmatrix} a_{11} & a_{12} & \cdots & a_{1n} \\ a_{21} & a_{22} & \cdots & a_{2n} \\ \vdots & & & \vdots \\ b_{i1}+c_{i1} & b_{i2}+c_{i2} & \cdots & b_{in}+c_{in} \\ \vdots & & & \vdots \\ a_{n1} & a_{n2} & \cdots & a_{nn} \end{vmatrix}$$

$$= \begin{vmatrix} a_{11} & a_{12} & \cdots & a_{1n} \\ a_{21} & a_{22} & \cdots & a_{2n} \\ \vdots & & & \vdots \\ b_{i1} & b_{i2} & \cdots & b_{in} \\ \vdots & & & \vdots \\ a_{n1} & a_{n2} & \cdots & a_{nn} \end{vmatrix} + \begin{vmatrix} a_{11} & a_{12} & \cdots & a_{1n} \\ a_{21} & a_{22} & \cdots & a_{2n} \\ \vdots & & & \vdots \\ c_{i1} & c_{i2} & \cdots & c_{in} \\ \vdots & & & \vdots \\ a_{n1} & a_{n2} & \cdots & a_{nn} \end{vmatrix} \qquad (36)$$

A similar theorem applies if the elements of a column are written as the sum of two terms.

Theorem 11. If $[a]$ and $[b]$ are square matrices of the same order then the determinant of the product $[a][b]$ is equal to the product of the determinants of $[a]$ and $[b]$. Thus

$$[c] = [a][b] \Rightarrow |c| = |a||b| \qquad (37)$$

Theorem 12. If $[a]$ is a square matrix, the sum of the products of the elements in any row (column) and the corresponding cofactors of the elements in another row (column) is zero. Thus

$$\sum_j a_{ij}|A_{kj}| = 0 \qquad i \neq k \qquad (38)$$

$$\sum_i a_{ij}|A_{ik}| = 0 \qquad j \neq k \qquad (39)$$

Theorem 13. If $[a)_1, [a)_2, \ldots, [a)_n$ are the columns of a square matrix $[a]$ of order n, then $|a| = 0$ if and only if the columns are linearly dependent, that is, if and only if there exists a set of constants $\lambda_1, \lambda_2, \ldots, \lambda_n$ not all zero such that

$$\sum_i \lambda_i [a)_i = [0) \qquad (40)$$

A similar result holds for the rows.

THE INVERSE OF A SQUARE MATRIX

Here we use the ideas developed in the preceding sections to define the inverse of a square matrix and to show how it can be evaluated. We state theorems without proof.

The *inverse* of a square matrix $[a]$ of order n is a square matrix $[a]^{-1}$ of order n such that

$$[a][a]^{-1} = [e] \tag{41}$$

and

$$[a]^{-1}[a] = [e] \tag{42}$$

where $[e]$ is the unit matrix of order n.

Theorem 14. If a square matrix has an inverse the inverse is unique.

A square matrix $[a]$ for which the determinant $|a| = 0$ is said to be *singular*. If $|a| \neq 0$ then $[a]$ is said to be *nonsingular* or *regular*.

Theorem 15. A square matrix has an inverse if and only if the matrix is nonsingular.

The *adjugate* or *adjoint* of a square matrix $[a]$, which we designate as adj$[a]$, is the transpose of the matrix that is obtained by replacing the elements a_{ij} in $[a]$ by their cofactors $|A_{ij}|$. Thus

$$\text{adj}[a] \equiv \left[|A_{ij}|\right]^T \tag{43}$$

Theorem 16. The inverse of a nonsingular square matrix is equal to the adjoint of the matrix divided by the determinant of the matrix. Thus if $[a]$ is a nonsingular square matrix then

$$[a]^{-1} = \frac{\text{adj}[a]}{|a|} \tag{44}$$

SUMMARY: The inverse of a square matrix $[a]$ is given by

$$[a]^{-1} = \frac{\text{adj}[a]}{|a|} \tag{45}$$

where

$$\text{adj}[a] \equiv \left[|A_{ij}|\right]^T \tag{46}$$

$$|A_{ij}| \equiv (-1)^{i+j} |M_{ij}| \tag{47}$$

and $|M_{ij}|$ is the determinant of the matrix formed by striking out the ith row and the jth column in $[a]$. Thus to obtain the inverse of a square matrix $[a]$ we perform the following operations:

a. Replace each element in $[a]$ by the determinant of the matrix obtained by striking out the row and column containing the given element.
b. In the resulting matrix change the signs of all the elements that are an odd number of steps from the leading element.
c. Transpose.
d. Divide by the determinant $|a|$.

The following theorems, which we state without proof, are useful in the evaluation and the manipulation of inverses.

Theorem 17. If by a sequence of operations in which a row is multiplied by a number other than zero or a multiple of one row is added to another row we can reduce a square matrix to a unit matrix, the same sequence of operations when performed on the unit matrix will produce the inverse of the original matrix. Thus we can find the inverse of a matrix by carrying out such a sequence of operations simultaneously on the matrix whose inverse we desire and the unit matrix. A similar theorem applies if the operations are applied to columns rather than rows.

Theorem 18. If $[a]$ and $[b]$ are square matrices and $[c] = [a][b]$, then $[c]^{-1} = [b]^{-1}[a]^{-1}$.

ORTHOGONAL MATRICES

A square matrix $[a]$ is defined to be *orthogonal* if and only if the transpose of $[a]$ is equal to the inverse of $[a]$. Thus if $[a]$ is orthogonal then

$$[a]^T = [a]^{-1} \tag{48}$$

Theorem 19. A square matrix $[a]$ is an orthogonal matrix if and only if the elements a_{ij} satisfy either of the following conditions:

$$\sum_k a_{ik} a_{jk} = \delta_{ij} \tag{49}$$

$$\sum_k a_{ki} a_{kj} = \delta_{ij} \tag{50}$$

Theorem 20. The determinant of an orthogonal matrix is $+1$ or -1.

Theorem 21. The product of orthogonal matrices is an orthogonal matrix.

Theorem 22. The inverse of an orthogonal matrix is an orthogonal matrix.

UNITARY MATRICES

A square matrix $[a]$ is defined to be *unitary* if and only if the transposed complex conjugate of $[a]$ is equal to the inverse of $[a]$. Thus if $[a]$ is unitary then

$$[a]^\dagger = [a]^{-1} \tag{51}$$

RANK OF A MATRIX

A matrix obtained by omitting some rows and columns from a given matrix $[a]$ is called a *submatrix* of $[a]$.

The determinant of a square submatrix is called a *minor* of the matrix.

A matrix $[a]$ is said to be of *rank* r if and only if all minors of order greater than r are singular, and at least one minor of order r is nonsingular.

Theorem 23. The rank of a matrix is equal to the maximum number of linearly independent rows or columns.

Theorem 24. The rank of a matrix does not change if:
a. We change the order of the rows (columns).
b. We multiply a row (column) by a nonzero number.
c. We add to one of the rows (columns) a linear combination of the remaining rows (columns).
d. We drop a row (column) that is a linear combination of the remaining rows (columns).
e. We interchange the rows and the columns.

APPENDIX 13

Systems of Linear Equations

In this appendix we consider some properties of systems of equations of the form

$$a_{11}x_1 + a_{12}x_2 + \cdots + a_{1n}x_n = b_1$$
$$a_{21}x_1 + a_{22}x_2 + \cdots + a_{2n}x_n = b_2$$
$$\vdots$$
$$a_{m1}x_1 + a_{m2}x_2 + \cdots + a_{mn}x_n = b_m$$
(1)

where the a_{ij} and the b_i are constants. This set of equations can be written in the form

$$\sum_{j=1}^{n} a_{ij}x_j = b_i \qquad i = 1, 2, \ldots, m \tag{2}$$

or in terms of matrices in any one of the following forms:

$$\begin{bmatrix} a_{11} & a_{12} & \cdots & a_{1n} \\ a_{21} & a_{22} & \cdots & a_{2n} \\ \vdots & \vdots & \vdots & \vdots \\ a_{m1} & a_{m2} & \cdots & a_{mn} \end{bmatrix} \begin{bmatrix} x_1 \\ x_2 \\ \vdots \\ x_n \end{bmatrix} = \begin{bmatrix} b_1 \\ b_2 \\ \vdots \\ b_m \end{bmatrix} \tag{3}$$

$$[a_{ij}][x_j) = [b_i) \tag{4}$$

$$[a][x) = [b) \tag{5}$$

The matrix

$$[a] \equiv \begin{bmatrix} a_{11} & a_{12} & \cdots & a_{1n} \\ a_{21} & a_{22} & \cdots & a_{2n} \\ \vdots & \vdots & \vdots & \vdots \\ a_{m1} & a_{m2} & \cdots & a_{mn} \end{bmatrix} \tag{6}$$

is called the *coefficient matrix*. The matrix

$$[a,b] \equiv \begin{bmatrix} a_{11} & a_{12} & \cdots & a_{1n} & b_1 \\ a_{21} & a_{22} & \cdots & a_{2n} & b_2 \\ \vdots & \vdots & \vdots & \vdots & \vdots \\ a_{m1} & a_{m2} & \cdots & a_{mn} & b_n \end{bmatrix} \qquad (7)$$

is called the *augmented matrix*.

Theorem 1. The system of m linear equations

$$\sum_{j=1}^{n} a_{ij} x_j = b_i \qquad i = 1, 2, \ldots, m \qquad (8)$$

in the n variables $x_1, x_2, \ldots, x_n$ possesses a solution if and only if the rank of the coefficient matrix $[a]$ is equal to the rank of the augmented matrix $[a,b]$. If the rank of the matrix $[a]$ and the matrix $[a,b]$ is r, then $n - r$ of the unknowns x_i can be chosen arbitrarily, after which the remaining r unknowns are determined uniquely. Any set of r of the unknowns x_i can be selected as those that are determined uniquely when the others have been given arbitrary values, provided the matrix of their coefficients is of rank r.

Corollary 1. The system of n linear homogeneous equations

$$\sum_{j=1}^{n} a_{ij} x_j = 0 \qquad i = 1, 2, \ldots, n \qquad (9)$$

in the n unknowns $x_1, x_2, \ldots, x_n$ will have a solution other than the solution $x_1 = x_2 = \ldots = x_n = 0$ if and only if the determinant $|a|$ is zero.

APPENDIX 14

Functional Dependence

In this appendix we define the functional dependence of two or more functions. We suppose throughout this appendix that we are dealing with a set of m quantities $f_1, f_2, \ldots, f_m$, which are functions of a set of n variables $\mathbf{x} \equiv x_1, x_2, \ldots, x_n$, where $m \leq n$, and we assume that each of the functions f_i and each of its partial derivatives $\partial f_i / \partial x_j$ is single valued and continuous everywhere in a region R.

Definition. The functions $f_1(\mathbf{x}), f_2(\mathbf{x}), \ldots, f_m(\mathbf{x})$ are said to *functionally dependent* in R if there exists a function $F(f_1, f_2, \ldots, f_m)$ such that

$$F[f_1(\mathbf{x}), f_2(\mathbf{x}), \ldots, f_m(\mathbf{x})] = 0 \tag{1}$$

for all values of $\mathbf{x}$ in R.

Theorem. If the functions $f_1(\mathbf{x}), f_2(\mathbf{x}), \ldots, f_m(\mathbf{x})$ are functionally dependent in R, the rank r of the matrix

$$\begin{bmatrix} \dfrac{\partial f_1}{\partial x_1} & \dfrac{\partial f_2}{\partial x_1} & \cdots & \dfrac{\partial f_m}{\partial x_1} \\ \dfrac{\partial f_1}{\partial x_2} & \dfrac{\partial f_2}{\partial x_2} & \cdots & \dfrac{\partial f_m}{\partial x_2} \\ \cdots & \cdots & \cdots & \cdots \\ \dfrac{\partial f_1}{\partial x_n} & \dfrac{\partial f_2}{\partial x_n} & \cdots & \dfrac{\partial f_m}{\partial x_n} \end{bmatrix} \tag{2}$$

is less than m, that is, $r < m$. Conversely if the rank r of the matrix in Eq. (2) is less than m, and if one of the r dimensional minors of this matrix is not equal to zero anywhere in the region R, the functions $f_1(\mathbf{x}), f_2(\mathbf{x}), \ldots, f_m(\mathbf{x})$ are functionally dependent.

PROOF: If the functions $f_1(\mathbf{x}), \ldots$ there exists an $F(f_1, f_2, \ldots, f_m)$ such [that the functions] are functionally dependent, [equation] is satisfied, and thus

$$dF \equiv \sum_{i=1}^{m} \frac{\partial F}{\partial f_i} df_i \equiv \sum_{i=1}^{m} \sum_{j=1}^{n} \cdots \equiv 0 \tag{3}$$

Equation (3) must be true for every set of values of [dx_j]; therefore

$$\sum_{i=1}^{m} \frac{\partial F}{\partial f_i} \frac{\partial f_i}{\partial x_j} = 0 \qquad j = 1, 2, \ldots, n \tag{4}$$

or in matrix form

$$\begin{bmatrix} \dfrac{\partial f_1}{\partial x_1} & \dfrac{\partial f_2}{\partial x_1} & \cdots & \dfrac{\partial f_m}{\partial x_1} \\ \dfrac{\partial f_1}{\partial x_2} & \dfrac{\partial f_2}{\partial x_2} & \cdots & \dfrac{\partial f_m}{\partial x_2} \\ \cdots & \cdots & \cdots & \cdots \\ \dfrac{\partial f_1}{\partial x_n} & \dfrac{\partial f_2}{\partial x_n} & \cdots & \dfrac{\partial f_m}{\partial x_n} \end{bmatrix} \begin{bmatrix} \dfrac{\partial F}{\partial f_1} \\ \dfrac{\partial F}{\partial f_2} \\ \cdots \\ \dfrac{\partial F}{\partial f_m} \end{bmatrix} = \begin{bmatrix} 0 \\ 0 \\ \cdots \\ 0 \end{bmatrix} \tag{5}$$

Equation (5) provides us with a set of n simultaneous equations in the m quantities $\partial F/\partial f_1, \partial F/\partial f_2, \ldots, \partial F/\partial f_m$. At least one of the quantities $\partial F/\partial f_i$ is not zero, and thus there is a nontrivial solution to the set of Eqs. (5). But from the results of Appendix 13, this is possible only if the rank r of the matrix (2) is less than m.

Conversely suppose the rank r of the matrix (2) is less than m, and let us suppose further that

$$\begin{vmatrix} \dfrac{\partial f_1}{\partial x_1} & \cdots & \dfrac{\partial f_r}{\partial x_1} \\ \cdots & \cdots & \cdots \\ \dfrac{\partial f_1}{\partial x_r} & \cdots & \dfrac{\partial f_r}{\partial x_r} \end{vmatrix} \neq 0 \tag{6}$$

anywhere in R. It then follows from Appendix 18 that the equations

$$f_1 = f_1(\mathbf{x})$$
$$f_2 = f_2(\mathbf{x})$$
$$\cdots$$
$$f_r = f_r(\mathbf{x}) \tag{7}$$

can be solved to obtain

$$x_1 = x_1(f_1, f_2, \ldots, f_r, x_{r+1}, \ldots, x_n)$$
$$\cdots$$
$$x_r = x_r(f_1, f_2, \ldots, f_r, x_{r+1}, \ldots, x_n) \tag{8}$$

462 Functional Dependence

If we substitute these equations

$$f_{r+1} = f_{r+1}(x_1, x_2, \ldots, x_n)$$
$$\vdots$$
$$f_m = f_m(x_1, x_2, \ldots, x_n) \tag{9}$$

we obtain

$$f_{r+1} = f_{r+1}(f_1, f_2, \ldots, f_r, x_{r+1}, \ldots, x_n)$$
$$\vdots \tag{10}$$
$$f_m = f_m(f_1, f_2, \ldots, f_r, x_{r+1}, \ldots, x_n)$$

We now show that the functions above are not explicit functions of $x_{r+1}, \ldots, x_n$. Let us consider in particular the function $f_{r+1}(f_1, f_2, \ldots, f_r, x_{r+1}, \ldots, x_n)$. The proof is the same for any one of the other functions. Since the matrix (2) is of rank r, every minor of order $r+1$ must vanish identically. In particular

$$\begin{vmatrix} \dfrac{\partial f_1}{\partial x_1} & \cdots & \dfrac{\partial f_{r+1}}{\partial x_1} \\ \cdots & \cdots & \cdots \\ \dfrac{\partial f_1}{\partial x_{r+1}} & \cdots & \dfrac{\partial f_{r+1}}{\partial x_{r+1}} \end{vmatrix} \equiv 0 \tag{11}$$

But if Eq. (11) is true then the columns of the determinant are linearly dependent, and thus for a given point x there exists a set of constants c_i such that

$$\sum_{i=1}^{r+1} c_i \frac{\partial f_i}{\partial x_j} \equiv 0 \tag{12}$$

Multiplying Eq. (12) by dx_j and summing over j we obtain

$$\sum_{j=1}^{n} \sum_{i=1}^{r+1} c_i \frac{\partial f_i}{\partial x_j} dx_j \equiv 0 \tag{13}$$

Noting that

$$\sum_j \frac{\partial f_i}{\partial x_j} dx_j \equiv df_i \tag{14}$$

we obtain

$$\sum_{i=1}^{r+1} c_i \, df_i \equiv 0 \tag{15}$$

It follows that the change in f_{r+1} at a given point depends only on the change in the quantities $f_1, f_2, \ldots, f_r$. Hence the function $f_{r+1}(f_1, f_2, \ldots, f_r, x_{r+1}, \ldots, x_n)$ is not an explicit function of $x_{r+1}, \ldots, x_n$, and thus

$$f_{r+1} = f_{r+1}(f_1, f_2, \ldots, f_r) \tag{16}$$

We can similarly show that
$$f_{r+2} = f_{r+2}(f_1, f_2, \ldots, f_r)$$
$$\ldots$$
$$f_m = f_m(f_1, f_2, \ldots, f_r) \tag{17}$$
These equations provide us with $m - r$ functional relations among the quantities $f_1, f_2, \ldots, f_m$. Hence the functions $f_1, f_2, \ldots, f_m$ are functionally dependent.

APPENDIX 15

The Method of Lagrange Multipliers

A function $f(\mathbf{x})$ of a set of variables $\mathbf{x} = x_1, x_2, \ldots, x_n$ is said to have a stationary value at a point $\mathbf{x}$ if and only if

$$df \equiv \sum_i \frac{\partial f(\mathbf{x})}{\partial x_i} dx_i = 0 \tag{1}$$

for all allowed values of the dx_i. If there are no constraints on the $\mathbf{x}$, the condition above is equivalent to the conditions

$$\frac{\partial f(\mathbf{x})}{\partial x_i} = 0 \quad i = 1, 2, \ldots, n \tag{2}$$

If the values of $\mathbf{x}$ are restricted to those satisfying some constraint condition, the set of Eqs. (2) is not equivalent to Eq. (1).

Theorem 1. If $f(\mathbf{x})$ is a function of the set of variables $\mathbf{x}$ and the values of $\mathbf{x}$ are restricted to those satisfying the constraint

$$g(\mathbf{x}) = 0 \tag{3}$$

Then $f(\mathbf{x})$ will have a stationary value at a point $\mathbf{x}$ if and only if there exists a nonzero constant λ such that

$$\frac{\partial}{\partial x_i}[f(\mathbf{x}) + \lambda g(\mathbf{x})] = 0 \quad i = 1, 2, \ldots, n \tag{4}$$

at the point. Equations (3) and (4) are sufficient to determine the values of λ and the values of $\mathbf{x}$ at which the function has a stationary value. It follows that we can determine the points at which $f(\mathbf{x})$ has a stationary value subject to the constraint $g(\mathbf{x}) = 0$ by simply finding the points $(\mathbf{x}, \lambda)$ at which the function

$$F(\mathbf{x}, \lambda) \equiv f(\mathbf{x}) + \lambda g(\mathbf{x}) \tag{5}$$

has a stationary point. The parameter λ is called a Lagrange multiplier.

PROOF 1: At a stationary point

$$df = \sum_i \frac{\partial f}{\partial x_i} dx_i = 0 \qquad (6)$$

However the values of dx_i are not arbitrary but must satisfy the constraint condition

$$dg = \sum_{i=1}^{n} \frac{\partial g}{\partial x_i} dx_i = 0 \qquad (7)$$

It follows that only $n-1$ of the dx_i can be chosen arbitrarily. Equations (6) and (7) can be rewritten

$$-\left(\frac{\partial f}{\partial x_j}\right) dx_j = \sum_{i \neq j} \left(\frac{\partial f}{\partial x_i}\right) dx_i \qquad (8)$$

$$-\left(\frac{\partial g}{\partial x_j}\right) dx_j = \sum_{i \neq j} \left(\frac{\partial g}{\partial x_i}\right) dx_i \qquad (9)$$

Dividing Eq. (8) by Eq. (9) we obtain

$$\frac{\partial f / \partial x_j}{\partial g / \partial x_j} = \frac{\sum_{i \neq j} (\partial f / \partial x_i) dx_i}{\sum_{i \neq j} (\partial g / \partial x_i) dx_i} \qquad (10)$$

There are only $n-1$ of the dx_i appearing in Eq. (10); hence Eq. (10) must be true for arbitrary choice of these dx_i. Setting all the dx_i equal to zero except dx_k we obtain

$$\frac{\partial f / \partial x_j}{\partial g / \partial x_j} = \frac{\partial f / \partial x_k}{\partial g / \partial x_k} \qquad (11)$$

It follows that at a stationary point

$$\frac{\partial f / \partial x_1}{\partial g / \partial x_1} = \frac{\partial f / \partial x_2}{\partial g / \partial x_2} = \cdots = \frac{\partial f / \partial x_n}{\partial g / \partial x_n} \qquad (12)$$

The set of $n-1$ equations (12) is equivalent to the following set of n equations:

$$\frac{\partial f}{\partial x_i} + \lambda \frac{\partial g}{\partial x_i} = 0 \qquad i = 1, 2, \ldots, n \qquad (13)$$

Conversely suppose there exists a λ and a point x that satisfy Eq. (3) and at which Eqs. (4) are true. It then follows that at this point

$$df = \sum_i \frac{\partial f}{\partial x_i} dx_i = -\lambda \sum_i \frac{\partial g}{\partial x_i} dx_i = -\lambda \, dg = 0 \qquad (14)$$

Hence the point is a stationary point. This completes the proof of the theorem.

PROOF 2: If we multiply Eq. (7) by an arbitrary parameter and add the result to Eq. (6) we obtain

$$\sum_i \left(\frac{\partial f}{\partial x_i} + \lambda \frac{\partial g}{\partial x_i}\right) dx_i = 0 \qquad (15)$$

The parameter λ and any $n-1$ of the dx_i can be chosen arbitrarily. If we choose λ in such a way that for a particular value of i, say $i=j$,

$$\frac{\partial f}{\partial x_j} + \lambda \frac{\partial g}{\partial x_j} = 0 \tag{16}$$

then

$$\sum_{i \neq j} \left(\frac{\partial f}{\partial x_i} + \lambda \frac{\partial g}{\partial x_i} \right) dx_i = 0 \tag{17}$$

There are only $n-1$ of the dx_i appearing in Eq. (17), hence Eq. (17) must be true for arbitrary choice of these dx_i. These can be true only if

$$\frac{\partial f}{\partial x_i} + \lambda \frac{\partial g}{\partial x_i} = 0 \quad \text{for} \quad i \neq j \tag{18}$$

Combining Eqs. (16) and (18) we obtain

$$\frac{\partial f}{\partial x_i} + \lambda \frac{\partial g}{\partial x_i} = 0 \quad i = 1, 2, \ldots, n \tag{19}$$

The rest of the proof proceeds as in Proof 1.

APPENDIX 16

Elliptic Functions

THE JACOBIAN ELLIPTIC FUNCTIONS

The Jacobian elliptic functions occur quite frequently in the solutions of the differential equations of mechanics. In this appendix we define them and consider some of their properties.

Consider the integral

$$x = \int_0^\phi \frac{du}{(1 - k^2\sin^2 u)^{1/2}} = \int_0^{\sin\phi} \frac{dv}{[(1 - v^2)(1 - k^2 v^2)]^{1/2}} \tag{1}$$

where k is a number between 0 and 1. Equation (1) defines a function $x(\phi, k)$. The function $x(\phi, k)$ is a monotonic increasing function of ϕ, and it satisfies the relation

$$x(\phi + \pi, k) = x(\phi, k) + 2K(k) \tag{2}$$

where

$$K(k) = \int_0^{\pi/2} \frac{du}{(1 - k^2\sin^2 u)^{1/2}} = \int_0^{\pi/2}\left[1 + \frac{k^2}{2}\sin^2 u + \cdots\right] du$$

$$= \frac{\pi}{2}\left\{1 + \sum_{n=1}^{\infty}\left[\frac{1 \cdot 3 \cdots (2n-1)}{2 \cdot 4 \cdots 2n}\right]^2 k^{2n}\right\} \tag{3}$$

The function $K(k)$ is plotted in Fig. 1. Since $x(\phi, k)$ is a monotonic increasing function of ϕ, it can be inverted to obtain $\phi(x, k)$. The function $\phi(x, k)$ satisfies the relation

$$\phi[x + 2K(k), k] = \phi(x, k) + \pi \tag{4}$$

The three elliptic functions are defined by the equations

$$\text{sn}(x, k) \equiv \sin\phi(x, k) \tag{5}$$

$$\text{cn}(x, k) \equiv \cos\phi(x, k) \tag{6}$$

$$\text{dn}(x, k) \equiv \left[1 - k^2\text{sn}^2(x, k)\right]^{1/2} \tag{7}$$

468   Elliptic Functions

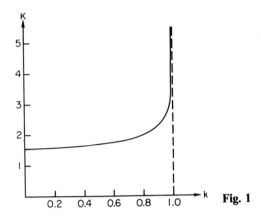

Fig. 1

where "sn" is spoken "ess en," and so forth. The functions sn(x, k) and cn(x, k) have a period 4K(k), and the function dn(x, k) has a period 2K(k). Figure 2 plots sn(x, k), cn(x, k), and dn(x, k), respectively, as functions of x for $k^2 = 0.7$.

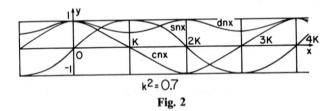

Fig. 2

SOME PROPERTIES OF THE JACOBIAN ELLIPTIC FUNCTIONS

The following properties of Jacobian elliptic functions are frequently useful:

$$\frac{d}{dx}[\text{sn}(x,k)] = \text{cn}(x,k)\text{dn}(x,k) \tag{8}$$

$$\frac{d}{dx}[\text{cn}(x,k)] = -\text{sn}(x,k)\text{dn}(x,k) \tag{9}$$

$$\frac{d}{dx}[\text{dn}(x,k)] = -k^2\text{sn}(x,k)\text{cn}(x,k) \tag{10}$$

$$\int \text{sn}(x,k)\,dx = k^{-1}\ln[\text{dn}(x,k) - k\,\text{cn}(x,k)] \tag{11}$$

$$\int \text{cn}(x,k)\,dx = k^{-1}\cos^{-1}[\text{dn}(x,k)] \tag{12}$$

$$\int \text{dn}(x,k)\,dx = \sin^{-1}[\text{sn}(x,k)] \tag{13}$$

$$[\text{sn}(x,k)]^2 + [\text{cn}(x,k)]^2 = 1 \tag{14}$$

$$[\text{dn}(x,k)]^2 = 1 - k^2\text{sn}^2(x,k) \tag{15}$$

SPECIAL CASES

When $k = 0$ the Jacobian elliptic functions reduce to circular functions. In particular

$$\text{sn}(x, 0) = \sin x \tag{16}$$

$$\text{cn}(x, 0) = \cos x \tag{17}$$

$$\text{dn}(x, 0) = 1 \tag{18}$$

When $k = 1$ the Jacobian elliptic functions reduce to hyperbolic functions. In particular

$$\text{sn}(x, 1) = \tanh x \tag{19}$$

$$\text{cn}(x, 1) = \text{sech } x \tag{20}$$

$$\text{dn}(x, 1) = \text{sech } x \tag{21}$$

DEFINITIONS OF THE JACOBIAN ELLIPTIC FUNCTIONS AS SOLUTIONS OF A DIFFERENTIAL EQUATION

From Eqs. (8), (14), and (15) we obtain

$$\frac{d}{dx}\left[\text{sn}(x, k)\right] = \left[1 - \text{sn}^2(x, k)\right]^{1/2}\left[1 - k^2\text{sn}^2(x, k)\right]^{1/2} \tag{22}$$

It follows that $\text{sn}(x, k)$ can be defined as the solution of the differential equation

$$\frac{dy}{dx} = (1 - y^2)^{1/2}(1 - k^2 y^2)^{1/2} \tag{23}$$

with the boundary condition

$$y(0) = 0 \tag{24}$$

It is possible in a similar fashion to define $\text{cn}(x, k)$ and $\text{dn}(x, k)$ as solutions to a differential equation.

APPENDIX 17

Coordinate Transformations

INTRODUCTION

Let $\mathbf{x} \equiv x_1, x_2, \ldots, x_n$ and $\mathbf{y} \equiv y_1, y_2, \ldots, y_n$ be two sets of numbers, and let us assume that there is a set of equations $\mathbf{y} = \mathbf{y}(\mathbf{x})$, that is,

$$\begin{aligned} y_1 &= y_1(x_1, x_2, \ldots, x_n) \\ y_2 &= y_2(x_1, x_2, \ldots, x_n) \\ &\vdots \\ y_n &= y_n(x_1, x_2, \ldots, x_n) \end{aligned} \tag{1}$$

which associates with each value of $\mathbf{x}$ a unique value of $\mathbf{y}$. The process by which we pass from the set of variables $\mathbf{x}$ to the set of variables $\mathbf{y}$ is called a transformation, and the equations $\mathbf{y} = \mathbf{y}(\mathbf{x})$ that define the transformation are called the transformation equations for the transformation.

From a geometrical point of view we may think of the variables x_1, $x_2, \ldots, x_n$ as the coordinates of a point referred to a set of rectangular axes in a space that we refer to as the space $S(\mathbf{x})$, and the variables $y_1, y_2, \ldots, y_n$ as the coordinates of a point referred to a set of rectangular axes in a space that we refer to as the space $S(\mathbf{y})$. The transformation $\mathbf{y} = \mathbf{y}(\mathbf{x})$ can then be interpreted as a mapping of the points in the space $S(\mathbf{x})$ onto the space $S(\mathbf{y})$. This point of view is represented symbolically in Fig. 1 for the case in which $n = 2$.

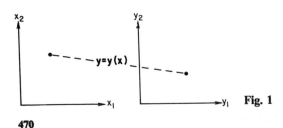

Fig. 1

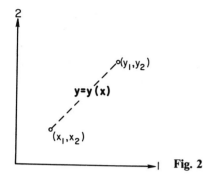

Fig. 2

For another possible geometrical point of view we can think of the variables $x_1, x_2, \ldots, x_n$ as the coordinates of a point referred to a set of rectangular axes in a space S and the variables $y_1, y_2, \ldots, y_n$ as the coordinates of a second point referred to the same set of axes. The transformation $\mathbf{y} = \mathbf{y}(\mathbf{x})$ can then be interpreted as a mapping of the space S onto itself. This point of view is represented symbolically in Fig. 2 for the case in which $n = 2$.

The transformation $\mathbf{y} = \mathbf{x}$ is called the *identity transformation* and is designated by the letter I.

If T_1 and T_2 are two transformations, the transformation T_3, which is obtained by first carrying out the transformation T_1 and then carrying out the transformation T_2, is called the *product* of T_2 and T_1 and is designated $T_2 T_1$.

Given a transformation T, if there exists a transformation T^{-1} such that $T^{-1}T = I$ then the transformation T^{-1} is called the *inverse* of the transformation T.

REGULAR TRANSFORMATIONS

A transformation defined for some region R of the space $\mathbf{x}$ by the transformation equations $\mathbf{y} = \mathbf{y}(\mathbf{x})$ is said to be a *regular transformation* if and only if the functions $y_i(\mathbf{x})$ and their first derivatives $\partial y_i / \partial x_j$ are continuous in the region R, and the Jacobian $J(\mathbf{y}/\mathbf{x})$ does not vanish at any point in the region R.

The set of regular transformations satisfies the following theorems.

Theorem 1. If T_1 and T_2 are two regular transformations, then the product $T_2 T_1$ of T_1 and T_2 is a regular transformation.

Theorem 2. If T_1, T_2, and T_3 are regular transformations, then $T_3(T_2 T_1) = (T_3 T_2) T_1$.

Theorem 3. There exists a regular transformation I called the identity transformation, such that if T is any regular transformation, then $TI = IT = T$.

Theorem 4. If T is a regular transformation, the inverse transformation T^{-1} exists and is regular.

NOTE 1: Theorem 4 is a very important theorem in dynamics. Because of its importance it is considered in greater detail in Appendix 18, in Volume 2.

NOTE 2: From Theorems 1–4 it follows that the set of all regular transformations forms a group.

AFFINE TRANSFORMATIONS

An *affine transformation* is a transformation of the form

$$y_i = \sum_j a_{ij} x_j + b_i \tag{2}$$

where the a_{ij} and the b_i are constants.

Theorem 5. A transformation $y = y(x)$ is an affine transformation if and only if for any two arbitrary values a and b of x and any constant α

$$y[\alpha a + (1 - \alpha)b] = \alpha y(a) + (1 - \alpha)y(b) \tag{3}$$

NOTE 1: From a geometrical point of view an affine transformation is a transformation that transforms lines into lines and preserves the ratios of distance between points along a line.

NOTE 2: The set of all regular affine transformations forms a subgroup of the group of all regular transformations.

LINEAR TRANSFORMATIONS

A *linear transformation* is a transformation of the form

$$y_i = \sum_j a_{ij} x_j \tag{4}$$

where the a_{ij} are constants.

Theorem 6 A transformation $y = y(x)$ is a linear transformation if and only if for any two arbitrary values a and b of x and any constant α.

$$y(a + \alpha b) = y(a) + \alpha y(b) \tag{5}$$

NOTE 1: From a geometrical point of view a linear transformation is a transformation that transforms the origin into the origin and lines into lines, and preserves the ratios of distances between points along a line.

NOTE 2: The set of all regular linear transformations forms a subgroup of the group of all regular affine transformations.

EUCLIDEAN TRANSFORMATIONS

A *Euclidean transformation* or *congruence transformation* is a transformation of the form

$$y_i = \sum_j a_{ij} x_j + b_i \qquad (6)$$

where the b_i are constants, and the a_{ij} are constants satisfying either of the equivalent conditions

$$\sum_k a_{ik} a_{jk} = \delta_{ij} \qquad (7)$$

$$\sum_k a_{ki} a_{kj} = \delta_{ij} \qquad (8)$$

Theorem 7. A transformation $y = y(x)$ is a Euclidean transformation if and only if for any two arbitrary values **a** and **b** of **x**

$$\sum_i [y_i(\mathbf{a}) - y_i(\mathbf{b})]^2 = \sum_i (a_i - b_i)^2 \qquad (9)$$

NOTE 1: From a geometrical point of view a Euclidean transformation is a transformation that preserves the distance between points. Thus a Euclidean transformation has the same effect on the coordinates of a set of points as does a rotation and/or translation and/or inversion of axes.

NOTE 2: The set of all Euclidean transformations forms a subgroup of the group of all regular affine transformations.

ORTHOGONAL TRANSFORMATIONS

An *orthogonal transformation* is a transformation of the form

$$y_i = \sum_j a_{ij} x_j \qquad (10)$$

where the a_{ij} are constants satisfying either of the equivalent conditions

$$\sum_k a_{ik} a_{jk} = \delta_{ij} \qquad (11)$$

$$\sum_k a_{ki} a_{kj} = \delta_{ij} \qquad (12)$$

Theorem 8. A transformation y(x) is orthogonal if and only if

$$y(0) = 0 \qquad (13)$$

and for any arbitrary values **a** and **b** of **x**

$$\sum_i [y_i(\mathbf{b}) - y_i(\mathbf{a})]^2 = \sum_i (b_i - a_i)^2 \qquad (14)$$

NOTE 1: From a geometrical point of view an orthogonal transformation is a transformation that transforms the origin into the origin and preserves the distance between points. Thus an orthogonal transformation has the same effect on coordinates of a point as does a rotation and/or inversion.

NOTE 2: The set of all orthogonal transformations forms a subgroup of the set of Euclidean transformations.

ROTATIONS

A *rotation* is a transformation of the form

$$y_i = \sum_j a_{ij} x_j \qquad (15)$$

where the a_{ij} are constants satisfying either of the equivalent conditions

$$\sum_k a_{ik} a_{jk} = \delta_{ij} \qquad (16)$$

$$\sum_k a_{ki} a_{kj} = \delta_{ij} \qquad (17)$$

and the determinant $|a_{ij}| = 1$.

Tables

Tables

Table 1 ABBREVIATIONS OF UNITS

Unit	Abbreviation
Joule	J
Kilogram	kg
Meter	m
Newton	N
Second	s
Watt	W

Table 2 CONSTANTS

Earth, angular velocity of	$= 7.292 \times 10^{-5}$ rad/s
Earth, mass of	$= 5.97 \times 10^{24}$ kg
Earth, mean radius of	$= 6.371 \times 10^{6}$ m
Earth to moon, mean distance	$= 3.85 \times 10^{8}$ m
Earth to sun, mean distance	$= 1.495 \times 10^{11}$ m
Electron mass	$= 9.1091 \times 10^{-31}$ kg
Gravitational constant, G	$= 6.673 \times 10^{-11}$ N m^2/kg^2
Gravity, acceleration due to gravity at sea level at latitude 45°, g	$= 9.80621$ m/s^2
Gravity, standard value of the acceleration due to gravity, g_n	$= 9.80665$ m/s^2
Mercury, mass of	$= 3.3 \times 10^{23}$ kg
Mercury, period of orbit	$= 7.6 \times 10^{6}$ s
Mercury to sun, mean distance	$= 5.8 \times 10^{10}$ m
Moon, mass of	$= 7.35 \times 10^{22}$ kg
Moon, radius of	$= 1.74 \times 10^{6}$ m
Neutron mass	$= 1.67482 \times 10^{-27}$ kg
Proton mass	$= 1.67252 \times 10^{-27}$ kg
Speed of light, c	$= 2.997925 \times 10^{8}$ m/s
Sun, mass of	$= 1.987 \times 10^{30}$ kg
Sun, radius of	$= 6.96 \times 10^{8}$ m

Table 3 CONVERSION FACTORS

1 acre	$= 4.0468564 \times 10^3$ m^2
1 ångstrom	$= 10^{-10}$ m
1 are	$= 10^2$ m^2
1 atmosphere	$= 1.01325 \times 10^5$ N/m^2
1 atomic mass unit (^{12}C = 12)	$= 1.660531 \times 10^{-27}$ kg
1 astronomical unit	$= 1.495 \times 10^{11}$ m
1 bar	$= 10^5$ N/m^2
1 British thermal unit	$= 1.05435 \times 10^3$ J
1 bushel (U.S.)	$= 3.5239 \times 10^{-2}$ m^3
1 calorie (g)	$= 4.184$ J
1 Calorie (kg)	$= 4.184 \times 10^3$ J
1 cubic foot	$= 2.8317 \times 10^{-2}$ m^3
1 cubic inch	$= 1.6387064 \times 10^{-5}$ m^3
1 day (mean solar)	$= 8.64 \times 10^4$ s
1 day (sidereal)	$= 8.61641 \times 10^4$ s
1 dyne	$= 10^{-5}$ N
1 electron volt	$= 1.60209 \times 10^{-19}$ J
1 erg	$= 10^{-7}$ J
1 fathom	$= 1.8288$ m
1 fermi	$= 10^{-15}$ m
1 foot	$= 0.3048$ m
1 foot-pound	$= 1.35582$ J
1 furlong	$= 2.01168 \times 10^2$ m
1 gallon (U.S. liquid)	$= 3.7854 \times 10^{-3}$ m^3
1 geepound	$= 14.5939$ kg
1 gram	$= 10^{-3}$ kg
1 hertz	$= 1$ cycle per second
1 horsepower	$= 7.45700 \times 10^2$ W
1 hour (mean solar)	$= 3.6 \times 10^3$ s
1 hour (sidereal)	$= 3.59017 \times 10^3$ s
1 inch	$= 2.540 \times 10^{-2}$ m
1 international nautical mile	$= 1.852 \times 10^3$ m
1 kilogram of force	$= 9.80665$ N
1 knot	$= 0.5144$ m/s
1 light year	$= 9.46055 \times 10^{15}$ m
1 liter	$= 1.000028 \times 10^{-3}$ m^3
1 liter atmosphere	$= 1.01328 \times 10^2$ J
1 micron	$= 10^{-6}$ m
1 mile	$= 1.609344 \times 10^3$ m
1 mile per hour	$= 0.44704$ m/s

Table 3 CONVERSION FACTORS (*continued*)

1 millimeter of mercury	$= 1.333224 \times 10^2 \text{ N/m}^2$
1 minute (mean solar)	$= 60 \text{ s}$
1 minute (sidereal)	$= 59.83617 \text{ s}$
1 parsec	$= 3.08374 \times 10^{16} \text{ m}$
1 pound (force)	$= 4.44822 \text{ N}$
1 pound (mass)	$= 0.45359237 \text{ kg}$
1 pound per square inch	$= 6.89476 \times 10^3 \text{ N/m}^2$
1 poundal	$= 0.1382550 \text{ N}$
1 quart (U.S. liquid)	$= 9.4635295 \times 10^{-4} \text{ m}^3$
1 rod	$= 5.0292 \text{ m}$
1 slug	$= 14.5939 \text{ kg}$
1 square foot	$= 9.290304 \times 10^{-2} \text{ m}^2$
1 square inch	$= 6.4516 \times 10^{-4} \text{ m}^2$
1 square mile	$= 2.5899881 \times 10^6 \text{ m}^2$
1 square yard	$= 0.836127 \text{ m}^2$
1 ton force (U.S.)	$= 8.89644 \times 10^3 \text{ N}$
1 ton mass (U.S.)	$= 9.0718474 \times 10^2 \text{ kg}$
1 ton (metric)	$= 10^3 \text{ kg}$
1 watt	$= 1 \text{ J/s}$
1 yard	$= 0.9144 \text{ m}$
1 year (calendar)	$= 3.1536 \times 10^7 \text{ s}$

Table 4 CENTERS OF MASS

Centers of Mass of Uniform Filaments

Circular Arc. Let A be an arc of a circle, r the radius of the circle, and 2θ the angle that the arc subtends at the center of the circle. The center of mass of A is located on the axis of symmetry of A at a distance $r \sin \theta / \theta$ from the center of the circle.

Semicircle. The center of mass of a semicircle of radius r is at a distance $2r/\pi$ from the center. This is a special case of the result stated for a circular arc.

Centers of Mass of Uniform Flat Laminae

Triangle. The center of mass of a triangular lamina is on the line joining the center A of any side with the opposite vertex B at the point that is one-third of the way from A to B. Hence it is at the intersection of the medians.

Table 4 CENTERS OF MASS (*continued*)

Rectangle. The center of mass of a rectangular lamina is at the intersection of the diagonals.

Quadrilateral. Let $ABCD$ be a quadrilateral lamina. From each vertex lay off segments equal to one-third of the length of the corresponding sides, meeting at this vertex. Draw extended lines through the ends of the segments associated with each vertex, respectively. These intersect to form a parallelogram. The intersection of the diagonals of this parallelogram is the center of mass of $ABCD$.

Sector of a Circle. Let S be a sector of a circle, r the radius of the circle, and 2θ the angle that the arc of the sector subtends at the center of the circle. The center of mass of S is located on the axis of symmetry at a distance $(2r\sin\theta)/(3\theta)$ from the center of the circle.

Segment of a Circle. Let S be a segment of a circle, r the radius of the circle, 2θ the angle subtended by the arc at the center of the circle, $l = 2r\sin\theta$ the length of the chord, and $A = r^2(2\theta - \sin 2\theta)/2$ the area of the segment. The center of mass of S is on the axis of symmetry of S at a distance $l^3/12A$ from the center of the circle.

Semicircle. The center of mass of a semicircular lamina of radius r is located on the axis of symmetry at a distance $4r/3\pi$ from the center of the circle. This is a special case of the result stated for a sector of a circle and the result stated for a segment of a circle.

Semiellipse. Let E be an elliptical lamina with axes A and B of lengths $2a$ and $2b$, respectively. The center of mass of one of the halves of E bounded by the axis B is on the axis A at a distance $4a/3\pi$ from the center of the ellipse.

Quarter of an Ellipse. Let E be an elliptical lamina with axes A and B of lengths $2a$ and $2b$, respectively. The center of mass of one of the quarters of E bounded by the axes A and B is at a point $4b/3\pi$ above axis A, and $4a/3\pi$ above axis B.

Parabolic Segment. Let S be a parabolic segment formed by a parabola P and a chord C perpendicular to the symmetry axis of P. Let h be the perpendicular distance of C from the vertex of P. The center of mass of S is on the axis of symmetry of S at a distance $2h/5$ from C.

Centers of Masses of Uniform Curved Laminae

Belt Cut from a Spherical Lamina by Two Parallel Planes. Let B be a belt cut from a spherical lamina by two parallel planes P and P'. The center of mass of B is on the axis of symmetry of B halfway between P and P'.

Spherical Cap. The piece cut from a spherical lamina by a plane is called a spherical cap. Let C be a spherical cap, and P the plane that cuts off the cap. The center of mass of C is located on the axis of symmetry of C halfway between the points at which the symmetry axis passes through P and C, respectively. This is a special case of the result stated for a belt cut from a spherical lamina by two parallel planes.

Hemisphere. The center of mass of a hemispherical lamina of radius r is located on the axis of symmetry at a distance $r/2$ from the center of the sphere. This is a special case of the result stated for a spherical cap.

Table 4 CENTERS OF MASS (*continued*)

Cone. Let C be a conical lamina, and h its height. The center of mass of C, excluding its base, is on the axis of symmetry at a distance $h/3$ from its base.

Centers of Mass of Uniform Solids

Pyramid. Let P be a solid pyramid having any plane figure as base and any point as vertex. Let A be the center of mass of the base, that is, the center of mass of a uniform lamina coinciding with the base, and let B be the vertex. The center of mass of P is on the line joining A and B at the point which is one fourth of the way from A to B.

Frustrum of a Pyramid. A frustrum of a pyramid is the section between the base and a plane parallel to the base. Let P be a pyramid, F a frustrum of P, B and B' the larger and smaller bases, respectively, of F, C and C' the centers of mass of B and B', respectively, A and A' the areas of B and B', respectively, and a the distance between C and C'. The center of mass of F is on the line joining C and C' at a distance

$$\frac{a\left[A + 2(AA')^{1/2} + 3A'\right]}{4\left[A + (AA')^{1/2} + A'\right]}$$

from C.

Cone. The center of mass of a solid cone C is on the line joining the center O of the base with the vertex V, at a point one-fourth of the way from O to V. This is a special case of the result stated for a pyramid.

Frustrum of a Cone. A frustrum of a cone is the section between the base and a plane parallel to the base. Let F be a frustrum of a solid cone, B and B' the larger and smaller bases, r and r' the radii of B and B', respectively, O and O' the centers of B and B', respectively, and a the distance between O and O'. The center of mass of F is on the line joining O and O' at a distance

$$\frac{A\left[(r + r')^2 + 2(r')^2\right]}{4\left[(r + r')^2 - rr'\right]}$$

from O. This is a special case of the result stated for the frustrum of a pyramid.

Spherical Sector. Let S be a sector of a sphere, r the radius of the sphere, and θ the semivertical angle of the cone by which the sector is bounded. The center of mass of S is located on the axis of symmetry at a distance $3r(1 + \cos\theta)/8$ from the center of the sphere.

Hemisphere. Let H be a solid hemisphere of radius r. The center of mass of H is on the axis of symmetry of H a distance $3r/8$ from the center of the sphere. This is a special case of the result stated for a spherical sector.

Spherical Segment. Let S be a segment of a sphere, r the radius of the sphere, C the center of the base of the segment, and h the maximum height of the segment. The center of mass of S is located on the axis of symmetry at a distance $[h(4r - h)]/[4(3r - h)]$ from C.

Table 4 CENTERS OF MASS (*continued*)

Octant of an Ellipsoid. Let E be the solid ellipsoid bounded by the surface $(x^2/a^2)+(y^2/b^2)+(z^2/c^2)=1$. The center of mass of the octant of E for which $x>0, y>0$, and $z>0$ is located at the point $x=3a/8, y=3b/8$, and $z=3c/8$.

Paraboloid of Revolution. Let P be a paraboloid of revolution, h its height, and r the radius of the base. The center of mass of P is located on the axis of symmetry at a distance $h/3$ from the base.

Hemispherical Shell. Let S be a hemispherical shell, a its inner radius, and b its outer radius. The center of mass of S is located on the axis of symmetry at a distance

$$\frac{3(a+b)(a^2+b^2)}{8(a^2+ab+b^2)}$$

from its geometrical center.

Table 5 MOMENTS OF INERTIA

Notation

We designate the moments of inertia of a body about axes that pass through the point P and are in the x, y, and z directions, respectively, as $I_x(P)$, $I_y(P)$, and $I_z(P)$, respectively, and the mass of the body as M.

Moments of Inertia of Uniform Filaments

Rod. Given a straight filament AB of length l lying along the x axis with its center at the origin O,

$$I_x(O)=0 \qquad I_y(O)=I_z(O)=\frac{Ml^2}{12}$$

$$I_x(A)=0 \qquad I_y(A)=I_z(A)=\frac{Ml^2}{3}$$

Circle. Given a circular filament of radius r lying in the x-y plane with its center at O,

$$I_x(O)=I_y(O)=\frac{Mr^2}{2} \qquad I_z(O)=Mr^2$$

Circular Arc. Given a circular arc of radius r and angle 2θ, lying in the x-y plane with the axis of symmetry on the x axis and the center of the circle at the origin O,

$$I_x(O)=\frac{Mr^2}{2}\left(1-\frac{\sin 2\theta}{2\theta}\right)$$

$$I_y(O)=\frac{Mr^2}{2}\left(1+\frac{\sin 2\theta}{2\theta}\right)$$

$$I_z(O)=Mr^2$$

Table 5 MOMENTS OF INERTIA (*continued*)

Moments of Inertia of Uniform Flat Laminae

Isosceles Triangle. Given an isosoceles triangular lamina of height h and base b with base on the x axis and axis of symmetry on the y axis,

$$I_x(O) = \frac{Mh^2}{6}$$

$$I_y(O) = \frac{Mb^2}{24}$$

$$I_z(O) = M\left(\frac{h^2}{6} + \frac{b^2}{24}\right)$$

Rectangle. Given a rectangular lamina of sides a and b, with side a parallel to the x axis and side b parallel to the y axis and center at the origin O,

$$I_x(O) = \frac{Mb^2}{12}$$

$$I_y(O) = \frac{Ma^2}{12}$$

$$I_z(O) = \frac{Ma^2 + Mb^2}{12}$$

Circle. Given a circular lamina of radius r lying in the x-y plane with its center at the origin O,

$$I_x(O) = I_y(O) = \frac{Mr^2}{4} \qquad I_z(O) = \frac{Mr^2}{2}$$

Annulus (Circular Ring). Given an annular lamina of inner radius a and outer radius b lying in the x-y plane with its center at the origin O,

$$I_x(O) = I_y(O) = \frac{M(a^2 + b^2)}{4}$$

$$I_z(O) = \frac{M(a^2 + b^2)}{2}$$

Ellipse. Given an elliptical lamina whose boundary is defined by the equations $z = 0$ and $(x^2/a^2) + (y^2/b^2) = 1$,

$$I_x(O) = \frac{Mb^2}{4}$$

$$I_y(O) = \frac{Ma^2}{4}$$

$$I_z(O) = \frac{M(a^2 + b^2)}{4}$$

Table 5 MOMENTS OF INERTIA (*continued*)

Moments of Inertia of Curved Laminae

Sphere. Given a spherical lamina of radius r with center at the origin,
$$I_x(O) = I_y(O) = I_z(O) = \frac{2Mr^2}{3}$$

Cylinder. Given a cylindrical lamina of length l and radius r with axis along the z axis and center at the origin O,
$$I_x(O) = I_y(O) = M\left(\frac{r^2}{2} + \frac{l^2}{12}\right)$$
$$I_z(O) = Mr^2$$

Moments of Inertia of Uniform Solids

Sphere. Given a solid sphere of radius r with center at the origin O,
$$I_x(O) = I_y(O) = I_z(O) = \frac{2Mr^2}{5}$$

Hemisphere. Given a solid hemisphere of radius r with center at the origin O,
$$I_x(O) = I_y(O) = I_z(O) = \frac{2Mr^2}{5}$$

Hollow Sphere. Given a hollow sphere of inner radius a and outer radius b with center at the origin O,
$$I_x(O) = I_y(O) = I_z(O) = M\frac{2(b^5 - a^5)}{5(b^3 - a^3)}$$

Ellipsoid. Given a solid ellipsoid bounded by the surface $(x^2/a^2) + (y^2/b^2) + (z^2/c^2) = 1$,
$$I_x(O) = \frac{M(b^2 + c^2)}{5}$$
$$I_y(O) = \frac{M(a^2 + c^2)}{5}$$
$$I_z(O) = \frac{M(a^2 + b^2)}{5}$$

Cylinder. Given a solid cylinder of radius r and length l with axis along the z axis and center at the origin O,
$$I_x(O) = I_y(O) = M\left(\frac{r^2}{4} + \frac{l^2}{12}\right)$$
$$I_z(O) = \frac{Mr^2}{2}$$

Table 5 MOMENTS OF INERTIA (*continued*)

Hollow Cylinder. Given a hollow cylinder of length l, inner radius a, and outer radius b, with axis along the z axis and center at the origin O,

$$I_x(O) = I_y(O) = M\left(\frac{a^2 + b^2}{4} + \frac{l^2}{12}\right)$$

$$I_z(O) = \frac{M(a^2 + b^2)}{2}$$

Elliptical Cylinder. Given a solid right elliptical cylinder of height h, and transverse axes $2a$ and $2b$ with longitudinal axis in z direction, axis a in the x direction, axis b in the y direction, and center at the origin O,

$$I_x(O) = M\left(\frac{b^2}{4} + \frac{h^2}{12}\right)$$

$$I_y(O) = M\left(\frac{a^2}{4} + \frac{h^2}{12}\right)$$

$$I_z(O) = \frac{M(a^2 + b^2)}{4}$$

Rectangular Parallelepiped. Given a solid rectangular parallelepiped with sides a, b, and c parallel to the x, y, and z axes, respectively, and with center at the origin,

$$I_x(O) = \frac{M(b^2 + c^2)}{12}$$

$$I_y(O) = \frac{M(a^2 + c^2)}{12}$$

$$I_z(O) = \frac{M(a^2 + b^2)}{12}$$

Cone. Given a solid right circular cone of height h and radius of base r with the symmetry axis in the z direction and the base in the x-y plane with center at the origin O,

$$I_x(O) = I_y(O) = \frac{M(3r^2 + 2h^2)}{20}$$

$$I_z(O) = \frac{3Mr^2}{10}$$

Spherical Sector. Given a solid spherical sector of radius r and semivertical angle θ, with the axis of symmetry on the z axis,

$$I_z(O) = \frac{Mr^2(1 - \cos\theta)(2 + \cos\theta)}{5}$$

Table 5 MOMENTS OF INERTIA (*continued*)

Spherical Segment. Given a solid spherical segment of radius r and height h with the axis of symmetry on the z axis,

$$I_z(O) = M\left(r^2 - \frac{3rh}{4} + \frac{3h^2}{20}\right)\frac{2h}{3r-h}$$

Frustrum of a Cone. Given a frustrum of a solid right circular cone, with radius of the smaller base a and radius of the larger base b, and symmetry axis coinciding with the z axis,

$$I_z(O) = \frac{3M(b^5 - a^5)}{10(b^3 - a^3)}$$

Torus or Anchor Ring. Given a solid torus whose equation in cylindrical coordinates ρ, ϕ, and z is given by $(\rho - b)^2 + z^2 = a^2$, where $b > a$,

$$I_x(O) = I_y(O) = \frac{M(4b^2 + 5a^2)}{8}$$

$$I_z(O) = \frac{M(4b^2 + 3a^2)}{4}$$

Table 6 VECTOR IDENTITIES

$\mathbf{A} \cdot (\mathbf{B} \times \mathbf{C}) = \mathbf{B} \cdot (\mathbf{C} \times \mathbf{A}) = \mathbf{C} \cdot (\mathbf{A} \times \mathbf{B})$	(1)
$\mathbf{A} \times (\mathbf{B} \times \mathbf{C}) = (\mathbf{A} \cdot \mathbf{C})\mathbf{B} - (\mathbf{A} \cdot \mathbf{B})\mathbf{C}$	(2)
$(\mathbf{A} \times \mathbf{B}) \cdot (\mathbf{C} \times \mathbf{D}) = (\mathbf{A} \cdot \mathbf{C})(\mathbf{B} \cdot \mathbf{D}) - (\mathbf{A} \cdot \mathbf{D})(\mathbf{B} \cdot \mathbf{C})$	(3)
$(\mathbf{A} \times \mathbf{B}) \times (\mathbf{C} \times \mathbf{D}) = [(\mathbf{A} \times \mathbf{B}) \cdot \mathbf{D}]\mathbf{C} - [(\mathbf{A} \times \mathbf{B}) \cdot \mathbf{C}]\mathbf{D}$	(4)
$\nabla(\phi + \psi) = \nabla\phi + \nabla\psi$	(5)
$\nabla(\phi\psi) = \phi\nabla\psi + \psi\nabla\phi$	(6)
$\nabla \cdot (\mathbf{F} + \mathbf{G}) = \nabla \cdot \mathbf{F} + \nabla \cdot \mathbf{G}$	(7)
$\nabla \times (\mathbf{F} + \mathbf{G}) = \nabla \times \mathbf{F} + \nabla \times \mathbf{G}$	(8)
$\nabla \cdot (\phi\mathbf{F}) = \mathbf{F} \cdot \nabla\phi + \phi\nabla \cdot \mathbf{F}$	(9)
$\nabla \times (\phi\mathbf{F}) = \nabla\phi \times \mathbf{F} + \phi\nabla \times \mathbf{F}$	(10)
$\nabla(\mathbf{F} \cdot \mathbf{G}) = (\mathbf{F} \cdot \nabla)\mathbf{G} + (\mathbf{G} \cdot \nabla)\mathbf{F} + \mathbf{F} \times (\nabla \times \mathbf{G}) + \mathbf{G} \times (\nabla \times \mathbf{F})$	(11)
$\nabla \cdot (\mathbf{F} \times \mathbf{G}) = \mathbf{G} \cdot (\nabla \times \mathbf{F}) - \mathbf{F} \cdot (\nabla \times \mathbf{G})$	(12)
$\nabla \times (\mathbf{F} \times \mathbf{G}) = \mathbf{F}(\nabla \cdot \mathbf{G}) - \mathbf{G}(\nabla \cdot \mathbf{F}) + (\mathbf{G} \cdot \nabla)\mathbf{F} - (\mathbf{F} \cdot \nabla)\mathbf{G}$	(13)
$\nabla \times (\nabla \times \mathbf{F}) = \nabla(\nabla \cdot \mathbf{F}) - (\nabla \cdot \nabla)\mathbf{F}$	(14)
$\nabla \times \nabla\phi = 0$	(15)
$\nabla \cdot (\nabla \times \mathbf{F}) = 0$	(16)

Answers

Answers

CHAPTER 5

5. 43.3 km/hr, 54.7° W of N
6. 15.8 km/hr, 18.4° E of S

CHAPTER 6

1. $2\mathbf{e}_1 - 40\mathbf{e}_2 + 17\mathbf{e}_3$
2. $\mathbf{A} = (6\cos t - 3\exp t)\mathbf{e}_{1'} + (6\sin t - 2\exp t)\mathbf{e}_{2'} + (3\sin t - 2\cos t + \exp t)\mathbf{e}_{3'}$
   $\dot{\mathbf{A}} = \cos t\, \mathbf{e}_{1'} + \sin t\, \mathbf{e}_{2'} + \exp t\, \mathbf{e}_{3'}$
   $\ddot{\mathbf{A}} = (-45\sin t + 6\cos t + 4\exp t)\mathbf{e}_{1'} + (-6\sin t + 40\cos t - 19\exp t)\mathbf{e}_{2'} + (14\sin t + 21\cos t - 12\exp t)\mathbf{e}_{3'}$
   $\dot{\mathbf{A}}' = -\sin t\, \mathbf{e}_{1'} + \cos t\, \mathbf{e}_{2'} + \exp t\, \mathbf{e}_{3'}$
3. a. $(\omega+1)\cos[(\omega+1)t]\mathbf{e}_1 + (\omega+1)\sin[(\omega+1)t]\mathbf{e}_2 + \exp t\, \mathbf{e}_3$
   b. $(\omega+1)\cos t\, \mathbf{e}_{1'} + (\omega+1)\sin t\, \mathbf{e}_{2'} + \exp t\, \mathbf{e}_{3'}$
   c. $\cos[(\omega+1)t]\mathbf{e}_1 + \sin[(\omega+1)t]\mathbf{e}_2 + \exp t\, \mathbf{e}_3$
   d. $\cos t\, \mathbf{e}_{1'} + \sin t\, \mathbf{e}_{2'} + \exp t\, \mathbf{e}_{3'}$

7. $\omega + (a/b)(\Omega - \omega)$

8. $\mathbf{v} = (u + a\dot{\alpha}\cos\alpha + d\dot{\beta}\cos\beta)\mathbf{e}_H - (a\dot{\alpha}\sin\alpha + d\dot{\beta}\sin\beta)\mathbf{e}_V$
   $\mathbf{a} = -(a\dot{\alpha}^2\sin\alpha + d\dot{\beta}^2\sin\beta)\mathbf{e}_H - (a\dot{\alpha}^2\cos\alpha + d\dot{\beta}^2\cos\beta)\mathbf{e}_V$
   where $a \equiv c + d$, $\alpha \equiv [(u/c) + (v/a)]t$, $\beta \equiv [(u/c) + (v/d)]t$
   $\mathbf{e}_H \equiv$ unit vector in the horizontal direction
   $\mathbf{e}_V \equiv$ unit vector in the vertical direction

9. $[2\pi n \sin(\alpha + \beta)/\sin\alpha][\sin\beta\, \mathbf{e}_{1'} + \cos\beta\, \mathbf{e}_{2'}]$, where S' is a frame with the $3'$ axis along the axis of the inner cone and the $1'$ axis in the plane determined by the axes of both cones.

CHAPTER 7

1. a. $-3v^2/4k$
   b. $(3v^2/4k)[2/(1 + \cos\theta)]^{1/2}$
   c. $v/(2kr)^{1/2}$
2. $\mathbf{v} = 2bt\mathbf{e}_r + bct^2\mathbf{e}_\phi \qquad \mathbf{a} = (2b - bc^2t^2)\mathbf{e}_r + 4bct\mathbf{e}_\phi$
3. $\mathbf{v} = ab\mathbf{e}_\phi + 2ct\mathbf{e}_z \qquad \mathbf{a} = -ab^2\mathbf{e}_\rho + 2c\mathbf{e}_z$
   $\alpha = \cos^{-1}\{4c^2/[(a^2b^2 + 4c^2)(a^2b^4 + 4c^2)]^{1/2}\}$
4. $\mathbf{r} = (\xi^2 + \eta^2)^{1/2}(\xi/2)\mathbf{e}_\xi + (\xi^2 + \eta^2)^{1/2}(\eta/2)\mathbf{e}_\eta + z\mathbf{e}_z$
   $\mathbf{v} = (\xi^2 + \eta^2)^{1/2}\dot{\xi}\mathbf{e}_\xi + (\xi^2 + \eta^2)^{1/2}\dot{\eta}\mathbf{e}_\eta + \dot{z}\mathbf{e}_z$

CHAPTER 8

1. $m_a : m_b : m_c :: 5 : 10 : 4$
2. $m = 3$
3. If we assume only that the force is along the line joining the two stars then $\mathbf{R} = \mathbf{A} + \mathbf{B}t - (m/M)C(\cos\omega t\, \mathbf{i} + \sin\omega t\, \mathbf{j})$, where M is the unknown mass of the invisible star. If we assume further that the magnitude of the force is given by GMm/d^2, where d is the distance between the stars and G is the gravitational constant then we can determine M. See Problem 2, Chapter 25.
4. b/a
5. The universe in Newton's time was believed to consist of particles, differing only in size and shape, that were moving in empty space. Hence the density of a body was a measure of the fraction of the body occupied by matter, and the product of the volume of a body and its density was a measure of the quantity of matter in the body. Newton's definition furthermore implied the additivity of mass.

CHAPTER 9

1. $\mathbf{r} = \mathbf{c}(t^4/12m) \qquad \mathbf{v} = \mathbf{c}(t^3/3m)$
2. $x = (mv_0/a^2)[1 - \exp(-a^2t/m)]$
3. $5f_0 t_0^2/2m$

Answers   491

4. $f(x) = -mp^2 x$
5. $(1/m)[ase_1 + bve_2 + et_0 e_3]$
6. $-5.48\,Ne_r + 20.9\,Ne_\phi$, where r and ϕ are the polar coordinates of the particle.
11. This argument was used by Newton.
12. This argument presupposes that mass is additive.
13. This argument presupposes that two forces are equal if the reaction they produce on the agent is the same.
14. a. $\mu mg[1 + (m/M)]$
    b. $\mu mg[1 + (M/m)]$
15. $(n/N)F$

CHAPTER 10

1. $(2, 0, 2)$
2. $3\mathbf{i} - 2\mathbf{j} - \frac{1}{2}\mathbf{k}$
5. $x(c) = 0 \quad y(c) = 3a/5$
6. $AC/CB = (a + 2b)/(2a + b)$
7. See Table 4.
8. See Table 4.
9. $(h^2 - 3a^2)/[4(h + 2a)]$ measured from the interface
10. $x(c) = \frac{1}{2} \quad y(c) = 1 \quad z(c) = 2$
11. $x(c) = 0 \quad y(c) = 0 \quad z(c) = 2a/3$
12. $h/4$ from the base, $2a/\pi$ from the flat face, where a is the radius of the base and h is the altitude
13. $(4b^2 + a^2)/2\!:\!b$ from the center of curvature
14. $a/6$ from the center of the large circle
15. On the axis of the cone at a distance $a\cos^2\alpha$ from the vertex
17. See Table 4.
18. $2a/3$ from the low density end
19. $5a/2$ from the center of the base
20. $64a\sqrt{2}/45\pi$ from the center of curvature

CHAPTER 11

2. $-4Gm\sigma \sin^{-1}\{ab/[(a^2 + z^2)^{1/2}(b^2 + z^2)^{1/2}]\}$

3. If $r > R$ then $f = -GMm/r^2$; if $r < R$ then $f = -GMmr/R^3$, where R is the radius of the sphere.

4. $\rho = C/r$

5. $2\pi G\rho\{[(b + z)^2 + a^2]^{1/2} - [(b - z)^2 + a^2]^{1/2} - 2z\}$ if the point is inside the cylinder; and $2\pi G\rho\{[(b + z)^2 + a^2]^{1/2} - [(b - z)^2 + a^2]^{1/2} - 2b\}$ if the point is outside the cylinder and z is positive.

6. $2G\sigma a/bc$

7. The force is directed toward the apex and is of magnitude $GMm/\rho s$, where m is the mass of the particle, M is the total mass on the surface of the cone, and ρ is the distance of the particle from the edge of the base of the cone.

8. $-2\pi\sigma GM$

9. $2\pi a\alpha\beta G\{[(h + b)^2 + a^2]^{-1/2} - (h^2 + a^2)^{-1/2}\}$

10. 1.72

CHAPTER 12

2. 490 N

3. $mgl/(9l^2 - 3s^2)^{1/2}$

4. To the point at which the line joining the bug and the center of curvature of the bowl makes an angle 14.0° with the downward vertical.

5. $\mu a(2m + M)/[M^2 + \mu^2(2m + M)^2]^{1/2}$

6. The force is of magnitude $mg \sin[(\tan^{-1}\mu) - \alpha]$ and is directed along a line making an angle $\tan^{-1}\mu$ with the plane.

7. $\cos^{-1}\frac{3}{5}$

8. $\alpha = \epsilon - \sin^{-1}(m\cos\epsilon/M)$

10. $0 \leqslant \theta < 10°$ $\quad$ $69.1° < \theta < 74.4°$ $\quad$ $155.8° < \theta \leqslant 180°$

11. The polar equation of the boundary of the region within which the particle can remain at rest is

$$(mg)^2(\sin^2\alpha - \mu^2\cos^2\alpha) + \frac{(r - a)^2\lambda^2}{a^2} + \frac{2(r - a)\lambda}{a}mg\sin\alpha\cos\theta = 0$$

where θ is the angle that the string makes with the line of greatest slope and r is the distance of the particle from the fixed point.

CHAPTER 13

2. a. mg
   b. $m(g + a\alpha)$
5. $x = [(v^2 + V^2)^{1/2} V]/(2\bar{\mu}g)$ $y = [(v^2 + V^2)^{1/2} v]/(2\bar{\mu}g)$
8. Let v be the initial speed of the snowball, V the speed of the truck, a the distance between the boy and his friend when they are abreast of each other, α the elevation angle at which the snowball must be thrown, and ϕ the lag angle at which the snowball must be thrown; then α and ϕ can be obtained from the equations $V = v \cos\alpha \sin\phi$ and $ga = v^2 \sin(2\alpha)\cos\phi$.
10. $\cos^{-1}(\frac{2}{3})$
12. $y = -(gx^2/2v_0^2\cos^2\alpha)[1 + (2kx/3)] + x \tan\alpha$
15. See Example 3.
16. If we assume that the steel balls attract each other as if they were point particles, they will collide after approximately 26 hours and their relative speed at collision will be approximately 2.6×10^{-4} m/s.
17. 1.96 m/s   2.35 N
18. $-(M - m)^2 g/(M + m)$
19. The following equations:
    $(M + m\sin^2\theta)a\ddot{\theta} + ma\dot{\theta}^2 \sin\theta \cos\theta + (M + m)g \sin\theta = 0$
    $T = (mg + ma\ddot{\theta}\sin\theta + ma\dot{\theta}^2 \cos\theta)/\cos\theta$
    where a is the length of the string and T the tension, together with the boundary conditions $\theta(0) = \theta_0$ and $\dot{\theta}(0) = 0$ will suffice.
20. $a = g \sin\alpha - [(\bar{\mu}'m' + \bar{\mu}m)/(m' + m)]g \cos\alpha$
    $F = (\bar{\mu} - \bar{\mu}')[m'm/(m' + m)]g \cos\alpha$
22. $F = -m\{A\beta^2 \sin\beta t + [\omega_0^2 r_0^4/(r_0 + A \sin\beta t)^3]\}$
23. $(mMg \cos\alpha)/(m \sin^2\alpha + M)$
25. $a\ddot{\theta} + b\omega^2 \sin\theta = 0$, where θ is the angle that the string makes with the line through O and P.

CHAPTER 14

1. a. $-8\mathbf{e}_1 + 31\mathbf{e}_2 + 36\mathbf{e}_3$
   b. $\mathbf{e}_1 + 33\mathbf{e}_2 + 25\mathbf{e}_3$
2. $bn - mc$
3. $l(bZ - cY) + m(cX - aZ) + n(aY - bX)$

4. $h_x = -mr^2(\dot\theta \sin\phi + \dot\phi \sin\theta \cos\theta \cos\phi)$
$h_y = mr^2(\dot\theta \cos\phi - \dot\phi \sin\theta \cos\theta \sin\phi)$
$h_z = mr^2\dot\phi \sin^2\theta$

5. $\mathbf{h} = (8t^4 + 36t^3 - 130t^2 + 64t - 104)\mathbf{i} + (-2t^6 - 48t^4 + 56t^3 + 18t^2 + 96)\mathbf{j} + (-6t^6 + 16t^5 - 90t^4 + 80^3 + 6t^2 - 48t + 102)\mathbf{k}$
$\mathbf{g} = \dot{\mathbf{h}}$

CHAPTER 15

1. 14
2. -28π
3. $2\pi\sigma Gm[(h^2 + a^2)^{1/2} + a - (2a^2/h)]$
4. $\pi\sigma aGM$
5. The minimum work the spider would have to do is $mg(a + b)/2$.
6. a. $[2gx(M - m)/(M + m)]^{1/2}$
   b. $\{2gx[M - m\exp(\bar\mu\pi)]/[M + m\exp(\bar\mu\pi)]\}^{1/2}$
7. a. $h = [(2mgb/K) + (mg/K)^2]^{1/2} - (mg/K)$
   b. $v = [2g(b - x) - (Kx^2/m)]^{1/2}$
   where $K \equiv kk'/(k + k')$
8. $v^2/(2\bar\mu g)$
9. $[2gs(\sin\alpha - \bar\mu \cos\alpha)]^{1/2}$
10. $b \geqslant (1/\lambda)\{\bar\mu mga + a[(\bar\mu mg)^2 + 2\lambda\bar\mu mg]^{1/2}\}$
12. $x = [(v^2 + V^2)^{1/2}V]/(2\bar\mu g)$ $y = [(v^2 + V^2)^{1/2}v]/(2\bar\mu g)$
13. $(1 - 2\mu^2)e^{\mu\pi} - 3\mu = 0$
14. $3mv_0^2/2$

CHAPTER 16

1. $(3czx/r^5)\mathbf{i} + (3cyz/r^5)\mathbf{j} + [c(3z^2 - r^2)/r^5]\mathbf{k}$
2. a. $-(\mathbf{i}x + \mathbf{j}y + \mathbf{k}z)/(x^2 + y^2 + z^2)$
   b. $(2a\cos\theta/r^3)\mathbf{e}_r + (a\sin\theta/r^3)\mathbf{e}_\theta$
   c. $-k_1 x\mathbf{i} - k_2 y\mathbf{j} - k_3 z\mathbf{k}$
   d. $(e^{-kr}/r^2)(1 + kr)\mathbf{e}_r$
4. $V = -4\pi Gm\rho[(b^2/2) - (a^2/2)]$ $r \leqslant a$
   $V = -4\pi Gm\rho[(b^2/2) - (a^3/3r) - (r^2/6)]$ $a \leqslant r \leqslant b$
   $V = -4\pi Gm\rho[(b^3/3r) - (a^3/3r)]$ $r \geqslant b$

5. $Gm\sigma \ln[(r_1 + r_2 + 2a)/(r_1 + r_2 - 2a)]$
6. $-xy^2z^3 + 3x^2z^2$
7. $-xe^y - z$
8. a. The first force is not conservative and the work done is $abB/2$.
   b. The second force is conservative and the work done is zero.
9. a. Not conservative
   b. $U = -\int r\phi(r)\,dr$
   c. Not conservative; see example in Appendix 11.
   d. $U = -x^2y^3z^4 + $ constant
10. a. $U = \int \phi_1(x)\,dx + \int \phi_2(y)\,dy + \int \phi_3(z)\,dz$
    b. Not conservative
    c. $U = -(x^2y^2z/2) + $ constant
11. a. $U = -(ax^2/2) - by^2x - ayz - (bz^3/3) + $ constant
    b. Not conservative
    c. Not conservative
    d. $U = -(\mathbf{a}\cdot\mathbf{r})^2/2 + $ constant
13. $U = -\int_r^\infty (k/r^n)\,dr$

CHAPTER 17

2. $v_1^2 = 2gf^2(z_2)(z_1 - z_2)/[f^2(z_1) - f^2(z_2)]$
   $v_2^2 = 2gf^2(z_1)(z_1 - z_2)/[f^2(z_1) - f^2(z_2)]$
4. $\dot{x}, \quad \dot{x}^2 + \dot{y}^2 + 2gy, \quad \dot{x}\dot{y} + gx, \quad \dot{y} + gt$
5. $(9\lambda a/2m)^{1/2}$
6. $v = [V^2 - 2g(z - a)]^{1/2} \quad z_0 = a \quad z_1 = V^2/2g$
7. $v_H = 0 \quad v_V = \pm(2ga\sin\theta)^{1/2}$
9. $(4mg/5\pi)(3\pi + 1)$

CHAPTER 18

1. a. $3mv/[2(M+m)]$
   b. $3mMv/[2\Delta t(M+m)]$
2. $[Nmv(1+e)t]/[Nm(1+e)t + (m+M)]$
3. $v_1' = \frac{1}{2}[(1-e)v_1 + (1+e)v_2]$ in same direction as v_1
   $v_2' = \frac{1}{2}[(1+e)v_1 + (1-e)v_2]$ in same direction as v_2
   Loss of kinetic energy $= (m/4)(1-e^2)(v_1-v_2)^2$

4. he^{2n}

5. $e = (b/a)^{1/2}$ loss of kinetic energy $= mga[1 - e^{2n}]$

6. $v'_2 = (v/2)(1 + e)\cos\alpha$ at an angle α with the direction of the incident ball
$v'_1 = (v/2)[(1 - e)^2\cos^2\alpha + 4\sin^2\alpha]^{1/2}$ at an angle β with the direction of the incident ball, where $\tan\beta = [(1 + e)\sin\alpha\cos\alpha]/[2 - (1 + e)\cos^2\alpha]$

7. If $\bar{\mu} \leq \tan\theta/(1 + e)$ the speed of the rebounding particle will be $v\{e^2\cos^2\theta + [\sin\theta - \bar{\mu}(1 + e)\cos\theta]^2\}^{1/2}$ and the angle its direction of motion makes with the normal to the surface will be $\tan^{-1}\{-(\tan\theta/e) + [\bar{\mu}(1 + e)/e]\}$. If $\bar{\mu} \geq \tan\theta/(1 + e)$ the particle will rebound in a direction normal to the wall with a speed $ev\cos\theta$.

8. 17.3 m/s at an angle of 30° to the horizontal

9. Let $\mathbf{e}_1$ be the direction of the initial velocity of mass m.
   a. $-\{[2M - m(1 + \sin^2\theta)]/[2M + m(1 + \sin^2\theta)]\}v\mathbf{e}_1$
   b. $\{2m/[2M + m(1 + \sin^2\theta)]\}v\mathbf{e}_1$
   c. $\sin^2\theta$

CHAPTER 19

1. $\ddot{x} = -bz - 2bt\dot{z} + b^2t^2x$
$\ddot{y} = 0$
$\ddot{z} = bx + 2bt\dot{x} + b^2t^2z$
where $x \equiv x_{1*}$ $y \equiv x_{2*}$ $z \equiv x_{3*}$

2. $\ddot{x} = -b\cos\omega t + 2\omega\dot{y} + \omega^2 x$
$\ddot{y} = b\sin\omega t - 2\omega\dot{x} + \omega^2 y$
$\ddot{z} = 0$
where $x \equiv x_{1*}$ $y \equiv x_{2*}$ $z \equiv x_{3*}$
Solution: $x = -(bt^2/2)\cos\omega t$ $y = (bt^2/2)\sin\omega t$

3. $\omega = [(\mu g - a)/r]^{1/2}$

4. The orbit is a circle of radius $u/(2\omega\sin\lambda)$ with the center at $x = 0$; $y = a - [u/(2\omega\sin\lambda)]$.

5. If we choose the axes as in Problem 4, $R = mg - 2m\omega\dot{y}\cos\lambda$.

7. $(2^{1/2}8\omega h^{3/2}\cos\lambda)/(3g^{1/2})$ to the west

8. Given x axis south, y axis east, and z axis vertically up, the rigorous solution to the equation $\ddot{\mathbf{r}} = \mathbf{g} - 2(\boldsymbol{\omega} \times \dot{\mathbf{r}})$ for a particle dropped from rest at the point $(x, y, z) = (0, 0, h)$ is:
$x = (g\sin 2\lambda/8\omega^2)\{2(\omega t)^2 - [1 - \cos(2\omega t)]\}$
$y = (g\cos\lambda/4\omega^2)\{2\omega t - \sin(2\omega t)\}$
$z = h - (g/4\omega^2)\{2(\omega t)^2\sin^2\lambda + [1 - \cos(2\omega t)]\cos^2\lambda\}$
An approximation can be obtained by expanding in powers of ωt.

Answers   497

11. If we choose the z axis as the axis of rotation and the x axis along the rod, then $\mathbf{R} = 2m\omega\dot{x}\mathbf{j}$, where $\dot{x}$ is determined from the solution of the equation $\ddot{x} = \omega^2 x$.

CHAPTER 20

1. $\ddot{r} - r\dot{\theta}^2 - r\sin^2\theta\,\dot{\phi}^2 = -(\lambda/ma)(r-a) + g\cos\theta$
   $r\ddot{\theta} + 2\dot{r}\dot{\theta} - r\sin\theta\cos\theta\,\dot{\phi}^2 = -g\sin\theta$
   $r\sin\theta\,\ddot{\phi} + 2\dot{r}\dot{\phi}\sin\theta + 2r\cos\theta\,\dot{\theta}\dot{\phi} = 0$

2. $\ddot{r} = r\sin^2\theta\,\alpha^2 t^2 - g\cos\theta - \mu[(g + r\cos\theta\,\alpha^2 t^2)^2 + \alpha^2(r + 2\dot{r}t)^2]^{1/2}\sin\theta\,(\dot{r}/|\dot{r}|)$

5. $R_{\max} = mg\sin\alpha\,(1 + 16\cot^2\alpha)$

CHAPTER 21

1. a. $x = (1/\alpha)\ln\{A - [A(A - |E|)]^{1/2}\cos[\alpha t(2|E|/m)^{1/2} + C]\}$
      $\quad - (1/\alpha)\ln|E| \qquad E < 0$
      $x = (1/\alpha)\ln\{\tfrac{1}{2} + (A\alpha^2/m)(t + C)^2\} \qquad E = 0$
      $x = (1/\alpha)\ln\{[A(A + E)]^{1/2}\cosh[\alpha t(2E/m)^{1/2} + C] - A\}$
      $\quad - (1/\alpha)\ln E \qquad E > 0$
      where $E = (m\dot{x}^2/2) + V$
   b. $\tau = (\pi/\alpha)(2m/|E|)^{1/2}$

2. $x = -a \qquad \tau = 2\pi(2ma^3/c)^{1/2}$
   a. $-(c/ma)^{1/2} < v < (c/ma)^{1/2}$
   b. $v < -(c/ma)^{1/2}$ or $(c/ma)^{1/2} < v < (2c/ma)^{1/2}$
   c. $v > (2c/ma)^{1/2}$

3. a. $x = (1/\alpha)\sinh^{-1}\{[(|E| + A)/|E|]^{1/2}\sin[\alpha t(2|E|/m)^{1/2} + C]\}$
      $E < 0$
      $x = \pm(1/\alpha)\sinh^{-1}\{[(E + A)/E]^{1/2}\sinh[\alpha t(2|E|/m)^{1/2} + C]\}$
      $E > 0$
      $x = \pm(1/\alpha)\sinh^{-1}[\alpha t(2A/m)^{1/2} + C] \qquad E = 0$
      where $E = (m\dot{x}^2/2) + V(x)$
   b. $x = (1/\alpha)\sin^{-1}\{[E/(E + A)]^{1/2}\sin[\alpha t(2(E + A)/m)^{1/2} + C]\}$,
      $E > 0$

4. The motion is oscillatory for $-\tfrac{1}{4} < E < 0$; $\tau = \pi m^{1/2}/2^{1/2}|E|^{3/2}$

5. $x(t) = x_0/[1 \pm tx_0(2A/m)^{1/2}]$, where $x_0 \equiv x(0)$ and the sign is $(+)$ when $\dot{x}(0) < 0$ and $(-)$ when $\dot{x}(0) > 0$.

6. The general oscillatory motion of a simple pendulum of length l is given by $\sin(\theta/2) = \sin(\alpha/2)\text{sn}[(g/l)^{1/2}(t - t_0)]$, where θ is the angular dis-

placement, α is the amplitude, t_0 is the time of passage through $\theta = 0$ in the positive direction, and the modulus of the elliptic function (see Appendix 16) is $k = \sin(\alpha/2)$.

10. $V = 2m\pi^2 x^2/\tau^2$, where τ is the period.

CHAPTER 22

1. a. $\tau = 0.284\,\text{s}$ $A = 0.18\,\text{cm}$
   b. $[12 + 0.181\sin(22.1t)]$ cm

2. a. $\omega^2 = (A\alpha^2/m)[1 - (B/A\alpha)^2]^{1/2}$; a minimum occurs when $B < A\alpha$
   b. $\omega^2 = (8\pi A\alpha^4 x_0^2/3m)[\Gamma(\frac{3}{4})/\Gamma(\frac{1}{4})]^2$, where $\Gamma(x)$ is the gamma function and the amplitude x_0 is determined from the equation $E = A\alpha^4 x_0^4/3$

4. $\tau = 2.012\,\text{s}$
   $A(1):A(2):A(2.5)::7.23 \times 10^{-2}:1:5.48 \times 10^{-1}$

5. $(M + m)g/4b$

7. $x = e^{-kt}\{C_1\exp[(k^2 - \lambda^2)^{1/2}t] + C_2\exp[-(k^2 - \lambda^2)^{1/2}t]\} + (g/\lambda) - a + \lambda a[(\lambda - p^2)\cos pt + 2kp\sin pt]/[(\lambda - p^2)^2 + 4k^2 p^2]$

8. $m(u\omega + \mu a \omega^2)$

12. $x(t) = 0 \quad t \leq 0$
    $x(t) = (a/\omega_0^3 \tau)(\omega_0 t - \sin\omega_0 t) \quad 0 \leq t \leq \tau$
    $x(t) = (a/\omega_0^3 \tau)[\omega_0 \tau - \sin\omega_0 t + \sin(\omega_0 t - \omega_0 \tau)] \quad t \geq \tau$

13. If $\omega \neq \omega_0$, then
    $x(t) = 0 \quad t \leq 0$
    $x(t) = (a/\omega_0)\{[\omega\sin(\omega_0 t) - \omega_0 \sin(\omega t)]/[\omega^2 - \omega_0^2]\} \quad 0 \leq t \leq \pi/\omega$
    $x(t) = (a\omega/\omega_0)\{(\sin[\omega_0 t - (\pi\omega_0/\omega)] + \sin[\omega_0 t])/(\omega^2 - \omega_0^2)\} \quad t \geq \pi/\omega$
    If $\omega = \omega_0$, then
    $x(t) = 0 \quad t \leq 0$
    $x(t) = (a/2\omega_0^2)[\sin(\omega_0 t) - \omega_0 t \cos(\omega_0 t)] \quad 0 \leq t \leq \pi/\omega$
    $x(t) = -(a\pi/2\omega_0^2)\cos(\omega_0 t) \quad t \geq \pi/\omega$

CHAPTER 23

3. The orbit of the comet is parabolic, hence the comet will escape. It crosses the earth's orbit at an angle of 45° with a speed $\sqrt{2}\,v_e$, where v_e is the orbital speed of the earth.

4. $\pi\eta/[(1 - \epsilon^2)\tau]$

6. $\{2[t^2 + (6)^{1/2}t + 2]\}^{1/2}$

8. $f(r) = mv^2a^2[(1/r^3) - (2c^2/r^5)]$
   $\phi - \coth\phi = (av/c^2)t + \phi_0 - \coth\phi_0$

11. $2a\omega/\sqrt{3}$

13. If $E = 0$ [assuming $V(\infty) = 0$] and $h^2 = km/(1 + b^2)$, the orbit will be of the desired form.

CHAPTER 24

1. a. $\sigma = 4r$
   b. $\sigma(\theta) = |r\sin(\theta/2)|$ if θ is allowed to range from $-\pi$ to π
      $\sigma(\theta) = 2r\sin(\theta/2)$ if θ is allowed to range from 0 to π

2. $\sigma(v,\theta) = \{k/[2mv^2\sin^2(\theta/2)]\}^2$

3. $\sigma(v,\theta) = \{[(ax/2)(1 + x)]/[(1 + 2x)\sin^2(\theta/2) + x^2]\}^2$
   where $x \equiv k/2Ea \equiv k/mv^2a$

6. $\pi nd[k\cot(\alpha/2)/(2E)]^2$

7. $In d\sigma(\alpha)A/r^2$

CHAPTER 25

4. Assuming that the force between the charges is given by ke^2/r^2:
   $r = 2ke^2(1 + \sqrt{2})/(mv^2)$
   $v_1 = (v/2)[3(2 - \sqrt{2})]^{1/2}$
   $v_2 = (v/2)[2 - \sqrt{2}]^{1/2}$

5. a. $6a$
   b. $5(5k/m)^{1/2}$
   c. $\pi(m/5k)^{1/2}$

7. The motion lies along a line. We assume the positive direction to be the initial direction of the velocity of m.
   a. $2mv/(m + M)$
   b. $mv/(m + M)$
   c. $2\pi\{[mM/(m + M)]/k\}^{1/2}$

8. $h_M(t) = -(gt^2/2) + [mv/(m + M)]t - [m/(m + M)](v/\omega)\sin\omega t + h_M(0)$
   $h_m(t) = -(gt^2/2) + [mv/(m + M)]t + [M/(m + M)](v/\omega)\sin\omega t + h_m(0) + a$
   where $\omega^2 = k(M + m)/(Mm)$ and $v < a/\omega$

9. $v = 3a(3k/m)^{1/2}$

CHAPTER 26

2. $E = 2\epsilon_0 \qquad \Theta = \theta/2$

3. 0.25

6. $r_0 = [k(M+m)/(Mmv^2)]\{[2M/(M+m)]^{1/2} + 1\}$, where $k \equiv Qq/(4\pi\epsilon_0)$

CHAPTER 28

1. $2mvn$

3. $-\pi a^2 v^2 \rho$

4. a. 81.6 m/s
   b. No

5. a. $h(t) = (uM_0/A)\{[(M_0 - At)/M_0]\ln[(M_0 - At)/M_0] + (At/M_0)\} - (gt^2/2) \qquad 0 < t \leq \Delta M/A$
   $h(t) = -g[t - (\Delta M/A)]^2/2 + \alpha[t - (\Delta M/A)] + \beta \qquad t \geq \Delta M/A$
   where α and β are the values of $dh(t)/dt$ and $h(t)$, respectively, at $t = \Delta M/a$
   b. $h_{max} = \beta + (\alpha^2/2g)$

6. a. 1.55×10^2 m/s 7.83×10^2 m/s
   b. 4.64×10^3 m

7. 372 s

8. $(g/2\alpha^2)[(\ln \beta)^2 + 2\ln \beta - 2\beta + 2]$

9. $M_1 = 39{,}800$ kg $M_2 = 1946$ kg (fuel + rocket, excluding payload)

10. $v_0 - (\mu m g/\rho A)^{1/2}$

11. 1.64 m/s

12. 7.16×10^2 N

13. $(gl/2)^{1/2}$

15. $F = -W + (8W/b^2)[x(b-x)]$

18. $F = 0 \qquad 0 < t < [2(h-l)/g]^{1/2}$
    $F = (3g^2\rho^2 t^2/2) - \rho g(h-l) \qquad [2(h-l)/g]^{1/2} < t < (2h/g)^{1/2}$
    $F = \rho l g \qquad t > (2h/g)^{1/2}$

19. $(m/\rho)\{[1 + (3\rho v_0^2/2mg)]^{1/3} - 1\}$

Answers 501

CHAPTER 29

5. a. $5\mathbf{i} + 5\mathbf{j} + 3\mathbf{k}$
   b. $-5\mathbf{i} + 3\mathbf{j} - 2\mathbf{k}$
   c. $5\mathbf{i} - 25\mathbf{j} + 3\mathbf{k}$
   d. $-5\mathbf{i} - 2\mathbf{j} - 7\mathbf{k}$

6. a. $m\omega a \mathbf{i} \quad -(3m\omega a^2/2)\mathbf{k}$
   b. $0 \quad -(m\omega a^2/2)\mathbf{k}$
   c. $m\omega a \mathbf{i} \quad -(3m\omega a^2/2)\mathbf{k}$
   d. $0 \quad -(m\omega a^2/2)\mathbf{k}$
   where $\mathbf{i}$ is parallel to the motion of the center of the disk and $\mathbf{j}$ is vertically up

7. $\mathbf{H} = (MR^2\omega/2)\mathbf{n}$, where $\mathbf{n}$ is a unit vector perpendicular to the disk

8. a. $\mathbf{H} = (-12t + 31)\mathbf{i} + (6t^2 - 10t - 12)\mathbf{j} + (5t^2 + 21)\mathbf{k}$
      $\mathbf{G} = (-12)\mathbf{i} + (12t - 10)\mathbf{j} + (10t)\mathbf{k}$
   b. $\mathbf{H} = (4t^2 - 12t + 137)\mathbf{i} + (8t^2 - 40t + 14)\mathbf{j} + (-5t^2 + 21)\mathbf{k}$
      $\mathbf{G} = (8t - 12)\mathbf{i} + (16t - 40)\mathbf{j} + (-10t)\mathbf{k}$

9. $(m\omega a^2/3)[-\mathbf{i}\sin\theta\cos\theta + \mathbf{k}\sin^2\theta]$

10. $T_B/T_A = (10a + 4x)/(9a - 2x)$

13. $\dot{\theta} = [(3g/a)(1 - \cos\theta)]^{1/2}$
    $F_r = (5mg\cos\theta/2) - (3mg/2) \qquad F_\theta = -(mg\sin\theta/4)$

15. $g/[1 + (4\pi^2 b^2/3p^2)]$

CHAPTER 30

1. a. $ma^2\omega^2/6$
   b. $ma^2\omega^2/24$

4. $2F\theta/ma$

5. $\{[3k(b-a)^2/2m] - 4\bar{\mu}gb\}^{1/2}$

CHAPTER 31

1. $m\ddot{x}_1 = -kx_1 + F_1 \qquad m\ddot{x}_2 = -kx_2 + F_2 \qquad m\ddot{x}_3 = F_3$
   $M\ddot{x}_4 = -F_1 \qquad M\ddot{x}_5 = -F_2 \qquad M\ddot{x}_6 = -Mg - F_3$
   $(x_4 - x_1)^2 + (x_5 - x_2)^2 + (x_6 - x_3)^2 = a^2$
   $F_1(x_5 - x_2) = F_2(x_4 - x_1) \qquad F_1(x_6 - x_3) = F_3(x_4 - x_1)$
   where F_1, F_2, and F_3 are components of the force exerted by the rod on

the particles. Note that there are nine equations in the nine unknowns x_1, x_2, x_3, x_4, x_5, x_6, F_1, F_2, F_3. The boundary conditions require at $t = 0$ that $x_1 = b$, $x_2 = b$, $x_3 = 0$, $x_4 = b$, $x_5 = b$, $x_6 = a$, and $\dot{x}_1 = \dot{x}_2 = \dot{x}_3 = \dot{x}_4 = \dot{x}_5 = \dot{x}_6 = 0$.

CHAPTER 32

2. a. $V = -(m_c g a^2 \theta^2 / l) + m_p g a (1 - \cos\theta)$
   b. If $2m_c a < m_p l$, there is a minimum at $\theta = 0$ and a maximum at the value of θ in the range $0 < \theta < \pi$ satisfying the equation $lm_p \sin\theta = 2m_c a\theta$.
3. $V = mg\{a\theta \sin\theta - [a + (d/2)](1 - \cos\theta)\}$
4. $V = nmga(4\cos^2\theta + 8\cos\theta - 5) + mga(2\sin\theta - \sqrt{3})$

CHAPTER 33

1. If $v(1) = 2v\mathbf{i}$ and $v(2) = 3v\mathbf{j}$, then $v(3) = -2v\mathbf{i} - 3v\mathbf{j}$.
4. $3\dot{\theta}^2 + \dot{\phi}^2 + 2\dot{\theta}\dot{\phi} + (k/m)[2\sin(\phi/2) - \sqrt{2}\,]^2 = v^2/2a^2$
   $3\dot{\theta} + \dot{\phi} = v/a$
8. a. $1.653a$
   b. $0.435 mv^2/a$
9. $a = [(M + 4m)/4m]^{1/2} R$

CHAPTER 35

1. $\mathbf{f} = R\mathbf{j} + 4R\mathbf{k}$ acting at the point $x = 4a/17$, $y = (4a/17) + t$, $z = -(a/17) + 4t$, where t is an arbitrary parameter
   $\mathbf{g} = -(aR/17)[\mathbf{j} + 4\mathbf{k}]$
2. The pitch of the equivalent wrench is $[(\pi/4) - 1]a \tan\alpha$.

CHAPTER 36

1. $T = [3mg(a - b)]/8(2ab - b^2)^{1/2}]$
2. $\theta = (\pi/2) - 2\epsilon$
3. $\tan^{-1}[(\sin\theta - \mu\cos\theta)/(2\mu\sin\theta)]$
4. $16°4'$ $2mg/(13)^{1/2}$ $6mg/(13)^{1/2}$

8. $(2ma\tan^{-1}\mu)/(b - 2a\tan^{-1}\mu)$

15. The force will be directed horizontally toward the vertical line passing through M.

CHAPTER 37

1. a. $ma^2/12$
    b. $ma^2/3$
3. See Table 5.
5. See Table 5.
7. $I = (4ma^2/5) + 2mx^2 - 4mbx + 4mb^2$, where x is the distance of the center of either sphere from the axis. The maximum angular momentum will occur when the axis passes through one of the extremities of the rod.
9. Assume that the velocity of O is $V\mathbf{i}$, the angular velocity is $\Omega\mathbf{k}$, and $\mathbf{j}$ is vertically up. $T = (mV^2/2) + (ma^2\Omega^2/5)$.
    a. $(7ma^2\Omega/5)\mathbf{k}$
    b. $[(2ma^2\Omega/5) - mVa]\mathbf{k}$
    c. $(2ma^2\Omega/5)\mathbf{k}$
    d. $(2ma^2\Omega/5)\mathbf{k}$
    e. The answer depends on the meaining of "highest point."
11. $(3g/2a)^{1/2}$
13. $mg/4$
14. $[2r^2/(R^2 + 2r^2)]g$
15. $a_2 = [2(m_2 - m_1)/(2m_1 + 2m_2 + M)]g$
    $T_1 = [(4m_2 + M)/(2m_1 + 2m_2 + M)]m_1 g$
    $T_2 = [(4m_1 + M)/(2m_1 + 2m_2 + M)]m_2 g$
16. $\ddot{\theta}_A = b^2 G/(I_A b^2 + I_B a^2)$ $\ddot{\theta}_B = abG/(I_A b^2 + I_B a^2)$
17. $(15/56)mg$ $(403/448)mg$
18. $(3/16)(15ga/2)^{1/2}$ in the horizontal direction
19. $mg\sin\theta/3$, where θ is the angle that the plane of the axes makes with the vertical
20. $\cos^{-1}(10/17)$
22. $2a^2\Omega^2/(49\mu g)$
24. $\mu = \{1 + [v^2/(2gh)]\}\tan\alpha$ $\omega_0 = \mu v/[a(\mu - \tan\alpha)]$
32. $7Mv/(7M + 2m)$ $2Mv/[a(7M + 2m)]$

CHAPTER 38

4. a.
$$ma^2 \begin{vmatrix} 18 & 5 & -3 \\ 5 & 18 & -1 \\ -3 & -1 & 18 \end{vmatrix}$$

b. $23ma^2$

5. a.
$$ma^2 \begin{vmatrix} 9 & 5 & 0 \\ 5 & 9 & 0 \\ 0 & 0 & 18 \end{vmatrix}$$

b. $I_{\bar{1}} = 4ma^2$ $\mathbf{e}_{\bar{1}} = (1/\sqrt{2})(\mathbf{e}_1 - \mathbf{e}_2)$
$I_{\bar{2}} = 14ma^2$ $\mathbf{e}_{\bar{2}} = (1/\sqrt{2})(\mathbf{e}_1 + \mathbf{e}_2)$
$I_{\bar{3}} = 18ma^2$ $\mathbf{e}_{\bar{3}} = \mathbf{e}_3$

6. a.
$$m/12 \begin{vmatrix} 4b^2 & -3ab & 0 \\ -3ab & 4a^2 & 0 \\ 0 & 0 & 4(a^2 + b^2) \end{vmatrix}$$

b. $ma^2b^2/[6(a^2 + b^2)]$
c. $I_{\bar{1}} = (5 + 3\sqrt{2})(ma^2/6)$ $\mathbf{e}_{\bar{1}} = (4 - 2\sqrt{2})^{-1/2}[\mathbf{e}_1 + (1 - \sqrt{2})\mathbf{e}_2]$
$I_{\bar{2}} = (5 - 3\sqrt{2})(ma^2/6)$ $\mathbf{e}_{\bar{2}} = (4 + 2\sqrt{2})^{-1/2}[\mathbf{e}_1 + (1 + \sqrt{2})\mathbf{e}_2]$
$I_{\bar{3}} = 5ma^2/3$ $\mathbf{e}_{\bar{3}} = \mathbf{e}_3$

7. $\tan(2\theta) = (12/5)[(24 - \pi)/(32 - \pi)]$ $\theta = 30.02°$

9. $ma^2/6$

10. c. $I_{\bar{1}} = 2Ma^2/3$ $\mathbf{e}_{\bar{1}} = (1/\sqrt{3})(\mathbf{e}_1 + \mathbf{e}_2 + \mathbf{e}_3)$
$I_{\bar{2}} = 11Ma^2/3$ $\mathbf{e}_{\bar{2}}$ can be any unit vector perpendicular to $\mathbf{e}_{\bar{1}}$
$I_{\bar{3}} = 11Ma^2/3$ $\mathbf{e}_{\bar{3}} = \mathbf{e}_{\bar{1}} \times \mathbf{e}_{\bar{2}}$

11. Let S be a coordinate system with origin at the vertex of the cone and 3 axis along the axis of the cone. The 1, 2, and 3 directions are principal directions for the point O and for the center of mass C.
$I_{11}(O) = I_{22}(O) = (3M/20)(a^2 + 4h^2)$ $I_{33}(O) = 3Ma^2/20$
$h/a = \frac{1}{2}$ $x_1(C) = x_2(C) = 0$ $x_3(C) = 3a/8$
$I_{11}(C) = I_{22}(C) = 51Ma^2/320$ $I_{33}(C) = 3Ma^2/10$

12. $9ma^2/4$ $ma^2/2$ $9ma^2/4$

13. $83Ma^2/320$ $83Ma^2/320$ $2Ma^2/5$

14. $5ma^2/3$ $(7 + 3\sqrt{2})(ma^2/6)$ $(7 - 3\sqrt{2})(ma^2/6)$

CHAPTER 39

3. $\theta = \tan^{-1}(19/12)$

5. $T = (m/10)[(b^2 + c^2)\dot{\alpha}^2 + (a^2 + c^2)\dot{\beta}^2\cos^2\alpha + (a^2 + b^2)\dot{\beta}^2\sin^2\alpha]$
   If the $\bar{3}$ axis is along the axis of length $2c$, the $\bar{1}$ axis along the axis of length $2a$, and the 3 axis in the direction of the axis about which the $\bar{3}$ axis is rotating, α is the angle between the $\bar{1}$ axis and the direction $\mathbf{e}_3 \times \mathbf{e}_{\bar{3}}$.

6. $T = (M/10)(a^2 + c^2)\dot{\beta}^2\sin^2\gamma + (Ma^2/5)(\dot{\beta}\cos\gamma + \dot{\alpha})^2$

CHAPTER 40

2. $(ma^2\omega^2\sin\alpha\cos\alpha)/4$ in a direction perpendicular to the plane containing the axis of symmetry and the axis of rotation.

3. $ma^2\Omega/2\tau$. The reaction couple has a component in the plane of the crankshaft of magnitude $2ma^2\Omega/\pi\tau$ and a component perpendicular to the plane of the crankshaft of magnitude $2ma^2\Omega^2 t^2/\pi\tau^2$.

4. $\Omega(1 + 3\cos^2\alpha)^{1/2}$

7. b. $2\pi/[\Omega(1 + 3\cos^2\alpha)^{1/2}]$
   c. $2\pi/(\Omega\cos\alpha)$

13. The motion is stable if the angular velocity exceeds $2(g/a)^{1/2}$.

15. $\theta, \dot{\phi}$, and $\dot{\psi}$ are constants and are related as follows:
    $(C - A)\cos\theta\,\dot{\phi}^2 + C\dot{\phi}\dot{\psi} - mgh = 0$

18. $\mu \geq |[2\cos\alpha(r - a\cos\alpha)]/[\sin\alpha(4r - 3a\cos\alpha)]|$

21. $\theta = 25.1°$

CHAPTER 41

6. $3a/2$

CHAPTER 43

1. $Q_r = -mg\cos\theta \quad Q_\theta = mgr\sin\theta \quad Q_\phi = 0$

2. $Q_r = -mg\cos(\omega t + \alpha)$; r is the distance from O to the particle P; α is the angle that OP makes with the upward vertical at $t = 0$

3. $Q_s = 0 \quad Q_\theta = -mga\sin\theta$

Answers

4. $Q_s = -mg\sin\alpha$, where s is the distance of the cylinder from the bottom of the plane

5. $Q_\theta = 0 \qquad Q_\phi = -mga\sin\phi$

6. $Q_x = -kx \qquad Q_\theta = -mg(b-a)\sin\theta$

7. $Q_s = -(M+m)g\sin\alpha \quad Q_r = mg\cos\theta - k(r-b) \quad Q_\theta = -mgr\sin\theta$,
   s is the distance of the hoop from the bottom of the plane, r is the length of the filament, θ is the angle that the filament makes with the downward vertical.

CHAPTER 44

1. $m\ddot{r} - mr\dot\theta^2 = -dV/dr$
   $(d/dt)(mr^2\dot\theta) = 0$

2. $\ddot\theta - \sin\theta\cos\theta\,\dot\phi^2 = -(g/a)\sin\theta$
   $(d/dt)(\dot\phi\sin^2\theta) = 0$

3. $(d/dt)[(M+m)\dot x + ma\cos\theta\,\dot\theta] = 0$
   $\ddot x\cos\theta + a\ddot\theta = -g\sin\theta$

4. $(m+M)a\ddot\theta + Mb\sin(\theta-\phi)\dot\phi^2 + Mb\cos(\theta-\phi)\ddot\phi = -(m+M)g\sin\theta$
   $b\ddot\phi - a\sin(\theta-\phi)\dot\theta^2 + a\cos(\theta-\phi)\ddot\theta = -g\sin\phi$

8. $\ddot\theta + \omega^2\sin\theta = 0$. If P is the particle, C the center of the circle, and A the point on the circle opposite the fixed point, then θ is the angle ACP.

10. $g/2 \qquad -g/6 \qquad -5g/6$

13. $L = (m/2)[\dot s^2 + a^2\dot\theta^2 - 2a\cos(\alpha+\theta)\dot s\dot\theta] - mgs\sin\alpha + mga\cos\theta$ where s is the distance up the plane at which the plate is located, and θ is the angle that the pendulum makes with the downward vertical, the positive sense being away from the rod.

17. $2a\ddot\theta = 3g\sin\theta$

18. $M\ddot x - Mb\sin\theta\,\ddot\theta - Mb\cos\theta\,\dot\theta^2 = -kx$
    $4b\ddot\theta - 3\sin\theta\,\ddot x = -3g\sin\theta$
    where x is the displacement of the point of suspension of the rod from its equilibrium position, and θ is the angle that the rod makes with the vertical.

19. $(3b^2 - 2a^2)\ddot\theta + (3b^2 - 4a^2)\sin\theta\cos\theta\,\dot\phi^2 = 3g(b^2-a^2)^{1/2}\cos\theta$
    $(d/dt)\{[m(b^2-a^2)\cos^2\theta + (ma^2\sin^2\theta/3) + (Mb^2/2)]\dot\phi\} = 0$
    where ϕ is the angle that the plane of the loop makes with a fixed vertical plane, and θ is the angle that the rod makes with the vertical.

20. $\phi = [a/(a^2+b^2)^{1/2}]\tan^{-1}[x/(a^2+b^2)^{1/2}]$

CHAPTER 45

1. $\omega_{\bar{1}} = (g/a)^{1/2}$ $\quad \omega_{\bar{2}} = (g/b)^{1/2}$
2. $\theta_0/\phi_0 = -\frac{2}{3}$
3. a. $\omega_{\bar{1}} = (2g/a)^{1/2}$ $\quad \omega_{\bar{2}} = (g/2a)^{1/2}$
   b. $\phi = -2\theta$ $\qquad\qquad \phi = \theta$
   c. $q_{\bar{1}} = \theta - \phi$ $\qquad\quad\; q_{\bar{2}} = 2\theta + \phi$
8. $[mv/(M+2m)][t - \omega^{-1}\sin\omega t]$, where $\omega^2 \equiv k(M+2m)/(Mm)$
10. $\omega_{\bar{1}} = (2k/m)^{1/2}$ $\quad \omega_{\bar{2}} = (3k/m)^{1/2}$ $\quad \omega_{\bar{3}} = (6k/m)^{1/2}$
    $c_{\bar{1}} = (\alpha, \alpha, \alpha)$ $\quad\;\; c_{\bar{2}} = (\beta, 0, -\beta)$ $\quad\;\;\; c_{\bar{3}} = (\gamma, -\gamma, \gamma)$
    $q_{\bar{1}} = x + 2y + z$ $\;\; q_{\bar{2}} = x - z$ $\qquad\qquad q_{\bar{3}} = x - 2y + z$

CHAPTER 46

2. Let x be the downward displacement of the end of the spring from its equilibrium position, and θ the angle that the rod makes with the vertical.
   $\dot{x} = (4ap_x + 3\sin\theta p_\theta)/(4ma - 3ma\sin^2\theta)$
   $\dot{p}_x = -kx$
   $\dot{\theta} = (3a\sin\theta\, p_x + 3p_\theta)/(4ma^2 - 3ma^2\sin^2\theta)$
   $\dot{p}_\theta = ma\cos\theta\, \dot{x}\dot{\theta} + mga\sin\theta$

3. Choosing ϕ as generalized coordinate and dropping noncontributing terms in the Lagrangian (see Chapter 54); we obtain:
   $L = (m/2)(a^2 + c^2)\dot\phi^2 - mgc\phi$
   $H = [p_\phi^2/2m(a^2 + c^2)] + mgc\phi$
   $\ddot\phi = -gc/(a^2 + c^2)$

6. $H(\theta, \phi, p_\theta, p_\phi, t) = -(m\dot r^2/2) + (p_\theta^2/2mr^2) + (p_\phi^2/2mr^2\sin^2\theta) + mgr\cos\theta$, where $r = r(t)$ and $\dot r = \dot r(t)$. The term $-(m\dot r^2/2)$ may be dropped because it does not affect the equations of motion.

7. a. $H = [(p_\theta^2/2m)/(r_0 - \alpha t)^2] - mg(r_0 - \alpha t)\cos\theta - (m\alpha^2/2)$
   b. No
   c. No

8. $s(t) = (g/2\omega^2)(\cosh\omega t - \cos\omega t)$

10. $H = (p_X^2 + p_Y^2 + p_Z^2)/[2(M+m)] + [p_r^2 + (p_\theta^2/r^2) + (p_\phi^2/r^2\sin^2\theta)][(M+m)/(2Mm)] + V(r)$;
    $p_X, p_Y, p_Z, H, p_\phi, p_\theta^2 + (p_\phi^2/\sin^2\theta)$

Index

Abelian group, 950
Absolute derivative, 926-927
Acceleration:
  in cylindrical coordinates, 45
  of gravity, 73, 74
  in moving frame, 38
  relative, 27
  in spherical coordinates, 45
Action and angle variables, 807-814
Action at distance, 56
Action and reaction, law of, 60, 66, 216
Amonton's law of friction, 75
Angular frequency, 160
  of driven oscillator, 163
Angular impulse, 127, 129-130
Angular momentum:
  and arbitrary point, 102, 226-227
  axial component, 271-272
  and center of mass, 227
  change due to impulse, 127, 129, 130, 338
  conservation, 123, 174, 248, 580, 585, 586, 598
  definition, 102, 226
  of incident particle, 191
  Poisson bracket of components, 830
  of rigid body, 271, 272, 312
  time rate of change, 103, 229
  and torque, 103, 229-230
Angular motion, 176-177, 600
Angular velocity, 33-36
  definition and properties, 33-38
  Euler angles, 310, 650
Anholonomic constraint, 532

Anholonomic system:
  definition, 554
  and Gibbs-Appell equations, 722-725
Aphelion, 184
Apocentron, 183
Appell's equations of motion, 722
Appell's function, 721
Apse, 177, 600
Apsidal angle, 177, 600
Apsidal distance, 177, 600
Apsidal radius, 600
Apsis, 177, 600
Associative law, 945
Automorphic group, 948
Axial direction, 269
Axis of rotation, instantaneous, 36

Basis vectors, 683
Bertrand's theorem, 606-613
Bilinear form, 928
Boundary conditions, 435
Bound orbit, 606, 607
Brachistochrone problem, 979

Calculus of variations, 968-980
Canonical transformation:
  definition, 755-756, 762-764
  inverse, 761
  Jacobian for, 760, 771-772, 794-795
  Lagrange bracket test, 757
  and modified Hamilton's principle, 846-852
  Poisson bracket test, 826
  product of two, 761-762
  properties, 759-762
Canonical variables, 773-774

Carrier space, 955
Cartesian configuration space, 240-241, 349, 514
Cartesian tensors, 417-425
Cartesian vector quantity, 421
Cartesian vectors, 420-421, 424
Center of mass:
  acceleration, 67, 217
  coordinates, 205, 624-627, 632-633
  definition, 66
  determination, 67
  motion, 67-68, 197, 216, 220, 254, 274, 320, 655
  position, 217
  properties, 65-68
  of rigid body, 269
  table of values, 479-482
  of two particles, 196, 622
  uniqueness, 66
  velocity, 217, 622
Central force, 172-174
Central force motion:
  angular motion, 176-177, 600
  circular, 606-607
  fundamental equations, 175, 598
  Hamilton-Jacobi treatment, 803-804
  Lagrangian treatment, 597-602
  Newtonian treatment, 172-184
  orbit in, 175, 599
  radial motion in, 176, 599-600
  stability of circular orbit in, 184
  symmetry of orbit, 600
  for two-particle system, 198
Centrifugal force, 138
Character:
  of matrix group element, 961
  of operator, 961
  of representations, 961-963
Characteristic equation, 931
Character table, 689-690
Christoffel symbols, 926
Circular motion in central force field, 606-607
Class of element in group, 950, 960, 962
Clocks, 9
Closed orbit, 607
Closed systems, 215
Closure law, 945
Coefficient of kinetic friction, 76
Coefficient of restitution, 131, 339
Coefficient of static friction, 75
Collision:
  change in energy in, 204, 623
  definition, 49-50, 203, 622, 875-876
  description, 203, 622-624
  elastic, 131, 204, 339, 623
  inelastic, 131, 339
  of particle with surface, 131-132
  possible, 50, 876
  properties, 50, 876
  quantities conserved in, 52-56, 876-883
  rigid body, 339
  two-particle, 203-209, 622-633
Colonization of space, 643
Column matrix, 449
Column matrix representation of vector, 683, 953
Complex numbers, 437
Complex of group, 949
Compression period, 131, 339
Configuration of harmonic system, 684
Configuration space:
  basis vectors in, 683
  Cartesian, 240-241, 349, 514
  decomposition, by group theory, 688
  definition, 241, 514
  for holonomic system, 539, 682
  modal directions in, 683
  vector in, 682-683
Configuration of system, 241, 513
Conjugate elements in group, 950
Conservation:
  of angular momentum, 123, 174, 248, 580, 585, 586, 598
  of energy, 123, 174, 249, 598, 888
  of four-momentum, 897
  of linear momentum, 57, 122-123, 248, 578, 883-884
  of mass, 57
  see also Constants of motion
Conservative force field, 114-115, 244
Conserved quantities, 50-58, 876-884
  see also Conservation; Constants of motion
Constants:
  of motion, see Constants of motion
  in solution of differential equation, 433-434
  table, 477
Constants of motion:
  in central force motion, 174
  definition, 120, 247, 575
  generalized coordinate as, 740

generalized momentum as, 577-578, 585, 740
Hamiltonian as, 576-577, 585, 739, 829
in Hamiltonian formulation, 390, 739-740
importance, 122, 247-248
in Lagrangian formulation, 364-365, 575-584
number of independent, 120-121, 247, 575
in Poisson formulation, 829
in restricted three-body problem, 638
and symmetry, 578-584
see also Conservation; Conserved quantities
Constraint, 349, 528
Constraint forces:
  anholonomic:
    definition, 532
    determination, 555-556
  definition, 349, 528
  generalized components, 532-534, 541
  geometric, 529
  holonomic:
    definition, 349-351, 531-532
    determination, 549-550
  ideal, 529-531
  kinematic, 529
Contact:
  perfectly rough, 76
  perfectly smooth, 76
Contraction of tensor, 422
Contravariant vector, 923
Conversion factors, table, 478-479
Coordinates:
  center of mass, 205, 624-625, 632-633
  generalized, 352-353, 514, 539
  laboratory, 205, 624-625, 632-633
  orthogonal curvilinear, 43-46, 427-431
  relative, 205, 624-625, 632-633
Coordinate transformations, 470-474
Coriolis force, 138
Coset, 949
Couple, 256-257
Coupled oscillators, 676
  see also Vibrating systems
Covariant derivative, 927
Covariant vector, 923
Cross product, 425

Cross section:
  differential scattering, see Differential scattering cross section
  hard sphere, 192-193, 619
  inverse square, 193, 619-620
  total scattering, 187-188, 614-615, 628
Curl, 408-409, 414, 416, 425
Curvilinear coordinates, orthogonal:
  equations of motion in, 144-145
  properties, 43-46, 427-431
Cyclic coordinate, 577, 747
Cyclic group, 948, 950
Cyclic system, 808
Cylindrical coordinates:
  definition and properties, 44-45, 429-430
  equations of motion in, 144
  relation to rotating frame, 45

d'Alambert's principle, 56
Damped harmonic oscillator, 160-162
Decay modulus, 161
Decrement, 161
Degeneracy, 686-687
Degrees of freedom, 352, 534
Del operator, 409
Derivative, absolute, 926-927
Descartes, René, 12
Determinant, 452-454, 930
Diagonalization of matrices, 935-936
Differential, exact, 442-446
Differential equations:
  linear, 436-440
  ordinary, 432-435
Differential scattering cross section:
  center of mass, 207-208, 627-629, 631-633
  definition, 188-189, 615-616
  determination, 190-192, 617-619
  hard sphere, 192-193, 619
  inverse square, 193, 619-620
  transformation between lab and center of mass, 208-209, 629-633
Differentiation:
  covariant, 927
  of integral, 441
  partial, 398-400
  of tensor, 424
Dimension:
  of representation, 955
  irreducible, 960
Disk, rolling, 534-536, 556
Displacement, virtual, 349, 354, 515, 540, 712-713

Index

Dissipation function, 560-561
Distance, shortest, 978
Divergence, 408-409, 414, 416, 424
Dot product, 403, 424, 425
Driven harmonic oscillator, 162-168
Dynamical state, 120
Dynamical system, 65, 215
Dynamics, aim of, 122
Dynamics problems, method of solution, 88

Effective potential, 176, 179, 181, 326, 599
Eigencolumn, 931
Eigenvalues of matrix, 931-934
Einstein, Albert, 12
Elastic collision, 339
Electromagnetic field, generalized potential, 561
Elementary system, 521
Ellipsoid:
  of inertia, 291
  Poinsot, 330
Elliptic functions, 330, 467-469
Energy:
  of acceleration, 721
  conservation, 123, 174, 249, 598, 888
  dissipation, 164
  kinetic, *see* Kinetic energy
  and momentum, 888-889
  potential, *see* Potential energy
  relativistic, 888
  and work, 107-108, 234-237, 888
Equations of motion:
  in configuration space, 241
  Euler's, 322, 656, 719, 730
  Gibbs-Appell, 720-725, 727-730
  Hamiltonian type, 744-747
  Hamilton's, *see* Hamilton's equations of motion
  Lagrange's, *see* Lagrange's equations of motion
  Newton's, *see* Newton's equation of motion
  in non-inertial frames, 136
  Poisson's, 828-830
  relativistic, 885, 901, 903
  of rigid body, 274, 321-323
  Routh's, 746-749
  for two-particle system, 197
Equilibrium points, 638
Equimomental systems, 323
Equivalence of matrices, 928

Equivalent systems of forces, 255, 258
Escape speed, 182-183
Euclidean set of coordinate systems, 417
Euclidean transformation, 473
Euler angles, 303-307, 649
Euler equation, 972-974
Euler's equations of motion, 322, 656, 719, 730
Euler's theorem, 914
Exact differentials, 442-446
Extended point transformation, 750-751, 759
External forces, 78, 215

F function, 763
Fictitious forces, 137-138
Field, 113, 243, 408, 409, 424
Fixed plane, definition, 269
Force:
  action at distance, 61, 78
  active, 528
  central, 172-174
  classification, 61
  components:
    generalized, *see* Generalized components of force
    ordinary, 518-519
    physical, 517
  conservative, 114, 244
  of constraint, *see* Constraint forces
  contact, 61, 78
  coplanar, 260
  definition, 59, 65, 885
  derivable from generalized potential, 559
  determination, 77
  electromagnetic, 61
  equivalent, 77, 255-258
  external, 65, 215, 219
  fictitious, 137-138
  field, 113, 243
  generalized components, *see* Generalized components of force
  given, 528
  gravitational, 60-61
  instantaneous action at distance, 61
  internal, 65, 215
  interparticle, 216
  inverse square, 178-184, 601-602
  irrotational, 113, 243-244
  line of action, 256
  and linear momentum, 218-220

Lorentz transformation, 885-886
net, 219
net external, 67, 216
Newtonian, 59-62
parallel, 261
passive, 528
point of application, 77, 255
reduction of systems, 258
relativistic, 885-886
representation, 78
specification, 77
Foucault pendulum, 138-139
Four-force, 898
Four-momentum, 897
Four-velocity, 896
Frames of reference:
 inertial, 11-12, 857-858
 relative motion, 32
 transformation between, 13-21, 859-865
Frequencies of system, 810
Frequency, resonance, 164
Friction, 75-76, 80
Function, 968
Functional:
 definition, 968
 in Hamilton's modified principle, 843
 in Hamilton's principle, 839
 stationary values, 972-975
Functional dependence, 460-463

Galilean transformation, 13, 20
Galileo, 12
Gauss's theorem, 412
Generalized components of force:
 Cartesian components, 355
 as covariant components, 572
 definition, 354, 515
 derived from potential function, 356
 determination, 355, 357, 516
 and Euler angles, 654-655
 for holonomic constraint, 356
 for holonomic system, 539-541
 for quasi-coordinates, 712
 transformation, 516-517
Generalized components of impulse, 588
Generalized coordinates, 352-353, 514, 539
Generalized force, *see* Generalized components of force
Generalized force functions, 558-562
Generalized momentum:
 conservation, 577, 578, 585
 definition, 387, 577, 737
Generalized potential, 558-560
Generalized velocities, 352, 514, 539
General tensors, 921-927
Generating functions, 789-795
Geometry of space, 8
Gibbs-Appell equations of motion, 720-725, 727-730
Gibbs-Appell function, 721, 725, 727-730
Gradient, 408-409, 414-416, 424
Gram-Schmidt orthogonalization, 674, 915-917
Gravitational constant, 73
Gravitational force:
 due to arbitrary mass distribution, 74
 due to spherically symmetric mass distribution, 74
 effective, 139
 of homogeneous disk, 78
 reduction to single force, 262
Gravity:
 acceleration due to, 73
 law, 73
Green's theorems, 410-412
Group:
 Abelian, 948
 automorphic, 948
 complete Lorentz, 893
 conjugate elements in, 950
 cyclic, 948
 definition, 945
 homomorphic, 948
 inhomogeneous Lorentz, 893
 isomorphic, 948
 Lorentz, 893
 matrix representation, 955
 operator representation, 955
 order, 945
 periodic, 948
 Poincaré, 893
 properties, 945-946
 special Lorentz, 894
 of symmetry operations, 685
Group theory, 945-950

Hamiltonian:
 canonical transformation, 774, 782
 conservation, 390-391, 576, 585, 739
 definition, 387-388, 737
 importance, 390
 of particle in electromagnetic field, 740

relativistic, 903
and total energy, 388, 738
transformation, 774, 791
Hamiltonian function, 738
Hamilton-Jacobi equation:
  generalization, 801-802
  for particle in central force field, 803
  for simple harmonic oscillator, 802-803
  solution by separation of variables, 800-801
  time-dependent, 797-799
  time-independent, 799-800
Hamilton's canonical equations of motion, 773-782
Hamilton's characteristic function:
  definition, 800
  for simple harmonic oscillator, 802
Hamilton's equations of motion, 387-391, 737-740
  in condensed notation, 769
  and Hamilton's principle, 840-841
  for particle in electromagnetic field, 740-741
  relativistic, 903
  of second kind, 744-745
Hamilton's modified principle, 843-852
Hamilton's principal function:
  definition, 798
  for particle in central force field, 804
  for simple harmonic oscillator, 802
Hamilton's principle, 838-841
  relation to modified Hamilton's principle, 845
Harmonic motion, 158-168, 373-378, 665-707
Harmonic oscillator, simple:
  amplitude, 161
  angular frequency, 161
  critically damped, 162
  damped, 160-162
  decay modulus, 161
  driven, 162
  effect of constant force, 166
  frequency, 161
  overdamped, 161
  period, 161
  relaxation time, 161
  solution by canonical transformation, 782-783
  solution by Hamilton-Jacobi technique, 802-803
  underdamped, 160
Harmonic system, 373, 665, 684

Holonomic constraints, 349-351, 356, 531-532
Holonomic systems, 351-352, 538
Homomorphic group, 948
Hooke's law, 74
Hooke's law potential, motion in, 602
Huygens, Christian, 12

Ideal constraint force, 529-531
Identity element in group, 945, 946, 950
Ignorable coordinate, 577
Impact parameter, 188, 615
Impulse:
  angular, 127, 129-130
  definition and properties, 126-132
  generalized components, 588
  on harmonic oscillator, 167
  instantaneous, 127, 128, 130, 338, 588
Impulsive equations of motion:
  Lagrange's, 588-590
  for particle, 126-132
  for rigid body, 338-341
Indices, raising and lowering, 925-926
Inertia:
  law of, 12
  moment of, see Moment of inertia
Inertia ellipsoid, 291
Inertial forces, 137
Inertial frames, 11-12, 857
Inertia tensor, 286-298, 647-648
Initial conditions, 435
Integral:
  differentiation, 441
  line, 409
  surface, 409-410
  volume, 409-410
Interaction between particles, 59, 875
Internal forces, 215
Interparticle forces, 216
Invariable line, 330
Invariable plane, 330
Invariant subgroup, 949
Invariant subspace, 686, 953, 963
Inverse of element in group, 945, 946, 950
Inverse square force field, motion in, 178-184, 601-602
Inverse transformations, 909-911
Irreducible representations:
  definition and properties, 959-963

and natural modes, 689-692
Irreducible subspace:
  of configuration space, 686-687
  properties, 963
Irrotational force field, 113
Isomorphic groups, 947-948

Jacobian:
  for canonical transformation, 760, 771-772, 794-795
  in change of integration variables, 912-913
  definition, 401
  and inverse transformation, 909
  properties, 401
  of transformation from lab to center of mass, 630
Jacobian elliptic functions, 330, 467-469
Jacobi's identity, 824

Kinematics:
  Newtonian, 7-46
  relativistic, 857-871
  of rigid body, 303-319, 647-652
Kinetic energy:
  average, for driven oscillator, 163
  change in collision, 623
  definition, 107, 234
  of dynamical system, 234
  generalized coordinates, 524-525
  of particle, 107, 234
  relative, 623
  of rigid body, 271-272, 312-314, 650-651
  of two-particle system, 198
  and work, 107-108, 234-237
Kronecker delta, 34, 405

Laboratory coordinates, 205, 624-625, 632-633
Lagrange bracket, 757, 770, 825
Lagrange multipliers, 53, 464-466, 533, 880, 970-972
Lagrange points, 639-643
Lagrange's equations of motion:
  for anholonomic systems, 554-556
  for elementary systems, 521-526
  and Hamilton's principle, 840
  for holonomic systems, 361-365, 538-546
  for impulsive forces, 588-590
  for Lagrangran systems, 564-566
  for quasi-coordinates, 715-719
  relativistic form, 901
  for rigid body, 653-655
  and tensor analysis, 570-572
Lagrangian:
  definition, 363, 564
  harmonic, 373, 665, 682
    approximately, 375-376, 668-670
  indeterminate nature, 566
  invariance, 578-581, 585-586
  normal coordinates, 674
  relativistic, 901
Lagrangian system, 564
Laplacian, 425
Latitude, definition, 140
Left-handed systems, 406
Legendre transformation:
  definition and properties, 918-920
  of kinetic energy, 744
  of Lagrangian, 737, 746
Length:
  contraction, 868
  definition, 7
  measurement, 7
Levi-Civita symbol, 35, 321, 407, 423-424, 562
Linear differential equations, 436-440
Linear equations, system, 458-459
Linear momentum:
  at arbitrary point, 101, 218-219
  at center of mass, 219
  conservation, 57, 122-123, 248, 578, 883-884
  definition:
    for particle, 57, 101
    relativistic, 883-884
    for system of particles, 218
  effect of impulse, 126-130, 338
  and force, 59, 218-220, 885
  Lorentz transformation, 884
  time rate of change, 59, 220, 885
Linear operator:
  matrix representation, 954
  on vector space, 953
Linear operator representation:
  of group, 955-966
  of symmetry group, 686
Linear transformations, 472, 676
Line integral:
  definition and properties, 409-414
  independence of path, 443-446
Line of nodes, 303
Logarithmic decrement, 161
Lorentz force, 561, 741
Lorentz group, 893-894

I-8 Index

Lorentz transformation:
  of accelerations, 863-864
  consequences, 867-869
  of force, 885-886
  four-dimensional formulation, 893-895
  of linear momentum, 884
  statement and derivation, 13-21, 859-860
  in vector form, 18-20, 860-861
  of velocities, 861-863

Mass, 56-57, 883
Matrix:
  adjoint, 455
  adjugate, 455
  antisymmetric, 450
  augmented, 459
  coefficient, 459
  column, 449
  complex conjugate, 449
  congruent, 929
  definition, 447
  diagonal, 450, 932
  diagonalization, 935-936
  eigenvalues, 931-934
  equivalent, 928
  inverse, 455-456
  non-negative, 450
  non-singular, 455
  order, 447
  orthogonal, 456, 930
  positive definite, 450, 932
  positive semidefinite, 450, 932
  properties, 451
  rank, 457, 459-460
  regular, 455
  row, 449
  scalar, 450
  similar, 929, 932
  singular, 455
  skew-symmetric, 450
  square, 450
  submatrix, 457
  symmetric, 450, 932
  trace, 961
  transposition, 448, 451
  unit, 450-451
  unitary, 457
  zero, 449, 451
Matrix addition, 447, 451
Matrix equality, 447
Matrix multiplication, 448, 451

Matrix representation:
  of group, 955
  of linear operator, 954
  of vector, 953
Matrix transformation:
  congruent, 929
  definition, 928
  equivalence, 928
  orthogonal, 930
  similarity, 929
Mechanics, definition, 1
Metric tensor, 896, 925
Modal direction, 374, 666, 671, 683
Modal frequency, 374, 666, 683, 688-689, 692-695
Modal vector, 374, 666, 673, 683
Modified Hamilton's principle, 843-852
Modified velocity, 865, 876
Modulus of elasticity, 75
Moment of inertia:
  definition and properties, 269-271, 286-287, 647
  principal, 292-295, 648-649
  table of values, 482-486
Moment of mass, 66
Momentum:
  angular, *see* Angular momentum
  and energy, relativistic relation between, 888-889
  linear, *see* Linear momentum
Motion:
  of center of mass, 67-68, 197, 216, 220, 254, 274, 320, 655
  central force, *see* Central force motion
  constants of, *see* Constants of motion
  equations, *see* Equations of motion
  harmonic, 158-168, 373-378, 665-707
  impulsive, 126-132, 338-341, 588-590
  one-dimensional, 151-156
  relative:
    of frames of reference, 32-39
    of particles, 27-28
  uniplanar, 269-275
Multiplication table for group, 947-948
Multiplicity:
  of degeneracy, 686-687
  of modal frequency, 688-689
Multiply periodic function, 810, 967

Natural coordinates, 376-378, 670-673
Natural modes of motion:
  definition and properties, 373-376, 665-670

determination, using group theory, 682-695
superposition, 377, 671
Newton, Isaac, 12
Newton's equation of motion:
  application to elementary problems, 88-89
  in configuration space, 240-242, 513-514
  in cylindrical coordinates, 144
  and d'Alembert's principle, 835
  impulsive form, 126
  and Lagrange's equations, 361-363, 522-524
  in non-inertial frames, 136-139
  in spherical coordinates, 145
  statement, 59-60, 102
  for system of particles, 216, 240-242, 513-514
  in tensor form, 570-571
  and torque-angular momentum theorem, 103
  and work-energy theorem, 108
Newton's law for collisions, 339
Newton's law of gravity, 73
Newton's laws:
  first law, 12
  second law, 59-60
  third law, 60-61
  see also Newton's equation of motion
Nodes, line of, 303
Noether's theorem, 581-586
Normal coordinates, 378, 673-676
  see also Natural coordinates
Normal divisor, 949
Nutation, 326

Operator:
  linear, see Linear operator
  projection, 690-691, 964-966
Orbit:
  bound, 606
  in central force motion, 175, 599
  in inverse square force field, 180, 182, 601-602
Order:
  of cyclic group, 948
  of element in group, 947
  of group, 945
Orthogonal curvilinear coordinates, see Curvilinear coordinates
Orthogonality condition:
  for irreducible representations, 961-962
  for modal vectors, 673

Orthogonal matrix, 456
Orthogonal transformation, 473
Oscillations, small, see Vibrating systems
Oscillators:
  coupled, 676
  harmonic, see Harmonic oscillator
  see also Vibrating systems

Parallel-axis theorem, 270, 290
Partial differentiation, 398-400
Particles:
  collisions between, 203-209, 622-623
  definition, 1-2
  equations of motion, see Equations of motion
  interaction, 49
  relative motion, 27-28
  statics, 83-85
  system of two, 196-200
Pericentron, 183
Perihelion, 183
Period:
  fundamental, 813, 916
  of harmonic oscillator, 159
  of inverse square orbit, 183
  of multiply periodic function, 813, 916
  in oscillatory one-dimensional motion, 153-156
Periodic group, 948
Permutation symbol, 35, 321, 407, 423-424, 562
Phase angle, 163
Poincaré group, 893
Poinsot ellipsoid, 330
Point transformation, 750-751
Poisson brackets, 759, 823-827
Poisson's equations of motion, 828-830
Poisson's hypothesis, 339
Poisson's theorem, 829
Position, 27, 38, 45, 196
Postulates of mechanics:
  classical:
    I, 11
    II, 12
    III, 20
    IV, 56
    V, 216
  relativistic:
    I, 857
    II, 857
    III, 858
    IV, 883

Potential:
  effective, 176, 179, 181, 326, 599
  generalized, 558-560
  power law, 600-602
  single-valued, 114, 244
Potential energy:
  average, for driven oscillator, 163
  definition and properties, 113-115, 243-244
  determination from orbit, 178
  gravitational, 117
  of harmonic oscillator, 158
  and work, 115, 244
Potential function, 114, 243, 560
Power of group element, 946-947
Power law potential, 600-602
Principal axis, 292-298, 648-649
Principal directions, 292-298, 648-649
Principal moments of inertia, 292-295, 648-649
Product:
  cross, 403, 406
  dot, 424
  scalar, 403, 424
  vector, 403, 424
Products of inertia, 287
Projection operator, 690-691, 964-966
Proper subspace, 952

Quadratic forms:
  definition, 937
  properties, 937
  reduction to sum of squares:
    by linear transformation, 939-941
    by orthogonal transformation, 941-943
    simultaneous, 943-944
  transformation, 938-939
  types, 937
Quasi-coordinates:
  definition and properties, 711-714
  and Gibbs-Appell equations, 720-725
  Lagrange's equation, 715-717
Quotient law, 925

Radial motion, 176, 599-600
Reduced mass, 198
Reducible representation, 959-960
Reduction of quadratic form, 939-943
Reference frames, 7, 9, 23, 136
Regular transformation, 471
Relative coordinates, 205, 624-625, 632-633
Relative motion, 27-28, 32-39, 197-200

Relativistic mechanics:
  basic postulates, 857-858, 883
  dynamics, 875-889
  four-dimensional formulation, 893-898
  Hamilton's equations in, 903-904
  kinematics, 857-871
  Lagrange's equations in, 901-902
Relativity:
  general theory, 1
  principle, 11-12
  special theory, 1
  see also Relativistic mechanics
Relaxation time, 161, 166
Representation:
  of operator by matrix, 954
  of quantity by tensor, 921
  of vector by matrix, 953
Representations of group:
  characters, 961-963
  dimensions, 955
  direct sum, 957
  equivalence, 962
  equivalent, 956
  generation, 957
  irreducible, 956, 959-961
  matrix, 955
  natural, 958
  one-dimensional, 957
  operator, 686, 955
  reducible, 959-960
  regular, 958-960
  unitary, 957
Resonance, 164-166
Restitution:
  coefficient, 131, 339
  period, 131, 339
Restricted three-body problem, 634-643
Riemannian space, 896, 925
Right-handed systems, 406
Rigid body:
  angular momentum, 271-273, 312-315
  collisions, 339-341
  configuration, 253-254, 271, 303
  definition, 253
  kinetic energy, 271-273, 312-315, 650-651
  statics, 264-266
Rigid-body motion:
  dynamics, 320-323, 653-656
  equations of motion:
    Gibbs-Appell, 727-730
    Lagrangian, 653-655
    Newtonian, 254, 261, 274-275, 320-323

impulsive, 338-341
kinematics, 303-315, 647-651
uniplanar, 269-275
Rocket, 220
Rotating frames, 33-39, 136-139
Rotation:
   instantaneous axis, 36-37
   as transformation, 474
Rotational motion, 33
Routhian, 746
Routh's equations of motion, 746-749

Scalar, 402, 419, 922-923
Scalar field, 408
Scalar potential, 470
Scalar product, 403, 424
Scattering angle, 188-192, 205-206, 618-619, 624-625
Scattering cross section, *see* Cross section
Separable systems, 807
Separation of variables, 800-801
Simple closed orbit, 607
Simple harmonic motion, 158-168
Simple harmonic oscillator, *see* Harmonic oscillator
Sine function, 397
Space:
   absolute, 22
   configuration, *see* Configuration space
   definition, 7
Speed of particle, upper limit, 16, 20, 858
Sphere, rolling, 332-333, 658, 730
Spherical coordinates:
   definition and properties, 45-46, 430-431
   equations of motion, 145, 525-526
   relation to rotating frame, 45-46
Spring constant, 74
Spring linear, 74
Statics:
   of particle, 83-85
   of rigid body, 264
Stationary value:
   of function, 464-466, 969-972
   of functional, 972-978
Steady-state solution, 163
Stokes' theorem, 413
Strings, 76-77
Subgroups, 949-950
Summation convention, 425, 768, 922
Surface:
   perfectly rough, 76
   perfectly smooth, 76

Symmetric top, *see* Top, symmetric
Symmetry:
   and constants of motion, 578-584
   and natural modes of motion, 682-707
Symmetry group, 685-686
Symmetry operations, 684-686

Tables, 477-486
Tensors:
   Cartesian, 288, 417-426
   general, 921-927
   inertia, 286-298, 647-648
   and Lagrange's equations of motion, 570-572
   metric, 896, 925
   in relativistic mechanics, 893-898
Three-body problem, restricted, 634-643
Thrust of rocket, 220
Time:
   absolute, 22
   definition, 8
   measurement, 8-9
Time dilation, 867-868
Time rate of change in different frames, 33-36
Top:
   asymmetric, 327-332
      stability of motion, 658-659
   spin, 325
   symmetric:
      freely rotating, 323-325
      in gravitational field, 325-327, 657, 748-749
Torque:
   and angular momentum, 103, 229-230
   at arbitrary point, 103, 227-228
   axial component, 273-274
   at center of mass, 228
   definition, 102-103, 227
   produced by couple, 256-257
Total cross section, 187-188, 614-615, 628
Trace of matrix, 930, 933-934
Transformation coefficients, 308, 418-419
Transformation matrices, 307, 308
Transformations:
   canonical, *see* Canonical transformation
   congruent, 929
   coordinate, 470-474
   equivalence, 928
   extended point, 750-751, 759
   Galilean, 13, 20
   inverse, 909-911
   lab to center of mass, 206-209, 625-633

Lorentz, *see* Lorentz transformation
matrix, *see* Matrix transformation
   orthogonal, 930
   point, 750-751
   of quadratic forms, 938-939
   similarity, 929
Transient solution, 163
Trojan asteroids, 643
Turning points, 152, 177, 180
Twin paradox, 869-870
Two-particle collisions, 203-209, 622-633
Two-particle systems, 196-200

Uniplanar motion of rigid body, 269-275
Unitary representation, 957
Units, abbreviations, 477

Vector:
   basis, 952
   Cartesian, 420
   components, 404-407, 952
   contravariant, 923
   covariant, 923
   definition, 402, 420
   derivative, 408
   linearly independent, 952
   negative, 402
   unit, 403
   zero, 403
Vector algebra, 402-407
Vector calculus, 408-416
Vector field, 409
Vector function, 408

Vector identities, 486
Vector potential, 740
Vector product, 403, 424
Vector space, 682, 951-954
Velocity:
   in cylindrical coordinates, 45
   Lorentz transformation, 861-863
   modified, 865
   in moving frame, 38
   relative, 27
   relativistic addition, 867
   in spherical coordinates, 45
Vibrating systems, 373-378, 665-676, 682-695
Virtual displacement, 349, 354, 515, 540, 712-713
Virtual work, 350, 354, 515, 540, 654-655, 713

Weightlessness, 138
Width of resonance curve, 165
Work:
   in conservative force field, 114, 244
   definition, 107, 235
   and kinetic energy, 108, 236-237, 888-889
   on particle, 107, 235
   and potential energy, 115, 244
   on system of particles, 235
   virtual, 350, 354, 515, 540, 654-655, 713
Wrench, 261